TH
7345
B63
1994

SIMPLIFIED DESIGN OF HVAC SYSTEMS

Other titles in the
PARKER-AMBROSE SERIES OF SIMPLIFIED DESIGN GUIDES

Harry Parker and James Ambrose
Simplified Design of Concrete Structures, 6th Edition

Harry Parker and James Ambrose
Simplified Design of Structural Wood, 4th Edition

Harry Parker, John W. MacGuire and James Ambrose
Simplified Site Engineering, 2nd Edition

James Ambrose
Simplified Design of Building Foundations, 2nd Edition

James Ambrose and Dimitry Vergun
Simplified Building Design for Wind and Earthquake Forces, 2nd Edition

Harry Parker and James Ambrose
Simplified Design of Steel Structures, 6th Edition

James Ambrose
Simplified Design of Masonry Structures

James Ambrose and Peter D. Brandow
Simplified Site Design

Harry Parker and James Ambrose
Simplified Mechanics and Strengths of Materials, 5th Edition

Marc Schiler
Simplified Design of Building Lighting

James Patterson
Simplified Design for Building Fire Safety

James Ambrose
Simplified Design of Wood Structures, 5th Edition

William Bobenhausen
Simplified Design of HVAC Systems

SIMPLIFIED DESIGN OF HVAC SYSTEMS

WILLIAM BOBENHAUSEN

Associate Professor of Architecture
City College of New York
School of Architecture and Environmental Studies

Associate Professor of Architecture
Pratt Institute
School of Architecture

A Wiley-Interscience Publication
JOHN WILEY & SONS, INC.
New York • Chichester • Brisbane • Toronto • Singapore

This text is printed on acid-free paper.

Copyright © 1994 by John Wiley & Sons, Inc.

All rights reserved. Published simultaneously in Canada.

Reproduction or translation of any part of this work beyond that permitted by Section 107 or 108 of the 1976 United States Copyright Act without the permission of the copyright owner is unlawful. Requests for permission or further information should be addressed to the Permissions Department, John Wiley & Sons, Inc., 605 Third Avenue, New York, NY 10158-0012.

This publication is designed to provide accurate and authoritative information in regard to the subject matter covered. It is sold with the understanding that the publisher is not engaged in rendering legal, accounting, or other professional services. If legal advice or other expert assistance is required, the services of a competent professional person should be sought.

Library of Congress Cataloging in Publication Data:
Bobenhausen, William, 1949-
    Simplified design of HVAC systems/William Bobenhausen.
      p.   cm.—(Parker-Ambrose series of simplified design guides)
    Includes index.
    ISBN 0-471-53280-0 (alk. paper)
    1. Heating—Equipment and supplies—Design and construction.
2. Ventilation—Equipment and supplies—Design and construction.
3. Air conditioning—Equipment and supplies—Design and construction. I. Title. II. Series.
TH7345.B63   1994
697—dc20   93-36664

Printed in the United States of America

10 9 8 7 6 5 4 3 2 1

CONTENTS

Preface   xiii

Introduction   xv

1 Human Comfort   1

    1.1 Conditions for Human Comfort / 1
    1.2 Effective Temperature / 3
    1.3 Mean Radiant Temperature / 3
    1.4 Air Movement that Increases Human Comfort / 5
    1.5 Air Movement that Decreases Human Comfort / 6
    1.6 Charting Climate for Human Comfort / 6

2 Climate   12

    2.1 Outdoor Temperature / 12
    2.2 Winter Design Temperatures / 13
    2.3 Heating Degree-Days (HDD) / 16
    2.4 Modified Degree-Day Method / 19
    2.5 Heating Degree-Hours (HDH) / 20
    2.6 Temperature Bins / 23
    2.7 Summer Design Conditions / 25

vi   CONTENTS

    2.8   Other Climatic Factors / 27
    2.9   Cooling Degree-Days (CDD) / 27
    2.10 Relative Humidity / 27
    2.11 Wind / 30
    2.12 Atmospheric Clearness / 32

3 Solar Basics   33

    3.1   Fundamentals / 33
    3.2   Altitude and Azimuth Angles / 37
    3.3   Profile (Shade) Angles / 43
    3.4   Clear Dry Solar Radiation on Surfaces / 45

4 Heat Gains Within Buildings   48

    4.1   Zone Type Designations / 49
    4.2   Heat from People / 49
    4.3   Heat from Equipment and Appliances / 54
    4.4   Heat from Lighting Systems / 57
    4.5   Heating and Cooling "Balance Points" / 60

5 Heat Loss from Buildings   63

    5.1   Heat Transfer Processes / 64
    5.2   Heat Loss Through Solid Building Elements / 65
    5.3   Air Spaces and Air Films / 65
    5.4   Construction Assemblies / 67
    5.5   Windows / 72
    5.6   Infiltration of Outside Air / 75
    5.7   Sizing Heating Systems / 80
    5.8   Calculations for a Simple Room / 80
    5.9   Below-Grade Walls / 81
    5.10 Below-Grade Floors / 83
    5.11 Slab-on-Ground Construction / 84
    5.12 Heat Loss to Adjacent Unheated Spaces / 85
    5.13 Annual Heating Needs / 86
    5.14 Heat Loss Calculation for "The House" / 86
    5.15 Checklist for Heat Loss Calculations / 92

6 Ventilation   94

    6.1   Need for Outside Air / 94
    6.2   Natural Ventilation / 95
    6.3   Ventilation Heat Loss / 96

CONTENTS vii

    6.4   Ventilation Heat Gain / 97
    6.5   Total Ventilation Heat Change / 99
    6.6   Indoor Air Quality (IAQ) / 100
    6.7   Heat Recovery Ventilation (HRV) / 101
    6.8   Power Ventilators / 102
    6.9   Stack Effect / 103
    6.10 Filters / 104

7 Solar Cooling Loads   105

    7.1   Solar Cooling Loads Through Glazing / 106
    7.2   Sun Control to Reduce Solar Heat Gain / 110
    7.3   Conduction of Heat Through Roofs, Walls, and Glazing / 115

8 Cooling Load Calculations   125

    8.1   Thermal "Zoning" of a Building / 125
    8.2   ASHRAE Calculation Procedure / 126
    8.3   Equations for External Heat Gains / 127
    8.4   Equations for Internal Heat Gains / 128
    8.5   Detailed Cooling Load Calculation for "The Office" / 128
    8.6   Peak Hour Cooling Loads / 132
    8.7   Peak Cooling Loads for Entire Buildings / 132
    8.8   Square Feet per Ton / 136
    8.9   Occupancy Diversity / 137

9 Air, Ducts, and Fans   138

    9.1   Air Properties / 138
    9.2   Properties of Moist Air / 139
    9.3   Air Pressure / 142
    9.4   Friction in Ductwork / 144
    9.5   Other Friction Losses / 147
    9.6   Duct Sizing Methods / 148
    9.7   HVAC Fans / 149
    9.8   Fan Performance / 151
    9.9   Fan Laws / 152

10 Water, Pipes, and Pumps   156

    10.1 Water Properties and Pressure / 156
    10.2 Using Water to Store and Deliver Heat / 158
    10.3 Pipe Data / 159
    10.4 Friction in Piping / 161

- 10.5 System Head Loss / 163
- 10.6 Centrifugal Pumps / 164
- 10.7 Pump Curves / 165
- 10.8 System Curves / 165
- 10.9 Pump Performance / 167
- 10.10 Pump Laws / 170

11 Selection of HVAC Systems — 173

- 11.1 Primary Selection Factors / 173
- 11.2 Secondary Selection Factors / 175
- 11.3 Systems for Residences / 177
- 11.4 Systems for Nonresidential Buildings / 180

12 Solar Heating Systems — 188

- 12.1 Active Solar Heating Systems / 188
- 12.2 Passive Solar Heating Systems / 189
- 12.3 Passive Solar Performance / 191
- 12.4 Performance for DG-A2 House Designs / 198
- 12.5 Indoor Air Temperature / 198
- 12.6 Productivity of Glazing / 202
- 12.7 Passive Solar Heating of "The House" / 205

13 Warm Air Heating Systems — 208

- 13.1 Basic System Description / 208
- 13.2 Early Warm Air Furnaces / 209
- 13.3 Modern Furnace Design / 209
- 13.4 Very High-Efficiency Furnaces / 213
- 13.5 Furnace Sizing / 214
- 13.6 Residential Air Distribution Options / 215
- 13.7 Amount of Warm Air Needed / 217
- 13.8 Sizing Ducts / 218
- 13.9 Duct Insulation / 220
- 13.10 Air Outlets / 220
- 13.11 Warm Air Heating of "The House" / 223

14 Steam Heating Systems — 227

- 14.1 Steam Properties / 227
- 14.2 Steam Boilers / 228
- 14.3 Steam Boiler Sizing / 233
- 14.4 Boiler-Related Piping / 234

14.5   System Pressure and System Velocity / 236
   14.6   One-Pipe Systems / 238
   14.7   Two-Pipe Systems / 239
   14.8   Pipe Sizing / 242
   14.9   Guidelines for Low-Pressure Steam Piping / 243
   14.10 Heaters (Radiators) / 244
   14.11 Temperature Control / 248

15 Hot Water Heating Systems — 249

   15.1   Basic System Description / 249
   15.2   Hot Water Flow / 250
   15.3   Piping / 252
   15.4   Boilers / 254
   15.5   Small Capacity Boilers / 257
   15.6   Boiler Sizing / 257
   15.7   Friction in Piping / 259
   15.8   Piping Arrangements and Temperature Control / 260
   15.9   Baseboard and Finned-Tube Radiation / 264
   15.10 Other Hydronic Terminal Devices / 268
   15.11 Hot Water Heating of "The House" / 270

16 Electric Heating Systems — 273

   16.1   Resistance Heat / 274
   16.2   Heat Pumps, General / 276
   16.3   Heat Pumps, Operational Performance / 278
   16.4   Heat Pump Sizing / 283
   16.5   Electric Boilers / 283

17 Heating Fuels and Combustion — 285

   17.1   Heat Content of Fuels / 285
   17.2   Steady-State Efficiency / 285
   17.3   Annual Fuel Utilization Efficiency (AFUE) / 286
   17.4   Measuring Improved Efficiency / 288
   17.5   Venting of Combustion Heating Appliances / 290

18 Cooling of Houses — 293

   18.1   Climatic Need for Cooling / 293
   18.2   Cooling Through Air Motion / 296
   18.3   Air-Circulation Fans / 300
   18.4   Whole-House Fans / 300

18.5 Mechanical Cooling / 301
18.6 Evaporative Cooling / 304
18.7 Residential Cooling Load Calculations / 305
18.8 Mechanical Cooling of "The House" / 313

19 Applied Psychrometrics 316

19.1 Air Conditioning Processes / 316
19.2 Total Heat (Enthalpy) / 317
19.3 Humidity Control / 319
19.4 Cooling Coils / 320
19.5 Heat Ratio Terminology / 321
19.6 Graphing HVAC Processes / 322

20 Cooling System Equipment 326

20.1 Basic Equipment Options / 326
20.2 Vapor Compression Cycle / 328
20.3 Unitary Air Conditioners / 330
20.4 Air-Cooled Condensers / 333
20.5 Other Condenser Types / 335
20.6 Cooling Towers / 336
20.7 Cooling Ponds and Sprays / 340
20.8 Evaporators / 342
20.9 Electric Chillers / 342
20.10 Types of Compressors / 344
20.11 Compressor Capacity Control / 347
20.12 Part-Load Performance (IPLV) / 348
20.13 Absorption Refrigeration / 349
20.14 Other Nonelectric Cooling Equipment / 351

21 Cooling Distribution and Delivery Systems 352

21.1 Local (Individual) Systems / 352
21.2 Air Supply Fundamentals / 355
21.3 Basic Air System Elements / 360
21.4 Air System Types / 363
21.5 Air Outlets / 368

22 Domestic Hot Water 373

22.1 Water Temperature Needed / 373
22.2 Quantity of Hot Water Required / 376
22.3 System Sizing / 377

22.4 Direct-Fired DHW Storage Heaters / 381
22.5 Indirect-Fired DHW Systems / 382
22.6 Heat Pump DHW Heaters / 384
22.7 Solar DHW / 385
22.8 DHW Piping Systems / 388

23 Control Systems   389

23.1 Basic Terminology / 389
23.2 Modes of Control / 391
23.3 Sensors, Controllers, and Actuators / 393
23.4 Summary of Control Systems / 394
23.5 Electric Control Systems / 395
23.6 Electronic Control Systems / 396
23.7 Pneumatic Control Systems / 396
23.8 Direct Digital Control (DDC) Systems / 398
23.9 Safety Controls / 398
23.10 Energy-Conserving Controls / 399
23.11 Smoke Controls / 400
23.12 Realistic Applications / 400

24 Applied Economics   402

24.1 Cost per Million BTUS (MMBtu) / 403
24.2 Simple Payback and Rate of Return / 404
24.3 Escalation of Energy Costs / 405
24.4 Discounted Payback / 407
24.5 Fuel Escalation and Discount Rates / 408
24.6 Mortgage Economics / 411
24.7 Conservation and "The Law of Diminishing Returns" / 413
24.8 Utility Rates for Electricity / 415
24.9 Dollars and Sense / 416

Index   417

PREFACE

It has been more than three years since Jim Ambrose asked me if I was interested in writing this book. After much thought (and some trepidation) I agreed. I agreed because it would be part of the "Simplified Design Series" of Parker/Ambrose books—books which I had used during my architectural schooling, and liked very much for their straightforward discussion of technical issues and their simple examples. The Simplified Design Series includes books that can be used during one's education and later during one's career. And so began my maiden journey into book creation.

The journey was often lonely and difficult (I suspect that only other authors can truly empathize), but along the way I have been assisted by many in numerous ways. Two individuals in particular were there throughout:

Catherine, my incredible wife, reviewed the entire manuscript several times, helped organize the sections, and translated some of my thoughts into English. She also provided a technical review since (as a toxicologist and industrial hygienist for Malcom Pirnie) she herself is very knowledgeable about solar heating, heat loss, industrial ventilation, indoor air quality, and the like. She also (occasionally) succeeded in pulling me away from my computer for an essential camping trip or walk in the park.

Alan Ballou, my good friend and fellow architect (Suffern, New York), reviewed the entire manuscript several times. Al grew up with the "simplified series" in the 1940s as an architectural student at Columbia. I enjoyed and benefitted from our many spirited discussions about what to include in the book and what was better suited for the cutting room floor (a lot).

Technical quality and appropriateness were essential in the creation of this book. The manuscript was improved immensely due to the contributions of the additional people:

Norbert Lechner, author of Wiley's *Heating, Cooling, Lighting*, reviewed the entire manuscript and offered some fine suggestions.

John Reynolds, co-author of Wiley's epic volume *Mechanical Electrical Equipment for Buildings*, reviewed the early chapters and helped me eliminate a few items which were not quite simplified enough for inclusion.

Frank Gerety, in New York, many think of him as "Mr. Steam," primarily reviewed the steam and hot water heating chapters. He offered such detailed knowledge and insight that it was sometimes difficult to draw the line as to what should be included in a simplified volume.

Michael Dexter, Associate Partner with Syska and Hennessey, reviewed Chapters 20 and 21 which pertain primarily to equipment and systems for commercial buildings. His comments in particular helped me develop material and examples which are appropriate for the contemporary design of large buildings.

Adrian Tuluca, Partner with Steven Winter Associates (with whom I often consult), read most of the manuscript with particular attention to load calculations and issues which concern the cutting-edge of technology.

And there were others...

Diane Serber, good friend and an AIA "fellow," provided an early review of the heat loss section which helped me to develop the book's basic organization.

James Callahan, engineer with Malcolm Pirnie, reviewed many sections pertaining to air distribution systems, ventilation, fans and pumps and offered sound advice.

I must also thank a few students: Joseph Michalak, a "life support systems" student of mine at Pratt Institute, read the entire manuscript and offered many good suggestions. Joe also helped check the math and data entry in tables. Jonathan Tham, a participant in my summer intern program at City College of New York made some good suggestions in reviewing the heat loss section.

And then, there were the illustrations...

Michael Koehler, Assistant Director of the Architectural Center at the City College of New York, did the wonderful freehand drawings which help to soften an otherwise technical book. The drawings in Chapter 11 in particular, bring to life the text on HVAC system selection for different building types.

Al Ballou also reviewed the concepts and drafts of the illustrations and produced many of the Autocad illustrations.

Victoria Ettlinger, of Kingston, New York, took on major responsibility for the computer drawings using her "Mac." I deeply appreciate Tory's flexibility and professionalism throughout a very difficult series of deadlines.

Sincere thanks to all of those noted above, and to my many contacts at Wiley. To Stephen Kliment, who initiated the project as Editor, and to his successors, first Everett Smethurst and now Amanda Miller. Much appreciation also to Allison Ort Morvay for smoothly managing the production process, and to Dean Gonzalez for his careful review of the artwork.

My journey in creating the first edition of this book is now complete. Now it is *your* turn to journey to a better understanding of HVAC systems regardless of your current level of knowledge. I have tried to demystify much of the information and to make it as logical, easy to understand, and fun as possible.

Good luck and enjoy!

WILLIAM BOBENHAUSEN

Hastings-on-Hudson, New York
February 1994

INTRODUCTION

This is a technical resource book—not a novel. Still, there is a definite sequential nature to the presentation of material. For example, before you can calculate passive solar performance (Chapter 12) you must be able to do heat loss calculations (Chapter 5) which apply climatic factors (Chapter 2). Therefore, users are urged to begin with Chapter 1 to be sure that basic fundamentals and underlying principles which will be applied and referred to in later chapters are understood.

And who will the "users" of this volume be? Who is it written for? Several groups. This book can easily serve as a text for general or concentrated courses in engineering, architectural, and technical training schools. Further, it should prove to be a ready reference to many within the architectural and engineering professions at various stages of experience and for those working in the heating, ventilating, and air-conditioning (HVAC) trades who wish to understand the theory behind these systems.

The topic of HVAC is broad, encompassing many issues and technologies. Some engineering concepts are very complex. And while this volume has been organized to live up to its title, it also covers the broad field of HVAC design. It is intended to give the reader enough of an understanding of the theory and concepts of HVAC design to facilitate basic design and selection of components for houses and simple commercial systems. It also will enable one to work knowingly with consultants on large and complex design applications.

Loads and systems for residences are presented in depth, including detailed examples of "The House" which first appears where heat loss calculations are described. Various heating and cooling systems for "The House" are then included at the end of the appropriate sections. With regard to larger buildings, "The Office" is used in various examples to compute cooling loads and size systems throughout the book.

The American Society of Heating, Refrigerating and Air Conditioning Engineers, Inc. (ASHRAE) is the organization in the United States that develops and publishes standards, procedures, and comprehensive technical handbooks for the HVAC industry. However, without a basic understanding of HVAC theory, using ASHRAE Handbooks can be intimidating and overwhelming. This book is not intended to impart all of the knowledge necessary to be a competent HVAC engineer. However, one fundamental goal is to serve as something of a "primer" for the latest editions of the voluminous ASHRAE Handbooks (*1993 Fundamentals*, *1992 HVAC Systems and Equipment*, and *1991 HVAC Applications*). Much of the data published in the tables contained herein are excerpted from the ASHRAE Handbooks, often in "simplified" form.

Beyond gaining the ability to perform heating or cooling load calculations, and to size equipment, you should learn to evaluate the factors that contribute to the calculation. You will then be in a position to manipulate building and site conditions to reduce the size of required equipment and the rate of fuel consumption. Selection of HVAC equipment requires a true understanding of fundamentals and operational characteristics. Bigger is not always better—and not only because bigger equipment takes up more space, and usually costs more to install and operate. For example, it is very important not to oversize central cooling equipment in residences. While a large capacity unit may produce physical and mental comfort on a very hot "design day," it will perform poorly on milder days with high humidity since the on/off cycling of the cooling coil may not remove enough moisture.

Throughout the book great attention is focused on the topic of "units." Historically some HVAC terms have been represented in a way that can be confusing to the uninitiated. For example, the term Btuh is widely used to mean Btus per hour, properly written as Btu/h. And "cfm" is universally used to refer to "cubic feet per minute," properly written as ft^3/m.

You will notice that most of the mathematical problems in this book use simple numbers which are often rounded and then referred to as "approximate." This is particularly true of any calculations based upon annual climate data since climate for a location may vary up or down 15% or more in a given year. You will also note that annual climate data tables in Chapter 2 are themselves rounded to avoid an appearance of exact precision given by values such as "7189" heating degree days—7200 will do just fine.

This volume avoids providing information only for "40 degrees North Latitude," so typical of many technical texts. The technical data and problems contained herein should be useful and illustrative for all 50 states.

In places readers may also detect humor and occasionally a degree of cynicism. I hope so. You are asked to think about why we now so often refer to "outside air" for ventilation needs rather than "fresh air" as we once did.

HVAC design is an ever-changing science and art. Not only must a designer understand principles and formulas, but societal issues as well, including issues of possible ozone depletion, "sick building syndrome," and application of new technologies.

Comments to improve future editions will be welcome.

1

HUMAN COMFORT

Human comfort . . . the principle reason for HVAC systems in buildings. Since the first fire was set at the mouth of a cave, people have created heating systems to remain comfortable and healthy when outside conditions turn cold. However, only within the last century has mechanical cooling equipment (originally developed for preservation of food) been used to provide and maintain comfort in buildings.

Comfort in buildings can be more than a luxury. A recent study by Rensselaer Polytechnic Institute (RPI) suggests that a comfortable work environment can lead to a 2% improvement in job performance. Others suggest even larger improvements and champion the use of personal environment module (PEM) work stations which have individual control of temperature, air flow, lighting, noise, and air quality. The PEM approach, while resulting in a comfortable work environment, can also be expensive and energy consuming.

In this chapter, the "human comfort zone" is defined by ranges of dry-bulb temperature and relative humidity. Also discussed are the concepts of "effective temperature" and "mean radiant temperature" (MRT).

"Uncomfortable" conditions can of course be changed by employing mechanical HVAC equipment. Nevertheless, the benefits of working with natural factors such as air movement and solar radiation should not be discounted, as will be shown. The issue of "long-term" comfort, or health, also has become an important HVAC design concern relative to indoor air quality, as discussed in Section 6.6.

1.1 CONDITIONS FOR HUMAN COMFORT

Human comfort is influenced by the climatic factors indicated in Figure 1.1.

An understanding of the following terms is required for a discussion of human comfort:

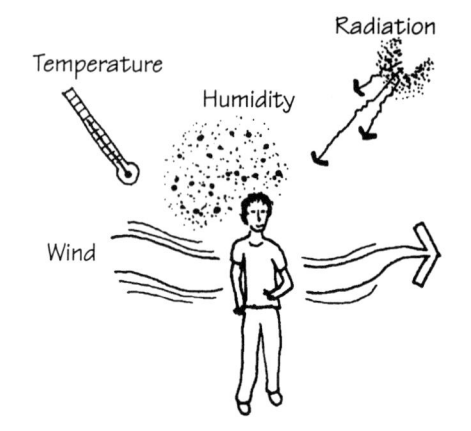

FIGURE 1.1 Factors influencing human comfort.

Dry-bulb temperature (db) The temperature one reads on a "normal thermometer" in degrees Fahrenheit (°F).

Wet-bulb temperature (wb) A temperature reading, also in °F, which takes into account the amount of moisture in the air. Its value can be obtained by spinning a thermometer with a wet wick encasing the bulb through the air (known as a sling psychrometer, shown in Figure 1.2). If the air is dry it will absorb moisture from the wick and indicate a low wet-bulb temperature. Conversely if the air is humid little moisture will evaporate from the wick and a high wet-bulb temperature will be indicated.

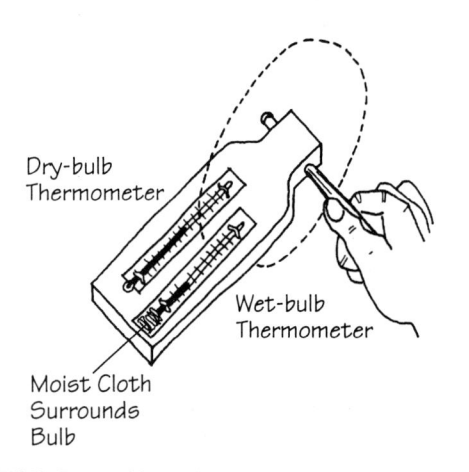

FIGURE 1.2 Measuring wet-bulb temperature with a sling psychrometer.

Relative humidity (RH) Measures the amount of water vapor (moisture) actually in the air compared to the total amount of moisture the air can hold at a specific temperature and pressure. Therefore, if air contains one-half of the total moisture it can hold under those conditions, the relative humidity is said to be fifty percent (50%).

The three values of dry-bulb temperature, wet-bulb temperature, and relative humidity are interconnected. Whenever any two of the values are shown, the third can be determined, as will be shown.

Human Comfort Zone

In general, most lightly clothed people are comfortable in "still" air when the dry-bulb temperature is between 68 and 78°F and the relative humidity ranges between 20 and 70%, although discomfort begins to result if both the temperature and humidity are near the upper limits of the ranges. The combination of these ranges of dry-bulb temperature and relative humidity is referred to as the "human comfort zone."

Figure 1.3 depicts the ASHRAE (American Society of Heating, Refrigerating and Air-Conditioning Engineers) comfort zone on a "psychrometric chart" used for winter and summer indoor design conditions. The psychrometric chart, a graphical way of plotting the properties of moist air, is explained in detail in Section 9.2. For now, realize that air temperature (as you would read it off a conventional thermometer) is indicated along the horizontal axis. And moisture in the air, known as humidity, is along the vertical axis. Curved lines through the chart indicate lines of "relative humidity."

Attempts to scientifically gauge human comfort have been many, and continue, including studies of people of various ages and nationalities. Older people, particularly those who are unable to move around much, generally require warmer temperatures to

1.3 MEAN RADIANT TEMPERATURE

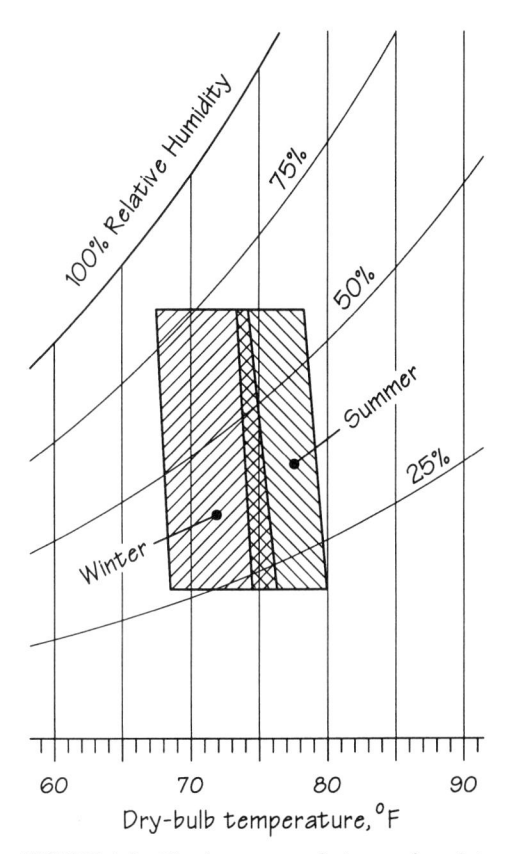

FIGURE 1.3 The human comfort zone for winter and summer.

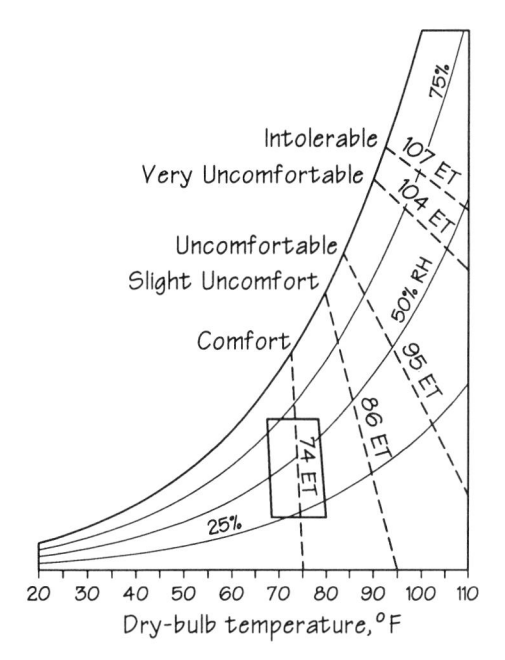

FIGURE 1.4 Chart of Effective Temperatures (ET). Adapted with permission from the *1993 ASHRAE Handbook of Fundamentals*, Chapter 8, Figure 10.

maintain comfort. On the other hand, activities such as rigorous exercise call for lower air temperatures.

An understanding of the requirements for comfort is fundamental to the selection and sizing of HVAC systems. For many occupancies, HVAC system performance is satisfactory if conditions fall anywhere within the human comfort zone. Other occupancies such as computer rooms and hospital operating rooms may have much more precise temperature and humidity requirements for reasons apart from human comfort.

1.2 EFFECTIVE TEMPERATURE

The "effective temperature" (ET) concept considers human comfort in response to factors of air temperature and relative humidity in still air. The effective temperature method, like most empirical comfort studies, is based on consideration of lightly clothed people who are seated at rest ("sedentary") in still air over a period of one hour.

Figure 1.4 is adapted from the ASHRAE ET chart. Indicated are lines of constant effective temperature, and applicable descriptors such as "uncomfortable" for the line representing 95°F ET. The effective temperature lines indicate that for a "constant" condition of comfort, higher air temperatures can be tolerated as relative humidities decrease.

1.3 MEAN RADIANT TEMPERATURE

On a very cold but bright winter day it is possible for body surfaces (face, hands) exposed to the sun to feel much warmer and more comfortable than expected for the actual air temperature. In a similar way, comfort can be attained indoors even if actual air temperatures are below the comfort zone, by using radiant heating systems.

Radiant effects can also produce discomfort. For example, one can feel very cool and uncomfortable when surrounded by cool surfaces in a cave or basement even when the air temperature is well within the normal comfort range. Most of us have experienced the chilling effect of sitting near a single-glazed window on a very cold day, as our body radiates heat to the cold surface.

Mean radiant temperature, or "MRT," accounts for the temperature impact of surrounding surfaces according to the angle of influence. The person located near the window in Figure 1.5 will have a 130-degree exposure to the cold glass. In contrast, the person located further from the window wall will have only a 94-degree exposure to the window and greater exposure to surrounding wall surfaces which will be much warmer in winter.

Figure 1.6 indicates combinations of air temperature and MRT which together produce comfort conditions which feel like 70°F db. For each degree of change in MRT from 70°F (either up or down), the air temperature must change 1.4°F in the opposite direction to compensate and maintain a feeling of 70°F comfort.

To use Figure 1.6 enter along the horizontal axis at the air temperature. Rise vertically until you hit the diagonal line, then move horizontally to the axis on the left to obtain the MRT required to provide a comfort level equal to having the air and all surrounding surfaces at 70°F. For example, an MRT of 75°F would yield the equivalent of 70°F comfort when the air temperature is only 63°F, as indicated by the dotted line on the figure.

A complete MRT calculation for a given point accounts for the heat radiating to and from all surfaces in three dimensions. However, typical applications within buildings normally allow for a much simpler two-dimensional approach. In either case, the MRT value is derived from a weighted-average of the influence of (i.e., radiation to or from) all surrounding surfaces and their temperatures, and, in the case of the customary two-dimensional analysis, considers a full, 360-degree exposure (either horizontal or vertical) through the surfaces which will have the greatest MRT impact. Two-dimensional MRT is calculated by

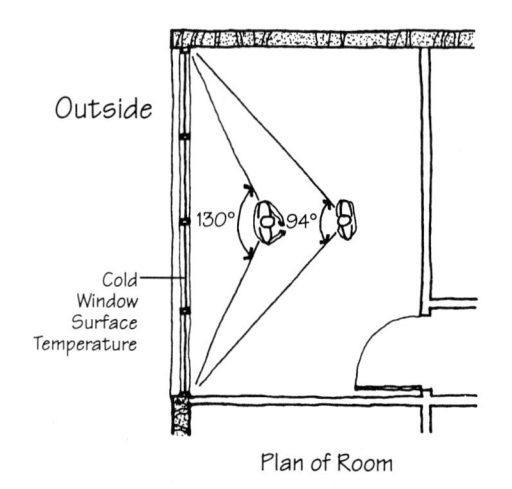

FIGURE 1.5 Influence of surrounding surfaces on mean radiant temperature (MRT).

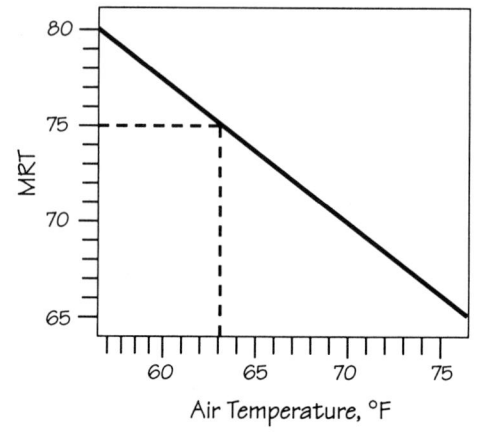

FIGURE 1.6 Air and MRT conditions for 70°F db comfort.

using Formula 1A.

$$MRT = \frac{<A \times t_1 + <B \times t_2 + <C \times t_3 + \cdots + <N \times t_N}{360} \quad \text{Formula 1A}$$

where

MRT is the mean radiant temperature in °F
$<A$ is the angle of exposure to surface 1
t_1 is the °F temperature of surface 1
$<B$ is the angle of exposure to surface 2
t_2 is the °F temperature of surface 2
$<C$ is the angle of exposure to surface 3
t_3 is the °F temperature of surface 3
$<N$ is the angle of exposure to surface N
t_N is the °F temperature of surface N

a heat-absorbing massive floor with a radiant heating system. Calculate the MRT at Point X above the floor.

Solution: Use Formula 1A. Note that $<A$ is 120 degrees and that t_1 is 80°F, the temperature of the radiant floor surface. Other surfaces are then considered in a clockwise manner.

$$MRT = \frac{<A \times t_1 + <B \times t_2 + <C \times t_3 + \cdots + <N \times t_N}{360} \quad \text{(Formula 1A)}$$

$$= \frac{120 \times 80 + 10 \times 65 + 40 \times 40 + 10 \times 65 + 120 \times 70 + 60 \times 70}{360}$$

$$= \frac{9600 + 650 + 1600 + 650 + 8400 + 4200}{360}$$

$$= \frac{25{,}100}{360} = 69.7°F, \text{ the MRT at Point X}$$

Example 1A: The living room in a residence (shown in section in Figure 1.7) has

This computation shows how the warming effect of radiant surfaces can offset the chilling effect of cold window surfaces.

Finally, be practical in applying the mathematics of MRT conditions. A person standing with one foot in a bucket of ice water while the other foot is in a bucket of steaming hot water is unlikely to feel comfortable even though the mathematical average may be right in the comfort zone!

1.4 AIR MOVEMENT THAT IMPROVES HUMAN COMFORT

Air movement impacts on human comfort. Air motion that is less than about 50 fpm

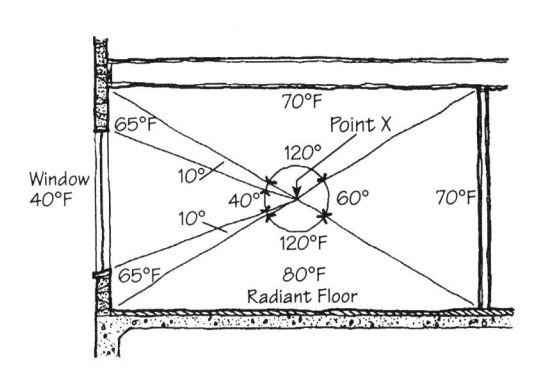

FIGURE 1.7 MRT calculation for a room.

6 HUMAN COMFORT

(0.57 mph) is basically unnoticed and considered "still air." Increased air motion of between about 50 and 250 fpm (slightly above 2 mph) is generally pleasant, but noticeable at the high end. Indoor air motion of speeds above 300 fpm are potentially drafty and annoying unless very warm weather makes it welcome.

1.5 AIR MOVEMENT THAT DECREASES HUMAN COMFORT

Winter air movement (winds) can, of course, add to outdoor discomfort due to drying and freezing of human skin as indicated by "wind chill" temperatures. For example, when the outside temperature is 0°F db it will feel like: −21 when a 10-mph wind blows, −39°F when the wind is 20 mph, −48°F in a 30 mph wind, and −53°F when the wind is blowing at 40 mph.

While not specifically used for HVAC design purposes, appreciation of wind chill temperatures should impress upon designers the importance of vestibules, protective landscaping, location of entries, and other architectural factors in cold climates.

1.6 CHARTING CLIMATE FOR HUMAN COMFORT

Bioclimatic Chart

The various elements required to provide for human comfort are perhaps most easily and graphically understood through use of the Bioclimatic Chart. It comes from what many consider to be one of the "bibles" on environmentally sensitive architectural design, *Design with Climate*, by Victor Olgyay, published initially in 1963 by Princeton University Press. Figure 1.8 provides a schematic representation of the Bioclimatic Chart referred to as the Bioclimatic Index.

In the center of Figure 1.8, one finds a relaxed person enjoying life in the human comfort zone which extends down to an air temperature of about 70°F. At lower temperatures, the schematic then suggests that solar radiation can provide the heat needed to remain comfortable. At temperatures above the comfort zone (about 80°F) winds are needed if comfort is to continue. The figure also indicates relative humidities along the horizontal axis, resulting human sensations, and the need to add moisture at low relative humidities (about 20%) and remove it at high relative humidities (over 75% or so).

Figure 1.9 presents the Bioclimatic Chart for moderate United States climates less than 1000 ft above sea level for persons dressed in ordinary light indoor clothing and doing light work in still air.

Note all the information that can be found in the Bioclimatic Chart. Below the comfort zone are lines which indicate the quantity of (solar) radiation needed in Btus per hour to restore comfort. For example, approximately 250 Btu/h of solar radiation is needed to provide comfort if the air temperature is about 50°F.

Also note that above the comfort zone, and to the right, wind speeds in feet per minute (fpm) are shown and suggest that air motion (from wind or a fan) of about 700 fpm (about 8 mph) is needed to produce comfortable conditions if the air temperature is about 90°F with a relative humidity of about 60%.

The Bioclimatic Chart also shows:

1. The "grains" of moisture needed per pound of dry air to improve the quality of overly dry warm air (Note: 7000 grains is equal to 1 lb of water, see Section 9.1).
2. Mean radiant temperature (MRT) values of up to 75°F to provide comfort when air temperatures are below 70°F as well as lower MRT values to

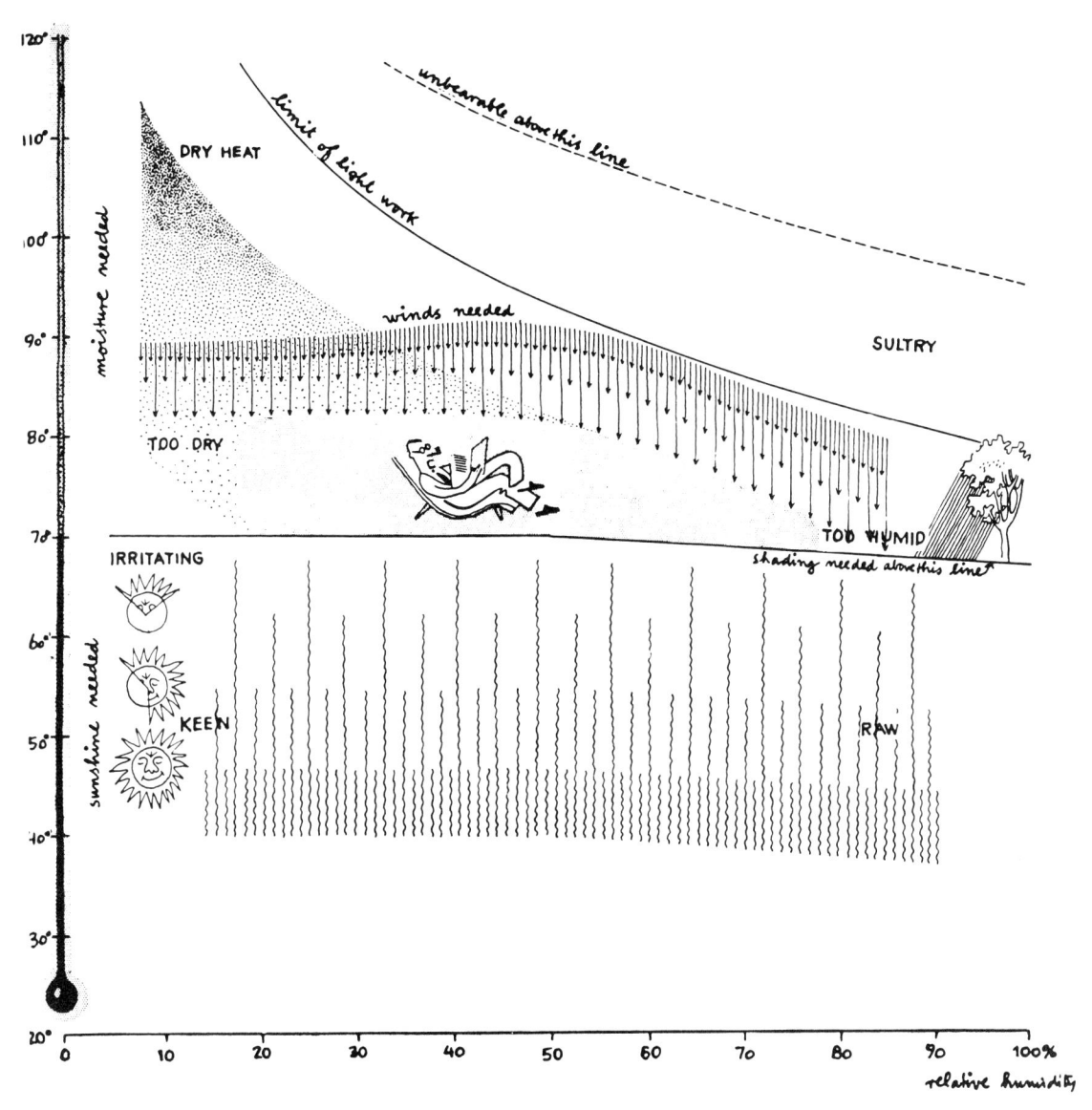

FIGURE 1.8 Schematic bioclimatic index.

provide comfort when air temperatures are above the upper limit of the comfort zone.
3. Curving lines to indicate temperature and humidity conditions which define the "limit of work of moderate intensity" and conditions for "possible sunstroke."

Building Bioclimatic Chart

In the 1970s two leading architectural educators/researchers, Murray Milne and Baruch Givoni, developed the concept of a "building bioclimatic chart" based on the psychrometric chart. Figure 1.10 illustrates such a chart in simplified form on the same

8 HUMAN COMFORT

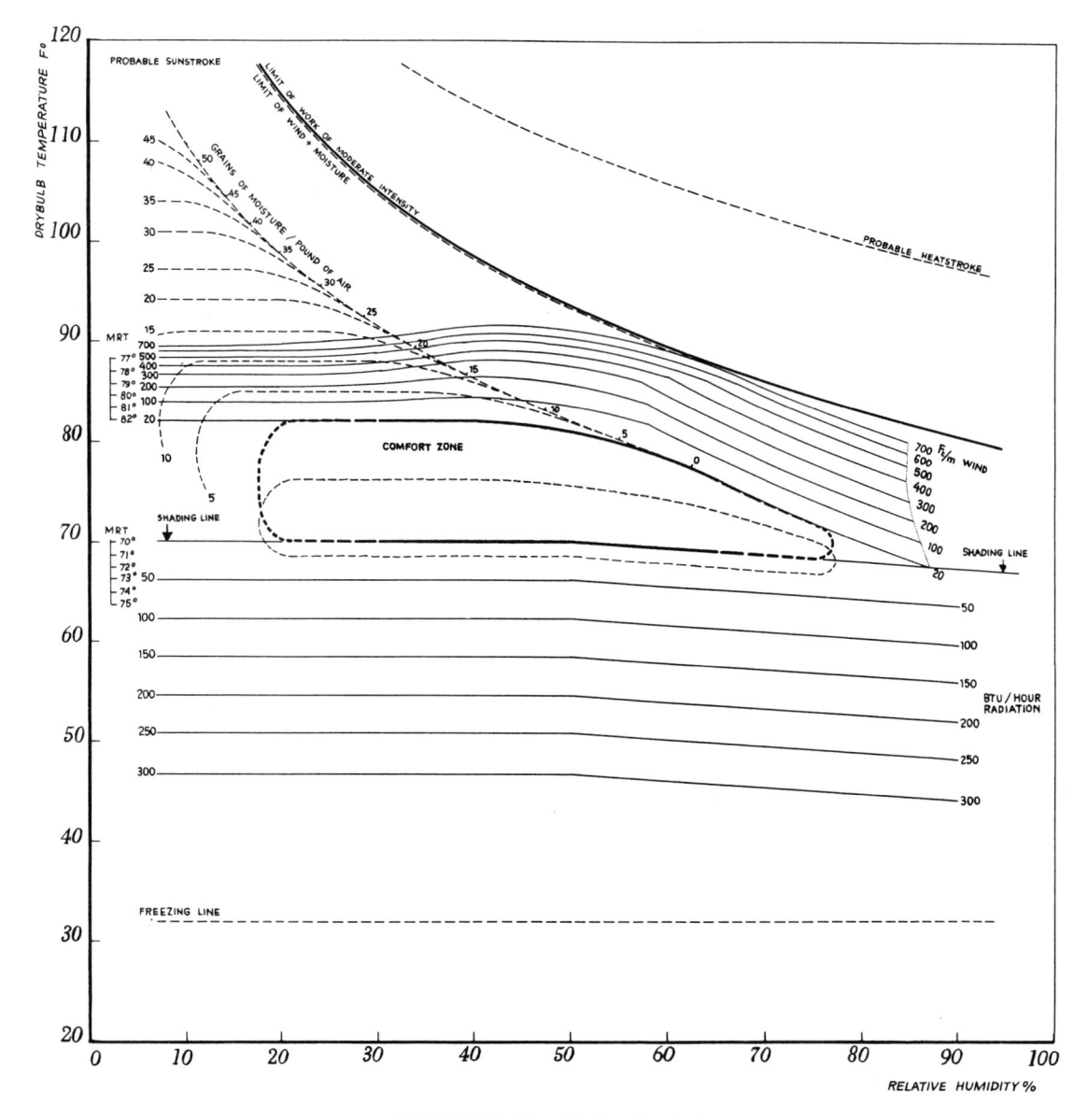

FIGURE 1.9 The bioclimatic chart.

psychrometric chart format used throughout this book. Noted on the chart are the design strategies that can be employed to achieve conditions comfortable for humans even if uncomfortable temperature and/or humidity conditions exist outside. As shown in the figure, if the temperature falls slightly below the comfort zone, "passive solar heating" is all that will be necessary until the temperature gets even lower when "active solar and conventional heating" will be required. Similarly, when the relative humidity is a little high, "natural ventilation" can provide comfort unless the air temperature is also high, in which case "conventional dehumidification" will be required. Other

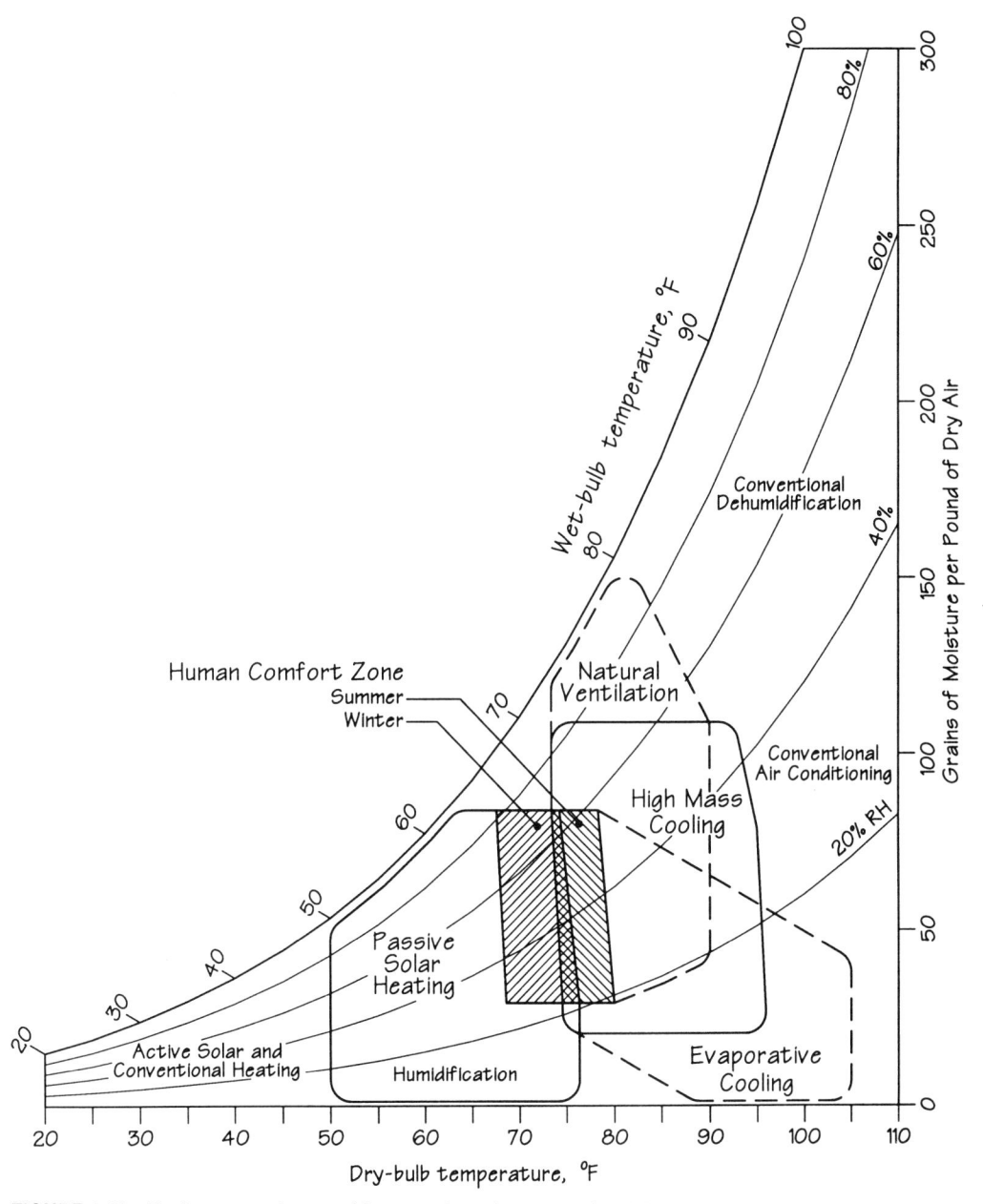

FIGURE 1.10 Design strategies to achieve comfort when uncomfortable outdoor temperature and humidity conditions exist.

design strategies noted include "evaporative cooling," "humidification" at low relative humidity levels, "high-mass cooling," and "conventional air conditioning."

By plotting temperature and humidity data representative of a particular climate on an overlay of the building bioclimatic chart one can see which design strategies are required to maintain comfort, particularly for residential buildings that have a

TABLE 1.1 ST. LOUIS, MISSOURI—NORMAL AND EXTREME TEMPERATURE AND HUMIDITY DATA

Month	Normal Daily Maximum	Normal Daily Minimum	Extreme Record High	Extreme Record Low	RH 00	RH 06	RH 12	RH 8
January	37.6	19.9	76	−18	77	81	66	69
February	43.1	24.5	85	−10	77	82	63	65
March	53.4	33.0	89	−5	74	81	59	59
April	67.1	45.1	93	22	70	78	54	53
May	76.4	54.7	93	31	75	82	55	55
June	85.2	64.3	102	43	76	82	55	54
July	89.0	68.8	107	51	77	84	56	55
August	87.4	66.6	107	47	79	87	56	57
September	80.7	58.6	104	36	80	88	57	60
October	69.1	46.7	94	23	76	84	55	60
November	54.0	35.1	85	1	77	83	63	67
December	42.6	25.7	76	−16	79	83	68	73

Source: Local Climatological Data, Annual Summary for 1992, National Oceanic and Atmospheric Administration.

relatively small amount of internal heat gain from lights, people, and equipment.

Table 1.1 provides "normal" long-term temperature and humidity data for St. Louis, Missouri.

This data can be used to determine two points (to form a line) representing the normal climate for each month. The highest daily temperature typically occurs at noon or several hours later. Similarly, the lowest daily temperature usually occurs before

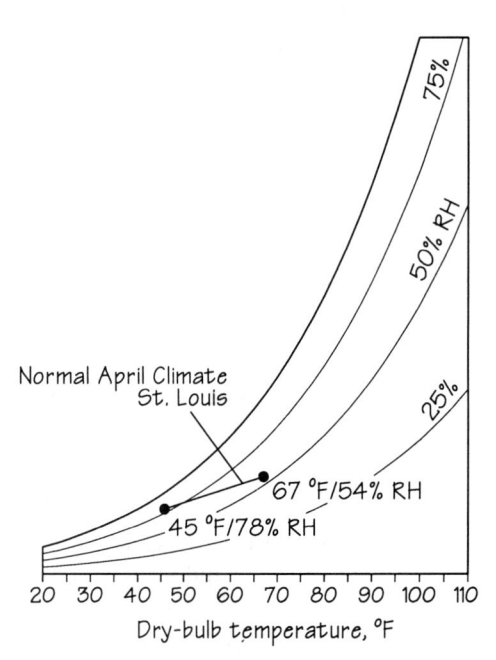

FIGURE 1.11 Plotting a line to represent a month's climate: April in St. Louis.

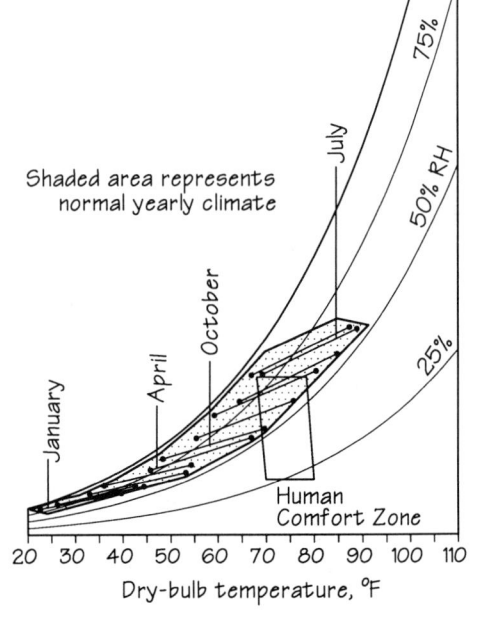

FIGURE 1.12 Plot of St. Louis's climate for a typical year.

sunrise. These occurrences of high and low daily temperature are used to obtain the corresponding relative humidities from the four hourly values given (note: for other cities data may be for four different hours such as 01, 07, 13, and 19).

For example, the normal high temperature in St. Louis in April of 67°F will normally occur near hour 12 (noon) when the relative humidity is 55%. Similarly, the normal low temperature in April of 46°F will normally occur near hour 6 (6 AM) when the relative humidity is 79%. These two values of temperature and corresponding relative humidity are used as points to draw a line representing the typical April climate in St. Louis as shown in Figure 1.11.

The normal climate for an entire year can be graphically seen once lines for all 12 months are plotted as shown in Figure 1.12.

By plotting a climate in such a manner we can quickly see how outdoor temperature and humidity conditions compare to the human comfort zone. Throughout this book we will see how HVAC systems can be selected and designed to in effect "bring us back into the comfort zone" regardless of the outdoor climate.

2

CLIMATE

In a few rare locations (e.g., Hawaii, Samoa, and the Virgin Islands), year-round weather conditions nearly always fall within the human comfort zone, and HVAC systems to heat or cool homes and small buildings are not needed. In most American locations, however, mechanical equipment is needed to offset uncomfortable climatic conditions. Good HVAC design is based on a thorough understanding of local conditions including normal and extreme temperatures, humidity, solar radiation, and wind.

A good source for basic climatic information is the National Oceanic and Atmospheric Administration (NOAA), a division of the Department of Commerce based in Asheville, North Carolina (phone 704-259-0682). NOAA has compiled over 50 years of detailed climate data for approximately 300 American cities, data which is excerpted and applied in many chapters of this book. In addition, the *ASHRAE Handbook of Fundamentals* provides data on climatic conditions which is needed for "design" purposes for over 600 locations, much of which is excerpted and contained herein for 40 United States cities.

2.1 OUTDOOR TEMPERATURE

In Figure 2.1 long-term normal and extreme dry-bulb air temperatures are given for Chicago. Note during which seasons that the average temperatures change markedly from month to month and the general shape of the line connecting monthly average temperatures. Looks a bit like a sine curve doesn't it? It does—see Section 3.1 for a full discussion. By the way, Chicago was only chosen for illustration because of its central location and population base. Almost all United States climates will display a similar pattern with variations in amplitude due primarily to humidity levels.

Before going into the details, a few words of philosophical advice regarding climate data (from whatever source) and computed results are in order. The values used, and results obtained, can easily be-

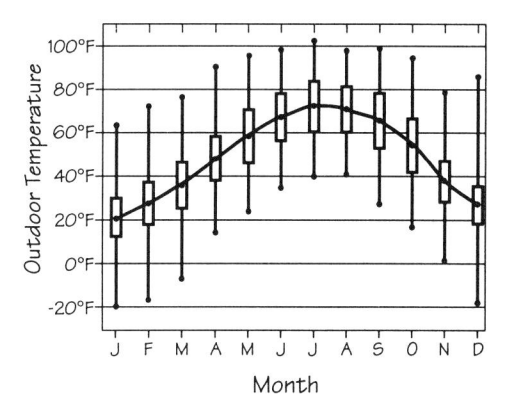

FIGURE 2.1 Normal and extreme temperatures for Chicago.

come a sea of numbers. Always keep the big picture in mind. Heating degree-days are a useful gauge of the severity of a winter climate, and resultant need for heating energy (fuel). If one source says that a given city experiences 4848 heating degree-days (see Section 2.3 for definition) and another says 5020, the goal normally is not to determine which is right. Assume they both are, perhaps at two airports in the same city. But say that the site you are currently concerned about in that city is somewhere between them on a small hill. The key is that the site experiences about 5000 heating degree-days per year (not 3000 and not 7000), and that design strategies and systems should be appropriate and cost effective for such a temperate climate. In this spirit, the data tables included below contain rounded numbers.

One must also realize that climatic conditions can change dramatically, even over very short distances. For example, Phoenix, Arizona correctly conjures up images of a hot desert city. It is, having an average yearly temperature of over 70°F, and record high temperatures of 88°F or higher during all 12 months. In stark contrast, Flagstaff, Arizona, which lies less than 200 mi to the north, but at an elevation of 7000 ft, has an average yearly temperature of 45°F and significant amounts of snow accumulation.

Significant changes also occur near large bodies of water as can be observed by comparing airport and downtown climate data for cities such as San Francisco and Washington, DC. Therefore, all climate data should be applied in calculations with as much understanding of actual site conditions and common sense as possible. If time and dedication allows, also consider that Frank Lloyd Wright reportedly would camp on a house site for several days to gain insight into its microclimate.

2.2 WINTER DESIGN TEMPERATURES

Almost all buildings on the mainland of the United States (except some in southern Florida) are built with some form of heating system. Historical climate information is needed to determine the required heating system output so that comfortable conditions can be maintained on very cold days.

How cold is a "very cold" day? Well, for heating system sizing purposes it's not the very coldest day recorded because that would result in a vastly oversized heating system. Instead, "design temperatures" are used; these are temperatures that are exceeded during either 99% or 97.5% of the hours in December, January, and February. The 97.5% winter design value is typically used, and even now required by some codes. The 99% winter design values are typically only three to five degrees lower and may be appropriate for design applications such as hospitals, where a fall in inside temperatures could be hazardous to health.

Climate information for cities indicated in Figure 2.2 will be presented.

Provided in Table 2.1 are the winter design dry-bulb temperatures, all-time low temperatures, heating degree-days, and latitudes for 40 United States cities.

As will be detailed elsewhere in this book, the capacity of heating equipment is

14 CLIMATE

FIGURE 2.2 Cities for which climate data is provided in this chapter.

determined by multiplying the calculated heat loss of a building or space by the difference in temperature between the desired indoor design temperature and the outdoor design temperature (typically the 97.5% value). The following example will be used to illustrate how the design temperature approach prevents gross oversizing of heating equipment.

Example 2A: In designing a heating system for a building in Albuquerque, New Mexico, how much larger would the heating system capacity have to be if the all-time extreme low temperature were used instead of the 97.5% winter design temperature?

Solution: The 97.5% design temperature approach means that during only 54 winter hours (mostly at night) will the temperature in a typical year fall below the design value. The 97.5% winter design temperature for Albuquerque is 16°F, whereas the all-time extreme low temperature is −17°F. Assume that the indoor design temperature is to be maintained at 70°F.

TABLE 2.1 WINTER DESIGN INFORMATION

City[a]	Lat.	All-Time Low (°F)	Winter (°F) Design (99%)	Winter (°F) Dry-Bulb (97.5)%	Heating Degree-Days (Base 65°F)
Albuquerque, NM AP	35.0	−17	12	16	4290
Anchorage, AK AP	61.1	−34	−23	−18	10,810
Atlanta, GA AP	33.6	−8	17	22	3100
Bismarck, ND AP	46.8	−44	−23	−19	9040
Boston, MA AP	42.4	−12	6	9	5620
Casper, WY AP	42.9	−41	−11	−5	7560
Charleston, SC Co	32.9	6	25	28	2150
Chicago, IL Co	41.8	−27	−3	2	6130
Cincinnati, OH Co	39.1	−25	1	6	5070
Dallas, TX AP	32.8	−1	18	22	2290
Denver, CO AP	39.7	−30	−5	1	6020
Detroit, MI	42.4	−21	3	6	6230
Hartford, CT AP	41.9	−26	3	7	6350
Honolulu, HI AP	21.2	53	62	63	0
Houston, TX Co	30.0	7	28	33	1430
Indianapolis, IN AP	39.7	−23	−2	2	5580
Jacksonville, FL AP	30.5	7	29	32	1330
Little Rock, AR AP	34.7	−5	15	20	3350
Los Angeles, CA Co	33.9	28	37	40	1820
Madison, WI AP	43.1	−37	−11	−7	7730
Memphis, TN AP	35.0	−13	13	18	3230
Miami, FL AP	25.8	30	44	47	200
Minn./St. Paul, MN AP	44.9	−34	−16	−12	8160
New Orleans, LA AP	30.0	11	29	33	1470
NYC, NY Central Park	40.8	−15	11	15	4850
Pensacola, FL Co	30.5	5	25	29	1570
Philadelphia, PA AP	39.9	−7	10	14	4870
Phoenix, AZ AP	33.4	17	31	34	1550
Pittsburgh, PA Co	40.5	−18	3	7	5930
Portland, OR Co	45.6	−3	18	24	4790
Reno, NV Co	39.5	−16	6	11	6020
St. Louis, MO Co	38.7	−18	3	8	4750
Salt Lake City, UT AP	40.8	−30	3	8	5980
San Antonio, TX AP	29.5	0	25	30	1570
San Diego, CA AP	32.7	29	42	44	1510
San Francisco, CA Co	37.6	28	38	40	3040
Seattle, WA Co	47.4	9	22	27	5190
Sioux Falls, SD AP	43.6	−36	−15	−11	7840
Topeka, KS AP	39.1	−26	0	4	5240
Washington, DC AP (Nat)	38.9	−5	14	17	5010

Sources: Winter design dry-bulb temperatures are extracted with permission from the *1993 ASHRAE Handbook of Fundamentals*, Chapter 24, Table 1. All-time low temperature data is extracted from NOAA Annual Summaries. Degree-day data is extracted from the *Passive Solar Design Handbook*, Volume Two, prepared for the U.S. Department of Energy and rounded.

[a]AP following the city name designates airport location. Co indicates office location within and affected by the urban area. Undesignated stations are semirural and may be compared to airport data. The ASHRAE *Handbook* lists two or more locations for many cities. The Co location data has been listed for cities whenever available.

16 CLIMATE

$$\text{Capacity increase if all-time low temperature is used} = \frac{\text{Indoor design temp} - \text{All-time low}}{\text{Indoor design temp} - 97.5\% \text{ design temp}}$$

$$= \frac{70 - (-17)}{70 - 16}$$

$$= 1.61$$

One can quickly see the validity of using the 97.5% design temperature approach. Use of all-time extreme low temperatures would greatly increase the capacity of the heating system, in this case by an additional 61%. This is unwise because oversized heating systems are more expensive to buy and to operate.

2.3 HEATING DEGREE-DAYS (HDD)

When the heating degree-day concept was first developed, it was assumed that heat would be needed for indoor comfort whenever outdoor temperatures fell below 65°F. As temperatures fall below 65°F (which is considered "base temperature" X), heating degree-days accrue based upon the mean (average) outdoor temperature, computed by use of Formula 2A.

$$\text{HDD}_X = X - \frac{(t_H + t_L)}{2} \quad \text{Formula 2A}$$

where

HDD_X is the total number of heating degree-days below base temperature X
X is the applicable base temperature in °F
t_H is a day's high dry-bulb temperature in °F
t_L is a day's low dry-bulb temperature in °F

It is useful to think of heating degree-days (for any given day) as "the temperature difference between a particular base temperature and the mean outdoor temperature, in degrees Fahrenheit, on that day." Therefore, the units are "°F·Day." Heating degree-days are often used in the procedure to estimate yearly heating energy use for houses and other buildings as shown below.

Example 2B: Determine how many heating degree-days (HDD) occur on a given day when the high dry-bulb temperature is 37°F and the low dry-bulb temperature is 25°F.

Solution: Use Formula 2A and assume the typical reference dry-bulb temperature of 65°F.

$$\text{HDD}_X = X - \frac{(t_H + t_L)}{2} \quad \text{(Formula 2A)}$$

$$= 65 - \frac{(37 + 25)}{2}$$

$$= 65 - 31$$

$$= 34 \text{ HDD}_{65} \text{ accrue for the day}$$

The temperatures used in this example are typical for a January day in Cincinnati, Ohio. Therefore, assuming that each of the 31 days of January were identical, a total of 1054 heating degrees-days would accrue for the month (31 × 34). And for the entire heating season, which generally begins in

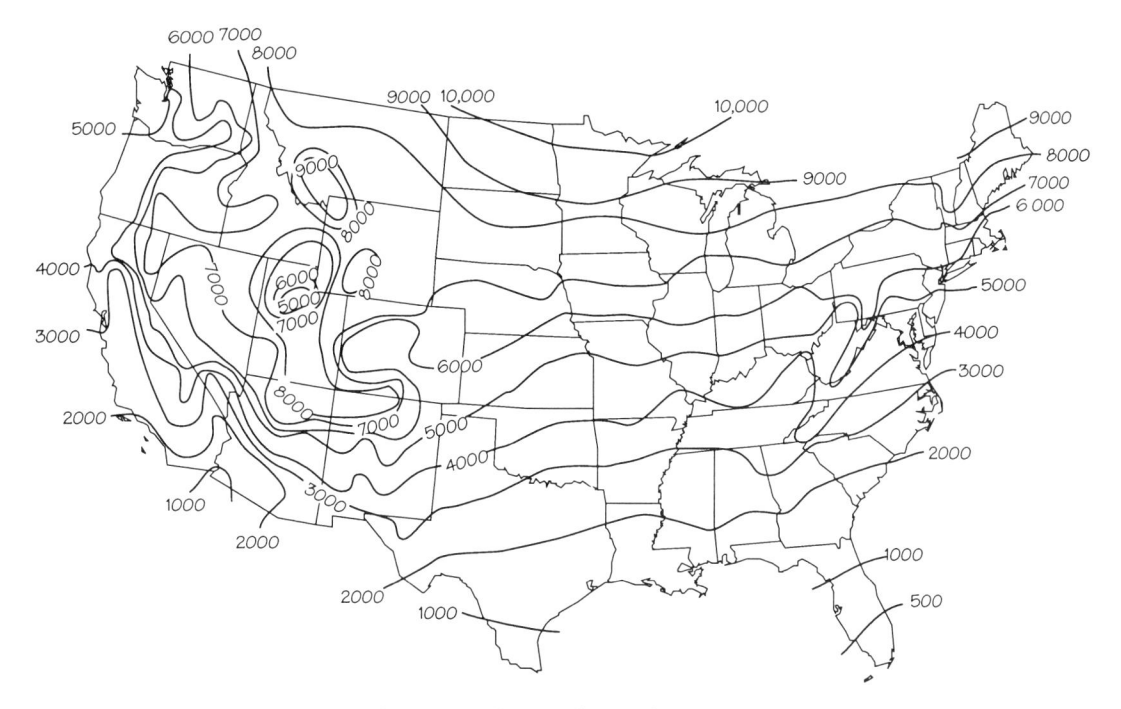

FIGURE 2.3 Heating degree-days, base 65.

September (when mean daily temperatures begin to fall below 65°F) and ends in May or June, the HDD$_{65}$ sum will total roughly the 5070 shown in Table 2.1 for a typical year. Illustrated in Figure 2.3 are heating-degree day (base 65) totals for the United States.

So, heating degree-days are a measure of the severity of climate for a given day, or they can be summed for a given time period or used (as is typical) for an entire heating season. An important use of heating degree-days is the estimation of heating energy requirements through use of a "Modified Degree-Day Method" as described below. Another use is by fuel oil companies which schedule deliveries based on historical fuel usage in relation to heating degree-days. Daily totals for heating degree-days, sums for the year to date, and comparisons to normal accumulations appear in many newspapers.

Despite the numerical precision of the degree-day method, do not forget that climate will vary greatly from year to year and deviations of up to 15% above or below heating degree-day totals (occasionally more) should be considered normal.

The simple heating degree-day methodology using 65°F as a base has proven to be very useful for houses built prior to the "energy crisis" of 1973. Since then though, insulation levels in houses are generally much higher, and better, tighter-fitting windows are used. Also, heat from internal sources (such as lights and various modern appliances) is greater than it used to be. As a result, many well-insulated houses do not need additional heat until the outdoor temperature falls to 60°F or below. The outdoor temperature at which the heat gains from internal sources (and solar gain) are no longer able to keep conditions at the indoor design temperature is known as the "heating balance point," which is fully explained in Section 4.5.

Variable-Base Degree-Days

Good estimates of yearly heating energy use can be made if the heating balance point of a building is known, using Formula 2B.

$$E_{\text{fuel}} = \frac{UA \times HDD_{bp} \times 24}{k \times V}$$

Formula 2B

where

E_{fuel} is the estimated fuel consumption for a time period which is usually a heating season (e.g., therms of gas, gallons of oil, or kilowatt hours of electricity)

UA is the total building heat loss (transmission plus infiltration) in Btu/°F · h

HDD_{bp} is the number of heating degree-days below the balance point base temperature

24 is the conversion of 24 hours in 1 day

k is a correction factor that includes the effects of rated full-load efficiency, part-load performance, oversizing, and energy conservation devices. Typical k factors are found in Table 2.2.

V is the heat value per fuel unit (e.g., Btus per therm, gallon, or kilowatt hour)

Typical k factors to be used in Formula 2B for various heating system types are listed in Table 2.2. See Section 16.3 for the estimation procedure for heat pumps.

Table 2.3 provides heating degree-day data for various base or "balance point" (bp) temperatures. Listed are large cities that experience at least 4000 (base 65°F) heating degree-days.

Example 2C: Estimate the amount of heating energy required each year by a house to be built based on the following:

It will be located in Albuquerque.

It will have a total heat loss of 600 Btu/°F·h (this is often referred to as the building UA—see Chapter 5 for detailed information on heat loss calculations).

The house has a computed balance point temperature of 60°F.

The house will be heated by a conventional gas-fired furnace ($k = 0.78$). The heating value of gas is 100,000 Btu per therm (V).

TABLE 2.2 TYPICAL k FACTORS

$k = 1.00$ for electric resistance heat
$k = 0.90$ for gas- or oil-fired boiler, fully condensing
$k = 0.85$ for gas- or oil-fired furnace, fully condensing
$k = 0.78$ for homes built after 1992 heated by gas- or oil-fired appliance
$k = 0.65$ for homes built from 1980 to 1992 heated by gas- or oil-fired appliance
$k = 0.55$ for homes built prior to 1980 heated by gas- or oil-fired appliance

2.3 HEATING DEGREE-DAYS (HDD)

TABLE 2.3 HEATING DEGREE-DAYS (HDD$_{bp}$) AT VARIOUS °F BASE TEMPERATURES

	50°F	55°F	60°F	65°F
Albuquerque	1500	2290	3220	4290
Bismarck	5240	6360	7630	9040
Boston	2370	3300	4380	5620
Casper	3720	4850	6130	7560
Chicago	2950	3880	4940	6130
Cincinnati	2120	2970	3950	5070
Denver	2590	3590	4730	6020
Detroit	2930	3890	4990	6230
Hartford	2970	3950	5080	6350
Indianapolis	2510	3400	4420	5580
Louisville	1820	2630	3560	4650
Madison	4090	5140	6350	7730
Minn./St. Paul	4580	5630	6820	8160
New York City	1930	2760	3740	4850
Philadelphia	1940	2780	3750	4870
Pittsburgh	2640	3570	4670	5930
Portland	1310	2230	3390	4790
Reno	2290	3350	4590	6020
St. Louis	1960	2760	3690	4750
Salt Lake City	2650	3610	4730	5980
Seattle	1390	2390	3660	5190
Sioux Falls	4320	5360	6530	7840
Topeka	2330	3180	4140	5240
Washington, DC	2000	2870	3860	5010

Source: Data extracted from *Passive Solar Design Handbook*, Volume Two, prepared for the U.S. Department of Energy, and then rounded.

Solution: Use Formula 2B to estimate the yearly heating requirement. Consult Table 2.3 to determine the total of 3220 heating degree-days for the balance point (base) temperature of 60°F.

$$E_{fuel} = \frac{UA \times HDD_{bp} \times 24}{k \times V}$$

(Formula 2B)

$$= \frac{600 \times 3220 \times 24}{0.78 \times 100{,}000}$$

= 594 therms of gas will be required per year to heat the house

2.4 MODIFIED DEGREE-DAY METHOD

When the heating balance point of a building is not known, the Modified Degree-Day Method can be used to estimate the need for yearly heating energy. This is accomplished by use of Formula 2C, which takes into account many of the same factors in Formula 2B. In addition, Formula 2C employs an empirical correction factor (C_D).

$$E_{fuel} = \frac{C_D \times UA \times HDD_{65} \times 24}{k \times V}$$

Formula 2C

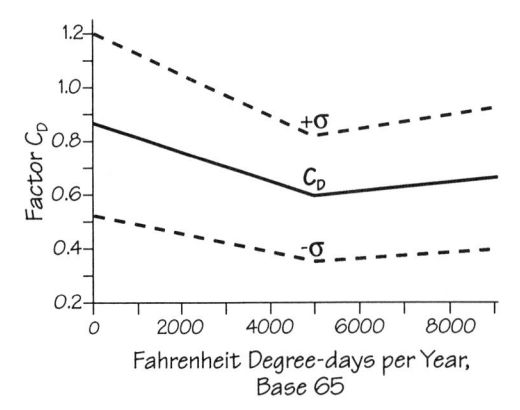

FIGURE 2.4 Correction factors (C_D) for various HDD$_{65}$ locations. Reproduced with permission from the *1989 ASHRAE Handbook of Fundamentals*, Chapter 28, Figure 1.

where

C_D is an empirical correction factor for heating effect versus HDD$_{65}$, whose value is obtained from Figure 2.4. The dashed lines in the figure indicate standard deviations.

Results obtained from use of Formula 2C must always be considered estimates. Indeed Chapter 28 of new *1993 ASHRAE Handbook of Fundamentals* no longer includes Formula 2C. Instead balance point temperature and bin methods are put forth as more reliable along with various simulation techniques.

Example 2D: Reestimate the amount of heating energy required annually for the house described in Example 2C. For this example assume that the heating balance point temperature of the house is not known.

Solution: Since the heating balance point temperature is not known, use Formula 2C to estimate the yearly heating requirement. Table 2.3 indicates a total of 4290 heating degree-days for base 65 in Albuquerque. The correction factor (C_D) is estimated from Figure 2.4 to be 0.62.

$$E_{\text{fuel}} = \frac{C_D \times UA \times HDD_{65} \times 24}{k \times V}$$

(Formula 2C)

$$= \frac{0.62 \times 600 \times 4290 \times 24}{0.78 \times 100{,}000}$$

$$= 491 \text{ therms of gas per year}$$

This estimate is a bit lower than the result of 594 therms found in Example 2C. Note the dashed line representing a positive standard deviation line in Figure 2.4. A C_D value of about 0.90 applies for the positive standard deviation for 4290 HDD$_{65}$. Substitution of that value of 0.90 in Formula 2C (instead of using 0.62) would result in a revised estimate of 713 therms.

2.5 HEATING DEGREE-HOURS (HDH)

Heating degree-day data will normally suffice for the evaluation of fuel use in houses and other simple buildings. However, reliable fuel use estimates for larger buildings will require the use of more detailed hourly data. The heating degree-hour (HDH) concept is similar to that for heating degree-days except values are obtained for each hour, not merely mean values for a particular day, as shown in Formula 2D.

$$HDH_X = X - t_h \quad \text{Formula 2D}$$

where

HDH$_X$ is the number of heating degree-hours below base temperature

X is the applicable base temperature in °F

t_h is the dry-bulb temperature during a particular hour in °F

Heating degree-hour evaluations are based on temperatures for all 24 hours each day. As a result, heating degree-hour totals for a given climate will be approximately 24 times as large as degree-day totals. Therefore, a climate with about 7000 heating degree-days will also experience approximately 168,000 heating degree-hours (7000 × 24). Keep in mind though that heating degree-hour totals will be slightly higher than 24 times the heating degree-day value since 24 hourly temperature values are considered each day. As a result, a day with a daytime temperature of 80°F, and a nighttime temperature of 50°F, would have a mean temperature of 65°F and result in a total of zero heating degree-days. Whereas heating degree-hours would accrue for all hours when the temperature was below 65°F.

The heating degree-hour concept is also useful in evaluating fuel use when indoor air temperatures are lowered or "set back" during night-time hours or during periods when a building is not occupied, as will be shown.

Example 2E: Determine how many heating degree-hours (HDH) occur during a particular hour when the outdoor dry-bulb temperature is 20°F relative to a night-setback base temperature (X) of 50°F.

Solution: Use Formula 2D.

$$\text{HDH}_X = X - t_h \quad \text{(Formula 2D)}$$
$$= 50 - 20$$
$$= 30 \text{ HDH}_{50}$$

And if this condition continues for an entire day, a total of 720 heating degree-hours (30 × 24) will accrue.

Table 2.4 provides day, night, and total heating degree-hours (HDH) for cities that experience at least 100,000 heating degree-hours for the 65°F base temperature. (Note: Data as excerpted, interpolated, and rounded from the source considers "day" to be a 12-hour period from roughly 7:30 AM to 7:30 PM with "night" being the remaining 12-h time period).

Revealed in Table 2.4 is the fact that temperature fluctuation from day to night is remarkably consistent across a wide range of climates. The data indicates that about 45% of a climate's heating degree-hours occur during the daytime, with the remaining 55% at night. The only exceptions are the very dry climates of Albuquerque, Denver, and Reno where a 40/60 split is more accurate.

Day and night climate data is particularly useful for estimation of energy use by buildings when temperatures are set back, or for unoccupied periods. Table 2.5 provides day and night HDH for various base temperatures in St. Louis, Missouri.

Example 2F: A St. Louis school board member has read that most commercial buildings actually use the majority of their heating energy when unoccupied. She has asked the architect of the new school to explain, and to estimate how much heating energy the building will require when unoccupied.

Solution: The architect knows the following:

- The heat loss for the school has been calculated to be 7500 Btu/°F·h.
- When unoccupied, the temperature of the building will be set back to 50°F.
- The heating balance point of the occupied building has been calculated to be 55°F.

First, the energy required during the daytime when the building is occupied is cal-

22 CLIMATE

TABLE 2.4 HEATING DEGREE-HOURS (HDH₆₅)—DAY VS. NIGHT

	Day	(% of Total)	Night	(% of Total)	Total
Albuquerque	43,300	(41)	62,880	(59)	106,180
Bismarck	98,560	(45)	119,610	(55)	218,170
Boston	71,170	(44)	89,980	(56)	161,150
Casper	78,830	(44)	101,490	(56)	180,320
Chicago	71,920	(45)	86,900	(55)	158,820
Cincinnati	57,260	(46)	66,990	(54)	124,250
Denver	79,270	(37)	136,200	(63)	215,470
Detroit	57,700	(45)	69,520	(55)	127,220
Hartford	72,000	(45)	88,500	(55)	127,220
Indianapolis	60,270	(45)	74,590	(55)	134,860
Madison	79,900	(46)	93,380	(54)	173,280
Minn./St. Paul	88,320	(46)	102,620	(54)	190,940
New York City	55,200	(46)	65,770	(54)	120,970
Philadelphia	53,440	(44)	67,460	(56)	120,900
Pittsburgh	66,360	(45)	80,300	(55)	146,660
Portland	50,850	(44)	65,940	(56)	116,790
Reno	65,060	(41)	94,060	(59)	159,120
Saint Louis	54,190	(44)	67,610	(56)	121,800
Salt Lake City	67,260	(45)	80,880	(55)	148,140
Seattle	52,370	(45)	62,760	(55)	115,130
Topeka	59,080	(45)	71,060	(55)	130,140
Washington, DC	50,640	(44)	63,950	(56)	114,590

Source: Derived from bin data in *Engineering Weather Data*, published by the Departments of the Army, Navy, and Air Force.

culated based upon 5/7th's (school open 5 days out of 7) of the heating degree-hours for the heating balance point temperature of 55°F.

Occupied HDH$_{55}$ = (5/7 × Day HDH$_{55}$)

= (5/7 × 32,860)

= 23,470 °F·h

which is then used to compute the heat needed when occupied:

Heat needed when occupied

= 23,470 °F·h × 7500 Btu/°F·h

= 176,025,000 Btu per year

Then the energy need for the unoccupied

TABLE 2.5 HEATING DEGREE-HOURS (HDH) FOR ST. LOUIS, MISSOURI

Base Temp. (°F)	Degree-Hours Below Base Temperature				
	Day	(% of Total)	Night	(% of Total)	Total
50	24,270	(43)	31,770	(57)	56,040
55	32,860	(44)	42,110	(56)	74,970
60	42,810	(44)	53,970	(56)	96,780
65	54,190	(44)	67,610	(56)	121,800
70	67,140	(45)	83,330	(55)	150,470

Source: Derived from bin data in *Engineering Weather Data*, published by the Departments of the Army, Navy, and Air Force.

two days and seven nights per week can be computed when the temperature is set back to 50°F:

Unoccupied HDH$_{50}$
$$= (2/7 \times \text{Day HDH}_{50})$$
$$+ (7/7 \times \text{Night HDH}_{50})$$
$$= (2/7 \times 24{,}270)$$
$$+ (7/7 \times 31{,}770)$$
$$= 6{,}935 + 31{,}770$$
$$= 38{,}705 \text{ °F·h}$$

which is then used to compute the heat needed when the building is unoccupied.

Heat needed when unoccupied
$$= 38{,}705 \text{ °F·h} \times 7500 \text{ Btu/°F·h}$$
$$= 290{,}288{,}000 \text{ Btu per year}$$

Therefore, the total heat needed by the building is estimated to be 466,313,000 Btu per year, 62% of which is needed when the building is unoccupied.

The architect presents these results to the school board along with some of the factors that will impact upon this estimate such as massiveness of construction, morning "warm up" of the building, solar heating benefits of some south-facing windows, and winter holidays. The architect also notes that the 12-h day/12-h night climate data used is appropriate since the school is normally actively used into the early evening hours.

2.6 TEMPERATURE BINS

It is clear that many engineering and energy applications require climatic temperature information which is far more detailed than simple base 65°F degree-day values.

To provide such data for various base temperatures, the bin method has been devised, which keeps track of average temperatures for each of the 8760 hours (365 × 24) that occur each year. The "bins," or groupings, typically used are 5-degree ranges centered on a reference value. Therefore, the 62-degree bin for a particular climate represents the number of hours during the year when temperatures fall between 59.5 and 64.5°F. Figure 2.5 graphically illustrates how temperatures fall into these bins.

Very detailed bin method temperature data is put forth in the latest edition of *Engineering Weather Data Manual*, published by the Departments of the Army, Navy, and Air Force and available from the U.S. Government Printing Office. A simplified listing of hourly weather (temperature) occurrences can also be found in the *ASHRAE Handbook of Fundamentals*. Table 2.6 shows data for three cities taken from that listing which can be used to compute heating degree-hours below any base temperature.

Example 2G: The heating balance point temperature of an office building to be built in Boston is 40°F. Using the bin method, determine the number of heating degree-hours which occur in Boston below 40°F.

Solution: Obtain the hours of occurrence for the 37°F bin (temperatures from 34.5 to 39.5), and all other bins as shown

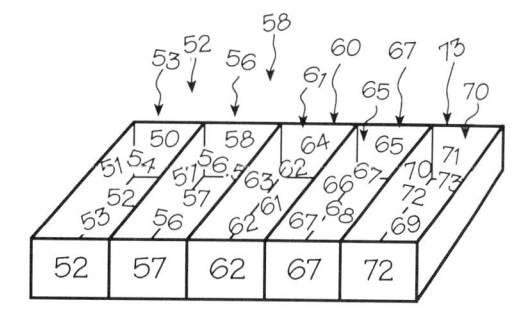

FIGURE 2.5 Temperatures falling into 5 degree "bins."

24 CLIMATE

TABLE 2.6 HOURLY TEMPERATURE OCCURRENCES (PER AVERAGE YEAR)

Outdoor Dry-Bulb Temperatures (°F)		Hours of Occurrence		
Bin	Range	St. Louis	Boston	Bismarck
72	69.5 to 74.5	823	676	454
67	64.5 to 69.5	728	819	566
62	59.5 to 64.5	646	804	614
57	54.5 to 59.5	575	781	606
52	49.5 to 54.5	585	766	563
47	44.5 to 49.5	578	757	520
42	39.5 to 44.5	620	828	518
37	34.5 to 39.5	671	848	604
32	29.5 to 34.5	650	674	653
27	24.5 to 29.5	411	429	550
22	19.5 to 24.5	219	256	474
17	14.5 to 19.5	134	151	371
12	9.5 to 14.5	77	74	338
7	4.5 to 9.5	40	35	292
2	−0.5 to 4.5	15	4	278
−3	−5.5 to −0.5	7	9	208
−8	−10.5 to −5.5	1	1	131
−13	−15.5 to −10.5			77
−18	−20.5 to −15.5			80

Source: Based on data extracted with permission from the *1989 ASHRAE Handbook of Fundamentals*, Chapter 28, Table 8.

below from Table 2.6. Then multiply the temperature difference of the bin from the 40°F reference temperature to obtain heating degree-hours for each bin.

This example shows that although somewhat tedious, it is easy to compute degree-hour information relative to any setpoint, setback, or heating balance point tempera-

Outdoor Dry-Bulb Temperatures (°F)		Degrees Below Ref. Temp. of 40°F		Hours of Occurrence		Degree-Hours (°F·h)
Bin	Range					
37	34.5 to 39.5	3	×	848	=	2544
32	29.5 to 34.5	8	×	674	=	5392
27	24.5 to 29.5	13	×	429	=	5577
22	19.5 to 24.5	18	×	256	=	4608
17	14.5 to 19.5	23	×	151	=	3473
12	9.5 to 14.5	28	×	74	=	2072
7	4.5 to 9.5	33	×	35	=	1155
2	−0.5 to 4.5	38	×	4	=	152
−3	−5.5 to −0.5	43	×	9	=	387
−8	−10.5 to −5.5	48	×	1	=	48
						25,408 °F·h

ture. And with the aid of a computer spreadsheet, a designer can quickly obtain results for their particular area.

2.7 SUMMER DESIGN CONDITIONS

Design conditions for hot summer weather are established using the same underlying principles discussed in Section 2.2 for winter. Namely, reasonable design conditions should be used, not the all-time worst case conditions. Such an approach will avoid wasteful and expensive oversizing of equipment and systems.

Table 2.7 provides Summer Design Conditions for 40 United States cities. Much of the data relates to humidity which is, of course, fundamental to the design and selection of cooling systems and equipment.

The percentages indicated under "Design Dry-Bulb and Mean Coincident Wet-Bulb" of 1%, 2.5%, and 5% refer to the percent of summer hours when the climate will exceed the temperatures indicated. There are a total of 2928 h during the cooling season months of June through September. Therefore, the 1% value represents only 29 h, the 2.5% value represents about 73 h, and the 5% value represents about 146 h during the summer months when the design conditions will be exceeded. In the table, the first number is the dry-bulb temperature and the second number is the mean coincident wet-bulb temperature used for design with that dry-bulb temperature. A selection is made from these values to compute cooling loads as discussed below.

The values in the "Mean Daily Range" refer to the difference between average daily maximum and minimum dry-bulb temperatures during the warmest summer month. This data will be used to illustrate the practicality of various design approaches requiring dry air, indicated by a large (over about 20°F) Mean Daily Range.

Finally the "Design Wet-Bulb" values indicated are wet-bulb temperatures which have been equaled or exceeded 1%, 2.5%, or 5% of the hours during the summer months. These values are normally only used to evaluate evaporative cooling options in dry climates, as discussed in Section 18.6.

Selection of Proper Design Conditions

How is the data in Table 2.7 used? Suppose one is designing a project near National Airport in Washington, DC. Find the numbers 93/75 in the "Design Dry-Bulb and Mean Coincident Wet-Bulb, 1%" column. This means that the dry-bulb temperature equals or exceeds 93°F in Washington during only 1% of the summer hours, and that the typical humidity level at those times will equate to a 75°F wet-bulb temperature.

So, are the 1% values for Washington, DC of 93°F dry-bulb and 75°F wet-bulb the climatic conditions which should be used for design purposes? Generally not. For most applications, designing for the 1% condition is excessive, and will increase system requirements and equipment sizing to satisfying conditions which will occur on average only 29 hours per summer.

Normal practice for most applications is to design for the 2.5% design condition. And for less demanding cooling applications, design for the 5% condition will result in initial cost and operational savings while yielding a system which may be unable to maintain indoor design conditions for about 145 hours each summer.

Ultimately, the design conditions to be employed will depend upon code requirements, the building type, professional judgment, and perhaps some direct discussion with the client of the cost ramifications of "oversizing" as well as the tolerance of the application to possible "undersizing" (inadequate cooling).

TABLE 2.7 SUMMER DESIGN CONDITIONS[a]

	Summer (°F) Design Dry-Bulb and Mean Coincident Wet-Bulb			Mean Daily Dry-Bulb Range (°F)	Design Wet-Bulb		
	1%	2.5%	5%		1%	2.5%	5%
Albuquerque, NM AP	96/61	94/61	92/61	27	66	65	64
Anchorage, AK AP	71/59	68/58	66/56	15	60	59	57
Atlanta, GA AP	94/74	92/74	90/73	19	77	76	75
Bismarck, ND AP	95/68	91/68	88/67	27	73	71	70
Boston, MA AP	91/73	88/71	85/70	16	75	74	72
Casper, WY AP	92/58	90/57	87/57	31	63	61	60
Charleston, SC Co	94/78	92/78	90/77	13	81	80	79
Chicago, IL Co	94/75	91/74	88/73	15	79	77	75
Cincinnati, OH Co	92/73	90/72	88/72	21	77	75	74
Dallas, TX AP	102/75	100/75	97/75	20	78	78	77
Denver, CO AP	93/59	91/59	89/59	28	64	63	62
Detroit, MI	91/73	88/72	86/71	20	76	74	73
Hartford, CT BF	91/74	88/73	85/72	22	77	75	74
Honolulu, HI AP	87/73	86/73	85/72	12	76	75	74
Houston, TX Co	97/77	95/77	93/77	18	80	79	79
Indianapolis, IN AP	92/74	90/74	87/73	22	78	76	75
Jacksonville, FL AP	96/77	94/77	92/76	19	79	79	78
Little Rock, AR AP	99/76	96/77	94/77	22	80	79	78
Los Angeles, CA Co	93/70	89/70	86/69	20	72	71	70
Madison, WI AP	91/74	88/73	85/71	22	77	75	73
Memphis, TN AP	98/77	95/76	93/76	21	80	79	78
Miami, FL AP	91/77	90/77	89/77	15	79	79	78
Minn./St. Paul, MN AP	92/75	89/73	86/71	22	77	75	73
New Orleans, LA AP	93/78	92/78	90/77	16	81	80	79
NYC, NY Central Park	92/74	89/73	87/72	17	76	75	74
Pensacola, FL Co	94/77	93/77	91/77	14	80	79	79
Philadelphia, PA AP	93/75	90/74	87/72	21	77	76	75
Phoenix, AZ AP	109/71	107/71	105/71	27	76	75	75
Pittsburgh, PA Co	91/72	88/71	86/70	19	74	73	72
Portland, OR Co	90/68	86/67	82/65	21	69	67	66
Reno, NV Co	96/61	93/60	91/59	45	64	62	61
St. Louis, MO Co	98/75	94/75	91/74	18	78	77	76
Salt Lake City, UT AP	97/62	95/62	92/61	32	66	65	64
San Antonio, TX AP	99/72	97/73	96/73	19	77	76	76
San Diego, CA AP	83/69	80/69	78/68	12	71	70	68
San Francisco, CA Co	74/63	71/62	69/61	14	64	62	61
Seattle, WA Co	85/68	82/66	78/65	19	69	67	65
Sioux Falls, SD AP	94/73	91/72	88/71	24	76	75	73
Topeka, KS AP	99/75	96/75	93/74	24	79	78	76
Washington, DC AP (Nat)	93/75	91/74	89/74	18	78	77	76

Source: Extracted with permission from the *1993 ASHRAE Handbook of Fundamentals*, Chapter 24, Table 1.

[a] AP following the city name designates airport location. Co indicates office location within and affected by the urban area. Undesignated stations are semirural and may be compared to airport data. The *ASHRAE Handbook* lists two or more locations for many cities. The Co location data has been listed for cities whenever available.

2.8 OTHER CLIMATIC FACTORS

Several other important climatic factors need to be understood in order to select appropriate HVAC systems and to estimate their performance and efficiency. Table 2.8 provides data for the representative cities on cooling degree-days (CDD), relative humidity (RH), mean annual wind speed, and several solar clearness indices.

2.9 COOLING DEGREE-DAYS (CDD)

Cooling degree-days can be thought of as the opposite of heating degree-days. Namely, when the average temperature for a day is above 65°F, cooling degree-days accrue as computed using Formula 2E.

$$CDD_X = \frac{(t_H + t_L)}{2} - X \quad \text{Formula 2E}$$

where

CDD_X is the total number of cooling degree-days above base temperature X
t_H is a day's high dry-bulb temperature in °F
t_L is a day's low dry-bulb temperature in °F
X is the base temperature in °F

Example 2H: The high dry-bulb temperature on a particular day is 90°F (t_H) and the low temperature is 70°F (t_L). Determine how many cooling degree-days accrue.

Solution: Assume the historical base temperature of 65°F and use Formula 2E.

$$CDD_X = \frac{(t_H + t_L)}{2} - X$$

(Formula 2E)

$$= \frac{(90 + 70)}{2} - 65$$

$$= 80 - 65$$

$$= 15 \; CDD_{65} \text{ occur for the day}$$

The actual need for cooling to provide human comfort after all does not begin at 65°F, but rather at an air temperature closer to 80°F. As discussed previously, heating degree-days can be used to predict heating consumption fairly well. However, the computation of required cooling energy from cooling degree-day data is far less direct because of the 65°F basis, and the influence of solar and internal heat gains. To estimate cooling energy use for larger buildings, one needs to use cooling degree-hour data (such as data in *Engineering Weather Data* compiled by the U.S. Air Force) along with a methodology that fully considers building orientation, local solar radiation, and many other factors best evaluated by computer.

The relative need for cooling is illustrated in annual totals (such as indicated in Table 2.8) of only 100 cooling degree-days for San Francisco; about 700 for Boston and Denver; over 2500 for Houston, Jacksonville, Dallas, and New Orleans; 3000 or more for San Antonio and Phoenix; and over 4000 for Miami and Honolulu.

2.10 RELATIVE HUMIDITY

The relative humidity values listed in Table 2.8 are average afternoon values for the month of July. The NOAA *Local Climatological Data, Annual Summaries* contain average values of relative humidity for 4 hourly benchmarks spaced 6 h apart using a 24-h military clock reference. Therefore,

TABLE 2.8 AVERAGE CLIMATIC FACTORS

	Cooling Degree-Days Above 65°F	% Relative Humidity July PM	Wet-Bulb Degree-Hours Above 65°F	Mean Wind Speed (mph)	Yearly % of Possible Sunshine	Number of Clear Winter Days
Albuquerque	1250	28%	0	9.0	76%	82
Anchorage	0	62%	0	7.0	42%	38
Atlanta	1670	60%	10,900	9.1	61%	60
Bismarck	470	47%	2250	10.2	59%	40
Boston	700	57%	1300	12.5	58%	52
Casper	460	26%	0	12.9	NA	46
Charleston	2090	62%	13,600	8.6	63%	60
Chicago	740	57%	3200	10.4	54%	38
Cincinnati	1040	57%	5100	9.1	51%	37
Dallas	2810	49%	8550	10.7	63%	66
Denver	680	34%	0	8.7	69%	60
Detroit	620	54%	3150	10.4	53%	30
Hartford	670	51%	3150	8.5	56%	43
Honolulu	4390	51%	34,200	11.3	69%	48
Houston	2760	58%	17,150	7.9	57%	50
Indianapolis	990	60%	3800	9.6	55%	40
Jacksonville	2520	58%	22,000	7.9	63%	56
Little Rock	2050	56%	10,900	7.8	62%	61
Los Angeles	1340	53%	0	6.2	73%	87
Madison	470	57%	1700	9.8	54%	42
Memphis	2070	57%	13,000	8.9	64%	57
Miami	4100	63%	46,900	9.3	72%	50
Minneapolis	660	54%	1300	10.6	58%	45
New Orleans	2690	66%	30,200	8.2	58%	55
New York City	1090	55%	4800	9.4	58%	55
Pensacola	2680	64%	33,800	8.4	60%	60
Philadelphia	1080	55%	4250	9.6	56%	47
Phoenix	3750	20%	5350	6.3	85%	94
Pittsburgh	650	54%	1350	9.1	46%	25
Portland	330	45%	650	7.9	48%	19
Reno	360	18%	0	6.6	79%	57
St. Louis	1470	56%	6050	9.7	57%	49
Salt Lake City	980	21%	0	8.8	66%	46
San Antonio	2980	52%	24,400	9.2	60%	58
San Diego	840	66%	50	6.9	68%	76
San Francisco	110	74%	0	8.7	66%	62
Seattle	200	49%	200	9.0	43%	21
Sioux Falls	750	53%	950	11.1	NA	47
Topeka	1380	59%	8400	10.0	60%	57
Washington, DC	1430	53%	3900	9.4	56%	49

NA = Not available in NOAA Summaries consulted.

Sources: Wet-bulb degree-hours derived from bin data in *Engineering Weather Data*, published by the Departments of the Army, Navy, and Air Force. Other data extracted from 1992 NOAA *Annual Summaries*.

for some cities, data will be published for hours 00, 06, 12, and 18 while the data for other cities may be for hours 02, 08, 14, and 20 with hour 20 of course referring to 8:00 PM.

The relative humidity values for afternoons in July indicated in Table 2.8 were obtained for the various cities for their early afternoon time (such as hours 12, 13, 14, 15, or 16) when dry-bulb temperatures will also typically be at or near their highest for the day. By contrast, at night, when air temperatures fall, the relative humidity will increase since air can hold less moisture at lower temperatures (see Section 9.1 for details).

As an example, the NOAA data for the very arid Reno, Nevada climate contains the following for average July conditions. At hour 16 (4 PM), the relative humidity is only 19% while the normal daily maximum temperature is 91.1°F. Then at hour 04 (4 AM) the relative humidity is 66% at a time when the temperature would be close to the normal daily minimum of 47.4°F.

Figure 2.6 shows these two conditions on a simplified psychrometric chart. Note that the two points are connected by a line which is close to horizontal, indicating that the absolute humidity (actual amount of moisture) in the air (on the right vertical axis of the chart) remains relatively constant.

Relative humidity can have a profound effect on design strategies and on equipment selection and performance. Climates that are considered to be "humid" in the continental United States are generally those in low coastal areas below 30°N latitude in the southeast and near the Gulf of Mexico. In such climates with a high year-round relative humidity, it is difficult or impossible to use outside air for free cooling (the "economizer cycle"), or cool night air to precool the thermal mass of a building. In addition, the performance of equipment which must reject heat to the air (air cooled

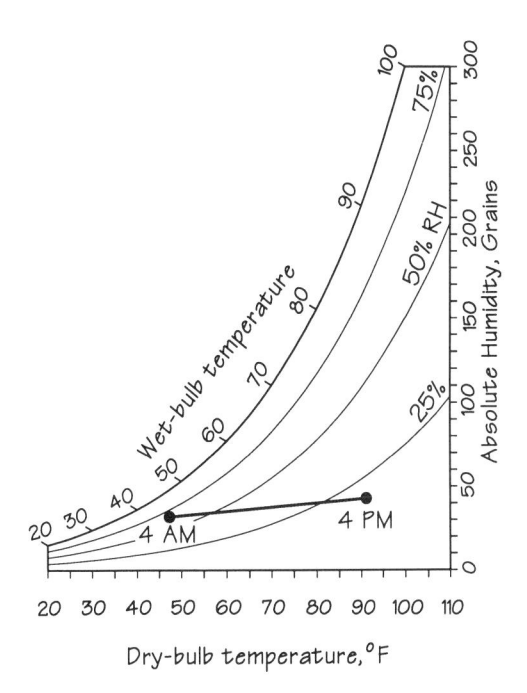

FIGURE 2.6 Plot of temperature/humidity conditions for a very dry climate (Reno, Nevada).

condensers and cooling towers—see tables in Chapter 20) will be reduced, and some equipment such as evaporative coolers will not perform adequately.

Wet-Bulb Degree-Hours

Another important indicator of the impact of humidity over the course of an entire cooling season is the total of wet-bulb degree-hours above 65°F, as represented in Table 2.8. Totals are obtained in the manner illustrated in the following example:

Example 2I: Determine the total number of wet-bulb degree-hours above 65°F which will occur in Boston on a given day assuming the following temperature conditions:

30   CLIMATE

Time Period	Dry-Bulb Temperature (°F)	Wet-Bulb Temperature (°F)
12 MID–9 AM	80 or less	60 or less
9 AM–10 AM	82	60
10 AM–11 AM	84	62
11 AM–NOON	86	63
NOON–1 PM	87	64
1 PM–2 PM	88	67
2 PM–3 PM	80	70
3 PM–4 PM	82	72
4 PM–5 PM	80	68
5 PM–6 PM	75	63
6 PM–12 MID	70 or less	62 or less

Solution: The data show that wet-bulb temperatures will exceed 65°F beginning at 1 PM and until 5 PM (4 h). The total for the day can then be totaled as follows:

Time Period	Wet-Bulb Temperature (°F)	−	Reference Temperature (°F)	=	Hourly Total
1 PM–2 PM	67	−	65	=	2
2 PM–3 PM	70	−	65	=	5
3 PM–4 PM	72	−	65	=	7
4 PM–5 PM	68	−	65	=	3
			Daily Total	=	17

This is not a large total for a day. Table 2.8 indicates a total for Boston of only about 1300 wet-bulb degree-hours exceeding 65°F each year. This is a relatively small total compared to the wet-bulb degree-hours exceeding 65°F which accrue in humid climates such as Atlanta (10,900), New Orleans (30,200), and Miami (46,900).

2.11 WIND

The average yearly wind speed information shown in Table 2.8 provides just an inkling of the monthly wind data published by NOAA. For most cities, NOAA data shows that winter winds are several miles per hour stronger than summer breezes. And for most weather stations, the prevailing direction is also indicated.

For example, note the normal NOAA wind data of mean speed in miles per hour and prevailing direction for Boston as follows:

Month	J	F	M	A	M	J	J	A	S	O	N	D
Mean Speed (m/h)	13.9	13.8	13.7	13.2	12.2	11.5	11.0	10.8	11.3	12.0	12.9	13.6
Dir	NW	WNW	NW	WNW	SW	SW	SW	SW	SW	SW	SW	WNW

2.12 ATMOSPHERIC CLEARNESS

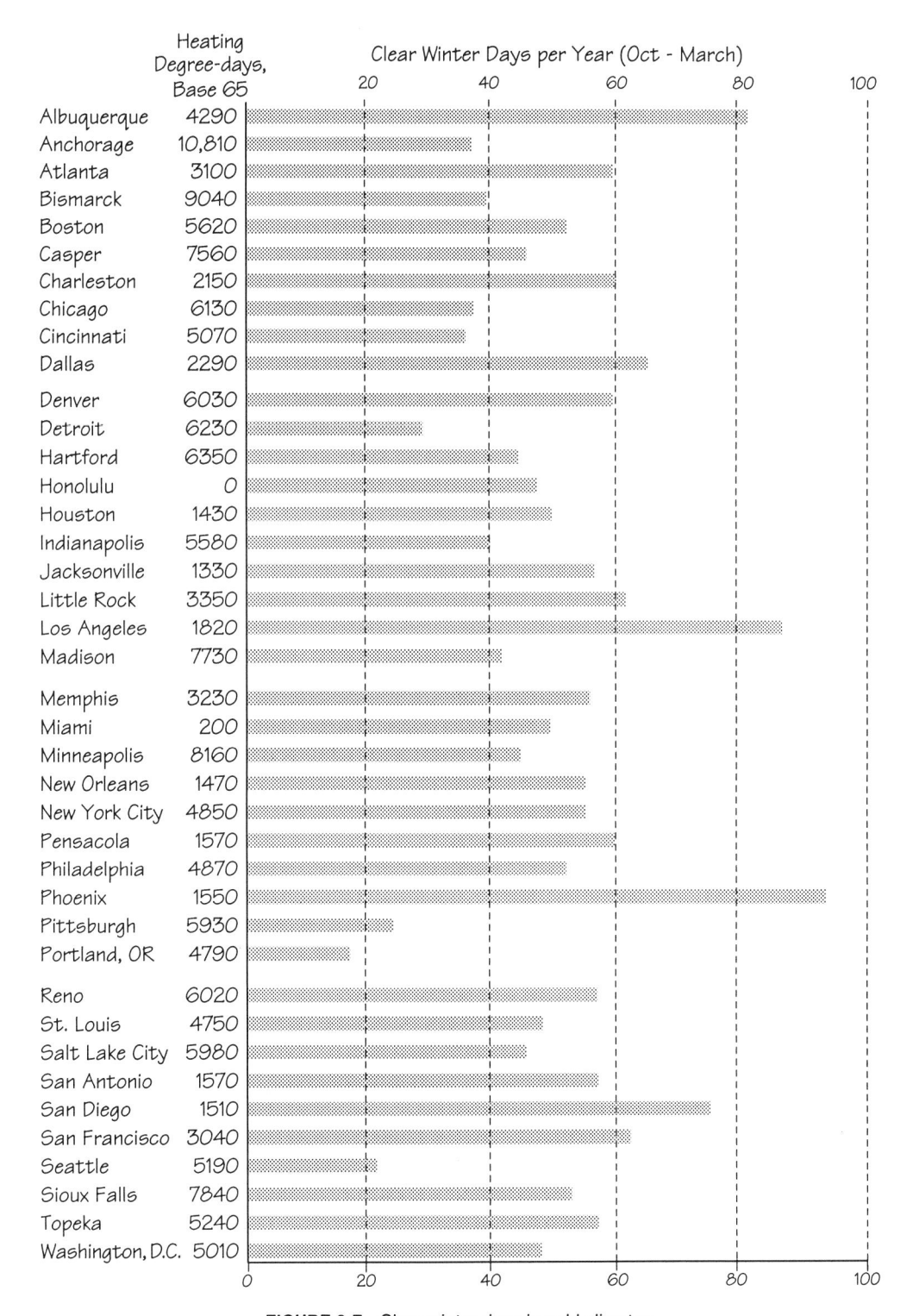

City	Heating Degree-days, Base 65	Clear Winter Days per Year (Oct - March)
Albuquerque	4290	80
Anchorage	10,810	37
Atlanta	3100	58
Bismarck	9040	40
Boston	5620	52
Casper	7560	45
Charleston	2150	58
Chicago	6130	38
Cincinnati	5070	36
Dallas	2290	64
Denver	6030	58
Detroit	6230	28
Hartford	6350	44
Honolulu	0	47
Houston	1430	50
Indianapolis	5580	40
Jacksonville	1330	55
Little Rock	3350	60
Los Angeles	1820	85
Madison	7730	42
Memphis	3230	55
Miami	200	50
Minneapolis	8160	45
New Orleans	1470	55
New York City	4850	55
Pensacola	1570	60
Philadelphia	4870	52
Phoenix	1550	93
Pittsburgh	5930	24
Portland, OR	4790	18
Reno	6020	57
St. Louis	4750	48
Salt Lake City	5980	45
San Antonio	1570	57
San Diego	1510	75
San Francisco	3040	62
Seattle	5190	22
Sioux Falls	7840	53
Topeka	5240	57
Washington, D.C.	5010	48

FIGURE 2.7 Clear winter days in cold climates.

Information on summer breezes is needed if natural cooling strategies are being considered (see Section 18.2). Winter wind information may also influence landscaping design (e.g., buffers) in harsh climates. Finally, U-factors used in heat loss calculations are based on a 15-mph winter wind. Note that the Boston wind data indicates a close agreement to 15 mph during the winter.

2.12 ATMOSPHERIC CLEARNESS

Atmospheric clearness is a primary climatic factor which affects energy consumption by HVAC systems and is fundamental to passive or active solar collection. Table 2.8 lists the "Percent of Possible Sunshine" for various cities on "average" each year as put forth in NOAA data. Note that values range from 85% for a very clear climate such as Phoenix to only 42% for the much more cloudy climate indicated for Anchorage.

NOAA data is also classified into days which are clear, partly cloudy, or cloudy. "Clear" includes 0, 1, 2, or 3 tenths of cloud cover (0/10, 1/10, 2/10, or 3/10). A designation of "Partly Cloudy" is given for conditions where the cloud cover is 4, 5, 6, or 7 tenths. Finally skies with 8, 9, or 10 tenths are considered "Cloudy." Listed in Table 2.8 is the number of "clear" days which occur each year on average out of the 182 winter days during the months of October through March.

Indicated in Figure 2.7 is number of clear winter days (October through March) for cities with over 3000 heating degree-days per year (HDD_{65}).

Note how cities with almost identical heating degree-day totals (e.g., Denver and Pittsburgh) experience markedly different winter clearness—conditions that will greatly affect the benefits derived from solar design strategies.

3

SOLAR BASICS

Humans have been fascinated by the movement of the sun through the sky since the beginning of time. Witness Stonehenge in England, Machu Picchu in the Peruvian Andes, and Mesa Verde in the southwestern United States. But the fascination was practical as well. When Aristotle and Socrates designed buildings, their choices for a heating system were not those contained in chapters of this book (steam, circulating hot water, forced hot air, or some form of electricity). They had to design for solar heating.

Reviewed herein are fundamentals about the sun, a large, but distant nuclear fusion reactor. This information on apparent solar motion, geometry, and resultant solar angles will help to provide the working knowledge to design passive solar heated buildings, as discussed in Chapter 12.

3.1 FUNDAMENTALS

Shown in Figure 3.1 is the earth's slightly elliptical orbit over 90 million miles from the sun. One complete 360 degree orbit around the sun takes approximately $365\frac{1}{4}$ days. This is why we have leap years every four years (exception, and final adjustment: there is no leap in centesimal years such as 1700, 1800, and 1900 unless the centesimal year is evenly divisible by 400, such as the year 2000, which is a leap year). Obviously Figure 3.1 is not even remotely to scale since the sun has a diameter of about 870,000 mi whereas the earth's is approximately 8000 mi.

The earth constantly spins, like a top, making one rotation in 24 h, on an axis which remains constantly tilted at approximately 23.5° to the plane in which the earth travels around the sun.

On the left side of Figure 3.1, the northern hemisphere is shown leaning into the sun, and at noon the sun falls directly over locations on earth that are 23.5° above the equator. And as the earth rotates, conceptually a circle is formed around the earth equidistant from the equator. This is known as a ''latitude'' (see below). In this case, 23.5°N latitude, also known as the Tropic

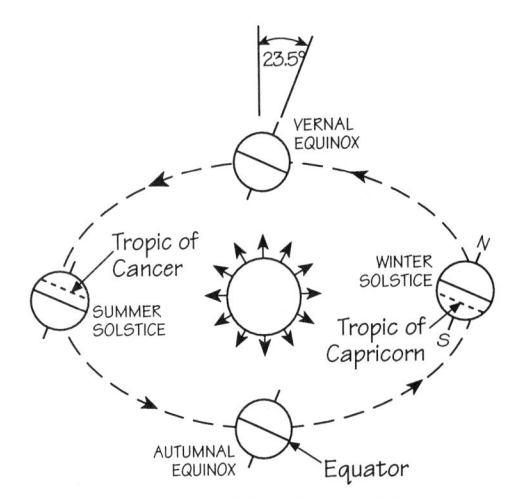

FIGURE 3.1 Earth's orbit around the sun.

of Cancer, denotes the northernmost latitude on earth where the sun is sometimes directly overhead (i.e., at a 90° altitude angle, see below). The day when this occurs is known as the summer solstice, the first day of summer; in the northern hemisphere, around June 21st.

Similarly, on the right side of the figure, the sun is directly over the southern hemisphere, at 23.5°S latitude (the Tropic of Capricorn), which is as far south as the sun can be seen directly overhead. This occurs around December 21st, and represents the first day of winter in the northern hemisphere (the winter solstice).

The two other earth positions in the middle of the figure represent intermediate positions between the two extremes. On the days after the winter solstice, the sun begins to "move north," falling directly over higher and higher latitudes until about March 21st when it falls directly over the equator. This is the spring (vernal) equinox, and all points on earth experience an equal amount of day and night (approximately 12 h with the sun above the horizon).

The sun continues to "travel north" until the summer solstice, and then recedes southward until the fall (autumnal) equinox (about September 21st) indicated by the lower middle earth in the figure, where the sun again falls directly over the equator.

The tilt of the earth's axis remains constant, at 23.5°. The effect of the tilt changes because of the earth's revolution around the sun, resulting in our changing seasons.

Solar Declination

The solar "declination" is equivalent to the latitude at which the sun is directly overhead (90° altitude). When the sun is directly over the equator, on the two equinoxes, the declination is 0°. On the summer solstice, the declination is +23.5°, whereas on the winter solstice, the declination is −23.5°. At all other times, the declination is between these limits and directly over locations between the Tropics of Cancer and Capricorn.

Listed in Table 3.1 are the standard declination values for the earth on the 21st day of each month.

TABLE 3.1 SOLAR DECLINATION, 21ST DAY OF MONTH			
Month	Declination	Month	Declination
January	−19.9°	July	+20.5°
February	−10.6°	August	+12.1°
March	0.0°	September	0.0°
April	+11.9°	October	−10.7°
May	+20.3°	November	−19.9°
June	±23.5°	December	−23.5°

The mathematical expression used to determine the changing position (declination) of a tilted object revolving in an elliptical orbit incorporates the sine function. The familiar sine curve can be used to illustrate changes in declination.

Think of one cycle of the sine curve (360 degrees) as representing a calendar year of 365 days, which is very nearly one degree for each day, with the four quadrants of the sine curve very nicely paralleling the four seasons. Then think of the amplitude of the sine curve which is equal to 1, as also being equal to one full measure of the earth's maximum declination of 23.5° as shown in Figure 3.2.

Figure 3.2 shows the sine curve with markings relative to the 21st day of each month and declination. Note that at the origin, where the cycle begins, declination is zero but becoming positive. This point equates to March 21st (the Spring Equinox). The sine curve then rises steeply between March 21st and April 21st, roughly 30° for 30 days. Trigonometrically, the sine of 30° is 0.5. Therefore, the declination has changed by 0.5 or one-half of the amplitude of the curve (0.5 × 23.5) equaling 11.75° of increased declination (and noon altitude) in just one month. Compare this result to the solar declination for April shown in Table 3.1 of +11.9°.

From April to May, the change is somewhat less dramatic. The sine of 60° is 0.866, thus the difference since April is 0.366 (0.866 − 0.5). Therefore, the solar declination only changes by 8.6° (0.366 × 23.5) between April and May. Therefore, the solar declination for May will be about 20.35° (11.75 + 8.6) which again agrees with the value shown in Table 3.1.

Then from May to June, the sine curve change is from 0.866 at 60°, to 1.0 at 90°, or a change of only 3.15° (1.0 − 0.866 = 0.134 × 23.5) from May 21st to June 21st. Therefore, the solar declination for June is 23.5° (11.75 + 8.6 + 3.15), just as it is listed in Table 3.1.

Solstice means "when the sun stands still." The sine curve dramatically illustrates why there is very little change in declination and resultant sun angles over the several month periods around the summer and winter solstices. It also shows why sun angles, day length, and indeed climate change rapidly around the equinoxes, as also indicated by Figure 3.2.

Latitude and Longitude

The sun's apparent location in the sky at a given point on earth depends upon the season (solar declination as described above), location (latitude), and time of day (hour). Figure 3.3a illustrates the term latitude, which is an imaginary circle around the earth "parallel" to the equator, as established by the angle above or below the equator from the earth's center.

Illustrated in Figure 3.3b is the term longitude, which can be thought of as "great circles" perpendicular to the equator connecting the north and south "poles" of the earth's axis. The earth rotates once, a full 360°, during each 24-h day, which forms the basis for the full time zones, with each time zone being approximately 15° (360/24) of longitude wide. This also can be thought of as 4 minutes per degree of longitude (60 min/15° of longitude).

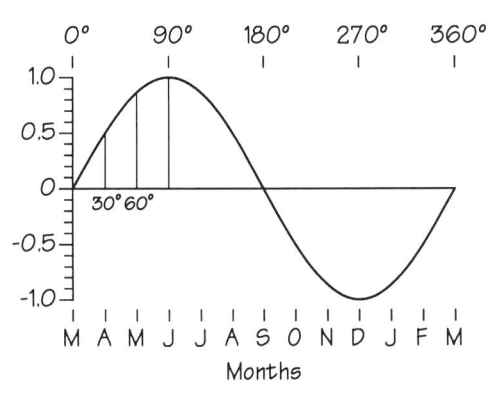

FIGURE 3.2 Solar declination and the sine curve.

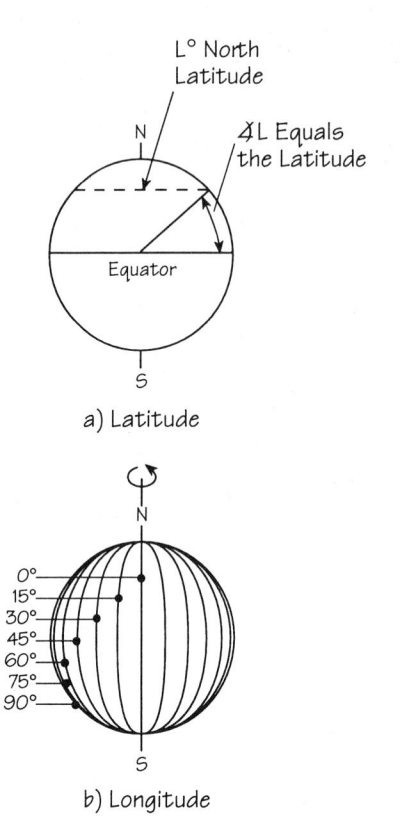

FIGURE 3.3 The meaning of latitude and longitude.

The starting point for longitudes of 0° is taken to be Greenwich (near London, England—not Connecticut). The longitudes which apply to time zones of the United States and nearby cities are: Eastern (75°, about Philadelphia); Central (90°, near New Orleans and Springfield, IL); Mountain (105°, close to Denver); Pacific (120°, near Los Angeles and Yakima, WA); and Alaska/Hawaii (150°, near Fairbanks and east of the Hawaiian Islands).

See Table 2.1 and Figure 2.2 for the latitudes of representative American cities. Information for other American and worldwide cities are contained in many sources, including almanacs and the *ASHRAE Handbook of Fundamentals*. Latitude information is fundamental to HVAC design since it directly correlates with basic climate and sun angles. Longitude information is only rarely used to adjust data for a given location relative to the basis of the time zone (see Figure 3.5).

The Solar Constant

After traveling through over 90 million miles of space, energy from the sun arrives at the earth. The quantity of solar energy which would fall just outside earth's atmosphere on a 1 ft^2 surface directly facing the sun in one hour is known as the "solar constant."

The solar constant represents the availability of solar energy before atmospheric absorption or reflection (see Figure 3.4). After scientific investigation, the value of the solar constant has been established at approximately 430 Btu/h · ft^2. In actuality, the solar constant varies slightly because the earth–sun distance changes by about 3 million miles during the elliptical orbit, with the earth being closer to the sun during winter in the northern hemisphere. As solar radiation travels through our atmosphere some of the heat is absorbed and reflected. The remainder then arrives on earth with some of it impacting upon building surfaces. In particular, solar heat gain

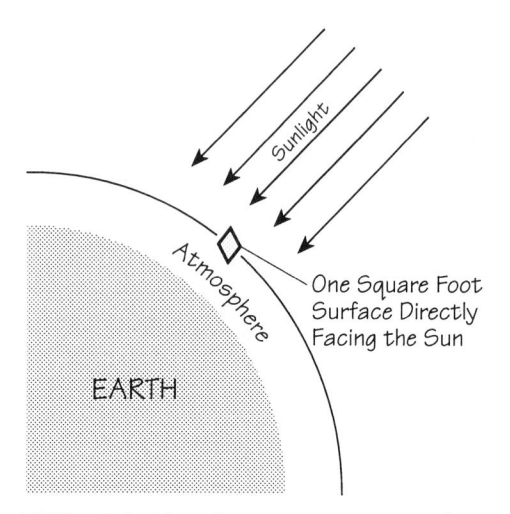

FIGURE 3.4 The solar constant as measured outside the earth's atmosphere.

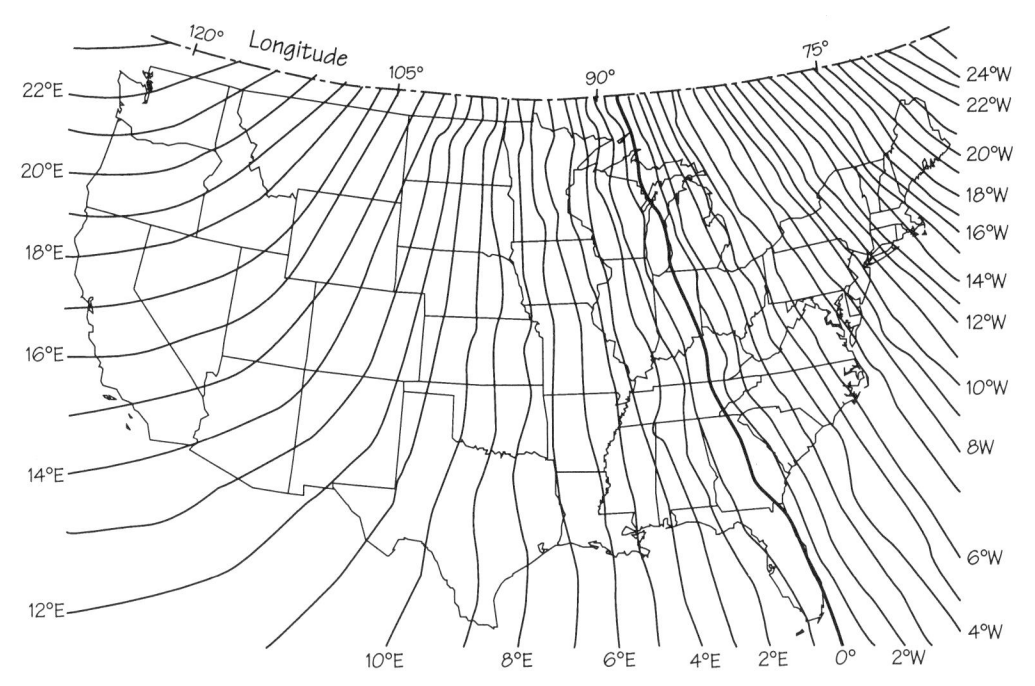

FIGURE 3.5 Magnetic deviation on an isogonic chart of the United States. Redrawn from the Isogonic Chart for the United States, U.S. Department of Commerce, Coast and Geodetic Survey, 1965.

through glazing in commercial buildings requires careful attention.

Magnetic Deviation

Because of the earth's magnetic field, there is a difference between true north and magnetic north for most locations in the United States which can be significant. For precise design applications (such as siting of a passive solar house), consult the "isogonic chart" in Figure 3.5 to identify the proper correction factor. For example, the isogonic chart indicates a value of 20°W for northern Maine. Therefore, solar (true) south will be 20° west of "south" on a magnetic compass.

Another way to determine solar south for a particular location is with the use of a local newspaper. On a sunny day find out what time sunrise and sunset occur (i.e., sunrise at 7:10 and sunset at 5:20). Solar noon will occur halfway between sunrise and sunset, or for the times stated at 12:15. Then at solar noon go to the site and determine true north by noting the direction of shadows cast by vertical objects.

3.2 ALTITUDE AND AZIMUTH ANGLES

An understanding of both altitude and azimuth angles as well as their combined effect is critical to various aspects of design including basic building orientation, selection of shading devices, location of thermal mass, tilt of solar collectors, and analysis of building self-shading.

The altitude angle (angle "A") is the angle of the sun above the horizon, achieving its maximum on a given day at solar noon (see Figure 3.6).

The azimuth angle (angle "B") is also known as the bearing angle. It is the directional angle of the sun's projection onto the

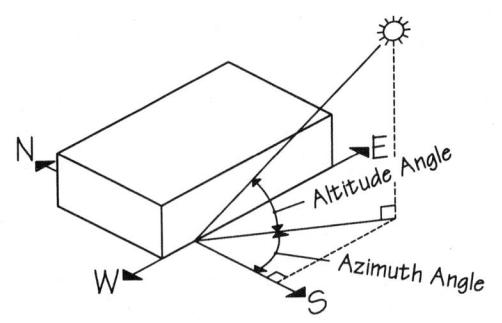

FIGURE 3.6 Definition of altitude and azimuth angles.

ground "plane" relative to north (historically for navigation) or south when building design and solar applications are being considered. In this volume, all references to azimuth angle are relative to solar south, also known as true south.

For North American locations, altitude angles are high in the summer and low in the winter. Also note that 47° (the absolute sum of the maximum positive and negative declination values of +23.5° and −23.5°) is the difference between highest summer and lowest winter noon altitudes for all locations above or below the tropics (23.5°N and 23.5°S latitudes). Therefore, if the noon altitude angle for a midwestern location is 28° on the winter solstice, one also knows that the noon altitude angle on the summer solstice will be 75° (28 + 47).

Table of Angles

Table 3.2 provides altitude and azimuth angles for various latitudes. Note that:

> Morning and afternoon solar times are listed; and there is symmetry around

TABLE 3.2 ALTITUDE AND AZIMUTH ANGLES (°)

Solar Time AM / PM	16° N. Lat. ALT. / AZ.	24° N. Lat. ALT. / AZ.	32° N. Lat. ALT. / AZ.	40° N. Lat. ALT. / AZ.	48° N. Lat. ALT. / AZ.	56° N. Lat ALT. / AZ.

December 21 Sun Angles

7 / 5	7 / 63	3 / 63				
8 / 4	19 / 57	15 / 55	10 / 54	5 / 53	1 / 53	
9 / 3	31 / 49	26 / 46	20 / 44	14 / 42	8 / 41	2 / 40
10 / 2	41 / 37	34 / 34	28 / 31	21 / 29	14 / 28	7 / 27
11 / 1	48 / 21	40 / 18	33 / 16	25 / 15	17 / 14	10 / 14
NOON	51 / 0	43 / 0	35 / 0	27 / 0	19 / 0	11 / 0

January and November 21st Sun Angles

7 / 5	8 / 66	5 / 66	1 / 65			
8 / 4	21 / 61	17 / 58	13 / 56	8 / 55	3 / 55	
9 / 3	33 / 52	28 / 49	22 / 46	17 / 44	11 / 43	5 / 42
10 / 2	43 / 40	37 / 36	31 / 33	24 / 31	17 / 29	10 / 28
11 / 1	51 / 23	44 / 20	36 / 18	28 / 16	21 / 15	13 / 14
NOON	54 / 0	46 / 0	38 / 0	30 / 0	22 / 0	14 / 0

February and October 21st Sun Angles

7 / 5	11 / 75	9 / 74	7 / 73	4 / 72	2 / 72	
8 / 4	25 / 70	22 / 66	18 / 64	15 / 62	11 / 60	7 / 59
9 / 3	38 / 62	34 / 57	29 / 53	24 / 50	19 / 47	13 / 46
10 / 2	50 / 50	45 / 44	38 / 39	32 / 35	25 / 33	19 / 31
11 / 1	59 / 30	52 / 25	45 / 21	37 / 19	30 / 17	22 / 16
NOON	63 / 0	55 / 0	47 / 0	39 / 0	31 / 0	23 / 0

TABLE 3.2 (Continued)

Solar Time AM	PM	16° N. Lat. ALT.	AZ.	24° N. Lat. ALT.	AZ.	32° N. Lat. ALT.	AZ.	40° N. Lat. ALT.	AZ.	48° N. Lat. ALT.	AZ.	56° N. Lat ALT.	AZ.
March and September 21 Sun Angles													
7	5	14	86	14	84	13	82	11	80	10	79	8	77
8	4	29	81	27	77	25	73	23	70	20	67	16	64
9	3	43	75	40	68	37	62	33	57	28	53	23	50
10	2	56	64	52	55	47	47	42	42	35	38	29	35
11	1	68	44	62	33	55	27	48	23	40	20	33	18
NOON		74	0	66	0	58	0	50	0	42	0	34	0
April and August 21st Sun Angles													
5	7											1	109
6	6	3	101	5	101	6	100	7	99	9	98	10	97
7	5	17	97	18	95	19	92	19	89	19	87	18	84
8	4	32	94	32	89	31	84	30	79	29	75	26	71
9	3	46	90	46	82	44	74	41	67	38	61	34	56
10	2	61	85	59	72	56	60	51	51	46	45	40	40
11	1	75	75	71	52	65	37	59	29	51	24	44	21
NOON		86	0	78	0	70	0	62	0	54	0	46	0
May and July 21st Sun Angles													
4	8											1	126
5	7							2	115	5	114	8	113
6	6	5	109	8	108	10	107	13	106	15	104	16	102
7	5	19	106	21	103	23	100	24	97	25	93	25	89
8	4	33	104	35	98	35	93	35	87	35	82	33	76
9	3	47	102	48	94	48	85	47	76	44	68	41	62
10	2	61	103	62	88	61	73	57	61	53	51	48	44
11	1	75	108	76	77	72	52	66	37	59	29	52	23
NOON		86	180	86	0	78	0	70	0	62	0	54	0
June 21st Sun Angles													
4	8											4	127
5	7					1	118	4	117	8	117	11	115
6	6	6	113	9	112	12	110	15	108	17	106	19	103
7	5	20	110	22	107	24	103	26	100	27	96	28	92
8	4	33	108	36	103	37	97	37	91	37	85	36	79
9	3	47	107	49	99	50	89	49	80	47	72	44	64
10	2	61	110	63	95	62	80	60	66	56	55	51	46
11	1	74	120	76	91	74	61	69	42	63	31	56	25
NOON		83	180	89	0	81	0	73	0	65	0	57	0

solar noon. As a result, the altitude and azimuth angles at 9 AM solar time will be identical to the angles at 3 PM solar time.

Spring and Fall sun angles around the two equinoxes (March 21st and September 21st) are equal and symmetrical.

Azimuth angles for 16°N latitude during the prime summer months are all over 100 degrees with the noon azimuth angle being 180° since the sun is ac-

40 SOLAR BASICS

tually north of this latitude and directly over 23.5°N latitude during that period.

Altitude and azimuth (bearing) angles for a given latitude can also be graphically presented as shown in Figures 3.7 and 3.8.

A few things to note about graphical representations of solar angles like Figure 3.7:

- The sun rises due east (90° azimuth) and sets due west (also 90° azimuth) on the equinoxes (March and September 21st). This is not only true for this latitude but for every point on earth on these days.
- The monthly curves are spaced relatively far apart around the equinoxes (March and September) when sun angles, and the seasons, change rapidly. Conversely the curves near the solstice months of June and December are closely spaced, indicating relatively small changes in altitude angles.

Sun angles are identical on October 21st and February 21st. This can create a challenge in designing sun control devices, since the climate and resultant HVAC needs of buildings in most locations are very different in October and February (due to the seasonal lag in climate resulting from absorption of summer heat and winter cold by land and water masses). The use of movable sun control devices (such as awnings) allows for seasonal (and daily) adjustment as needed.

Indicated in Figure 3.8 is a "Sun Path Diagram" for 40°N latitude which depicts the path of the sun as projected on a horizontal plane.

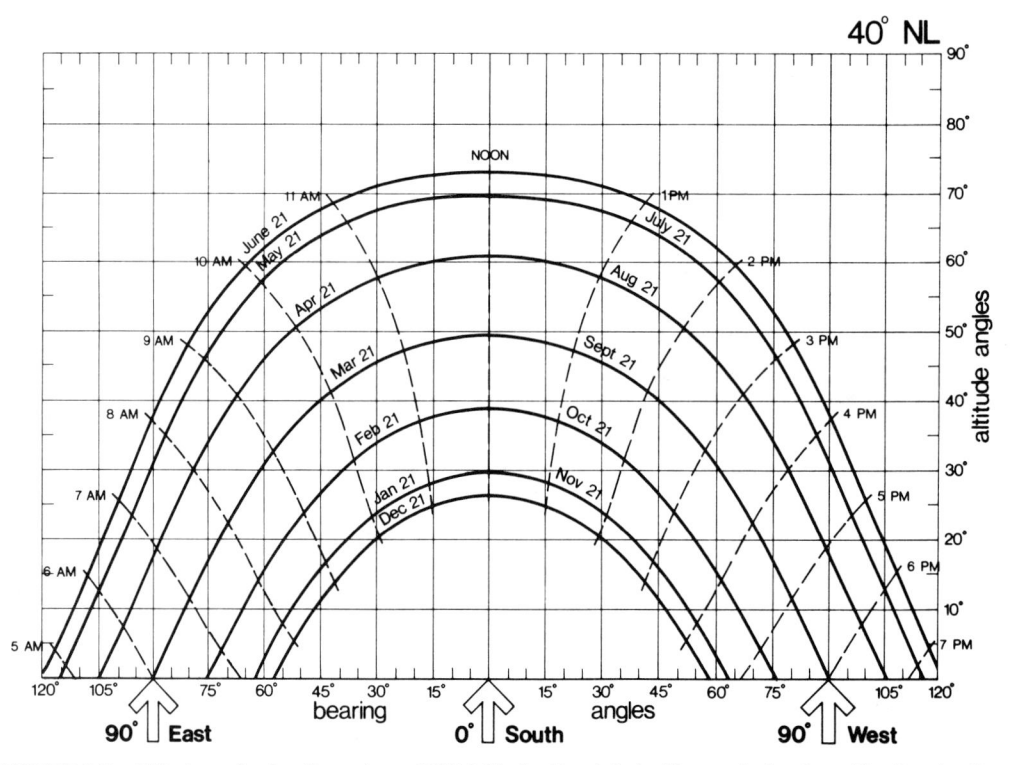

FIGURE 3.7 Altitude and azimuth angles—40°N latitude. Reprinted with permission from *The Passive Solar Energy Book* by Edward Mazria.

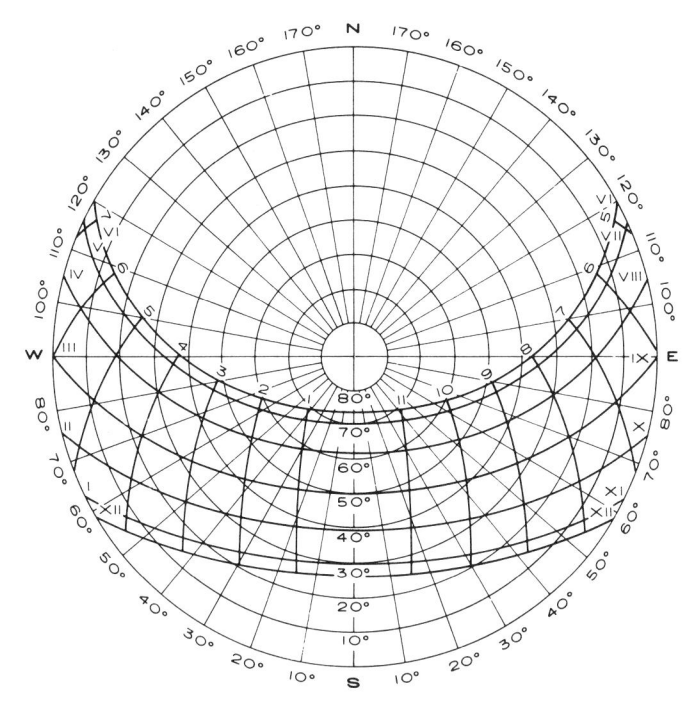

FIGURE 3.8 Sun path diagram—40°N latitude. From *Architectural Graphic Standards* Ramsey/Sleeper, 8th ed., John R. Hoke, ed. Copyright John Wiley, 1988.

A few things to note about Sun Path diagrams:

- True North is at the top of the diagram, east at the right, and so on. These directions are used to read off azimuth angles from solar south at the bottom of the circle.
- The large outer circle represents an altitude angle of 0°. On this circle one can determine the azimuth angle at sunrise.
- The smaller concentric circles represent increasing altitude angles in 10° increments. The point in the center represents an altitude angle of 90° when the sun would be directly overhead.
- The "apparent" path of the sun for the 21st day of each month (noted in Roman numerals) is indicated by curve lines which sweep from right (sunrise) to left (sunset) on the diagram.
- Solar time is indicated by lines which are vertical (and due south) at noon and slightly curving before and after noon.

Other Sun Path diagrams for eight latitudes (ranging from 24° to 52°N in 4° increments) are contained in Ramsey and Sleeper's *Architectural Graphic Standards*, which also contains information on how they can be used to evaluate shade conditions.

Altitude Angles at Noon (Only)

Solar *noon* altitude angles are particularly interesting since they represent the highest altitude angle for that day. As the geometry of Figure 3.9 shows, the solar noon altitude is found using Formula 3A.

$$A_{\text{noon}} = 90° - L + (D) \quad \text{Formula 3A}$$

42 SOLAR BASICS

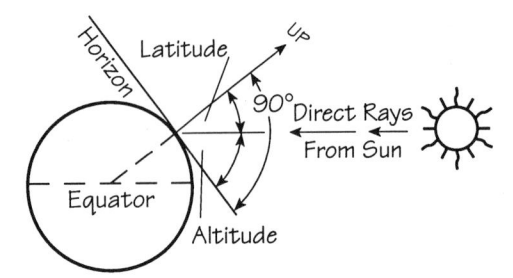

FIGURE 3.9 Determination of noon altitude angles.

where

A_{noon} is the solar noon altitude angle in degrees
90° is the angle between the horizon and the zenith (directly overhead)
L is the latitude of the location in degrees
D is the solar declination in degrees for the date (see Table 3.1). Note that declination is sometimes negative.

Example 3A: Determine the noon altitude angle for August 21st at 32°N latitude.

Solution: Obtain the declination (D) on August 21st of +12.1° from Table 3.1. Then use Formula 3A to determine the Solar Noon altitude angle for 32°N latitude (L) as follows:

$$A_{noon} = 90° - L + (D)$$

(Formula 3A)

$$= 90° - 32° + (+12.1)°$$
$$= 70.1° \text{ for August 21st at } 32°N \text{ latitude}$$

Be sure to realize that Formula 3A determines the altitude angle *at noon only* when the sun will be at its highest angle on that day.

Calculation of Altitude and Azimuth Angles

Altitude and azimuth angles can easily be computed (i.e., using a computer spreadsheet) for any specific location on earth using Formulas 3B and 3C, which take into account the factors that vary: location (latitude), season (declination), and time (hour of day).

The altitude angle (angle A), is calculated using Formula 3B:

$$\sin A = (\cos L \times \cos D \times \cos H) + (\sin L \times \sin D) \quad \text{Formula 3B}$$

where

A is the altitude angle in degrees
L is the latitude in degrees
D is the solar declination in degrees
H is the hour angle in degrees based upon 15 degrees per hour from solar noon

The azimuth (angle B) is then computed using Formula 3C:

$$\sin B = \frac{\cos D \times \sin H}{\cos A} \quad \text{Formula 3C}$$

where

B is the azimuth angle from solar (due) south in degrees
D is the solar declination in degrees
H is the hour angle in degrees based upon 15 degrees per hour from solar noon
A is the altitude angle in degrees

Example 3B: Compute the altitude and azimuth angles for 7 AM solar time for a location at 32°N latitude on March 21st.

Solution: First compute the altitude angle (A) using Formula 3B. The declination

(D) on March 21st is 0° as shown in Table 3.1. The hour angle (H) is 75 degrees since 7 AM is 5 hours from solar noon (5 × 15).

$$\sin A = (\cos L \times \cos D \times \cos H)$$
$$+ (\sin L \times \sin D)$$
(Formula 3B)

$$= [\cos(32) \times \cos(0) \times \cos(75)]$$
$$+ [\sin(32) \times \sin(0)]$$

$$= (0.848 \times 1 \times 0.259)$$
$$+ (0.530 \times 0)$$

$$= 0.220 + 0$$

= 0.220, or an altitude angle A equal to 12.7°

Now compute the azimuth (bearing) angle using Formula 3C.

$$\sin B = \frac{\cos D \times \sin H}{\cos A}$$
(Formula 3C)

$$= \frac{\cos(0) \times \sin(75)}{\cos(12.7)}$$

$$= \frac{1 \times 0.966}{0.975}$$

= 0.991, or an azimuth angle of 82° from solar south.

Compare these results with Table 3.2. Such calculations may at first glance appear tedious. They can be. But today the use of a computer spreadsheet can enable one to quickly determine a full series of altitude and azimuth angles for any location very quickly.

3.3 PROFILE (SHADE) ANGLES

It is often useful to precisely calculate the shading effect of horizontal projections. The angle of shade cast on a particular surface resulting from the combined effect of solar altitude and solar-wall azimuth is known as the "profile" or shade angle.

Profile angles can be calculated for any latitude, date, solar time, and surface orientation. Illustrated in Figure 3.10 is the profile angle (angle C). It is an angle in a plane perpendicular to the surface being evaluated.

The profile angle (C) is determined using Formula 3D. By knowing this angle, shadow heights can be determined based upon the width of the projection.

$$\tan C = \frac{\tan A}{\cos BW} \quad \text{Formula 3D}$$

where

C is the profile angle in degrees
A is the altitude angle in degrees
BW is the solar-wall azimuth angle (angle in the horizontal plane between the solar azimuth angle and the direction in which the surface is facing) in degrees

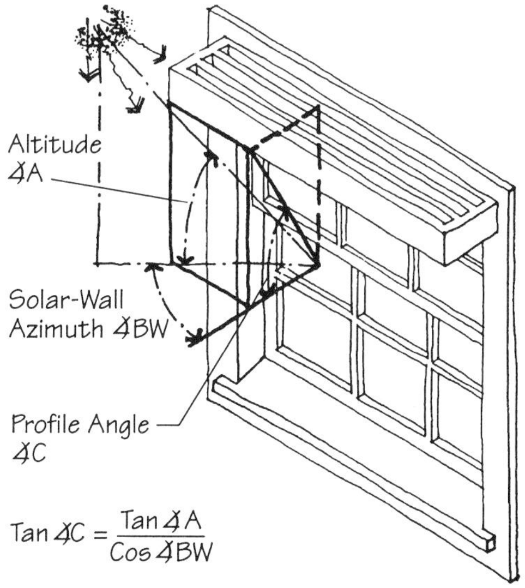

FIGURE 3.10 Definition of the profile (or "shade") angle.

44 SOLAR BASICS

Example 3C: Compute the profile angle at 11 AM on December 21st for the wall of a building which is oriented toward the southeast (45° azimuth) at 40°N latitude as shown in Figure 3.11. Indicate how shadows cast by projections from this building face would affect any glass areas below.

Solution: Obtain from Table 3.2 the 40°N latitude solar angles for 11 AM on December 21st. They are an altitude angle of 25° and an azimuth angle of 15°.

A solar azimuth angle of 15° from south, and a wall orientation for southeast of 45° from south (angle W) are indicated in Figure 3.11. Angle BW (the solar-wall azimuth) is equal to 30°. Apply Formula 3D to calculate the profile angle.

$$\tan C = \frac{\tan A}{\cos BW} \quad \text{(Formula 3D)}$$

$$= \frac{\tan(25)}{\cos(30)}$$

$$= \frac{0.466}{0.866}$$

$$= 0.538, \text{ yielding a profile angle of } 28°$$

The result means that the shadow cast by any projections (such as an overhang) would create a shade angle of 28° perpendicular to the wall orientation as indicated in Figure 3.12.

Tables of profile angles are contained in Section 7.2 where examples also illustrate how they are used to determine shading of glass areas. Perhaps the best way to obtain and appreciate the relative significance of any profile angle is to use a graphic method. Indicated in Figure 3.13 are profile angles based upon solar altitude (curves within the figure in 10° increments), and solar-azimuth angles (along the horizontal axis). Note how profile angles for altitude angles are only slightly more than the altitude angle itself unless there is a fairly large solar-wall azimuth angle.

Example 3D: Graphically determine the profile angle for the same southeast-facing wall and conditions noted in Example 3C.

Solution: As noted in Example 3C, the applicable angles are an altitude angle of 25° and a solar-wall azimuth angle of 30°. Find the interpolated line within Figure 3.14 representing an altitude angle of 25°.

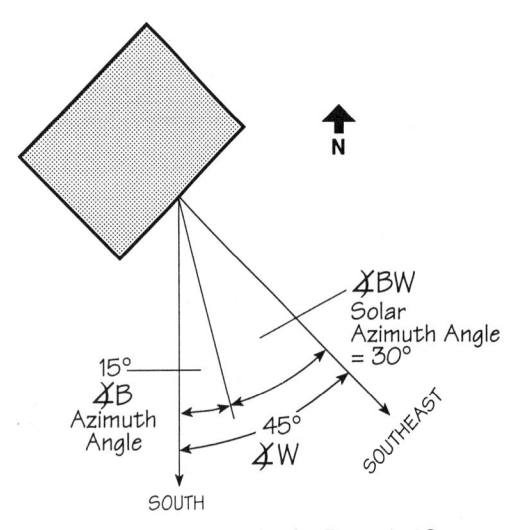

FIGURE 3.11 Angles for Example 3C.

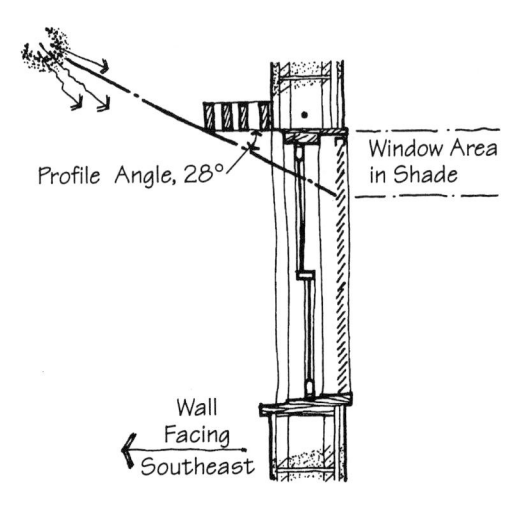

FIGURE 3.12 Shading on the southeast facing wall in Example 3C.

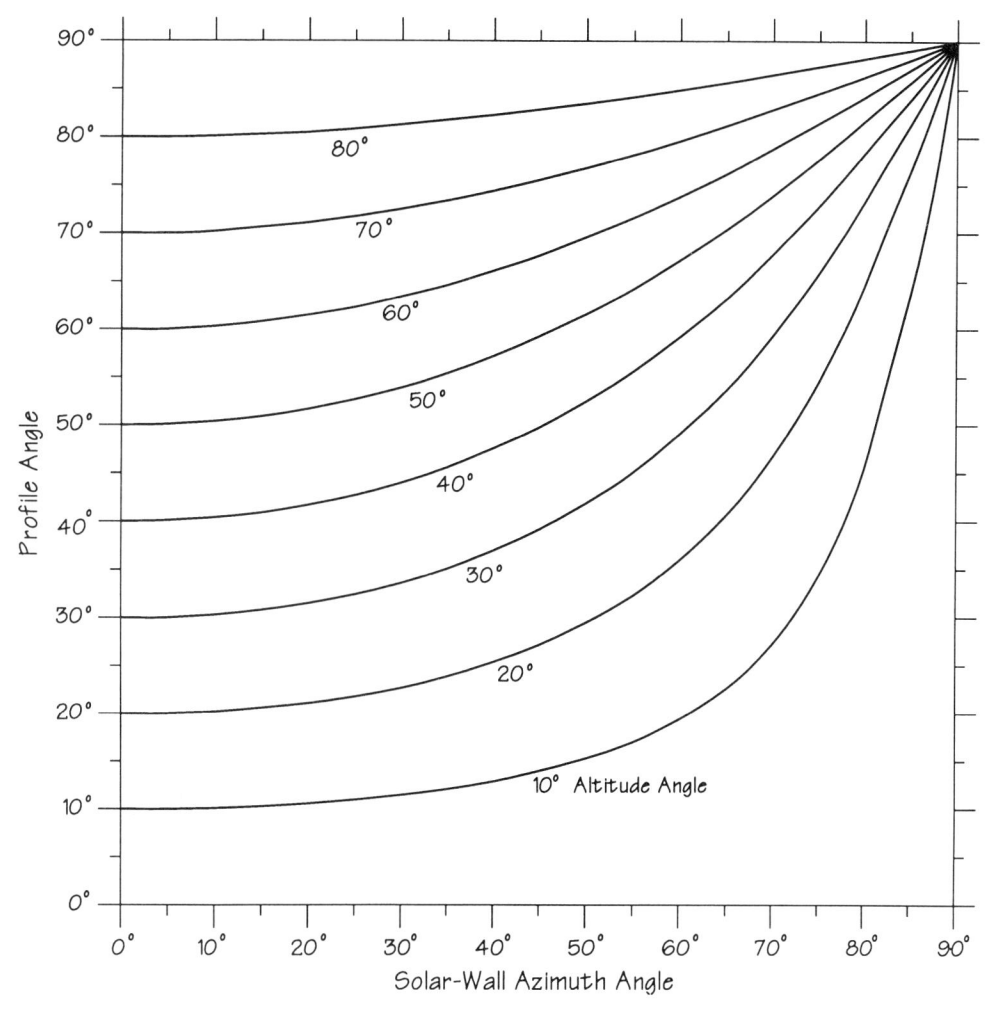

FIGURE 3.13 Graphic determination of profile angles.

Then locate along the horizontal axis a solar-wall azimuth angle of 30°. Rise vertically to intercept with the altitude curve (or interpolated curve when necessary). The answer is 28°, which is in agreement with the previous example.

3.4 CLEAR DAY SOLAR RADIATION ON SURFACES

Illustrated in Figure 3.15 is the total clear day solar radiation (for various north latitudes) which arrives on one square foot of the following surfaces:

Horizontal surfaces
South-facing vertical surfaces
South-facing surfaces which are tilted at an angle equal to the latitude (L).

Intensity of radiation is of prime importance to the performance of passive and active solar systems. Note how the radiation is maximum on south-facing vertical surfaces during the winter months. Therefore, vertical south-facing surfaces (such as win-

46 SOLAR BASICS

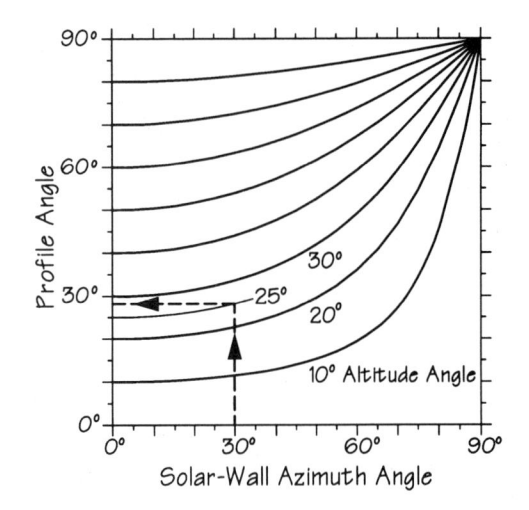

FIGURE 3.14 Determination of profile angle, Example 3D.

dows) are appropriate for solar collection to offset space heating needs in winter.

In contrast, radiation is much more constant year round for south-facing surfaces tilted at the latitude angle. This is why active solar collectors are mounted at a tilt angle equal to the latitude (or latitude plus an additional 10 to 15 degrees) when heat is needed year round such as for domestic hot water.

Finally, note the magnitude of the radiation on horizontal surfaces in the summer months when solar altitude angles are at their maximum. Notice that extreme care should be exercised in incorporating horizontal glazing (skylights) in building design if cooling loads are to be minimized.

a. Horizontal Surfaces

b. South-facing Vertical Surfaces

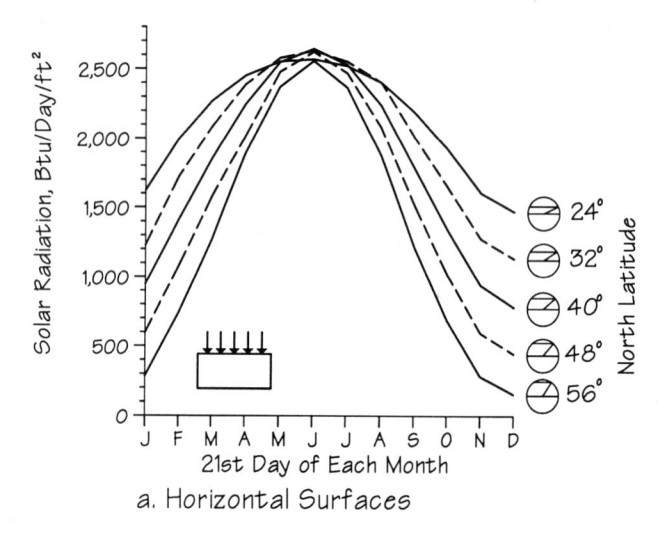

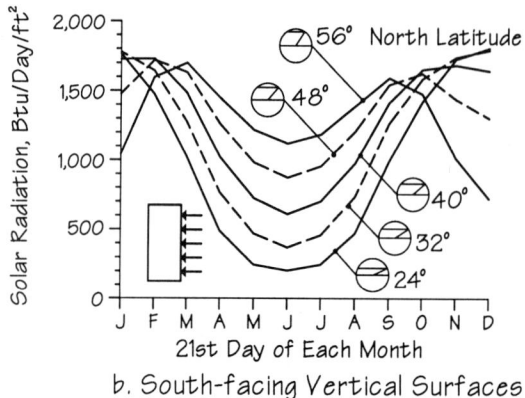

FIGURE 3.15 Daylong clear dry direct solar radiation.

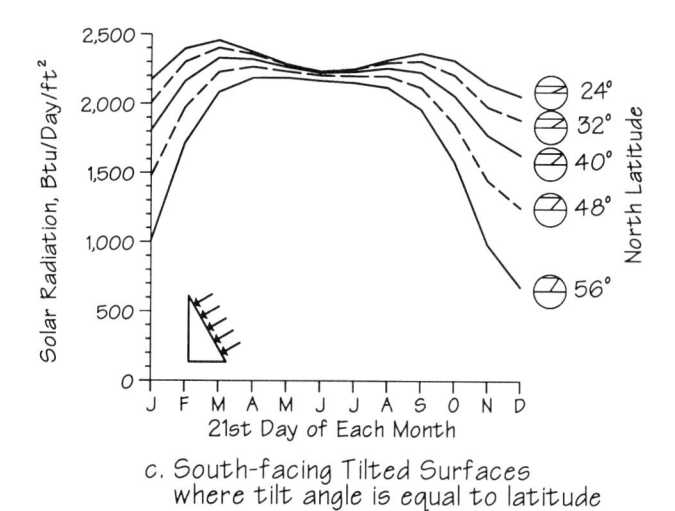

c. South-facing Tilted Surfaces
where tilt angle is equal to latitude

FIGURE 3.15 (Continued)

The data used to construct Figure 3.15 was obtained with permission from the *1987 ASHRAE Handbook—HVAC Systems and Applications*, Chapter 47, Table 3.

The figures are significant since they show how direct clear day solar radiation which arrives on surfaces here on earth varies seasonally at different locations. Of direct application to HVAC design is the effect of radiated or conducted solar heat which enters through glass and opaque building surfaces as discussed in Chapter 7 (Solar Cooling Loads) and Chapter 12 (Solar Heating Systems).

4

HEAT GAINS WITHIN BUILDINGS

Internal heat gain from people, lights, and equipment often has a significant impact on HVAC system selection and performance, and can be thought of as creating the "interior climate" of a building. This is particularly true for commercial buildings where people are often clustered in bright, artificially lit, highly automated environments.

There are two forms of heat which are of concern, sensible heat and latent heat. Sensible heat is heat which results in a change of the dry-bulb temperature of air. Often, a portion of the sensible heat gain being released in a building is stored in the structure or furnishings. As a result, the cooling load from this heat gain may be reduced and/or delayed. Latent heat (moisture) given off by people or processes such as cooking must also be quantified to assure the selection of equipment with adequate capacity to dehumidify air to within the comfort range.

Historically, the ASHRAE methods to compute cooling loads for internal heat gains have been rigorous. The *1993 ASHRAE Handbook of Fundamentals* incorporates a revised methodology (known as the CLTD/SCL/CLF method) to calculate space design cooling loads. With regard to the acronyms: CLTD refers to cooling load temperature differences which account for conductive heat gains of solar heat; SCL stands for solar cooling loads through glazing; and CLF refers to cooling load factors which account for the heat storage effect of building construction and furnishings.

The calculations of the CLTD/SCL/CLF method which quantify internal cooling loads due to people, equipment, and lights are illustrated below in a series of easy to understand examples. These calculations will then form the basis of much of the cooling load calculation for a complete room which is presented in Chapter 8. First, though, it is important to become familiar with "Zone Type" designations used by ASHRAE so that simple calculations can be modified for a wide range of conditions.

4.1 ZONE TYPE DESIGNATIONS

Not all internal heat gain becomes a cooling load which must be addressed instantaneously. Instead, a portion of the sensible heat is absorbed through radiation by surrounding construction and furnishings. In an attempt to quantify this heat storage effect, ASHRAE's revised methodology utilizes four Zone Types (A, B, C, and D). To qualify for a Zone Type designation, certain construction features must be met.

In simple terms Zone Type A represents the least amount of heat storage capability. Zone Types B, C, and D then offer increasingly massive construction and furnishings for additional storage of sensible heat.

Table 4.1 is used to determine the Zone Type(s) for interior rooms based upon type of construction and furnishings. These Zone Type designations are indicative of the heat storage capability of the construction.

Zone Type designations will be used in various tables included below to obtain the CLFs needed to compute cooling loads due to internal heat gain factors. In a similar manner, Zone Type designations for exterior spaces of single-story buildings with various zone parameters are determined by using Table 4.2.

Be sure to note that different Zone Type designations often apply for a room of a given description. For example, an exterior room with 2 walls, vinyl floor covering, gypsum partitions, and full inside shading is assigned Zone Type C when cooling loads are being computed for people and equipment, or lights. However, the same room will be designated as Zone Type B when cooling loads are computed due to solar heat gain through glass, as discussed in Section 7.1.

4.2 HEAT FROM PEOPLE

People, perhaps unflatteringly, can be thought of as "little boilers" which give off heat (both sensible and latent) to the environments they inhabit. The amount of heat

TABLE 4.1 ZONE TYPES—INTERIOR ROOMS

Room Location	Zone Parameters Middle Floor	Ceiling Type	Floor Covering	Zone Type People and Equipment	Lights
Single story	N/A	N/A	Carpet	C	B
	N/A	N/A	Vinyl	D	C
Top floor	2.5 in. Conc.	With	Carpet	D	C
	2.5 in. Conc.	With	Vinyl	D	D
	2.5 in. Conc.	Without	b	D	B
	1.0 in. Wood	b	b	D	B
Bottom floor	2.5 in. Conc.	With	Carpet	D	C
	2.5 in. Conc.	b	Vinyl	D	D
	2.5 in. Conc.	Without	Carpet	D	D
	1.0 in. Wood	b	Carpet	D	C
	1.0 in. Wood	b	Vinyl	D	D
Midfloor	2.5 in. Conc.	N/A	Carpet	D	C
	2.5 in. Conc.	N/A	Vinyl	D	D
	1.0 in. Wood	N/A	b	C	B

Source: Reprinted with permission from the *1993 ASHRAE Handbook of Fundamentals*, Chapter 26, Table 35A.
[b]The effect of inside shade is negligible in this case.

TABLE 4.2 ZONE TYPES—SINGLE-STORY BUILDING (EXTERIOR ROOMS)

	Zone Parameters			Zone Type		
No. Walls	Floor Covering	Partition Type	Inside Shade	Glass Solar	People and Equipment	Lights
1 or 2	Carpet	Gypsum	b	A	B	B
1 or 2	Carpet	Conc. blk.	b	B	C	C
1 or 2	Vinyl	Gypsum	Full	B	C	C
1 or 2	Vinyl	Gypsum	Half to None	C	C	C
1 or 2	Vinyl	Conc. blk.	Full	C	D	D
1 or 2	Vinyl	Conc. blk.	Half to None	D	D	D
3	Carpet	Gypsum	b	A	B	B
3	Carpet	Conc. blk.	Full	A	B	B
3	Carpet	Conc. blk.	Half to None	B	B	B
3	Vinyl	Gypsum	Full	B	C	C
3	Vinyl	Gypsum	Half to None	C	C	C
3	Vinyl	Conc. blk.	Full	B	C	C
3	Vinyl	Conc. blk.	Half to None	C	C	C
4	Carpet	Gypsum	b	A	B	B
4	Vinyl	Gypsum	Full	B	C	C
4	Vinyl	Gypsum	Half to None	C	C	C

Source: Reprinted with permission from the *1993 ASHRAE Handbook of Fundamentals*, Chapter 26, Table 35B.
[b]The effect of inside shade is negligible in this case.

depends upon the activity level, ranging from sitting at rest to strenuous work or exercise. Provided in Table 4.3 are heat gains from people in Btus per hour per person (Btu/h·per) for various activity levels in spaces maintained at 78°F. These values have been adjusted to account for the normal occupancy split of men, women, and children for the activity.

Notice that for the full range of activities listed in Table 4.3 the sensible heat gains vary by a factor of about 3 (225 to 710 Btus per hour per person). Also note that the latent heat gains can increase much more—by a factor of more than 10 (105 to 1090 Btus per hour per person) during strenuous activity. At such times the analogy to a "boiler" seems very appropriate indeed.

TABLE 4.3 HEAT GAIN FROM OCCUPANTS IN CONDITIONED SPACES

Activity	Sensible Heat (Btu/h·per)	Latent Heat (Btu/h·per)
Seated at theater	225	105
Seated, very light work	245	155
Moderately active office work	250	200
Standing, light work (retail store)	250	200
Light bench work (factory)	275	475
Walking, 3 mph (factory)	375	625
Bowling	580	870
Heavy machine work, lifting	635	965
Athletics, gymnasium	710	1090

Source: Excerpted by permission from the *1989 ASHRAE Handbook of Fundamentals*, Chapter 26, Table 3. (Note: Latent heat gain values published in the first printed edition of the *1993 ASHRAE Handbook* are incorrect.)

Occupancy Density

Density also determines the impact of heat gain from people. In Table 4.4 some guidance as to approximate square footage requirements by occupancy type is provided, however, the ultimate occupant density for many applications is uncertain during design. Values from Table 4.4 can be used to approximate the number of people likely to occupy a given space.

Sensible Cooling Load

The mass of a building's structure and furnishings has the potential to store heat and thus defer a portion of the instantaneous sensible heat gain from people. However, this potentially beneficial factor can only be assumed for light occupancies and when there is certainty that any stored heat from the day before has been removed before a new day's occupancy begins.

In many climates, particularly dry ones such as in the Southwest, large diurnal (day/night) temperature swings will assure adequate loss of any stored heat either through simple nighttime heat loss or by the introduction of cool nighttime air through the ventilation system. In other climates, cooling systems can be operated either for a period after or before normal building occupancy to ensure that all stored heat has been removed, and that the storage potential of the building structure and furnishings has been reestablished. In such cases a "Cooling Load Factor" adjustment to the sensible heat gain from people can be made to the cooling load calculation. The advantage of such an approach is the potential reduction in capacity of the refrigeration equipment and cooling distribution system.

Cooling Load Factor Adjustment—People

As indicated above, a rigorous calculation of sensible cooling loads due to heat gain from people takes into account heat storage. This is because about 70% of the heat output from people is radiant, and thus is first transferred to, and absorbed by, the building structure and furnishings.

For many applications, this heat storage effect can reduce the sensible cooling load substantially, and particularly during the first few hours of building occupancy. Then, over a period of time, some of the stored heat gain will reradiate, and contribute to the space cooling load for hours (and often after occupancy has ceased). This effect is graphically shown in Figure 4.1.

Cooling load factors (CLF) are used to account for this radiation and heat storage phenomenon. Table 4.5, excerpted from a much larger ASHRAE table, provides CLFs for people who occupy a space 8 or 10 hours

TABLE 4.4 TYPICAL OCCUPANCY DENSITY RANGES

Building Type	ft^2/Person
Office	100 to 200
Educational	100 to 175
Medical treatment	50 to 100
Assembly	15 to 20
Restaurant	20 to 30
Retail	30 to 75
Warehouse	500 to 1000
Apartment house	300 to 500
Single-family house	500 to 1000

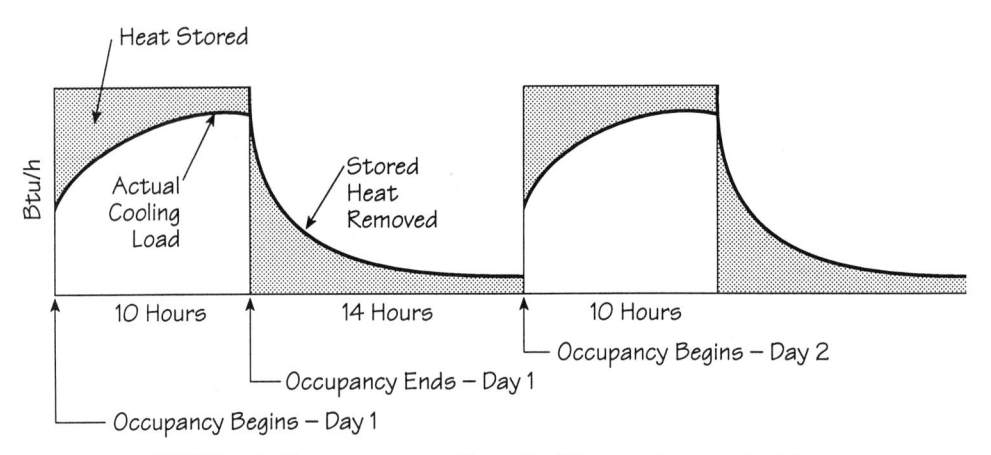

FIGURE 4.1 The heat storage effect of building structure and furnishings.

(which will cover almost all peak cooling load calculations for buildings) for the four Zone Types.

In Table 4.5 the first line lists CLFs for Zone Type A and 8 total hours of occupancy in the space. Note the following:

The value after 1 h in the space means that 0.75 or 75% of the sensible heat gain from people must be considered as part of the cooling load during that hour. Where did the rest of the heat gain go? It is being absorbed and

TABLE 4.5 SENSIBLE HEAT COOLING AND LOAD FACTORS FOR PEOPLE AND UNHOODED EQUIPMENT (CLF$_p$)[a]

Total Hours People Are in Space	Hours after People Enter Space or Equipment Turned on									
	1	2	3	4	5	6	7	8	9	10
Zone Type A										
8	0.75	0.88	0.93	0.95	0.97	0.97	0.98	0.98	0.24	0.11
10	0.75	0.88	0.93	0.95	0.97	0.97	0.98	0.98	0.99	0.99
Zone Type B										
8	0.65	0.75	0.81	0.85	0.89	0.91	0.93	0.95	0.31	0.22
10	0.65	0.75	0.81	0.85	0.89	0.91	0.93	0.95	0.96	0.97
Zone Type C										
8	0.61	0.69	0.75	0.79	0.83	0.86	0.89	0.91	0.32	0.26
10	0.62	0.70	0.75	0.80	0.83	0.86	0.89	0.91	0.92	0.94
Zone Type D										
8	0.62	0.69	0.74	0.77	0.80	0.83	0.85	0.87	0.30	0.24
10	0.63	0.70	0.75	0.78	0.81	0.84	0.86	0.88	0.89	0.91

Source: Excerpted by permission from the *1993 ASHRAE Handbook of Fundamentals*, Chapter 26, Table 37.
[a]Note: The sums shown are inserted for illustrative purposes only and are not actually part of the ASHRAE table.

stored by the spaces' structure and furnishings.

Then, the next value of 0.88 (for 2 h after people entered the space) indicates that a sensible cooling load calculation for the second hour of occupancy must consider an increased load of 88% of the hourly sensible heat gain from people. This is because as time goes on the mass of the structure and furnishings becomes less available to absorb additional heat, and in fact absorbed heat may begin to reradiate.

Finally, note the value for 9 h after people entered the space of 0.24 (the equivalent of 24% of one hour's instantaneous heat gain from people). This heat gain occurs in the hour after occupancy ends, and represents the release of heat stored during earlier hours.

You may be wondering, why are the values in Table 4.5 different for 1 h after people enter the space for the various Zone Types? This is because the cooling load factors are designed to simulate long-term heat storage and assume the same occupancy load the day before. As a result, some of the cooling load often is coming from heat stored the day before, and particularly for the heavier weight Zone Types C and D.

Be sure that you understand the basic methodology of the CLF approach since it will be applied in a similar manner to heat gains from other internal sources, as well as solar heat gains through glazing.

Cooling loads due to sensible heat gain from people ($q_{p\text{-sen}}$) are calculated by using Formula 4A.

$$q_{p\text{-sen}} = N \times SHG_p \times CLF_p \quad \text{Formula 4A}$$

where

$q_{p\text{-sen}}$ is the sensible cooling load due to people in Btu/h

N is the number of people in the space

SHG_p is the sensible heat gain per person in Btu/h per person based upon type of activity (see Table 4.3)

CLF_p is the applicable cooling load factor for people from the appropriate table (e.g., Table 4.5)

Example 4A: Determine the sensible cooling load at 2 PM for "The Office" space shown in Figure 4.2 and based on the following:

14 people doing moderately active office work.

Occupancy from 8 AM until 4 PM (8 hours).

Space temperatures are maintained on a 24 hour basis so that the mass of the building and furnishings is available for heat storage.

Zone Type Information: Carpeted floor covering, gypsum board partitions, and a negligible effect from interior shade.

Solution: The sensible heat gain per person (SHG_p) is 250 Btu/h for people doing

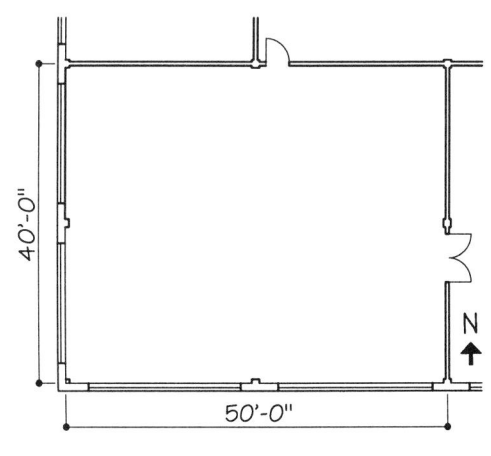

FIGURE 4.2 "The Office."

moderately active office work (from Table 4.3). Table 4.2 is then used to determine the Zone Type based upon the information given. In this case a space with 2 walls, carpeting, gypsum walls, and no interior shade is designated as Zone Type B for people (and equipment).

Table 4.5 is then consulted to determine the cooling load factor for Zone Type B, 8 hours of total occupancy, and 6 hours after people entered the space (2 PM). The applicable CLF_p is 0.91. Use Formula 4A.

$$q_{p\text{-sen}} = N \times SHG_p \times CLF_p$$

(Formula 4A)

$$= 14 \times 250 \times 0.91$$

$$= 3185 \text{ Btu/h}$$

Latent Cooling Load

For cooling load purposes, latent heat gain ($q_{p\text{-lat}}$) is computed using Formula 4B.

$$q_{p\text{-lat}} = N \times LHG_p \quad \text{Formula 4B}$$

where

$q_{p\text{-lat}}$ is the latent cooling load due to people in Btu/h
N is the number of people in the space
LHG_p is the latent heat gain in Btu/h per person based upon type of activity (see Table 4.3)

Note that Formula 4B assumes that all latent heat gain (moisture) is contained in the air in the building. This is not truly the case. Research work may soon result in an acceptable method to also quantify moisture storage in the building structure and furnishings.

Example 4B: Determine the latent cooling load in a factory if 100 people (N) are doing light bench work.

Solution: The latent cooling load is found by using Formula 4B. A value of 475 Btu/h per person for the activity of light bench work is obtained from Table 4.3.

$$q_{p\text{-lat}} = N \times LHG_p$$

$$= 100 \text{ persons} \times 475 \text{ Btu/h} \cdot \text{per}$$

$$= 47,500 \text{ Btu/h} \quad \text{(Formula 4B)}$$

4.3 HEAT FROM EQUIPMENT AND APPLIANCES

During the 1980s, an internal heat gain revolution took place as personal computers, printers, and other supporting electronic machines spread like wildfire through many work environments. With the equipment came heat, sometimes beneficial, though most often part of the cooling load for much of the year.

Provided in Table 4.6 is a list of typical equipment heat gains for a variety of work environments. The values shown in the two columns are rounded for simplicity and to remove any suggestion of precision since work environments vary greatly. Also remember the conversion of 1 watt equaling 3.413 Btu per hour.

To determine heat gains from equipment, be sure to evaluate each work environment individually. For example, the electronic equipment used by a sales force which is regularly and predictably out of the office will yield much less heat than a constantly operated "computer room" (which is likely to have its own dedicated cooling system). With regard to manufacturing plants, the general trend toward automation will result in ever increasing equipment heat gains.

Cooling Load Factor Adjustment—Equipment and Appliances

Table 4.5 is also used to determine CLFs for the sensible heat gain of "unhooded"

4.3 HEAT FROM EQUIPMENT AND APPLIANCES

TABLE 4.6 EQUIPMENT HEAT GAIN

Type of Work Environment	Watts/ft^2	Btu/h·ft^2 (max.)
General office with only a few typewriters, computers and other electrical items	0.25 to 1.0	1 to 4
Offices where most workers have personal computers	1.0 to 3.0	3 to 10
Rooms dedicated to large "main-frame" computers	15 to 50	50 to 175
Laboratories	5 to 20	15 to 70
Manufacturing plants	5 to 45	15 to 150

equipment. Unhooded equipment includes office equipment (i.e., computers, typewriters, and copy machines) and electric cooking appliances (microwaves, coffee makers, etc.) which are not locally ventilated or exhausted.

The values for 8 and 10 h of total operation are shown since they are the ones likely to be used for a commercial building load application. Use a CLF of 1.00 when calculating cooling loads for equipment and appliances which are operational for very long hours such as at a fast-food restaurant.

Sensible cooling loads due to equipment and appliance heat gain (q_{eq}) can be calculated using Formula 4C.

$$q_{eq} = EQ_{wsf} \times A \times 3.413 \times CLF_{eq}$$

Formula 4C

where

q_{eq} is the sensible cooling load due to equipment and/or appliances in Btu/h

EQ_{wsf} is the power density of equipment and appliances in the space in watts per square foot

A is the space's floor area in square feet

3.413   is the conversion of 3.413 Btu per hour per watt

CLF_{eq} is the applicable cooling load factor for equipment from the appropriate table (i.e., Table 4.5 or Table 4.8 for hooded cooking appliances)

Example 4C: Determine the sensible cooling load for "The Office" space shown in Figure 4.2 at 11 AM due to office equipment and based on the following:

An equipment power density of 1.0 W/ft^2 (i.e., a few personal computers).

Office hours are from 8 AM until 4 PM (8 hours).

Space temperatures are maintained on a 24-h basis so that the mass of the building and furnishings is available for heat storage.

Zone Type Information: Carpeted floor covering, gypsum board partitions, and a negligible effect from interior shade.

Solution: Table 4.2 is used to determine the Zone Type based upon the information given. In this case a space with 2 walls, carpeting, gypsum walls, and no interior shade is designated as Zone Type B for equipment (and people).

Table 4.5 is then consulted to determine the cooling load factor for Zone Type B, 8 h of total occupancy, and 3 h after people entered the space (11 AM). The applicable CLF_{eq} is 0.81. Formula 4C can now be used to calculate the sensible cooling load due to office equipment for the 2000 ft^2 office area.

TABLE 4.7 RECOMMENDED RATE OF HEAT GAIN FOR RESTAURANT EQUIPMENT LOCATED IN AIR-CONDITIONED AREA

	Rate of Heat Gain (Btu/h)		
	Without Hood		With Hood
Appliance	Sensible	Latent	All Sensible
Gas griddle/grill (large), per ft² of cooking surface	1140	610	460
Gas range (burners), per 2 burner section	Exhaust hood required		6590
Gas fryer (deep fat), per pound of fat capacity	Exhaust hood required		160
Elec. coffee brewer, 12 cup/2 burners	3750	1910	1810
Elec. toaster, 10 slice	9590	8500	5800

Source: Excerpted by permission from the *1993 ASHRAE Handbook of Fundamentals*, Chapter 26, Table 8.

$$q_{eq} = EQ_{wsf} \times A \times 3.413 \times CLF_{eq}$$

(Formula 4C)

$$= 1.0 \times 2000 \times 3.413 \times 0.81$$

$$= 5530 \text{ Btu/h}$$

Commercial Cooking Appliances

In restaurants and other buildings and spaces where food is cooked, the heat gains from commercial cooking appliances will represent significant heat gains. Table 4.7 provides a summary of heat gains (both sensible and latent) for the probable operation of a few items of commercial cooking equipment. See the *1993 ASHRAE Handbook of Fundamentals* and consult with manufacturers of equipment directly, for more complete information.

Often kitchen air is merely exhausted and not cooled, and the kitchen is under negative pressure relative to the dining room. Nevertheless, for applications where cooling of the space occurs, CLFs for sensible heat gain from "hooded" appliances which have exhaust hoods or direct ventilation should be applied. Table 4.8 provides CLFs for Zone Type C which is typical construction for many restaurant kitchens (i.e., vinyl floor, concrete block partitions).

Sensible cooling loads from "hooded" appliances should be computed using Table 4.8 and Formula 4C as described above.

TABLE 4.8 SENSIBLE HEAT COOLING LOAD FACTORS (CLF) HOODED EQUIPMENT

Total Operational Hours	Number of Hours after Equipment Turned On									
	1	2	3	4	5	6	7	8	9	10
Zone Type C										
2	0.43	0.54	0.20	0.16	0.13	0.10	0.09	0.07	0.06	0.04
4	0.43	0.54	0.63	0.70	0.33	0.26	0.20	0.17	0.14	0.11
6	0.44	0.56	0.63	0.70	0.76	0.80	0.40	0.31	0.26	0.21
8	0.44	0.56	0.64	0.70	0.76	0.80	0.84	0.87	0.46	0.37
10	0.46	0.57	0.64	0.71	0.76	0.80	0.84	0.87	0.89	0.91

Source: Excerpted by permission from the *1993 ASHRAE Handbook of Fundamentals*, Chapter 26, Table 39.

4.4 HEAT FROM LIGHTING SYSTEMS

The heat gain from artificial light sources can be substantial. So substantial, in fact, that building designers in the 1950s and 1960s sometimes consciously lit buildings with "luminous" ceilings to very high footcandle levels not only to improve visual acuity, but also to get the secondary benefit of the heat from the lights. Rising charges for electricity put an end to that mistaken design practice long ago.

Various artificial light sources are available, ranging from very inefficient incandescent lamps, to fluorescent lamps which are about four times as efficient, to low pressure sodium lamps which use electricity six to ten times as efficiently as incandescent bulbs. Inside homes, lighting density is usually low, and incandescent light sources have historically been used for their color rendition and flexibility of fixture selection. Now, though, compact fluorescent lamps are increasingly being found in houses and especially in areas where the local utility encourages their use by offering rebates and special purchase plans. In larger buildings, the great majority of contemporary lighting systems use fluorescent lamps.

Power requirements for light sources are expressed in watts such as 60 W for household incandescent and 40 W for a basic 4-ft long fluorescent tube. Throughout this book, the commonly used unit of heat is the British thermal unit (Btu). Commit to memory the conversion of 3.413 Btus per hour per watt.

Lighting design is an art, and while interrelated, lighting quality and lighting quantity should not be linked simplistically. Still, guidelines exist in codes and standards for lighting power densities in terms of watts per gross square foot of building area.

Provided in Table 4.9 are typical lighting power densities for lighting systems (including the power consumed by ballasts) in various building types. Use these values only as a rough guideline since building types vary greatly in design as do lighting requirements of specific tasks. In addition, many jurisdictions have adopted ASHRAE Standard 90.1-1989 "Energy Efficient Design of New Buildings Except Low-Rise Residential Buildings" which includes a computer program to determine allowable lighting power densities for various building and space types.

Lighting power densities such as those listed in Table 4.9 are in general 25 to 40% lower than lighting power densities which were typical over 20 years ago (before the initial energy crisis of 1973). This has been brought about by improvements in lamp and fixture efficiency along with some reductions of recommended/required footcandle levels.

TABLE 4.9 TYPICAL LIGHTING POWER DENSITIES (L_{wsf})

Building Type	Watts/Gross Square Foot (watts/ft^2)
Office	1.7 to 2.2
Store spaces	2.5 to 3.0
Shopping mall concourse	1.0 to 1.5
Fast food restaurant	1.3 to 2.0
Health/hospital	2.3 to 2.6
Warehouse/storage	1.0 to 1.5
Library	2.2 to 2.6
Public assembly	1.5 to 2.0

Source: Based upon author research and experience.

Cooling Load Factor Adjustment—Lights

Cooling load calculations for lighting also are based on the cooling load factor (CLF) concept. The *ASHRAE Handbook of Fundamentals* contains a series of large Cooling Load Factor tables for lights depending upon whether they are on for 8, 10, 12, 14, or 16 h (and assuming that conditioned space temperatures remain constant all 24 h). Table 4.10 contains an excerpt of the complete ASHRAE table for lights.

Values in Table 4.10 are based on a radiative/convective ratio of 0.59/0.41. Note that fluorescent lights have about a 0.50/0.50 ratio.

Sensible cooling loads due to lighting heat gain (q_{lgt}) are calculated using Formula 4D.

$$q_{lgt} = L_{wsf} \times A = 3.413 \times CLF_{lgt}$$

Formula 4D

where

q_{lgt}	is the sensible cooling load due to the lighting system in Btu/h
L_{wsf}	is the lighting system power density (including ballasts if any) in watts per square foot
A	is the space's floor area in square feet
3.413	is the conversion of 3.413 Btu per hour per watt
CLF_{lgt}	is the applicable cooling load factor for lights from the appropriate ASHRAE table (e.g., Table 4.10)

Example 4D: Determine the sensible cooling load for "The Office" space shown in Figure 4.2 for the lighting system at 1 PM and based on the following:

A lighting system power density of 2.0 W/ft².

TABLE 4.10 SENSIBLE HEAT COOLING LOAD FACTORS (CLF$_{lgt}$) FOR LIGHTS

Hours Lights Are On	\multicolumn{10}{c}{Number of Hours After Lights Turned On}
	1	2	3	4	5	6	7	8	9	10
Zone Type A										
8	0.85	0.92	0.95	0.96	0.97	0.97	0.97	0.98	0.13	0.06
10	0.85	0.93	0.95	0.97	0.97	0.97	0.98	0.98	0.98	0.98
Zone Type B										
8	0.75	0.85	0.90	0.93	0.94	0.95	0.95	0.96	0.23	0.12
10	0.75	0.86	0.91	0.93	0.94	0.95	0.95	0.96	0.96	0.97
Zone Type C										
8	0.72	0.80	0.84	0.87	0.88	0.89	0.90	0.91	0.23	0.15
10	0.73	0.81	0.85	0.87	0.89	0.90	0.91	0.92	0.92	0.93
Zone Type D										
8	0.66	0.72	0.76	0.79	0.81	0.83	0.85	0.86	0.25	0.20
10	0.68	0.74	0.77	0.80	0.82	0.84	0.86	0.87	0.88	0.90

Source: Excerpted by permission from the *1993 ASHRAE Handbook of Fundamentals*, Chapter 26, Table 38.

Office hours are from 8 AM until 4 PM (8 h). However, lights will be on for 10 h to allow for office cleaning.

Space temperatures are *not* maintained on a 24-h basis so that the mass of the building and furnishings are *not* available for heat storage.

Zone Type Information: Carpeted floor covering, gypsum board partitions and a negligible effect from interior shade.

Solution: Use Formula 4D to calculate the cooling load. Since the mass is *not* available for heat storage, the time period of occupancy is immaterial, and the cooling load factor to be used is 1.0.

$$q_{lgt} = L_{wsf} \times A \times 3.413 \times CLF_{lgt}$$

(Formula 4D)

$$= 2.0 \times 2000 \times 3.413 \times 1.0$$

$$= 13{,}650 \text{ Btu/h}$$

In contrast, if heat storage was available then adjustment could be made, as shown in the next example.

Example 4E: Recompute the sensible cooling load for the lighting system described in Example 4D except for the following:

Space temperatures are maintained on a 24-h basis so that the mass of the building and furnishings are available for heat storage.

Zone Type Information: Carpeted floor covering, gypsum board partitions, and a negligible effect from interior shade.

Solution: Table 4.2 is then used to determine the Zone Type based upon the information given. In this case, a space with 2 walls, carpeting, gypsum walls, and no interior shade is designated as Zone Type B for lights.

Table 4.10 is then consulted to determine the cooling load factor (CLF_{lgt}) for Zone Type B, lights scheduled to be on for 10 hours, and 5 hours after lights are turned on (1 PM). The applicable CLF_{lgt} is 0.94. Now calculate the load using Formula 4D.

$$q_{lgt} = L_{wsf} \times A \times 3.413 \times CLF_{lgt}$$

(Formula 4D)

$$= 2.0 \times 2000 \times 3.413 \times 0.94$$

$$= 12{,}830 \text{ Btu/h}$$

Note that a CLF_{lgt} of 0.94 indicates that 0.06 (or 6%) of the heat gain from lights that hour will be stored, reducing the cooling load which must be met during that hour.

Discussion

The lighting load is the largest relatively constant cooling load in most buildings. As a result, lighting system heat gain impacts upon the cooling distribution system, the capacity of the cooling plant and cooling towers or condensers, electrical service needs, and even the supporting structure. Therefore, every design effort should be made to install efficient lighting systems.

In buildings designed to use natural daylight, automatic dimmer controls can reduce the amount of artificial light output when daylight is available. Nevertheless, it should be pointed out that daylighting is not without potential pitfalls. Design must be carefully balanced to allow for the controlled entry of daylight without also increasing solar heat gain significantly. If this is not done, any potential lighting energy "savings" may be more than offset by larger cooling loads due to unnecessary solar heat gain.

Remember that many cooling load calculations (either manual or computerized) employ CLFs of 1.0 because of design or operational uncertainty, or simply as something of a safety factor. The procedures required to select lower CLFs have been emphasized here to help illustrate that economy is possible without loss of comfort if buildings are engineered rigorously but realistically, rather than for worst-case conditions.

4.5 HEATING AND COOLING "BALANCE POINTS"

First, a definition. The heating balance point of a space is the outside temperature at which the heat gain from internal factors (people, lights, and equipment) and solar radiation which enters a building is no longer enough to offset any heat losses occurring in the space to maintain comfort at a given room temperature (see Figure 4.3). It is the "theoretical" temperature at which additional heat is needed to maintain the thermostat set-point temperature.

Picture heating balance points in this way. The heating thermostat in a room is set to maintain a temperature of 70°F in a room with a calculated heating balance point temperature of 38°F. For such a case, when the outside temperature is 50°F the internal heat gains and solar gains will be more than enough to offset heat losses. Additional heat will not be needed to maintain comfortable conditions. In fact, the true concern is whether the internal heat gains will produce an interior temperature which is excessive, and thus uncomfortable. The temperature then falls to 40°F outside. The heat gains are just about enough to maintain comfort. Finally, the outside temperature falls to 38°F, which is the calculated heating balance point. At this point the internal and solar heat gains can no longer provide enough heat to maintain the set-point temperature and the mechanical heating system is activated.

Formula 4E is used to calculate heating balance points.

$$t_b = t_s - \frac{q_{\text{p-sen}} + q_{\text{eq}} + q_{\text{lgt}} + q_{\text{sol}}}{\text{HL}}$$

Formula 4E

where

t_b is the heating balance point temperature in °F

t_s is the thermostat set point temperature in °F

$q_{\text{p-sen}}$ is the sensible heat gain from people in Btu/h

q_{eq} is the heat gain from equipment in Btu/h

q_{lgt} is the heat gain from lights in Btu/h

q_{sol} is the heat gain from solar radiation getting into the space in Btu/h

HL   is the heat loss of the space including transmission, and infiltration and/or ventilation in Btu/°F·h

Examine the units of Formula 4E:

$$°F = °F - \frac{\text{Btu/h}}{\text{Btu/°F·h}}$$

The part after the minus sign is used to determine the ratio of the heat gain (in the numerator) to heat loss (in the denominator)

HEAT GAINS	HEAT LOSSES
People	Transmission
Lights	Infiltration
Equipment	Ventilation
Solar	

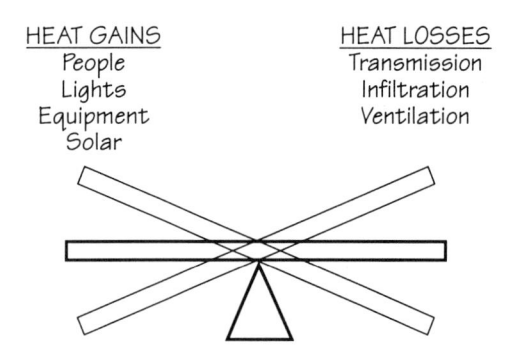

FIGURE 4.3 The meaning of "heating balance points."

as measured in units of °F. Put another way, this part of the formula is answering the question: How many degrees of heat loss are being made up by internal and solar heat gains?

Houses

Historically, data on outdoor temperatures below 65°F was collected because below that temperature it was assumed that heat would be needed to maintain comfort in houses (see the discussion on heating degree-days in Chapter 2). Therefore, 65°F has been the presumed balance point temperature for houses. But this is not always the case. Very well-insulated houses may have balance points of 60°F or lower and thus require less heating energy each year.

Example 4F: Calculate the average heating balance point temperature for a house assuming the following:

- A heating set-point temperature of 68°F.
- Two people occupy the house, each producing an average sensible heat gain of 250 Btu/h for a total of 500 Btu/h ($q_{\text{p-sen}}$).
- A total of 650 W of equipment energy (refrigerator, tv, etc.) provide a heat gain (q_{eq}) of 2220 Btu/h.
- The assumed heat gain from lights (q_{lgt}) is 225 W or about 770 Btu/h.
- The average solar heat gain per hour (q_{sol}) during the heating season is 500 Btu/h.
- The total heat loss (HL) for the well-insulated and constructed house (including infiltration) has been calculated to be 500 Btu/°F·h.

Solution: Apply Formula 4E.

$$t_b = t_s - \frac{q_{\text{p-sen}} + q_{\text{eq}} + q_{\text{lgt}} + q_{\text{sol}}}{\text{HL}}$$

(Formula 4E)

$$= 68°F - \frac{500 + 2220 + 770 + 500}{500}$$

$$= 68°F - \frac{3990 \text{ Btu/h}}{500 \text{ Btu/°F·h}}$$

$$= 68°F - 8°F$$

$$= 60°F$$

Therefore, 60°F is the heating balance point for this house. This result means that the internal and solar heat gains within the house will provide all of the heat required to maintain at least 68°F within the house when it is occupied, until the outdoor temperature goes down to 60°F. Below 60°F, some form of supplemental heat will be required to maintain the set-point temperature.

How accurate or useful are balance point temperatures? Will two people always be in a house with 225 W of lights turned on? Of course not. In addition, a lower set-point temperature may be maintained overnight. Still, the concept of balance points is very useful in determining which degree-day base should be used to estimate the need for auxiliary heat. The formula also shows how the balance point can be lowered by reducing the house heat loss.

As shown above, the computation of heating balance point temperatures using Formula 4E includes solar heat gain. Later, in Chapter 12, the required heating energy needs of passive solar houses will be estimated by using heating degree-day data relative to a "base temperature" which acknowledges internal heat gains (people, lights, and equipment) only.

Commercial Buildings

Balance point analysis can also be applied to commercial buildings where heat gains from internal factors are much greater. As a result, heating balance points in many commercial buildings are frequently in the

62 HEAT GAINS WITHIN BUILDINGS

range of 20 to 40°F. With such low balance points, such buildings (even in cold climates) will require very little heating fuel when occupied. In fact, most commercial buildings consume the great majority of heating energy (typically 75% or more) when unoccupied on nights and weekends.

Understand that the use of a single balance point for a house or building is a great simplification. Consider an entire two-story office building which has a computed heating balance point (assuming average winter solar heat gain) during daytime hours of 25°F. Such a total is actually little more than the weighted average of the heating balance points for the various building zones (rooms) which are shown in Figure 4.4. Also realize that excess heat in one part of a building is only useful in offsetting heat losses in another part of the building if the HVAC system is designed to transfer heat, such as with a closed-loop heat pump.

Cooling Balance Points

Formula 4E is also used to compute cooling balance point temperatures, which indicate the temperature above which heat removal (cooling) will be required to maintain human comfort. Cooling balance points are very important with regard to commercial buildings since they will only be eight to ten degrees higher than the heating balance point (the difference between the cooling and heating set-point temperatures). As a

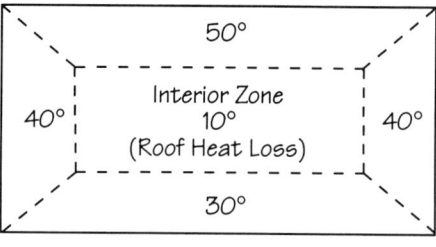

```
        50°
   ┌ ─ ─ ─ ─ ─ ─ ─ ┐
40° │ Interior Zone │ 40°
   │      10°      │
   │(Roof Heat Loss)│
   └ ─ ─ ─ ─ ─ ─ ─ ┘
        30°
```
Second (Top) Floor

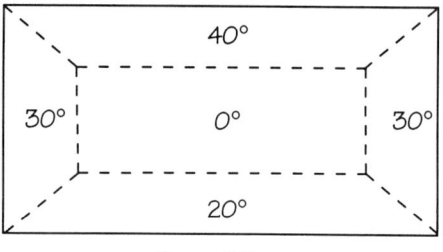

```
        40°
   ┌ ─ ─ ─ ─ ─ ─ ─ ┐
30° │      0°       │ 30°
   │               │
   └ ─ ─ ─ ─ ─ ─ ─ ┘
        20°
```
Ground Floor

FIGURE 4.4 Heating balance points by zone for a two-story office building.

result, the need for heat removal can exist when outdoor temperatures are as low as 30°F or even lower.

Therefore, in many commercial buildings, refrigeration air conditioning systems operate almost all year. Other, more efficient buildings remove heat when outdoor temperatures are low by increasing the quantity of cool outdoor air in the ventilation air stream (known as the "economizer cycle").

5

HEAT LOSS FROM BUILDINGS

Before designing a heating system, one must estimate the amount of heat which will be required to replace the heat which is lost from the building. This chapter will discuss the various routes of building heat loss and provide methods to quantify them.

The unit of heat commonly used is the British thermal unit (Btu). Since a Btu is a very small unit of heat (about equal to the heat from a wooden kitchen match), many thousands or even millions of Btus are required to heat a house or building each hour when it is cold outside.

Buildings lose heat in two fundamental ways:

1. Transmission through any solid elements of the building envelope (walls, windows, doors, roof, floors, etc.) which come in contact with cold outside air or the ground.
2. Infiltration/exfiltration as the wind drives cold outside air through cracks (around windows, sill plates, etc.) and openings (wall outlets, open doors, etc.) which in turn pushes warm inside air out through openings and cracks on the leeward sides of the building. Infiltration is the flow of cold air inward. Exfiltration is the flow of warm air out.

An understanding of this chapter will enable you to perform two very useful tasks:

1. Size heating systems (central equipment and individual delivery devices) for local climates.
2. Estimate how much heating fuel will be needed to maintain comfort each heating season.

Heat loss computations are easy if one takes time, keeps track of units, uses careful bookkeeping, and has learned the basic fundamentals as presented here in a step by step building block fashion.

This chapter concludes with an example of a complete heat loss calculation for "The House." The same house will then be used to illustrate the sizing and layout of various heating (passive solar, warm air, hot water, heat pump) and cooling systems.

5.1 HEAT TRANSFER PROCESSES

The flow of heat out of buildings must be quantified in order to size heating systems adequately. Similarly, heat which is produced within or enters into buildings often results in a need to remove it by some form of natural or mechanical cooling system. Therefore, the heat transfer processes of conduction, convection, and radiation need to be clearly understood as illustrated in Figure 5.1.

Conduction

Conduction is the transfer of heat through solid materials. The flow of heat is always from a warmer to a cooler material (object) such as the heat flow through a pan in cooking. Conductive heat loss through the solid portions of the envelope of a house or building (windows, walls, roof, etc.) normally accounts for most of the heat loss.

Convection

Convection, or convective heat transfer, takes place within the flow of gas or liquid. Natural convection results when air is warmed and obtains a higher energy level. This means that the molecules move more and expand into a larger volume while creating a flow.

Natural convection can be employed to warm houses using various systems including convectors, radiators, and even passive solar Trombe walls. Natural convection can also cool buildings passively. When convection is caused by the wind or a fan or pump it is known as forced convection (or forced circulation).

Convective (and wind driven) heat loss through cracks and openings in a building's envelope is known as air leakage (or infiltration). Air leakage can be responsible for 25% or more of the heat loss of a typical house in a cold climate.

Radiation

Radiation is the transfer of heat from a warmer object to a cooler object across an open space when the objects are in direct "sight" of each other. Sometimes the distance can be great such as the 91,410,000–94,460,000 mi which radiation from the sun travels through airless space. Within buildings, the distances are normally short, with heat sources from within the room, such as from a wood stove.

Radiant heat exchanges can either be a beneficial part of a heating system, or cre-

Conduction　　　　Convection　　　　Radiation

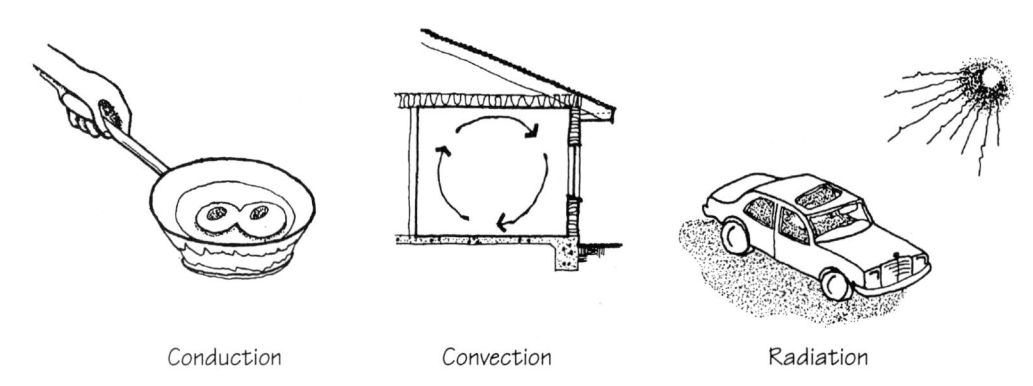

FIGURE 5.1 Heat transfer processes.

ate uncomfortable conditions such as from cold windows. Also see the discussion of "mean radiant temperature" (MRT) in Section 1.3.

5.2 HEAT LOSS THROUGH SOLID BUILDING ELEMENTS

Solid building elements which resist the flow of heat outward help maintain winter comfort within. The relative resistances of various construction materials are measured in the laboratory and assigned "R-values" (R for resistance). The higher the R-value, the more the material will resist heat flow, reducing heat loss in winter as well as resisting heat gain in the summer.

The most familiar products with well-advertised R-values are conventional fiberglass batts. At the local building supply center, "R-11" is emblazoned on $3\frac{1}{2}$-in. thick insulation, "R-19" on 6-in. thick batts, "R-30" on 9-in. thick blankets, and so on. If you are wondering about the units of R-values, they will be defined a bit later. For the moment, just remember, the higher the R-value, the less heat lost.

Construction Assemblies

Walls, floors, and roofs are constructed of finish and support materials (framing, siding, etc.), each of which provides some resistance to heat flow. However, in order to obtain assemblies with a high resistance to heat flow, insulation (e.g., batts, rigid boards, blown-in material) must be incorporated. The type and thickness of the insulation will depend on the climate and the degree of energy conservation required by local codes or desired to reduce operational costs.

In temperate and cold climates, houses should be built with walls which have total R-values of R-20 or more. Roofs often are built with even more insulation, resulting in total R-values of R-30 or greater. In contrast, most commercial buildings have less-insulated envelopes because they are often unoccupied (with less heat then required) and when they are occupied there is often useful heat gain during the winter from lights, people, and office equipment (see balance points in Section 4.5). The economics associated with the choice of various types and thicknesses of insulation and construction assemblies are discussed in Chapter 24.

Resistance (R) Values of Construction Materials

The resistance of a building element, such as a wall, is determined by summing the R-values of all elements which contribute to the total resistance. This will include all solid elements such as wood framing, sheathing, siding, interior finishes, and most importantly the insulation. And it will also include the resistance to heat loss provided by any "air spaces" in the construction, as well as "air films" on both the inside and outside surfaces.

Table 5.1 shows R-values for a simplified listing of building construction materials. Most values shown are per inch of thickness of the material. Other R-values shown are for the indicated thickness. In doing actual heat loss calculations, be sure to consult a complete listing of materials and manufacturers' product literature.

5.3 AIR SPACES AND AIR FILMS

"Air spaces" have insulating value, a point which is well-recognized by outdoor enthusiasts who wear down jackets and other insulated clothing with dead air spaces. In fact, fiberglass insulation derives its insulating value from its structural framework, which contains countless numbers of air pockets.

Air spaces provide thermal resistance depending upon their width and position,

TABLE 5.1 R-VALUES OF COMMON CONSTRUCTION MATERIALS

Materials	R-value
Insulating Materials	
Fibrous blankets or batts	3.00 to 4.00 per inch
Loose-fill, cellulose	3.13 to 3.70 per inch
Polyurethane foam	5.26 to 6.25 per inch
Cellular glass board	2.86 per inch
Glass fiber, rigid board	4.00 per inch
Extruded polystyrene	5.00 per inch
Cellular polyisocyanurate	7.20 per inch
Common Building Materials	
1/2 in. Gypsum board	0.45
1/2 in. Plywood	0.62
1/2 in. Fiberboard sheathing	1.32
Hardboard/particleboard	0.85 to 1.85 per inch
1/2 in. Gypsum plaster, lightweight (lw) aggregate	0.32
3/4 in. Wood subfloor	0.94
Hardwoods	0.80 to 0.94 per inch
Softwoods (construction lumber)	0.89 to 1.48 per inch
Face Brick	0.20 per inch
8 in. Concrete block, lw aggregate	2.00
Concrete, stone aggregate	0.08 per inch
Concrete, lightweight aggregate	0.19 to 2.00 per inch
Stucco	0.20 per inch
Wood shingles	0.87 to 1.40
Hardboard siding	0.67 per inch
Beveled wood siding, lapped	0.81
Aluminum siding, hollow-backed	0.61
Alum. siding, 3/4 in. insul. backed	1.82 to 2.96
Built-up roofing	0.33
Asphalt shingles	0.44
1/2 in. Slate	0.05
Wood shingles	0.94

the direction of heat flow, air temperature, and a property known as "emittance," which is the ability of a surface to radiate heat. For example, an air space where both of the facing surfaces are bright aluminum foil has a very low effective emittance of 0.03, whereas air spaces between normal nonreflective building materials (i.e., wood, masonry, gypsum wallboard, or nonmetallic paint) have much higher effective emittances of about 0.82 to 0.90.

Table 5.2 is a simplified listing of R-values for air spaces when both of the "sandwiching" surfaces are nonreflective (effective emittance of 0.82) for typical temperature conditions. As a rule of thumb, the data in Table 5.2 suggests that a resistance value of "R-1" is about right for air spaces if the surrounding surfaces are nonreflective with air space width being a minor factor.

Width is of increased importance for air

TABLE 5.2 AIR SPACE RESISTANCE VALUES (ASSUMING NONREFLECTIVE FACING SURFACES)[a]

Position of Air Space	Width of Air Space (in.)	Direction of Heat Flow	Applicable Component	R-value
Horizontal	1/2	Up	Ceiling	0.73 to 1.00
Horizontal	3-1/2	Up	Ceiling	0.80 to 1.12
Horizontal	1/2	Down	Floor	0.77 to 1.16
Horizontal	3-1/2	Down	Floor	1.00 to 1.64
Vertical	1/2	Horizontal	Walls	0.77 to 1.15
Vertical	3-1/2	Horizontal	Walls	0.85 to 1.23

[a] For surfaces with an effective emittance of 0.82.
Source: Excerpted with permission from the *1993 ASHRAE Handbook of Fundamentals*, Chapter 22, Table 2.

spaces enclosed by bright reflective surfaces with a low effective emittance (e.g., foil-faced gypsum board or insulation). For such conditions, resistance values range from about R-2 to almost R-4 when the direction of heat flow is either up or horizontal. And when the direction of heat flow is down, wide ($3\frac{1}{2}$ in.) horizontal air spaces enclosed by a bright aluminum surface can be R-8 or greater, a fact which can be well applied in attic design to resist summer heat gain through the use of radiant barriers.

"Air films" are thin layers of "insulating" air which naturally attach by static electricity of building surfaces both outside and inside. The R-values of outdoor air films are much lower than those for indoor air films because of the dislodging effect of the wind (see Figure 5.2).

Air film R-values are shown in Table 5.3 for nonreflective surfaces.

For exterior air films, a wind speed of 15 mph is typically used for winter heat loss calculations. The 7.5-mph wind condition is used for summer conditions and heat loss through ventilated attics and crawl spaces.

For most construction assemblies which incorporate insulation, the relatively small R-values of air films will have a minor impact on overall heat loss. An exception is the increased heat loss through overhead horizontal or sloped glazing which is discussed below.

5.4 CONSTRUCTION ASSEMBLIES

To obtain the total resistance (R_t) of an assembly of any number of individual elements and components, simply sum their individual resistances (R_1, R_2, R_3, etc.), as indicated in Formula 5A.

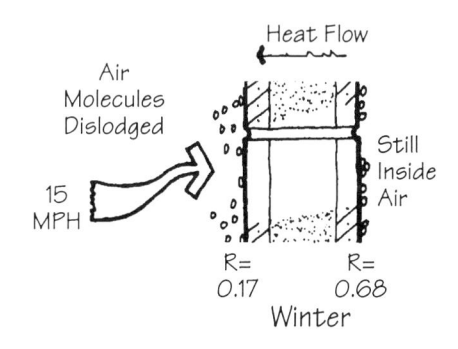

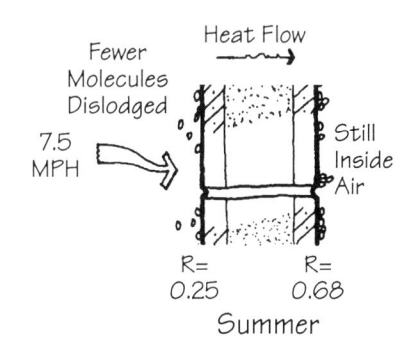

FIGURE 5.2 Heat flow resisting "air films" for various conditions.

68 HEAT LOSS FROM BUILDINGS

TABLE 5.3 AIR FILMS (ASSUMING NONREFLECTIVE SURFACE[a])

Position of Surface	Direction of Heat Flow	R-Value
Still air (inside):		
Horizontal	Upward	0.61
Sloping-45°	Upward	0.62
Vertical	Horizontal	0.68
Sloping-45°	Downward	0.76
Horizontal	Downward	0.92
Moving air (outside):		
15 mph Wind	Any	0.17
7.5 mph Wind	Any	0.25

[a]For surfaces with an effective emittance of 0.90.
Source: Excerpted with permission from the *1993 ASHRAE Handbook of Fundamentals*, Chapter 22, Table 1.

$$R_t = R_1 + R_2 + R_3 + \cdots$$

Formula 5A

where

R_t is the total resistance of the entire construction assembly

R_1 is the individual resistance of the first element (e.g., outside air film) of an entire assembly

R_2 is the individual resistance of the second element (e.g., concrete block) of an entire assembly

R_3 is the individual resistance of the third element of an entire assembly, and so on

Example 5A: The construction of a warehouse building wall is shown in Figure 5.3. Determine the total resistance rating of the wall.

Solution: This is a simple example. There are five R-values which must be summed. They are the R-values for the concrete wall (with lightweight aggregate), extruded polystyrene insulation, and face brick added to R's of 0.68 for the inside air film and 0.17 for the outside air film. Use Formula 5A.

$$R_t = R_1 + R_2 + R_3 + \cdots$$

(Formula 5A)

$$= 4.00 + 5.00 + 0.80 + 0.68 + 0.17$$

$$= 10.65$$

U-Factors and Units of Heat Loss

Until now, no units have been applied to R-values. The units of R-values are follows:

$$\frac{ft^2 \cdot °F \cdot h}{Btu}$$

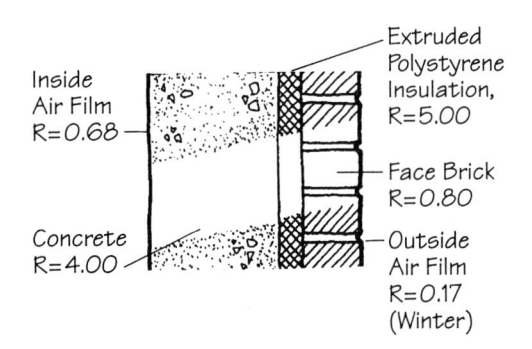

FIGURE 5.3 Thermal resistance of a simple wall.

Inside Air Film R=0.68
Concrete R=4.00
Extruded Polystyrene Insulation, R=5.00
Face Brick R=0.80
Outside Air Film R=0.17 (Winter)

Inspection of these units indicates that the

resistance to heat flow in Btus depends upon the area of the component (ft^2), the temperature difference (°F), and a time period (h).

The opposite of resistance to heat flow is the conductance of heat flow.

Conductance of heat through a construction assembly is represented by the overall coefficient of heat transmission known as the "U-factor" which have units as follows:

$$\frac{Btu}{ft^2 \cdot °F \cdot h}$$

U-factors are used to compute the flow of heat through solid building components. A U-factor (U_o) is the inverse, or reciprocal, of the total R-value (R_t) as indicated by Formula 5B.

$$U_o = \frac{1}{R_t} \quad \text{Formula 5B}$$

where

U_o is the overall heat transfer coefficient of an entire assembly in Btu/ft$^2 \cdot$°F$\cdot$h

R_t is the total R-value of an entire assembly

Example 5B: The wall in Example 5A had a total thermal resistance (R_t) of 10.65. Determine the U-factor for the wall.

Solution: Use Formula 5B.

$$U_o = \frac{1}{R_t} \quad \text{(Formula 5B)}$$

$$= \frac{1}{10.65}$$

$$= 0.094 \quad \frac{Btu}{ft^2 \cdot °F \cdot h}$$

Normally two decimal place accuracy for U-factors will suffice. This means that 0.09,

or about 1/11 of a Btu will be conducted through each square foot of the wall when there is a difference in temperature between inside and outside of 1°F over a time period of 1 h.

Be sure to compute U-factors by taking the reciprocal of the total resistance (R_t) of the entire assembly. Some make the mistake of adding the sums of individual reciprocals. Never do that!

Complex Assemblies

Most building components such as frame walls are more complex. Not only do they contain several layers of construction and finishes, but a fraction of the wall contains only wood for studs, plates, and framing around openings.

Example 5C: Determine the total R-values for the 2 × 6 wood frame wall illustrated in Figure 5.4 for sections where the

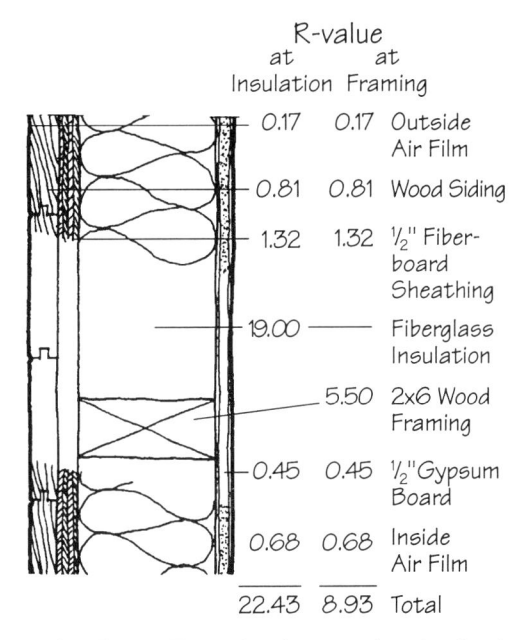

	R-value at Insulation	at Framing	
	0.17	0.17	Outside Air Film
	0.81	0.81	Wood Siding
	1.32	1.32	½" Fiberboard Sheathing
	19.00		Fiberglass Insulation
		5.50	2x6 Wood Framing
	0.45	0.45	½" Gypsum Board
	0.68	0.68	Inside Air Film
	22.43	8.93	Total

FIGURE 5.4 Thermal resistance of an insulated wood frame wall.

insulation occurs and for sections where there is wood framing instead. The R-ratings shown for the individual wall elements were obtained from Tables 5.1 and 5.3 with an R-value of 1.00 per inch assumed for the actual $5\frac{1}{2}$ in. width of the 2 × 6 wood framing.

Solution: Using Formula 5A, the total R-values for both the insulated and framed portions of wall can be calculated.

Insulated portion:

$$R_t = R_1 + R_2 + R_3 + \cdots$$

(Formula 5A)

$$= 0.17 + 0.81 + 1.32 + 19.00$$
$$+ 0.45 + 0.68$$
$$= 22.43$$

And for the framed portion (with the R-value for the wood framing replacing the R-value of the insulation):

$$R_t = R_1 + R_2 + R_3 + \cdots$$

(Formula 5A)

$$= 0.17 + 0.81 + 1.32 + 5.50$$
$$+ 0.45 + 0.68$$
$$= 8.93$$

Table 5.4 provides an approximation of percentages of solid wood area for various wood frame construction types allowing for studs and sills, rafters, and framing for openings. It shows that for a 2 × 6 wall with 24 in. on-center (o.c.) spacing of studs, the fraction of framing would be about 0.20 (20%). In general, framing percentages for walls are greater than for floors or roofs/ceilings because of the framing around openings. Heat loss calculations for a building component with two or more types of cross-sectional construction such as a wall containing areas of insulation ("A") and framing ("B") must be area-weighted using Formula 5C.

$$U_{aw} = \left(\frac{1}{R_{t-A}} \times F_A\right) + \left(\frac{1}{R_{t-B}} \times F_B\right)$$

Formula 5C

where

U_{aw} is the area weighted U-factor of an assembly made up of two or more cross-sectional construction assemblies

R_{t-A} is the total R-value of cross-section type "A"

F_A is the fraction of the total assembly constructed like cross-section type "A"

R_{t-B} is the total R-value of cross-section type "B"

F_B is the fraction of the total assembly constructed like cross-section type "B"

TABLE 5.4 WOOD FRAME CONSTRUCTION FRAMING FRACTION

	Approximate Framing Fraction
Wall framing:	
16 in. o.c.	0.25
24 in. o.c.	0.20
Floor or roof/ceiling framing:	
12 in. o.c.	0.13
16 in. o.c.	0.10

Of course, the sum of the fractions (F_A, F_B, etc.) must always equal 1.

Example 5D: Determine the area-weighted U-factor for the insulated wood frame wall shown in Figure 5.4.

Solution: Use Formula 5C. The fraction of framing (B) of 0.20 is obtained from Table 5.4 for 2 × 6 construction 24 in. o.c.

$$U_{aw} = \left(\frac{1}{R_{t-A}} \times F_A\right) + \left(\frac{1}{R_{t-B}} \times F_B\right)$$

(Formula 5C)

$$= \left(\frac{1}{22.43} \times 0.80\right)$$
$$+ \left(\frac{1}{8.93} \times 0.20\right)$$
$$= 0.036 + 0.022$$
$$= 0.058$$

Calculation of Component Heat Loss

Heat losses are usually designated by the symbol q followed by a qualifying subscript. For example, transmission heat loss (q_{tr}) for a building element is calculated by multiplying the U-factor by the area of construction, as indicated by Formula 5D.

$$q_{tr} = U_{aw} \times A \quad \text{Formula 5D}$$

where

q_{tr} is the transmission heat loss of a building component with units of Btu/°F·h

U_{aw} is the U-factor of the wall or other component (area weighted if applicable for frame walls or other) in Btu/ft²·°F·h

A is the area of the component in ft²

Example 5E: A small warehouse building has 2000 ft² of exterior concrete block wall having a U-factor of 0.09. Calculate the transmission heat loss (q_{tr}) of the wall.

Solution: Use Formula 5D:

$$q_{tr} = U_{aw} \times A \quad \text{(Formula 5D)}$$
$$= 0.09 \times 2000$$
$$= 180 \text{ Btu}/°F·h$$

Therefore, 180 Btus of heat will be lost for each degree of temperature difference (between inside and outside) per hour, also known as "Btu per degree-hour," as will be developed later. And many thousands and even millions of Btus will be lost as temperatures fluctuate over the many hours and days of an entire winter.

In building heat loss calculations, Formula 5D, which multiplies U-factors by areas, is used repeatedly to compute the heat loss through the various solid elements which form the envelope of the building. Such calculations are used to obtain the total transmission heat loss of the building by summing the results of all items of the envelope.

Thermal Bridges

It is clear that to minimize the loss of heat through opaque building elements (i.e., wall and roof assemblies) it is necessary to incorporate insulation materials which have high R-values. Yet that is not all that must be done. Attention must also be given to detailing and construction material selection relative to "thermal bridging." Thermal bridges result from having construction materials which are highly conductive of heat (i.e., heavyweight concrete and steel), penetrate or bypass the insulation, and act as a thermal short circuit. Several common details which exhibit thermal bridging are shown in Figure 5.5.

Particular attention should be paid to large areas of exterior walls constructed

72 HEAT LOSS FROM BUILDINGS

Heat Conducted by Metal Framing

Heat Conducted Through Edge of Concrete Slab

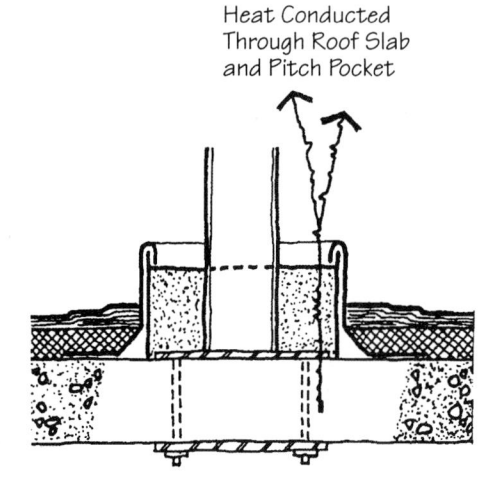

Heat Conducted Through Roof Slab and Pitch Pocket

FIGURE 5.5 Typical details with heat-losing thermal bridges.

with steel studs since steel has a thermal conductivity approximately 1000 times that of fibrous insulation. An example of this would be a wall built with 6-in. wide steel studs, R-19 cavity insulation, and simple interior and exterior finishes (which total to an R-value of perhaps 3 when added to the resistance of the air films). Such a wall would appear to have a value of R-22 and a U-factor of about 0.045 (1/22). However, laboratory tests have shown that such a metal stud wall would actually have a U-factor of about 0.09. This represents a doubling of heat transmission.

When construction details which are conducive to thermal bridging are used (such as steel studs), be sure to decrease the effective R-value of the insulation by approximately 50% in calculations. Also consider changing your detail! For more detailed information see *Catalog of Thermal Bridges in Commercial and Multi-Family Residential Construction* published by Oak Ridge National Laboratory.

5.5 WINDOWS

Windows are responsible for a considerable amount of the transmission heat loss which occurs in most houses and buildings. Before discussing glazing performance in detail let's start with some approximate rules of thumb. In simple terms, single-glazed windows have an R-value equal to about R-1, while the rating of double-glazed windows is about R-2, and you guessed it, triple-glazed windows and the relatively new windows with high performance "low-E" (low emittance) coatings are conveniently about R-3.

The U-factors, and resultant heat losses for window assemblies, vary depending upon the size of the opening and design of the window fame. Small windows have a relatively large percentage of frame area (sometimes 30–40%) whereas the percentage of frame area for large windows will be only 10–20%.

Tests developed by the National Fenestration Rating Council (NFRC) are currently used to determine two U-factors for glass. The first value refers to the U-factor at the "center" of glass (see Figure 5.6) where it is not thermally affected by edge spacers (which are used in glazing with multiple layers). The second value then provides the U-factor at the "Edge" of the glass (a strip assumed to be $2\frac{1}{2}$ in. wide) where it is affected by the edge spacer (either metal or an insulating sealant). Overall U-factors are then published for the entire window assembly based on frame type and window size.

Glass often is coated to enhance performance in some manner. Low-E (low emissivity) coated glass reduces heat loss and is now very commonly used for houses in climates that experience cold winters. Reflective coatings limit the amount of solar heat gain that will penetrate from the outside.

The surfaces of a glazing assembly are referenced in numerical order from the outside, as indicated in Figure 5.7.

The *1993 ASHRAE Handbook of Fundamentals* contains a comprehensive listing of U-factors for 54 configurations of glazing in various types of windows. Table 5.5 provides an excerpt of this table for common glazings and window types (aluminum and wood/vinyl). "Operable" windows are assumed to be 3 ft wide by 5 ft high double-hung windows. "Fixed" windows are assumed to be one large plate of glass 4 ft wide by 6 ft high.

A few things to consider for glazings shown in Table 5.5:

1. Filling spaces between glazings with inert gases (argon or krypton) has become relatively inexpensive. Still, questions persist regarding the longevity of enhanced performance by these gases.

2. The relatively high U-factors for aluminum windows indicate that their use should be very limited in temperate or cold climates. In addition, rarely will one couple poorly performing aluminum windows with high-performance low-E glazing, or an inert gas filler (argon or krypton).

3. The excellent performance of triple

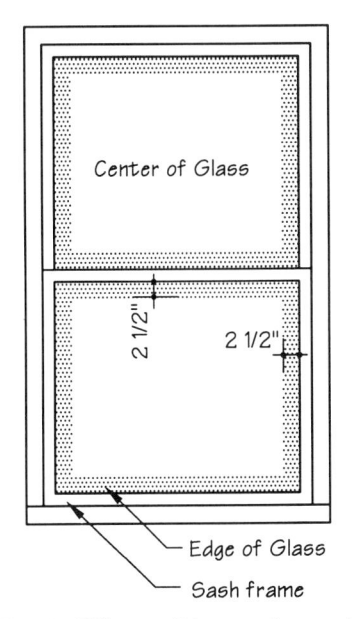

FIGURE 5.6 Different U-factors for portions of glazing.

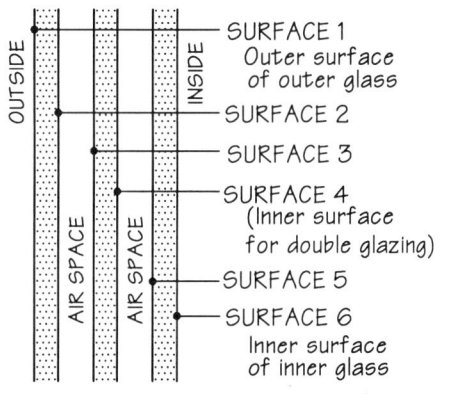

FIGURE 5.7 Numbering system for glazing surfaces.

TABLE 5.5 U-FACTORS FOR COMMONLY USED GLAZING AND WINDOWS

Glazing Type	Glazing Only, Center of Glass	Alum., Metal Thermal Break, Operable	Wood or Vinyl — Metal Thermal Break, Operable	Wood or Vinyl — Insul. Thermal Break, Operable	Wood or Vinyl — Insul. Thermal Break, Fixed
Single 1/8 in. glass	1.11	1.07	0.94	—	—
Double glazing					
1/4 in. air space	0.57	0.67	0.56	0.54	0.56
1/2 in. air space	0.49	0.62	0.51	0.48	0.49
1/4 in. argon space	0.52	0.64	0.53	0.50	0.51
1/2 in. argon space	0.46	0.59	0.49	0.46	0.46
Double glazing $E = 0.20$ on surface 2 or 3					
1/4 in. air space	0.46	0.59	0.49	0.46	0.46
1/2 in. air space	0.35	0.52	0.42	0.39	0.36
1/4 in. argon space	0.38	0.54	0.44	0.41	0.39
1/2 in. argon space	0.30	0.48	0.39	0.36	0.32
Triple glazing $E = 0.20$ on surfaces 3 and 5					
1/4 in. air spaces	0.29	0.47	0.37	0.33	0.31
1/2 in. air spaces	0.20	0.41	0.31	0.27	0.23
1/4 in. argon spaces	0.23	0.43	0.33	0.29	0.25
1/2 in. argon spaces	0.16	0.38	0.29	0.24	0.19
Quadruple glazing $E = 0.10$ on surfaces 3 and 5					
1/4 in. air spaces	0.23	0.43	0.33	0.29	0.25
1/2 in. air spaces	0.15	0.37	0.28	0.23	0.19
1/4 in. argon spaces	0.17	0.39	0.29	0.25	0.20
1/2 in. argon spaces	0.12	0.35	0.26	0.22	0.16
1/4 in. krypton spaces	0.12	0.35	0.26	0.22	0.16
1/2 in. krypton spaces	0.11	0.34	0.25	0.21	0.15

Source: Excerpted with permission from the *1993 ASHRAE Handbook of Fundamentals*, Chapter 27, Table 5.

glazing and quadruple glazing with low-E coating will generally be reserved for use in very cold climates (at least 7000 heating degree-days, base 65) and/or areas with high heating fuel costs.

4. Windows constructed of reinforced vinyl or aluminum-clad wood tend to have U-factors just slightly higher than those shown in the table for "wood or vinyl" windows.

Adjustment for Sloped and Horizontal Glazing

Owing to the lower R-value of air films for heat flow up through sloped or horizontal surfaces, it may be prudent to adjust the U-value for large areas of sloped or horizontal glazing. Table 5.6 provides adjustment factors to apply to commonly available glazing products where published U-

TABLE 5.6 U-FACTOR ADJUSTMENT FACTORS SLOPED AND HORIZONTAL GLAZING FOR UPWARD HEAT FLOW

	\multicolumn{6}{c}{U-Factor of Vertical Glazing}
	0.30	0.40	0.50	0.60	1.00	1.10
U-factor for sloped (45°) glazing	0.36	0.47	0.57	0.68	1.11	1.22
U-factor for horizontal (0°) glazing	0.40	0.51	0.61	0.72	1.14	1.25

Source: Excerpted with permission from the *1989 ASHRAE Handbook of Fundamentals*, Chapter 27, Table 13, Part B.

factors refer to the more typical vertical glazing application.

Adjustment for Wind Speed

U-factors for assemblies, and published values for manufactured products, are typically based on a winter wind of 15 mph. A complex mathematical expression is used to adjust for other wind speeds. In practical terms, though, the only adjustment which may be warranted is for glazing.

For example, a double-glazed window with a published winter U-factor of 0.50 (assuming a 15-mph wind) will have a summer U-factor of 0.47 when only 7.5 mph winds are assumed. This window will also have a U-factor of 0.45 in zero wind, and a U-factor of 0.52 with a 30-mph wind. And a single-glazed window with a winter U-factor of about 1.10 (at 15 mph) has a lower U-factor of 0.94 in zero wind.

Probably the only application where adjustment is warranted would be large glazed areas entirely shielded from the wind, in which case values indicated above for zero wind should be used in calculations.

5.6 INFILTRATION OF OUTSIDE AIR

During the winter, cold outside air can enter or "infiltrate" a building through cracks or openings, and in the process cause heated house air to "exfiltrate" through other building cracks and openings.

Although the heat is actually lost with the air that exfiltrates, by convention this type of heat loss is referred to as infiltration heat loss. Infiltration can account for a significant percentage of the heat loss of a building (generally somewhere between 25 and 50%). There are two commonly used methods of calculating infiltration heat loss.

The first method, referred to as the "Air Change" method, is based on an estimate of the fraction of a building's total air volume (in cubic feet) that will be replaced by infiltration in one hour.

The second and more complex method, known as the "Leakage Area" method, quantifies the infiltration through all of the cracks and openings in a house depending upon the tightness of the construction. Both methods are presented below.

Compared to transmission heat loss, the estimation of infiltration heat loss is relatively imprecise, with many hard to define variables. For example, the effect of wind on a particular building is very site specific. Also, the quality of construction will be instrumental in determining the leakiness of cracks.

Building "tightness" can only be determined accurately after construction has been completed. At that time, a "blower door" test to determine air leakage can be conducted using a fan to pressurize the house.

Blower door tests can be effectively used

on existing buildings (especially old houses) to determine locations of excessive leakage needing caulking or sealing. In Sweden, blower door testing is even used to verify the quality of new construction as a precondition for occupancy.

Air Change Method

For the Air Change method, typical values for contemporary average-sized residences (about 1700 ft² in floor area), based upon "tightness of construction," are presented in Table 5.7.

Although workmanship can and will vary greatly, most contemporary residential construction tends to fall at the low end of the "Medium" category, with an air change rate of about 0.7 air changes per hour on average during the winter.

To achieve a "Tight" air change designation, the use of top quality windows, careful caulking and sealing, and a vapor-permeable air barrier may be required. The potential negative side of tight construction is the possibility of poor indoor air quality, as discussed in Section 6.6

Calculation of infiltration heat loss (q_{inf}) by the air change method merely entails multiplying the heat capacity of air by the quantity of air changed, as illustrated in Formula 5E.

$$q_{inf} = C \times ACH \times 0.018 \quad \text{Formula 5E}$$

where

q_{inf} is the infiltration heat loss of a space in Btu/°F·h
C is the volume of the space in ft³
ACH is the infiltration air change rate per hour
0.018 is the heat capacity of air (Btu/°F·ft³)

Some people find the units of "air changes" confusing. To keep the units straight it helps to think of the volume of a building in cubic feet as being equal to 1 air change, as shown in the following example.

Example 5F: Determine the rate of infiltration heat loss by a cabin of "Medium" construction and a volume of 2380 ft³. Assume that the cabin has 0.7 air changes per hour.

Solution: Calculate the infiltration heat loss rate (q_{inf}) using Formula 5E.

$$q_{inf} = C \times ACH \times 0.018$$
$$\text{(Formula 5E)}$$
$$= 2380 \times 0.7 \times 0.018$$
$$= 30 \text{ Btu/°F·h}$$

Therefore, 30 Btus will be lost because of infiltration for each 1°F of temperature difference for each hour of occurrence. And it

TABLE 5.7 AIR CHANGES PER HOUR (ACH) ESTIMATED BY CONSTRUCTION TIGHTNESS

Tightness of Envelope Construction	Average Winter Air Changes per Hour
Tight	0.2 to 0.6 ACH
Medium	0.6 to 1.0 ACH
Loose	1.0 to 2.0 ACH

should be realized that over the months of an entire heating season many thousands of Btus of heat will be lost as temperature differences between inside and outside vary.

The air change method should be applied with judgment and common sense. This is particularly true for larger houses and commercial buildings. For example, Figure 5.8 shows House A with a volume of 15,000 ft^3 and House B with a volume which is twice as large at 30,000 ft^3. Strict application of the air change method based solely on volume would yield an infiltration heat loss total for House B which is twice that for House A.

However, the wall surface area of House B is only 39% greater than the wall surface of House A. Therefore, since most infiltration is related to cracks in walls (openings around windows, doors, and framing), the infiltration heat loss for House B is likely to be closer to 39% greater than that for House A, rather than double.

When using the air change method to size heating system components, remember to take into account the largest reasonable air change rate which might occur. For example, a house assumed to have an average winter air change rate of 0.7 ACH might well be designed for peak conditions of 1.0 to 1.5 ACH when system sizing is the concern. Whereas, the 0.7 ACH value would be used to calculate probable fuel consumption.

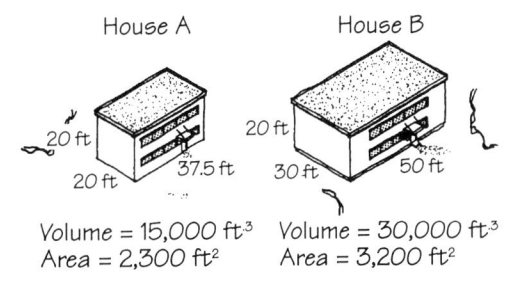

FIGURE 5.8 Relationship of building volume and surface area.

Effective Leakage Area Method

Computing air infiltration based upon openings and cracks goes back to the beginning of building heat loss calculations. Until recent years, the "crack method" was limited to a tabulation of air leakage based upon lengths of cracks around windows and other openings as well as a factor for type of wall construction.

In recent years, ASHRAE and others have attempted to quantify air leakage more rigorously by using a method which computes the infiltration air flow based upon the leakage area of the building construction. The method considers factors such as building height, exposure, outdoor temperature, and wind conditions. Because it is based on specific building components and conditions, the method provides more reliable results than the air change method.

Table 5.8 provides an abbreviated form of "best estimate" component leakage areas. For some items, such as sills and windows, leakage calculations are based upon square inches of leakage area per linear foot of crack (in.^{2}/lftc). For others, such as exterior walls, calculations are based upon square inches of leakage area per square foot of the element (in.^{2}/ft^2). For even others, such as furnaces, vents, and fireplaces, leakage calculations are based on leakage area for each item (in^2./ea).

Some items have a very large leakage area. For example, note that just one weatherstripped access door to an attic or crawl space may have as much leakage area (2.8 in.^{2}) as 1273 ft^3 of exterior wall with a continuous air infiltration barrier (2.8/0.0022). Other items of particular concern are fireplaces, fans, pipe penetrations, and furnaces without sealed combustion.

The ASHRAE calculation procedure for determining infiltration heat loss by the "Effective Leakage Area Method" is generally as follows:

STEP 1: Calculate the total building leakage area (*L*) by summing up the leakage area of all the components using factors such as those shown in Table 5.8.

STEP 2: Select Stack Coefficient A, related to building height, from Table 5.9. (Note: This calculation procedure provides values for buildings of one, two, or three stories, only. In taller buildings "stack effect" can be very pronounced and lead to increased infiltration as discussed in Section 6.9.)

STEP 3: Select Wind Coefficient B, related to building exposure (ranging from full exposure to winds with no obstructions, to very heavy shielding from winds by surrounding buildings, trees, and shrubs), from Table 5.10.

STEP 4: Determine the air flow rate due to infiltration in cubic feet per minute (generally referred to as cfm) with the appropriate coefficients A and B using Formula 5F. Try not to worry about the "crazy" units, which will be resolved.

TABLE 5.8 EFFECTIVE LEAKAGE AREAS OF BUILDING COMPONENTS (*L*) (LOW-RISE RESIDENTIAL APPLICATIONS)

Ceiling, general	0.026 in.^{2}/ft^2
Doors (weatherstripped):	
— Single	1.9 in.^{2}/ea
— Access to attic or crawl space	2.8 in.^{2}/ea
Fireplace w/o insert:	
— Damper closed	0.62 in.^{2}/ft^2
— Damper open	5.04 in.^{2}/ft^2
Floors over crawl space, general	0.032 in.^{2}/ft^2
Furnace	
— Sealed (or no) combustion	0.0 in.^{2}/ea
— Retention head or stack damper	4.6 in.^{2}/ea
Gas water heater in conditioned space	3.1 in.^{2}/ea
Joints	
— Ceiling-wall	0.76 in.^{2}/lftc
— Top plate, band joist	0.05 in^2/lftc
Piping/plumbing/wiring penetrations	
— Uncaulked	0.9 in.^{2}/ea
— Caulked	0.3 in.^{2}/ea
Sill	0.11 in.^{2}/lftc
Vents, dampers closed:	
— Kitchen	0.8 in.^{2}/ea
— Bathroom	1.6 in.^{2}/ea
— Dryer vent	0.46 in.^{2}/ea
Walls (exterior)	
— Lightweight concrete block, painted or stucco	0.016 in.^{2}/ft^2
— Continuous infiltration air barrier	0.0022 in^2/ft^2
— Rigid sheathing	0.005 in.^{2}/ft^2
Window framing, wood, caulked	0.004 in.^{2}/ft^2
Windows (weatherstripped):	
— Casement	0.12 in.^{2}/lftc
— Double hung	0.33 in.^{2}/lftc
— Double horiz. slider, wood	0.28 in.^{2}/lftc

Source: Excerpted with permission from the *1993 ASHRAE Handbook of Fundamentals*, Chapter 23, Table 3.

5.6 INFILTRATION OF OUTSIDE AIR

TABLE 5.9 STACK COEFFICIENT A

	Number of Stories		
	One	Two	Three
Stack coefficient	0.0156	0.0313	0.0471

Source: Excerpted with permission from the *1993 ASHRAE Handbook of Fundamentals*, Chapter 23, Table 6.

TABLE 5.10 WIND COEFFICIENT B

	Number of Stories		
	One	Two	Three
Shielding class (for building with):			
1 (No obstructions)	0.0119	0.0157	0.0184
2 (Few obstructions)	0.0092	0.0121	0.0143
3 (Moderate shielding)	0.0065	0.0086	0.0101
4 (Heavy shielding)	0.0039	0.0051	0.0060
5 (Very heavy shielding)	0.0012	0.0016	0.0018

Source: Excerpted with permission from the *1993 ASHRAE Handbook of Fundamentals*, Chapter 23, Table 8.

$$I_{cmf} = L \times [(A \times TD) + (B \times V^2)]^{1/2}$$

Formula 5F

where

- I_{cmf} is the infiltration air flow rate in cubic feet per minute (cfm)
- L is the infiltration leakage area in in.^{2}
- A is the stack coefficient in cfm^2/(sq. in.)$^2 \cdot °F$ (from Table 5.9)
- TD is the difference between the indoor and outdoor temperatures in °F ''for the time period''
- B is the wind coefficient in cfm$^2 \cdot$ mph^2 (from Table 5.10)
- V is the average velocity of the wind in miles per hour ''for the time period''

Note the phrase ''for the time period.'' When sizing a heating system, use winter outside design temperature and a reasonable maximum wind velocity. When estimating energy consumption, use the average winter temperature and wind velocity. Calculations for each will be illustrated later.

STEP 5: Heat loss due to infiltration (q_{inf}) can now be computed using Formula 5G.

$$q_{inf} = I_{cfm} \times 1.08 \quad \text{Formula 5G}$$

where

- q_{inf} is the infiltration heat loss in Btu/°F·h
- I_{cfm} is the infiltration air flow rate in cubic feet per minute (cfm)
- 1.08 is a ''constant'' which is the product of 60 minutes per hour and 0.018 Btu/°F·h (the heat capacity of air)

This leakage area calculation procedure will

be fully illustrated in the detailed heat loss example for "The House" in Section 5.14 below.

5.7 SIZING HEATING SYSTEMS

The most important design function played by heat loss calculations is the sizing of equipment and systems to provide adequate heat output under "design conditions." Simply put, the calculations are done to size the system so that it provides enough heat to maintain comfort when it gets very cold outside. Heating systems are normally sized to meet loads imposed by temperatures which are nearly the coldest of the year (95% winter design temperature—see Section 2.2). Systems are not normally sized to meet all-time record low temperatures because they would be substantially oversized at all other times.

Once the total heat loss of a room or entire building has been computed, the need for heat to satisfy design conditions (q_{design}) can be calculated using Formula 5H.

$$q_{design} = q_{tot} \times (t_i - t_o) \qquad \text{Formula 5H}$$

where

q_{design}	is the design heat loss in Btu/h
q_{tot}	is the total loss (transmission and infiltration) of a room or building in Btu/°F·h
t_i	is the inside design temperature in °F
t_o	is the outside winter design temperature in °F

Example 5G: Determine the quantity of heat required by a house in Cincinnati, Ohio which has a total design heat loss of 750 Btu/°F·h. The inside temperature is to be maintained at 70°F (t_i).

Solution: Obtain the 97.5% winter design dry-bulb temperature for Cincinnati of 6°F (t_o) from Table 2.1. Use Formula 5H.

$$q_{design} = q_{tot} \times (t_i - t_o) \quad \text{(Formula 5H)}$$
$$= 750 \text{ Btu}/°F·h \times (70°F - 6°F)$$
$$= 48{,}000 \text{ Btu/h}$$

Therefore, the heating system must provide an output of 48,000 Btus per hour to maintain inside temperatures when design conditions exist.

5.8 CALCULATIONS FOR A SIMPLE ROOM

Example 5H: Calculate the design heat loss in Btu/h for a bedroom with only one exposure on the middle floor of a 3-story townhouse in Memphis, Tennessee. Room parameters and U-values are as shown in Figure 5.9. The outside design temperature of 18°F for Memphis is found in Table 2.1, and the indoor temperature to be maintained is 65°F. The assumed infiltration heat loss for design purposes is 1.25 air changes per hour. For this example there is no heat loss through the floor, ceiling, and side walls because they abut heated spaces.

Solution: Calculate the transmission heat loss for each element by multiplying its U-factor by its area (A) in square feet (sim-

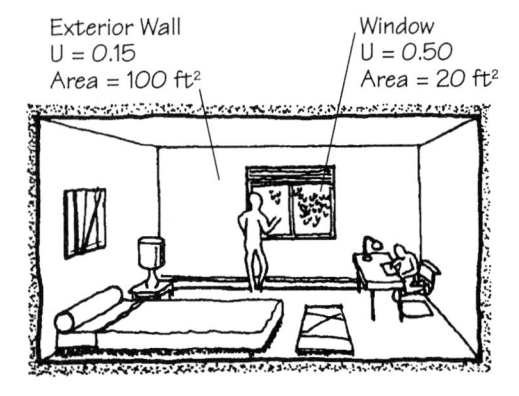

Exterior Wall
U = 0.15
Area = 100 ft²

Window
U = 0.50
Area = 20 ft²

FIGURE 5.9 Transmission heat loss calculation for a simple room.

ilar to Formula 5D). Sum these results to obtain the total transmission heat loss (q_{tr}). Then calculate infiltration heat loss (q_{inf}) using Formula 5E. Add the results together to get the total heat loss of the room (q_{tot}) in units of Btu/°F·h using Formula 5I.

Transmission Heat Loss:

$$q_{tr} = U \times A \quad \text{((Formula 5D)}$$

Surface	U		A		UA
Opaque Wall	.15	×	100	=	15
Window	.50	×	20	=	10
			q_{tr}	=	25 Btu/°F·h

Infiltration Heat Loss:

$$q_{inf} = C \times ACH \times 0.018$$

$$\text{(Formula 5E)}$$

$$= 1200 \times 1.25 \times 0.018$$

$$= 27 \text{ Btu/°F·h}$$

Total Heat Loss:

$$q_{tot} = q_{tr} + q_{inf} \quad \text{Formula 5I}$$

$$= 25 + 27$$

$$= 52 \text{ Btu/°F·h}$$

To determine the design heat loss of the room (q_{design}) multiply the room heat loss by the difference between the desired indoor temperature and the outdoor design temperature using Formula 5H.

$$q_{design} = (q_{tot}) \times (t_i - t_o)$$

$$\text{(Formula 5H)}$$

$$= 52 \times (65 - 18)$$

$$= 2444 \text{ Btu/h}$$

Therefore, the heating system must deliver 2444 Btu/h to this room if it is to maintain the desired temperature when design conditions exist outdoors.

5.9 BELOW-GRADE WALLS

Calculation of heat loss through below-grade (basement) walls is more complex than for above-grade walls which are exposed to a uniform temperature of outside air. In contrast, the deep earth soil temperature depends primarily on the mean annual air temperature or "steady state" temperature in a location, as shown in Figure 5.10.

Soil temperatures closer to the surface then vary depending upon time of year and depth below grade, as shown in Figure 5.11.

The quantity of heat loss through below-grade walls also depends on soil type, amount of moisture in the soil, and possible snow cover. In addition, construction characteristics can also be a factor, such as the cores in concrete blocks which can promote vertical convection of heat.

Obviously a great many variables impact upon heat loss from below-grade walls. If a heat loss calculation of great precision is required, as may be the case for a residence with substantial living space below grade, consult the *ASHRAE Handbook of Fundamentals* or the *Building Foundation Design Handbook* published by Oak Ridge National Laboratory. Nevertheless, approximate methods are usually adequate for most design applications. One approximate method modifies the heat loss which would occur if an assembly was an above grade wall by an "earth factor," as shown in Formula 5J.

$$q_{em} = U \times EF \times A_{bw} \quad \text{Formula 5J}$$

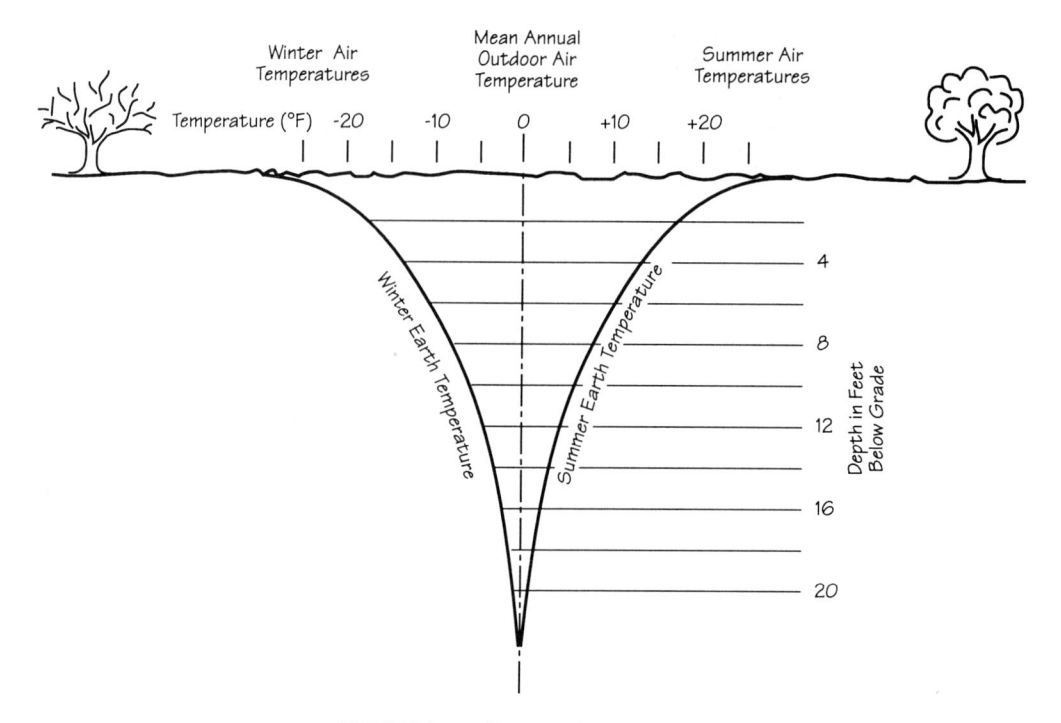

FIGURE 5.10 Deep earth temperature.

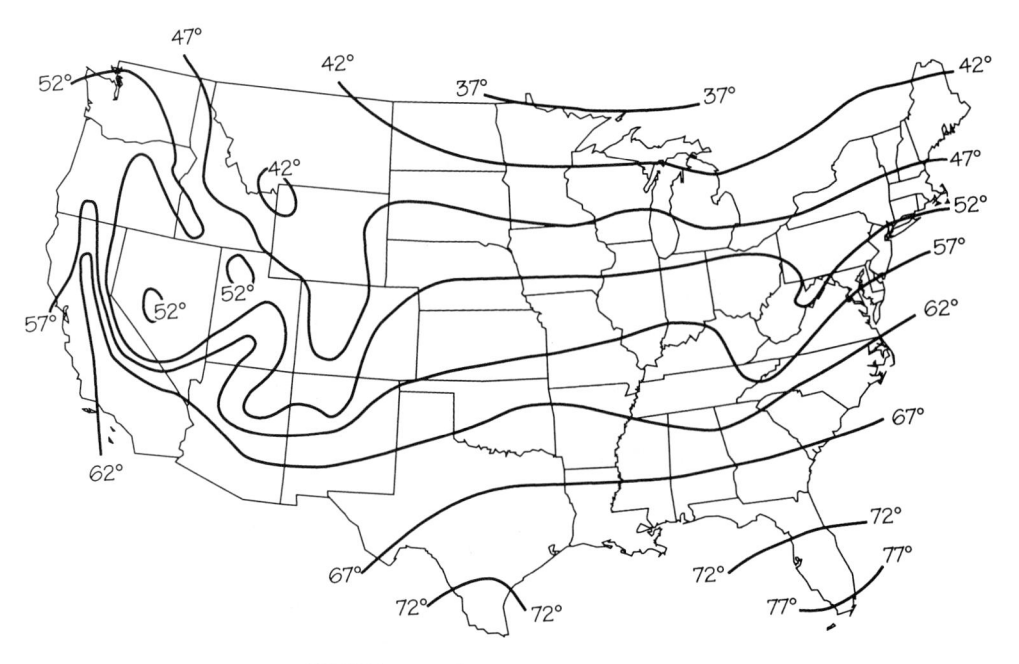

FIGURE 5.11 Soil temperature variation.

where

q_{em} is the earth modified heat loss in Btu/°F·h
U is the U-factor of the basement wall when outside and above grade
EF is the earth factor (typically 0.5)
A_{bw} is the area of basement wall below grade in ft^2

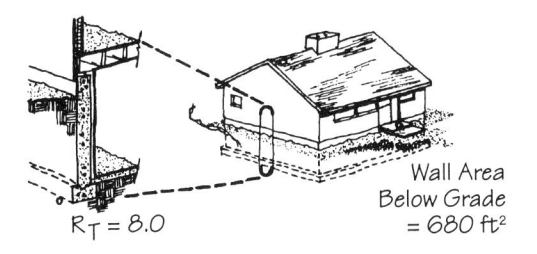

FIGURE 5.12 Heat loss by below-grade walls.

Example 5I: Estimate the below-grade wall heat loss for the house shown in Figure 5.12 using the "earth factor" approach. Assume that the basement wall has a total R-value of 8.0.

Solution: First compute the U-factor of the wall. Since the total R-value of the wall is 8, it is 0.125 (1/8). Then use Formula 5J. The entire area of below grade wall (A_{bw}) is 680 ft.

$$q_{em} = U \times EF \times A_{bw} \quad \text{(Formula 5J)}$$
$$= 0.125 \times 0.5 \times 680$$
$$= 42.5 \text{ Btu/°F·h}$$

Of course the earth factor approach is approximate, but it does acknowledge that the earth provides an insulating effect. It is usually justifiable, though, when one considers that basement heat loss is normally a minor component of total heat loss plus the fact that soil conditions vary greatly. Higher earth factors of 0.6 or 0.7 can be used to provide a more conservative estimate for sites with poor drainage, or relatively shallow embedment in the earth.

5.10 BELOW-GRADE FLOORS

Floors below grade also lose some heat to the earth below although this heat loss is relatively small since the temperature of the earth below a basement floor is moderated. In fact, for central floor areas located many feet in from the perimeter, the earth temperatures below approach the year-round deep earth temperature (see Figure 5.10).

Table 5.11 presents average heat loss values through each square foot of a concrete floor based upon depth of the foundation wall below grade, and the minimum width of the house.

TABLE 5.11 HEAT LOSS THROUGH BASEMENT FLOORS—EFFECTIVE U-FACTORS (Btu/ft^2·°F·h)

Depth of Foundation Wall Below Grade (ft)	Shortest Width of House (ft)			
	20	24	28	32
5	0.032	0.029	0.026	0.023
6	0.030	0.027	0.025	0.022
7	0.029	0.026	0.023	0.021

Source: Excerpted with permission from the *1993 ASHRAE Handbook of Fundamentals*, Chapter 25, Table 15.

Heat loss from basement floors is computed by using Formula 5K.

$$q_{bf} = U_{eff} \times A_{bf} \quad \text{Formula 5K}$$

where

- q_{bf} is basement floor heat loss in Btu/°F·h
- U_{eff} is the effective U-factor of the basement floor obtained from Table 5.11
- A_{bf} is the area of the basement floor in ft^2

Example 5J: Calculate the heat loss through the basement floor of the house shown in Figure 5.13. The basement floor area is 1120 ft^3 (28 × 40).

Solution: Since the basement floor is 5 ft below grade, and the shortest width of the house is 28 ft, the factor obtained from Table 5.11 is 0.026 Btu/ft^2·°F·h. This is the average heat loss for the area (A_{bf}) of the basement floor. Use Formula 5K.

$$q_{bf} = U_{eff} \times A_{bf} \quad \text{(Formula 5K)}$$
$$= 0.026 \times 1120$$
$$= 29.1 \text{ Btu/°F·h}$$

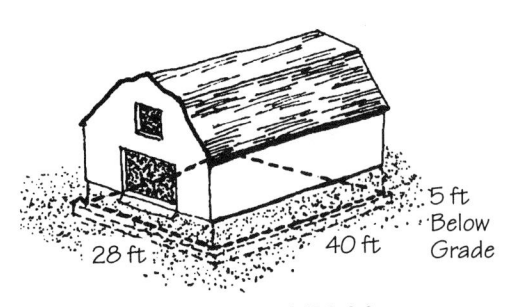

FIGURE 5.13 Heat loss through basement floors.

5.11 SLAB-ON-GROUND CONSTRUCTION

The majority of heat loss from slab-on-ground construction occurs at the perimeter where the edge of the slab is only a short distance from the surface soil and the effect of the cold outside air. As a result, calculation methods for slab-on-ground heat loss are based upon the perimeter length of slab edge in linear feet, and not on the area of floor slab. Heat loss is reduced if insulation is placed at the edge of the slab on the entire perimeter to a depth of at least 2 ft, as indicated in Figure 5.14.

Slab edge transmission heat loss (q_{se}) is then quantified using Formula 5L and what is referred to as "F2" factors. Slab edge insulation is often in direct contact with the earth and must be made of material that will not degrade. Extruded polystyrene insulation is commonly used. The F2 factors for such insulation are listed in Table 5.12 based upon R-5 per inch. Be sure to note that F2 factors provide heat loss per linear foot of perimeter (rather than per square foot for U-factors).

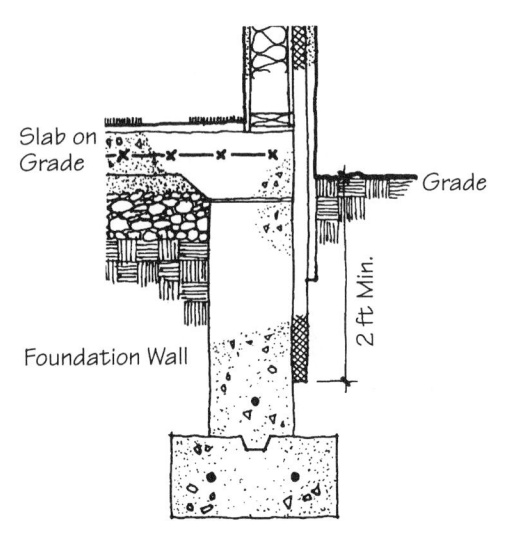

FIGURE 5.14 Heat loss by slab-on-ground construction.

TABLE 5.12 SLAB-ON-GROUND HEAT LOSS

R-Value of Insulation	Slab Edge Insulation Thickness	Approximate F2 Factor (Btu/ft·°F·h)
R-0	None	0.76
R-5	About 1 in.	0.47
R-10	About 2 in.	0.40
R-15	About 3 in.	0.36

$$q_{se} = F2 \times P \quad \text{Formula 5L}$$

where

q_{se} is the slab edge transmission heat loss in Btu/°F·h

F2 is the transmission heat loss per linear foot of slab edge in Btu/ft·°F·h

P is the perimeter of the slab edge in linear feet

As with the procedures for below-grade walls, precise calculations of slab-on-grade heat losses can be complex and depend to a degree on the type of construction, the quality of surrounding soil and sometimes even the type of heating system (if radiant slab heat is employed). For most applications, values obtained from manufacturers' literature or those shown in Table 5.12 are sufficient.

Use only materials for slab edge insulation which will not degrade structurally or thermally after long-term exposure to the elements. Although valuable and effective, many homebuilders refrain from including thick slab edge insulation due to detailing difficulties.

Example 5K: Determine the slab edge transmission heat loss (q_{se}) if the house in Example 5J were instead built on grade with R-5 insulation at the edge. The rectangular house is 28 ft wide by 40 ft long yielding a 136-ft perimeter (P).

Solution: From Table 5.12 obtain the F2 factor of 0.47 Btu/ft · °F · h for R-5 slab edge insulation (1-in. thick). Use Formula 5L.

$$q_{se} = F2 \times P \quad \text{(Formula 5L)}$$
$$= 0.47 \text{ Btu/ft·°F·h} \times 136 \text{ feet}$$
$$= 63.9 \text{ Btu/°F·h}$$

5.12 HEAT LOSS TO ADJACENT UNHEATED SPACES

Unheated spaces will be at a temperature which is above the outside air temperature when they abut heated interior spaces. One such type of space is a ventilated attic over an insulated ceiling where air temperatures typically will be 5 to 10 degrees warmer than the outside air temperature. Normally such a minor difference is ignored in heat loss calculations and the attic air temperature is assumed to be equal to the outside air temperature.

Other "tempered" spaces include unheated basements and garages. The possibility of the temperature of an unheated basement falling below freezing may be of concern regarding possible pipe freezing. In fact, most "unheated" basements are kept well above freezing because they are the location for the furnace or boiler (which would produce heat often during very cold weather), and other heat-producing equip-

86 HEAT LOSS FROM BUILDINGS

ment such as the hot water heater and laundry equipment. Some designers also use less insulation in the floor assembly over unheated basements so that the house will lose more heat to the basement.

When great precision is needed regarding air temperatures in adjacent unheated spaces the *1993 ASHRAE Handbook of Fundamentals* (page 25.8) contains a detailed methodology. For most applications, however, use of a "rule of thumb" that says that the temperature in the unheated space will be approximately halfway between the indoor and outdoor temperatures will suffice.

5.13 ANNUAL HEATING NEEDS

One of the prime uses of heat loss calculations is to be able to estimate the amount of heat needed annually (and resultant fuel cost) for a building of a particular design in a specific climate.

Remember that U-factors have units of Btu/ft$^2 \cdot$°F$\cdot$h. Therefore, when U-factors are multiplied by component areas in ft^2, the units "°F" and "h" remain in the denominator and combine to form "degree-hours." As discussed in Chapter 2, winter climates are commonly described in terms of "heating degree-hours" (HDH) and "heating degree-days" (HDD) relative to various base (heating balance point) temperatures.

Although something of a misnomer, "UA" is commonly used to represent total heat loss (transmission plus infiltration). Once the UA of a building is known it is possible to estimate building heat loss over the course of a heating season by Formula 5M (an adaptation of Formula 2B).

$$Q_{yr} = UA \times HDD_{bp} \times 24 \quad \text{Formula 5M}$$

where

Q_{yr} is the total amount of heat required to maintain the indoor design temperature in Btus per year

UA is the total design heat loss (transmission plus infiltration) in Btu/°F$\cdot$h

HDD_{bp} is the number of heating degree-days below a balance point base temperature

24   is the conversion of 24 h in 1 day

Example 5L: Determine the amount of heat which must be supplied to an old house in Chicago to maintain comfort conditions. The house has a total heat loss (UA) of 1000 Btu/°F $\cdot$ h and a computed balance point temperature of 65°F.

Solution: As indicated in Table 2.3, Chicago experiences approximately 6130 HDD for a base temperature of 65°F (the house balance point temperature). Use Formula 5M.

$$Q_{yr} = UA \times HDD_{bp} \times 24$$
$$\text{(Formula 5M)}$$
$$= 1000 \times 6130 \times 24$$
$$= 147{,}120{,}000 \text{ Btus of heat is needed per year}$$

When balance point temperatures are not known, the annual need for heat can be estimated by Formula 5N (an adaptation of Formula 2C), which employs an empirical correction factor (C_D) obtained from Figure 2.3.

$$Q_{yr} = C_D \times UA \times HDD_{65} \times 24$$
$$\text{Formula 5N}$$

5.14 HEAT LOSS CALCULATION FOR "THE HOUSE"

Example 5M: Determine the amount of heat needed by "The House," as shown in Figure 5.15, when winter design conditions

5.14 HEAT LOSS CALCULATION FOR "THE HOUSE" 87

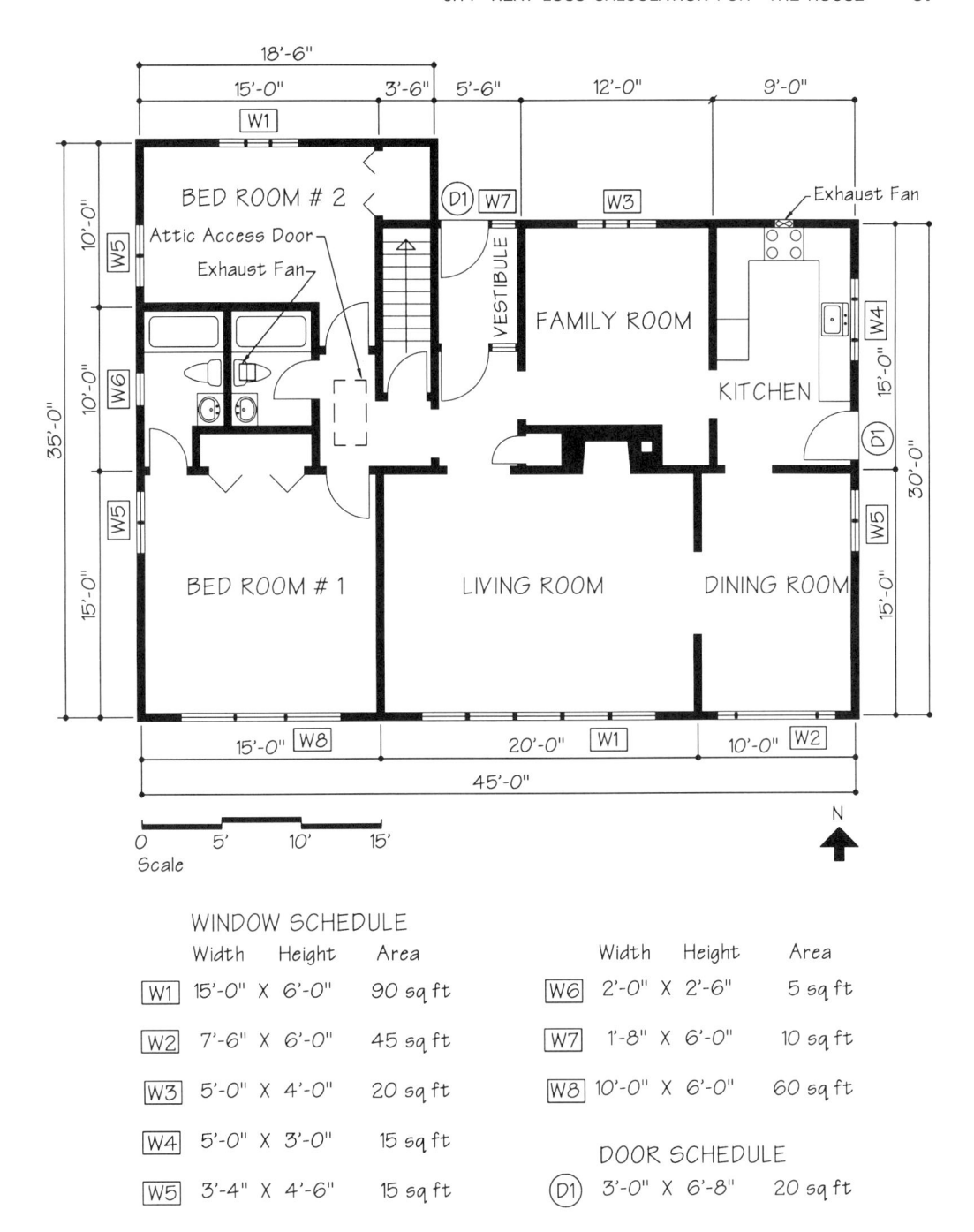

FIGURE 5.15 Floor plan for "The House."

WINDOW SCHEDULE

	Width	Height	Area
W1	15'-0" X 6'-0"		90 sq ft
W2	7'-6" X 6'-0"		45 sq ft
W3	5'-0" X 4'-0"		20 sq ft
W4	5'-0" X 3'-0"		15 sq ft
W5	3'-4" X 4'-6"		15 sq ft
W6	2'-0" X 2'-6"		5 sq ft
W7	1'-8" X 6'-0"		10 sq ft
W8	10'-0" X 6'-0"		60 sq ft

DOOR SCHEDULE

	Width	Height	Area
D1	3'-0" X 6'-8"		20 sq ft

88 HEAT LOSS FROM BUILDINGS

exist and using the leakage area method to compute infiltration heat loss. Also estimate the amount of heat needed annually.

It is a simple one-story house with the North "half" built over an unheated basement. The South "half" is built directly on grade because it will incorporate heavy thermal mass to improve passive solar heating potential.

U-factors of building components are as follows:

Walls:	0.05
Windows:	0.38
Doors:	0.20
Roof/ceiling:	0.03
Floors:	0.08
Slab edge:	0.47 (F2 factor)

The house has insulation directly over the flat room ceilings, and a ventilated peaked attic which will be assumed to be at the same temperature as the outside air. The height of walls on the first floor is 8 ft. In the interest of simplicity, this example does not reduce overall dimensions because of items such as exterior wall thicknesses. If all such construction conditions are carefully accounted for, room areas will be reduced slightly (typically by about 3–5%). Assume that the house will be built in St. Louis, Missouri with the following winter design and climate conditions:

- An indoor design temperature of 68°F.
- A 97.5% winter design temperature of 8°F (see Table 2.1).
- A yearly total of 4750 heating degree-days, base 65 (see Table 2.1).
- An assumed design wind speed of 25 mph, and an average winter wind speed of 11 mph (from NOAA data).
- An average heating season temperature of 40°F (from NOAA data).

Solution

PART A—ROOM-BY-ROOM DESIGN HEAT LOSS. A room-by-room calculation of total house heat loss (transmission plus infiltration) must be prepared in order to determine the required output of heat-producing equipment, and to size heat-releasing terminal devices (i.e., air outlets or baseboard heaters).

Areas and Lengths of Construction. First calculate the area of each component of the house which will lose heat. The areas for each component are summarized below for the vertical surfaces. As a check be sure that the gross wall area equals the sum of components.

SUMMARY: AREAS OF VERTICAL COMPONENTS FOR "THE HOUSE"

Space	Gross Wall Area (ft^2)	Window Area (ft^2)	Door Area (ft^2)	Net Wall Area (ft^2)
Dining room	200	60	0	140
Living room	160	90	0	70
Bedroom #1	240	75	0	165
Bathroom	80	5	0	75
Bedroom #2	268	35	0	233
Vestibule	44	10	20	14
Family room	96	20	0	76
Kitchen	192	15	20	157
Totals:	1280 ft^2	310 ft^2	40 ft^2	930 ft^2

Then calculate the areas of other house surfaces which will lose heat.

The area of the ceiling for the entire house is:

$(30 \text{ ft} \times 45 \text{ ft}) + (5 \text{ ft} \times 18.5 \text{ ft})$
$= 1442 \text{ ft}^2$

The area of floor over the unheated basement is:

$(15 \text{ ft} \times 45 \text{ ft}) + (5 \text{ ft} \times 18.5 \text{ ft})$
$= 768 \text{ ft}^2$

For the section of slab on grade construction, the heat loss calculation will be based upon perimeter. For "The House," the perimeter is 15 ft along the east wall of the Dining Room, 45 ft for the length of the South facade, and an additional 15 ft along the west wall of Bedroom #1 for a total of 75 ft.

Transmission Heat Loss. The transmission heat loss for each room is computed by applying Formula 5D for each component subject to transmission heat loss (and Formula 5L for slab edge heat loss). The complete calculation for the Dining Room is shown below.

SUMMARY: ROOM-BY-ROOM TRANSMISSION HEAT LOSS

Room	Btu/°F·h
Dining room	46.3
Living room	55.0
Bedroom #1	63.3
Exterior bathroom	9.7
Bedroom #2	40.0
Vestibule/hall	14.0
Family room	23.4
Kitchen	26.6
Total	278.3

U-factor of the floor over the basement (0.08) was assumed to be only half (or 0.04) in the calculations.

Infiltration. Infiltration heat loss attributed to each room must be computed. The following steps are needed to compute the infiltration heat loss of the Dining Room for winter design conditions.

TRANSMISSION HEAT LOSS: DINING ROOM

Transmission Heat Loss:		$q_{tr} = U \times A$		(Formula 5D)
Surface	U		A	Btu/°F·h
Opaque wall	0.05	×	140	= 7.0
Window	0.38	×	60	= 23.0
Ceiling	0.03	×	150	= 4.5
Floor	0.47 (F2)	×	25 ft	= 11.8
			q_{tr} =	46.3

Results for the other rooms which were calculated by the same procedure are shown in the following summary.
Ceiling and floor heat loss of interior spaces (Interior Bathroom, Inner Hall, Stair, and Closets) were assigned equally to the two Bedrooms. In addition, the unheated basement space was assumed to be at a temperature about halfway between indoor and outdoor temperatures. Therefore, the actual

A. First the leakage area of the room must be calculated. Note how the areas and lengths of all components of the construction which may be the source of air leakage are evaluated as shown in Figure 5.16. Leakage factors for the various items are then obtained from Table 5.8. Since the floor of the Dining Room is a slab on ground, no leakage through the floor is assumed.

90 HEAT LOSS FROM BUILDINGS

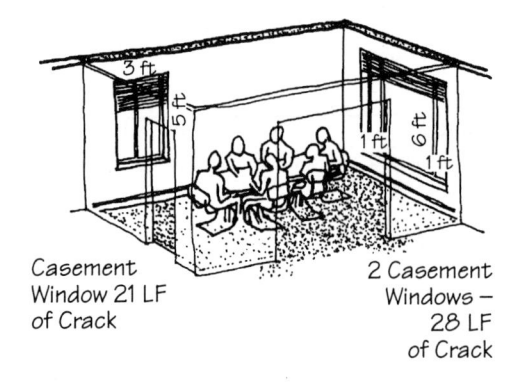

Casement Window 21 LF of Crack

2 Casement Windows — 28 LF of Crack

FIGURE 5.16 The "Dining Room."

The wind speed for design purposes is assumed to be 25 mph (V).

$$I_{cfm} = L \times [(A \times TD) + (B \times V^2)]^{1/2}$$

(Formula 5F)

$$= 14.4 \times [(0.0156 \times 60) + (0.0119 \times [25]^2)]^{1/2}$$

$$= 14.4 \times [0.936 + 7.44]^{1/2}$$

$$= 41.7 \text{ ft}^3/\text{m (cfm)}$$

DINING ROOM—CALCULATION OF LEAKAGE AREA

	Units		Factor/Unit		Product
Sill	25 lftc	×	0.11 in.²/lftc	=	2.8 in.²
Weatherstripped windows, casement	49 lftc	×	0.12 in.²/lftc	=	5.9 in.²
Window framing, wood, caulked	60 ft²	×	0.004 in.²/ft²	=	0.2 in.²
Walls, continuous infil. air barrier	140 ft²	×	0.0022 in.²/ft²	=	0.3 in.²
Joints, top plate band joist	25 lftc	×	0.05 in.²/lftc	=	1.3 in.²
Ceiling, general	150 ft²	×	0.026 in.²/ft²	=	3.9 in.²
		Dining room leakage area (L)		=	14.4 in.²

B. With the leakage area now known, formula 5F can be used to obtain the infiltration air flow rate in cfm (I_{cfm}) for winter design conditions. The following additional factors also apply:

- The stack coefficient A for the one-story house is 0.0156 (from Table 5.9).
- The wind coefficient B for the fully exposed one-story house (Shielding Class 1) is 0.0119 (from Table 5.10).
- The temperature difference (TD) is 60°F since the indoor design temperature is 68°F and the outdoor design temperature is 8°F.

C. The infiltration heat loss (q_{inf}) can now be computed by using Formula 5G.

$$q_{inf} = I_{cfm} \times 1.08 \quad \text{(Formula 5G)}$$

$$= 41.7 \times 1.08$$

$$= 45.0 \text{ Btu/°F·h}$$

In a similar manner, the total leakage area of the entire house was found to be 87.9 in.² and infiltration heat losses for each room were computed as summarized below. (Note: the effect of the attic access door was divided and evenly assigned to the two bedrooms).

SUMMARY: ROOM-BY-ROOM INFILTRATION HEAT LOSS

	Infiltration Heat Loss (Btu/°F·h)
Dining room	45.0
Living room	80.0
Bedroom #1	44.1
Exterior bathroom	12.7
Bedroom #2	38.0
Vestibule	10.4
Family room	20.2
Kitchen	23.9
Total:	274.3

Totals for all rooms of "The House" can also be computed as shown in the following summary. Remember, 33,150 Btu/h is the quantity of heat which must be delivered to the various rooms within the house to maintain the indoor design temperature when outdoor temperatures reach design condition. The energy input to combustion heating equipment such as a furnace or boiler will be greater to account for system losses, as discussed in Chapter 17.

SUMMARY: ROOM-BY-ROOM HEAT LOSS FOR "THE HOUSE" (WINTER DESIGN CONDITIONS)

	Room Transmission Heat Loss (Btu/°F·h)		Room Infiltration Heat Loss (Btu/°F·h)		Subtotal		Design Temp. Diff. (°F)		Room Design Heat Loss (Btu/h)
Dining room	(46.3	+	45.0)	=	91.3	×	60	=	5480
Living room	(55.0	+	80.0)	=	135.0	×	60	=	8100
Bedroom #1	(63.3	+	44.1)	=	107.4	×	60	=	6440
Ext. Bath.	(9.7	+	12.7)	=	22.4	×	60	=	1340
Bedroom #2	(40.0	+	38.0)	=	78.0	×	60	=	4680
Vest./hall	(14.0	+	10.4)	=	24.4	×	60	=	1460
Family room	(23.4	+	20.2)	=	43.6	×	60	=	2620
Kitchen	(26.6	+	23.9)	=	50.5	×	60	=	3030
							Total	=	33,150

Total Room-by-Room Design Heat Loss. The total design heat loss of each room (q_{design}) can now be determined. Sum the transmission and infiltration heat loss for each space and multiply the result by the design temperature difference (Formula 5H). The result of 4650 Btu/h for the Dining Room is shown below.

Design heat loss (q_{design}) for Dining Room:

$$q_{design} = (q_{tot}) \times (t_i - t_o)$$

(Formula 5H)

$$= (46.3 + 45.0) \times (68 - 8)$$

$$= 5480 \text{ Btu/h}$$

PART B—ESTIMATE OF HEAT REQUIRED ANNUALLY. A calculation of house heat loss forms the basis of an estimate of the amount of heat needed to maintain indoor design conditions during the heating season. Calculated in part A was a transmission heat loss of 278.3 Btu/°F·h (q_{tr}). It is now necessary to calculate infiltration heat loss for "average" winter temperature and wind conditions (instead of the more severe design conditions assumed in Part A). The following steps are needed to compute the infiltration heat loss of "The House" for average winter conditions.

A. The leakage area of the entire house has been determined to be 87.9 in^2.

B. With the house leakage area (*L*) known, Formula 5F can be used to obtain the infiltration air flow in cfm (I_{cfm}) for average winter conditions based upon the following additional factors which apply:

The house remains the same as does the amount of shielding from the wind. Therefore, coefficients *A* and *B* remain the same, namely, 0.0156 for coefficient *A* and 0.0119 for coefficient *B*.

Average climate conditions (from NOAA data) result in an average winter temperature difference (*TD*) of 28°F (68 − 40), and an average winter wind speed (*V*) of 11 mph.

$$I_{cfm} = L \times [(A \times TD) + (B \times V^2)]^{1/2}$$

(Formula 5F)

$$= 87.9 \times [(0.0156 \times 28)$$
$$+ (0.0119 \times [11]^2)]^{1/2}$$
$$= 87.9 \times [0.437 + 1.440]^{1/2}$$
$$= 120.4 \text{ cfm}$$

C. With the infiltration rate of the house for average winter conditions now known, the infiltration heat loss (q_{inf}) can be computed by using Formula 5G.

$$q_{inf} = I_{cfm} \times 1.08 \quad \text{(Formula 5G)}$$
$$= 120.4 \times 1.08$$
$$= 130.0 \text{ Btu}/°F \cdot h$$

We now know the infiltration heat loss of "The House" for average winter conditions. Therefore, Formula 5I can be used to determine total house heat loss (for average winter conditions), also known as the *UA*, as follows:

$$q_{tot} = q_{tr} + q_{inf} \quad \text{(Formula 5I)}$$
$$= 278.3 + 130.0$$
$$= 408.3 \text{ Btu}/°F \cdot h$$

With the house *UA* now known, we can estimate the annual amount of heat which must be delivered. The heating balance point of this well-insulated house has been computed to be 65°F (due to the large area of south glass). In St. Louis, a total of 4750 heating degree-days occur on average relative to a 65°F base temperature (see Table 2.3). Use Formula 5M.

$$Q_{yr} = UA \times HDD_{bp} \times 24$$

(Formula 5M)

$$= 408.3 \times 4750 \times 24$$
$$= 46,546,200 \text{ Btus are needed per year}$$

To get a feel for operational cost, assume that this heat is provided by a natural gas-fired warm air furnace with an Annual Fuel Utilization Efficiency (AFUE) of 80%. Then based upon an assumed fuel cost of $1.00 per therm, the annual operational cost would be about $580. Operational cost estimates to heat "The House" by other systems are contained in other chapters.

5.15 CHECKLIST FOR HEAT LOSS CALCULATIONS

Accurate calculations of building heat loss are needed to size elements of the heating system and to estimate yearly heating energy requirements. Calculations will be much clearer and easier if the following approach is taken:

A. Use a neat and organized method of computing the areas, lengths, and volumes of all required building components and elements. Accounting "analysis pads" and computer spreadsheets lend themselves nicely to these computations.

B. Compute the U-factors of all constructed components (walls, roof/ceiling, etc.), and consult technical data for manufactured items such as windows and special types of insulation or finishes.

C. Don't get lazy. Write down all units such as Btu/ft$^2 \cdot$°F$\cdot$h for U-factors.

D. If a computer spreadsheet is used, check all equations and definitions of ranges. Be reasonable in the number of significant digits you carry and round results when appropriate. Check all math. Better yet, have someone else check it!

E. Apply as much common sense to results as your experience allows. A house may have a total heat loss of 500 Btu/°F$\cdot$h, but it would have to be a mansion to have heat loss of 5000 Btu/°F$\cdot$h. Also be mindful of, but keep in perspective, building features such as the door and stair opening to an unheated basement which will lose a bit more heat than an insulated floor.

F. Keep in mind the purpose of the calculation. As shown in the detailed heat loss example for an entire house, different data is needed depending upon whether you are concerned about design conditions or yearly heating energy needs.

G. Check to see which items are major contributors to heat loss. For most houses, these will be windows and infiltration.

H. Use the best climate data available, but apply it using common sense and experience to account for local microclimate. Remember that climate can vary greatly from year to year.

I. Realize that although the examples put forth herein are simple and residential in scale, heat loss calculations for very large buildings are done in exactly the same manner.

J. Be sure to adhere to all codes regarding required conservation levels of building components.

6

VENTILATION

Before the air in buildings begins to get stale it must be removed and replaced by outside air (in some areas of the country still accurately referred to as "fresh air"). For this purpose, building codes establish criteria generally based on ASHRAE Standard 62-1989, "Ventilation for Acceptable Indoor Air Quality."

A very important consequence of ventilation is the need to heat, cool, and humidify or dehumidify incoming outside air to maintain comfortable indoor conditions. This chapter contains the various formulas used to compute these sensible and latent heat losses and gains.

Occupants of houses and other buildings are increasingly concerned about the quality of indoor air. This includes potential exposure to hard-to-detect hazards such as radon. HVAC systems should be designed with sensitivity to avoid problems which in commercial buildings have become known as "sick building syndrome."

6.1 NEED FOR OUTSIDE AIR

Figure 6.1 shows how recommended ventilation rates have varied over the years beginning with a requirement for 4 cubic feet per minute (cfm) per person of outside air in 1824 to reduce CO_2 (carbon dioxide) levels. The figure also shows that a value of 30 cfm per person was adopted (by the American Society of Heating and Ventilating Engineers) in 1893. Later, in 1936, the recommendation was reduced to 10 cfm per person, which was considered to provide adequate dilution to the threshold detection limit of human body odors. In response to the first "energy crisis" in 1973, the outside air requirement for ventilation was set at 5 cfm per person to reduce the amount of energy needed to heat or cool it. Then in the 1980s, complaints about poor indoor air quality led to arguments for increased use of outside air (15 cfm per person or more). This "tug of war" on ventilation rates is

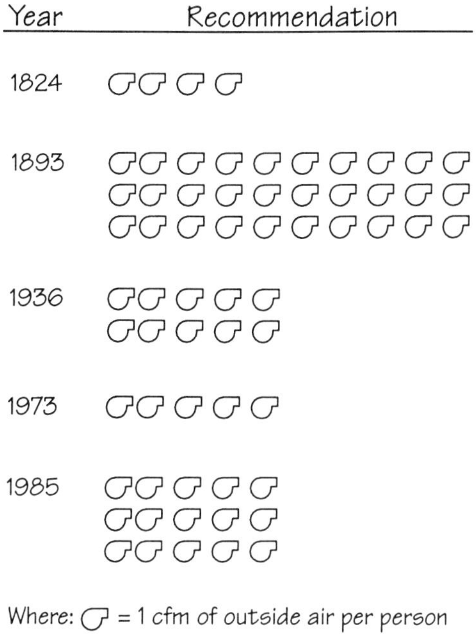

Year	Recommendation
1824	ᗅᗅ ᗅ ᗅ
1893	ᗅᗅ ᗅ ᗅ ᗅ ᗅ ᗅ ᗅ ᗅ ᗅ ᗅᗅ ᗅ ᗅ ᗅ ᗅ ᗅ ᗅ ᗅ ᗅ ᗅᗅ ᗅ ᗅ ᗅ ᗅ ᗅ ᗅ ᗅ ᗅ
1936	ᗅᗅ ᗅ ᗅ ᗅ ᗅᗅ ᗅ ᗅ ᗅ
1973	ᗅᗅ ᗅ ᗅ ᗅ
1985	ᗅᗅ ᗅ ᗅ ᗅ ᗅᗅ ᗅ ᗅ ᗅ ᗅᗅ ᗅ ᗅ ᗅ

Where: ᗅ = 1 cfm of outside air per person

FIGURE 6.1 Historical ventilation rate recommendations.

probably not over, but design approaches have now been developed which allow heat to be recovered from the air being expelled, thus minimizing energy demands while allowing for sufficient exchange.

ASHRAE 62-1989 specifies

> ... minimum ventilation rates and indoor air quality that will be acceptable to human occupants ... to minimize the potential for adverse health effects. This standard applies to all indoor or enclosed spaces that people may occupy, except where other applicable standards and requirements dictate larger amounts of ventilation than this standard. Release of moisture in residential kitchens and bathrooms, locker rooms, and swimming pools is included in the scope of this standard.

For houses and other residential facilities, ASHRAE 62-89 requires 15 cfm per person of outside air to account for odors, moisture, and carbon dioxide generated by people. This is normally satisfied by infiltration and natural ventilation. For kitchens, either 100 cfm of intermittent or 25 cfm of continuous outdoor air or operable windows are required. For baths and toilets, either 50 cfm of intermittent or 20 cfm of continuous outdoor air or operable windows are required.

Provided in Table 6.1 is an excerpted listing of outdoor air requirements for commercial facilities (offices, stores, shops, hotels, and sports facilities). Note that "positive engine exhaust systems" required for garages are exhaust hoses attached to tailpipes.

6.2 NATURAL VENTILATION

Before the age of mechanical cooling, buildings of all types depended on operable windows to satisfy the need for fresh air, and building design was different (see Section 18.2 for detailed information on cooling through air motion). Rooms in commercial buildings often had high ceilings and tall double-hung windows to facilitate convective air flow along with the entry of natural light (see Figure 6.2). Many large buildings were designed with courtyards to divide the building into narrow wings so that all rooms were on the exterior with access to the light and air. Schools and other buildings employed high "borrowed lights" in the walls between rooms and corridors to ventilate and light interior spaces as well. Today, most houses and many small commercial buildings still rely on windows and infiltration for ventilation of most spaces. However, owing to a concern for indoor air quality, more and more houses are being designed with mechanical ventilation systems, including those that recover heat, as discussed below.

TABLE 6.1 OUTDOOR AIR REQUIREMENTS FOR VENTILATION OF COMMERCIAL FACILITIES

Application	cfm/person	cfm/ft²	cfm/room
Dry cleaners, laundries			
Commercial laundry	25		
Coin-operated laundries	15		
Food and beverage service			
Dining rooms, fast food cafeterias	20		
Bars, cocktail lounges	30		
Garages, repair, service stations			
Enclosed parking garage, Auto repair rooms (*running engines must have positive engine exhaust syustems)		1.50*	
Hotels, motels, resorts, dorms			
Bedrooms and living rooms			30
Baths			35
Lobbies	15		
Assembly rooms	15		
Offices			
Reception areas	15		
Office space, data entry areas, conference rooms	20		
Public spaces			
Corridors and utilities		.05	
Public restrooms (cfm/water closet or cfm/urinal)	50		
Retail stores			
Basement and street floors		0.30	
Upper floors		0.20	
Theaters			
Lobbies	20		
Auditorium	15		
Transportation			
Waiting rooms, platforms, vehicles	15		

Source: Excerpted from ASHRAE Standard 62-1989 "Ventilation for Acceptable Indoor Air Quality," Table 2.

6.3 VENTILATION HEAT LOSS

During the heating season, outside air introduced into a building for ventilation purposes results in heat losses, both sensible and latent, estimated below.

Sensible Heat Loss

Formula 6A is used to calculate sensible heat loss due to ventilation.

$$q_{v\text{-sen}} = 1.08 \times \text{cfm} \times (t_i - t_o)$$

Formula 6A

where

- $q_{v\text{-sen}}$ is sensible heat loss of ventilation air in Btu/h
- 1.08 is the product of the heat capacity of air (0.018 Btu/°F·ft³) and 60 min/h in Btu·m/°F·h
- cfm is the flow rate of outside air in ft³/m
- t_i is the inside air temperature in °F
- t_o is the outside air temperature in °F

Example 6A: Determine the sensible heat loss for a ventilation flow rate of 250

is used to compute the latent heat loss of ventilation air.

$$q_{\text{v-lat}} = 4840 \times \text{cfm} \times (W_i - W_o)$$

Formula 6B

where

$q_{\text{v-lat}}$ is the amount of latent heat removed from ventilation air in Btu/h

4840   is the product of 0.075 lb/ft^3 (the density of air), 60 min/h, and 1076 Btu/lb (the approximate heat content of 50% relative humidity vapor at 75°F, less the heat content of water at 50°F)

cfm   is the flow rate of outside air entering the building in ft^3/m

W_i is the humidity ratio of the inside air in lb/lb$_{\text{dry air}}$

W_o is the humidity ratio of the outside air in lb/lb$_{\text{dry air}}$

Example 6B: Determine the latent heat loss due to ventilation for an air flow rate of 500 cfm when the humidity ratio of inside air (W_i) is 0.006 (based upon 70°F and 40% relative humidity), and the humidity ratio of the outside air (W_o) is 0.002 (based upon 30°F and 60% relative humidity).

Solution: Use Formula 6B.

$$q_{\text{v-lat}} = 4840 \times \text{cfm} \times (W_i - W_o)$$

(Formula 6B)

$$= 4840 \times 500 \times (0.006 - 0.002)$$

$$= 9680 \text{ Btu/h}$$

Therefore, 9680 Btus of heat will be required each hour to add water vapor to the ventilation air.

6.4 VENTILATION HEAT GAIN

Of prime importance to the design of mechanical cooling systems and equipment are

---

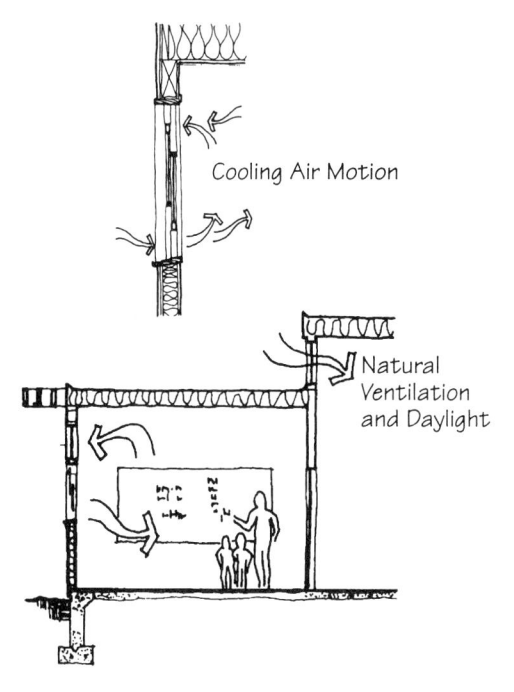

FIGURE 6.2 Natural ventilation of buildings.

cfm when the inside air temperature (t_i) is 70°F and the outside air temperature (t_o) is 30°F.

Solution: Use Formula 6A.

$$q_{\text{v-sen}} = 1.08 \times \text{cfm} \times (t_i - t_o)$$

(Formula 6A)

$$= 1.08 \times 250 \times (70 - 30)$$

$$= 10,800 \text{ Btu/h}$$

When ventilation is measured in air changes per hour (ACH), Formula 5A can be used and multiplied by the applicable °F temperature difference to determine sensible heat loss in Btu/h.

Latent Heat Loss

In winter, the loss of humidity due to the entry of cold dry outside air is of concern, particularly in houses and buildings such as hospitals and nursing homes. Formula 6B the latent and sensible cooling loads imposed by ventilation (or infiltration) heat gain. Depending upon the application, sometimes the total heat gain (enthalpy) is of interest. At other times either the sensible or latent heat gain alone needs to be determined to satisfy a design need.

Sensible Heat Gain

To ensure agreement with the equations contained in the current *ASHRAE Handbook of Fundamentals*, Formula 6C is used to compute sensible heat gain of ventilation (or infiltration) air in Btu per hour. This formula is similar to Formula 6A for sensible heat loss. The slight difference is the value of "1.10" instead of "1.08," due to the higher humidity ratios of air at temperatures when cooling for human comfort may be required. Note, though, that some designers simply use 1.08 for both heat gain and heat loss.

$$q_{oa\text{-}sen} = 1.10 \times cfm \times (t_o - t_i)$$

Formula 6C

where

$q_{oa\text{-}sen}$	is the sensible heat gain from outside air in Btu/h
1.10	is a constant based upon the product of the heat capacity of air (assuming a humidity ratio of 0.01 lb/lb$_{dry\ air}$) and 60 min/h in units of Btu·m/°F·ft^3·h
cfm	is the flow rate of outside air entering the building in ft^3/m
t_o	is the outside air temperature in °F
t_i	is the inside air temperature in °F

Example 6C: Determine the sensible heat gain for a ventilation flow rate of 10,000 cfm when the outside air temperature (t_o) is 88°F and the inside air temperature (t_i) is 78°F.

Solution: Use Formula 6C.

$$q_{oa-sen} = 1.10 \times cfm \times (t_o - t_i)$$

(Formula 6C)

$$= 1.10 \times 10{,}000 \times (88 - 78)$$

$$= 110{,}000 \text{ Btu/h}$$

Latent Heat Gain

The heat gain of increased moisture in outside ventilation air, known as latent heat, is calculated by using Formula 6D when information regarding grains of moisture is known.

$$q_{oa\text{-}lat} = 0.68 \times cfm \times (G_o - G_i)$$

Formula 6D

where

$q_{oa\text{-}lat}$	is the amount of latent heat which must be removed from the outside ventilation air in Btu/h
0.68	is a factor based upon the following: $\dfrac{60 \text{ min/h}}{13.5} \times \dfrac{1076}{7000}$
13.5	is the specific volume of air (at 70°F and 50% relative humidity) in ft^3/lb
1076	is the quantity of Btu required to extract one pound of water vapor from air
7000	is the number of grains of moisture in a pound
cfm	is the flow rate of outside air in ft^3/m
G_o	is the number of grains of moisture in the outside air per pound of dry air
G_i	is the number of grains of moisture in the inside air per pound of dry air

6.5 TOTAL VENTILATION HEAT CHANGE

The total (sensible and latent) heat change of ventilation air is known as the enthalpy difference. It can be determined by using Formula 6E.

$$q_{\text{oa-tot}} = 4.5 \times \text{cfm} \times (H_o - H_i)$$

Formula 6E

where

$q_{\text{oa-tot}}$ is the total heat gain (both sensible and latent) in Btu/h
4.5 is the product of 0.075 lb/ft³ (the density of air) and 60 min/h
cfm is the flow rate of outside air in ft³/m
H_o is the enthalpy of the outside air in Btu/lb$_{\text{dry air}}$
H_i is the enthalpy of the inside air in Btu/lb$_{\text{dry air}}$

Example 6E: Determine the total heat gain due to ventilation for 10,000 cfm of air when the enthalpy of outside air (W_o) is 36.7 (based on 88°F and 50% relative humidity), and the enthalpy of the inside air (W_i) is 30.0 (based on 78°F and 50% relative humidity).

Solution: Use Formula 6E.

$$q_{\text{oa-tot}} = 4.5 \times \text{cfm} \times (H_o - H_i)$$

(Formula 6E)

$$= 4.5 \times 10{,}000 \times (36.7 - 30)$$

$$= 301{,}500 \text{ Btu/h}$$

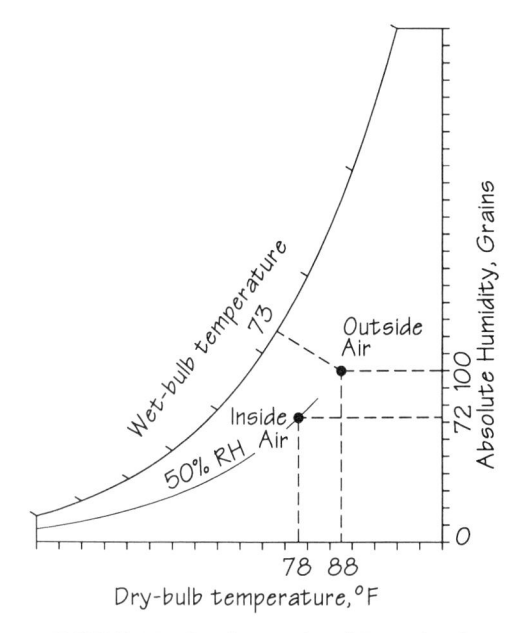

FIGURE 6.3 Psychrometrics of Example 6D.

Example 6D: Determine the latent heat gain due to ventilation for an air flow rate of 10,000 cfm for the indoor and outdoor conditions indicated on the psychrometric chart in Figure 6.3.

Solution: Use Formula 6D. As shown on Figure 6.3, the outside air contains 100 grains of moisture per pound of dry air while the inside air contains 72 grains per pound of dry air.

$$q_{\text{oa-lat}} = 0.68 \times \text{cfm} \times (G_o - G_i)$$

(Formula 6D)

$$= 0.68 \times 10{,}000 \times (100 - 72)$$

$$= 190{,}400 \text{ Btu/h}$$

Note that the outside conditions used in this example represent a wet-bulb temperature of 73°F which is the 2.5% summer design condition for cities such as Hartford and Minneapolis (see Table 2.7).

This total compares very closely with the total of the individual sensible and latent heat gains computed in Examples 6C and 6D for the same conditions.

6.6 INDOOR AIR QUALITY (IAQ)

HVAC systems must do more than provide conditions for immediate human comfort: they also must perform in a manner which avoids creating or exacerbating indoor air quality problems.

Residential Buildings

Many houses have living and sleeping quarters located below grade. Recently, attention has been focused on the possible entry of radon gas into these spaces through cracks in the foundation or drains in the basement floor. Radon is an odorless, colorless radioactive gas which has been implicated in the contraction of lung cancer by uranium miners, and may be a risk (albeit a much lower one) to humans exposed to low levels over many years.

Sources of radon in typical homes may be from the geological formation on which the house is sited, from ground water or the use of deep well water, or rarely, from bricks or concrete used in construction. Radon levels, while being relatively easy to monitor using detectors, are difficult to predict and may vary greatly from house to house within a single neighborhood. Radon releases are also strongly correlated to barometric pressure. The United States Environmental Protection Agency (USEPA) has set a standard for radon in residential buildings at 4 picocuries per liter (pCi/L). This "action" level also appears in ASHRAE 62-1989.

Owners of houses with radon levels above 4 pCi/L from external sources should have basement cracks and openings sealed and basement air positively pressurized to prevent negative stack effect, as discussed below.

Other indoor air contaminants which are of concern in residences and other buildings include cleaning products, asbestos, formaldehyde (prevalent in common building materials such as particle board, plywood, and carpeting), and the products of combustion including the potentially deadly carbon monoxide.

Commercial Buildings

Indoor air quality in commercial buildings is increasingly a concern and has led to use of the term "sick building syndrome" (SBS). The effects of IAQ problems are often nonspecific symptoms rather than clearly defined illnesses. Symptoms commonly attributed to IAQ problems include: headache, fatigue, shortness of breath, sinus congestion, cough; sneezing, eye, nose, and throat irritation, skin irritation, dizziness, and nausea. "Building-related illness" (BRI) is a term used to describe diagnosed cases such as Legionnaire's disease or Pontiac fever.

"Health" and "comfort" are used to describe a spectrum of physical sensations. For example, when the air in a room is slightly too warm or slightly too dry, a person may experience mild discomfort. And if these conditions get worse, discomfort increases and symptoms such as fatigue, stuffiness, and headaches can appear. Often high or low temperatures or humidities are improperly diagnosed as an IAQ problem. Some complaints by building occupants are clearly related to the discomfort end of the spectrum. One of the most common IAQ complaints is that "there's a funny smell in here." Odors are often associated with a perception of poor air quality, whether or not they cause symptoms. Environmental stress factors such as improper lighting, noise, vibration, overcrowding, poor ergonomics, and job-related stress can produce symptoms that are similar to those associated with poor air quality.

SBS can result from:

- Control strategies that "cycle" required ventilation or exhaust in order to save energy and reduce electrical peak demands.
- Use of variable air volume (VAV) and other air supply systems which reduce the quantity of ventilation air excessively.
- Interference with HVAC system operations (e.g., blocking of diffusers to avoid drafts, turning off exhaust to eliminate noise).
- Biological contaminants that grow within the HVAC system such as in stagnant water in drain pans.
- "Second hand" (passive) cigarette smoke.
- Psychological reactions to modern "sealed glass" design.
- Building finishes and furnishing, including carpeting.
- Excessive relative humidity in summer (over 70%) which can encourage mold growth, or low relative humidity in winter (below 40%), causing respiratory irritation.
- Spillage of combustion gases from heating equipment (e.g., inadequately vented appliances or leakage from cracked heat exchangers).
- Contaminants entering buildings from outdoors. These can include entry of soil gases (radon), outdoor air intakes near sources of exhaust fumes, and outdoor air intakes which capture cooling tower mist.
- Improper use and maintenance of personal equipment such as humidifiers brought in by building occupants.
- Contaminants from non-HVAC equipment (i.e., copiers, drawing reproduction machines).
- Inadequate or incorrect housekeeping practices. Cleansers and other products can produce odors and emissions.
- Locations where high surface humidity promotes condensation.

It is essential that the source of the problem(s) in sick buildings be identified, or, at least, corrective measures be taken based upon a realistic appraisal of the situation. Whenever possible design solutions for IAQ problems should be implemented that both solve the problem and are also sensitive to energy consumption and operational costs. New buildings should be designed proactively, to prevent potential indoor air quality problems. For detailed discussion and useful checklists see *Building Air Quality* as published by the Indoor Air Division of the USEPA, Washington, DC 20460.

6.7 HEAT RECOVERY VENTILATION (HRV)

The desire to minimize energy usage in the 1970s led to the development of new equipment known originally as "air-to-air heat exchangers." Then in the mid-1980s, the industry renamed them "heat recovery ventilators" or HRVs.

The concept is simple: to assure air quality in an energy-conserving manner. As shown in Figure 6.4, large quantities of indoor air are expelled from a building through the HRV while outside air is introduced, coming in contact with the heat exchange surface area (usually through a counterflow arrangement as shown). In the process, a portion (typically 40 to 60%) of the sensible heat (and for some equipment, latent heat as well) is transferred to the incoming air stream rather than being wasted.

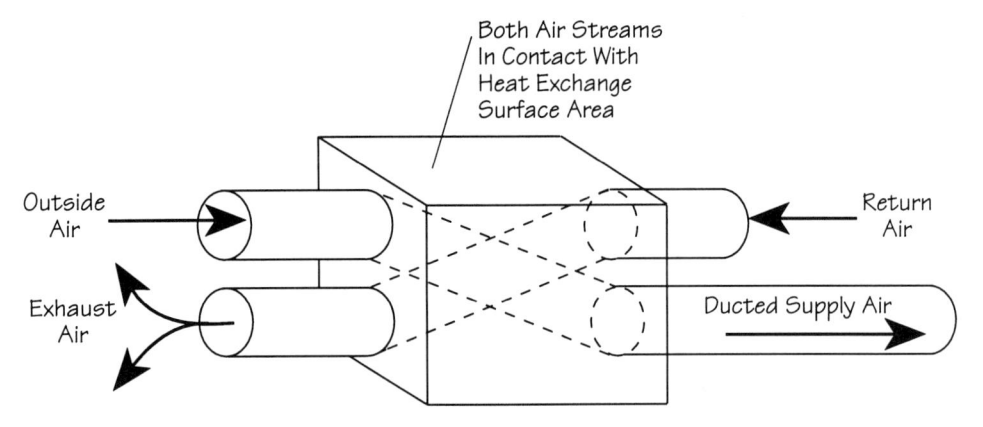

FIGURE 6.4 Heat recovery ventilator (HRV).

HRVs save energy and are most cost-effective in houses with intentionally high outdoor air ventilation rates. Nevertheless some designers take a different approach to maintaining indoor air quality. They prefer to limit outdoor ventilation air to that prescribed by the code for the house occupants only. In these houses the levels of contaminants are minimized, an approach known as "source control."

HRVs are also used for ventilation systems in larger buildings which frequently use large quantities of outside air (e.g., swimming pools, fitness centers). HRVs are most cost effective in buildings with long operating hours in cold climates where the cost of electricity is relatively low.

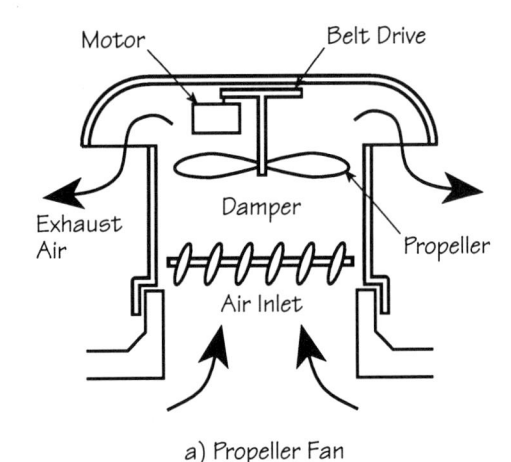

6.8 POWER VENTILATORS

Ventilation fans must be capable of moving high volumes of air to exhaust spaces such as toilets, kitchens, and warehouses. Fans can be either a propeller or centrifugal type, as shown in Figure 6.5 and discussed more fully in Section 9.7. Centrifugal fans operate more quietly and are suitable for ducted applications. Fans can be wall or roof mounted in housings which protect them

a) Propeller Fan

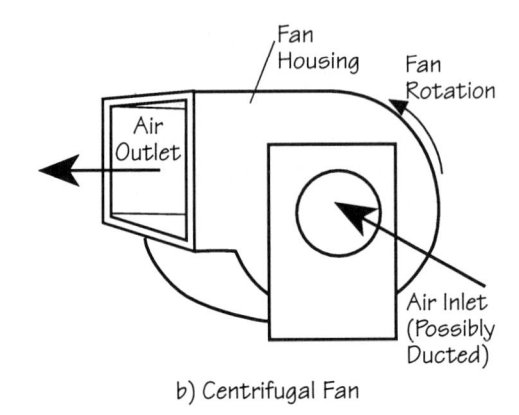

b) Centrifugal Fan

FIGURE 6.5 Power ventilators.

from the elements and assure safe operation.

6.9 STACK EFFECT

Pressures from wind, and pressures caused by differences between indoor and outdoor temperatures, result in air entering (infiltrating) through portions of a building while other air exits (exfiltrates). These pressures (both positive and negative) can be substantial in tall spaces and buildings where openings and shafts allow for vertical air movement; this phenomenon is known as the "stack effect."

Commercial buildings are normally designed to overcome stack and wind effects by employing fans to "pressurize" the building and thus prevent entry of infiltration air carrying unwanted humidity and unfiltered impurities.

Figure 6.6 shows how the stack effect can affect a building or large space which is not designed to prevent vertical air movement and air infiltration. Note how warm air will rise during the heating season and be replaced by cold outside air on the lower floors. In the summer, the denser cool air will fall and exfiltrate from the lower floors.

In Figure 6.6, note the NPL (Neutral Pressure Level) where there is no pressure either inward or outward. The NPL will be located at the midpoint of a building if cracks and openings are evenly distributed vertically and the air is free to move up and down from floor to floor. In new buildings this should not normally be the case since openings should be sealed to satisfy fire code requirements.

Stack effect can be estimated by using Formula 6F.

$$PD = 0.00027 \times \text{Feet}_{NPL} \times (t_i - t_o)$$

Formula 6F

where

PD	is the pressure difference due to stack effect, in inches of water column (in. wc)
0.00027	is a typically used estimate of stack effect in inches of water per foot from NPL per °F
Feet_{NPL}	is the vertical distance from the Neutral Pressure Level in feet
t_i	is the inside air temperature in °F
t_o	is the outside air temperature in °F

Example 6F: Determine the pressure difference due to stack effect at the bottom of a 100-ft tall atrium when the inside temperature is 70°F and the outside temperature is 20°F. Assume that openings are uniformly distributed vertically and that there is no resistance to vertical air movement within the building.

Solution: Assume that the NPL will occur at the midpoint of the atrium's height, or 50 ft up. Formula 6F can now be used.

$$PD = 0.00027 \times \text{Feet}_{NPL} \times (t_i - t_o)$$

(Formula 6F)

$$= 0.00027 \times 50 \times (70 - 20)$$

$$= 0.7 \text{ in. wc}$$

This is the pressure at which cold air will

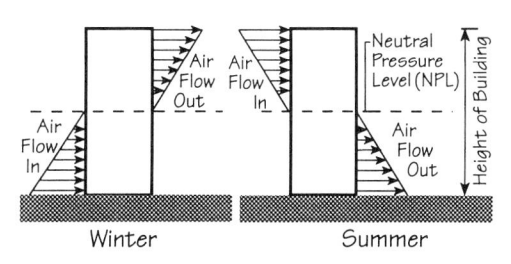

FIGURE 6.6 Winter and summer stack effect in tall buildings.

infiltrate at the base of the atrium unless fans are designed to overcome this pressure and wind pressure as well.

For more information on stack effect see Chapter 23 of the *1993 ASHRAE Handbook of Fundamentals*. Also see Chapter 9 of this book for information of air pressure.

6.10 FILTERS

Particles are usually removed from the air with filters. The type and design of a filter determine its removal efficiency for different size particles and the amount of energy needed by a fan to pull or push air through it. Filter ratings are based on their performance under test methods stipulated in ASHRAE Standard 52-76, "Method of Testing Air-Cleaning Devices Used in General Ventilation for Removing Particulate Matter."

Low efficiency filters with an ASHRAE Dust Spot rating of 20% or less are often used to keep lint and dust from clogging the heating and cooling coils of a system. In order to maintain clean air in occupied spaces, filters must also remove bacteria, pollen, insects, soot, dust, and dirt with an efficiency suited to the use of the building. Medium efficiency filters (ASHRAE Dust Spot rating of 30–60%) provide much better filtration than low efficiency filters. To maintain the proper airflow and minimize the amount of additional energy required to move air through these higher efficiency filters, bag- or pleated-type extended surface filters are typically used. In buildings that are designed to be exceptionally clean, it may be warranted to employ both medium efficiency prefilters and high efficiency filters (ASHRAE Dust Spot rating of 85–95%). "HEPA" (high efficiency particulate air) filters are used where over 99% of dust particles (over 0.3 microns in size) need to be removed such as in clean rooms and laboratories.

Air filters, regardless of their design or efficiency rating, require regular maintenance (cleaning for some and replacement for most). As a filter loads up with particles, it becomes more efficient at particulate removal but increases the frictional resistance (pressure drop) in the system, therefore reducing airflow. Filter manufacturers can provide information on the pressure drop through their products under different conditions of clogging.

Air handlers that are located in difficult-to-access places (e.g., in places that require ladders for access or are located on roofs with no roof hatch access) need special attention to maintain filter quality. For these exceptional instances, consider using automatic roll filters which expose new filtering material as resistance to air flow is sensed.

7

SOLAR COOLING LOADS

Residential buildings are sometimes referred to as "envelope dominated" since their energy performance depends on the ability of the outer envelope to resist heat loss. Most houses in North America have to be heated over some part of the year to maintain human comfort. For such houses, solar heat is useful during the heating season and design measures to improve passive solar heating are beneficial (see Chapter 12).

In contrast, larger commercial buildings are "load dominated" by internal heat gains from lights, people, and equipment (as discussed in Chapter 4) and heat removal is often required nearly year round. Therefore, special glazings (tinted and reflective), and sun control devices (overhangs, fins, etc.) are often employed to limit solar heat gain and reduce cooling loads.

In this chapter, the solar heat that reaches interior spaces through both glazing and opaque building elements on clear days will be quantified for use in cooling load calculations. Do not use any values from tables in this chapter to estimate building energy use since they provide information about clear days only! As will be seen, solar heat gain varies greatly depending upon latitude, date, time, orientation, and type of glazing and construction.

Throughout this chapter, solar time and the 24-hour military clock are used. Solar hour 11 refers to a time one hour before solar noon and solar noon (hour 12) is when the sun is due south for all North American locations. Similarly, solar hour 15 refers to a time three hours after solar noon.

Solar time does not directly correspond to clock time but it is usually adequate to think of a solar hour such that hour 12 is equivalent to 12 noon standard clock time. Still, remember that when the clock says it is 12 noon in the eastern time zone (based on 75° longitude) the sun is already west of south in Maine while still being very far east of south in Michigan. Also remember that during the warmer months solar noon (hour 12) corresponds to 1 PM clock time when daylight savings time is in effect.

Also try not to be numbed by the sea of numbers found in the various tables in this and other chapters. Try to think about using tables just like you use a huge telephone directory. The number of entries really doesn't matter—you just have to find the right one.

7.1 SOLAR COOLING LOADS THROUGH GLAZING

The quantity of Btus of solar radiation which will penetrate one square foot of clear, double-strength (1/8 in. thick) sheet glass in one hour on clear days is termed the "solar heat gain factor" (SHGF). Until publication of the *1993 ASHRAE Handbook of Fundamentals*, SHGFs were used directly in the ASHRAE CLTD/CLF procedure to calculate space design cooling loads. The new *ASHRAE Handbook*, however, introduces a new term, "solar cooling load" (SCL), for the solar gain through glazing. The SCLs are based upon the heat storage capacity of a building's structure and furnishings and directly lead to computation of cooling loads without the need for an additional cooling load factor (CLF).

Four basic Zone Types (A, B, C, and D) are defined in the revised ASHRAE methodology which is detailed in Section 4.1 (for internal heat gains from people, equipment, and lighting systems). Zone Type A represents the least amount of heat storage capability. Zone Type D is most massive and offers the greatest amount of heat storage.

Table 7.1 provides SCLs for sunlit glass for 40°N latitude and the 21st day of July. Values in the table represent the cooling load in Btus/h through each square foot of 1/8 in. clear glass during that solar hour for the applicable Zone Type. The compass points (N, E, S, W, etc.) represent vertical surfaces facing in the indicated direction.

"HOR" refers to glazed horizontal surfaces.

Graphic Depiction of SCLs

SCL tables should provide a sense of how solar time and heat storage capability (as represented by Zone Types) impact upon building cooling loads. Shown in Figure 7.1 are SCL values from Table 7.1 through square foot glazing areas for orientations which vary by 45-degree increments. Solar cooling loads in Btu/h per square foot of 1/8 in. clear glazing are shown for solar hours 9 and 15 and the four Zone Types ranging from A (lightest weight) to D (heaviest).

Shading Coefficients

The reference glass assumed for the SCL table is clear, double-strength (1/8 in.) sheet glass. For other types of glass, solar cooling loads are simply multiplied by what are known as shading coefficients (SC). Shading coefficients represent the fraction of solar heat which will get through a type of "fenestration" compared to how much gets through the reference glass. Fenestration is an architectural term which means the design of windows including glazing, mullions, muntins, the frame, and any shading elements within the rough opening of the window. Therefore, if a type of bronze tinted glass allows only 60% as much solar heat as the reference clear double-strength sheet glass, it has a shading coefficient of 0.60.

Provided in Table 7.2 are typical shading coefficients for various types of fenestration. For manufactured products, manufacturers' literature should be used to obtain actual shading coefficients.

Formula 7A is used to calculate the cooling load due to solar heat gain through glazing with consideration given to the shading coefficient and possible heat storage.

TABLE 7.1 JULY SOLAR COOLING LOAD (SCL) FOR SUNLIT GLASS AT 40° NORTH LATITUDE (Btu/h·ft²)

Glass Facing	8	9	10	11	12	13	14	15	16	17
Zone Type A										
N	28	32	35	38	40	40	39	36	31	31
NE	134	112	75	55	48	44	40	37	32	26
E	185	183	154	106	67	53	45	39	33	26
SE	131	150	150	131	97	63	49	41	34	27
S	25	41	64	85	97	96	84	63	42	31
SW	24	30	35	39	64	101	133	151	152	133
W	24	30	35	38	40	65	114	158	187	192
NW	24	30	35	38	40	40	50	84	121	143
Hor	120	169	211	241	257	259	245	217	176	125
Zone Type B										
N	24	28	32	35	37	38	37	35	32	31
NE	116	101	73	58	52	48	45	41	36	30
E	159	162	143	105	74	63	55	48	41	34
SE	112	131	134	122	96	69	58	49	42	35
S	21	36	56	74	86	87	79	63	46	37
SW	22	27	31	36	58	89	117	135	138	126
W	22	27	31	35	37	59	101	139	166	173
NW	22	27	31	34	37	37	46	76	108	128
Hor	104	147	185	214	233	239	232	212	180	137
Zone Type C										
N	24	27	30	33	34	35	34	32	29	29
NE	107	88	61	49	47	45	43	40	36	31
E	148	145	124	89	62	56	52	47	43	37
SE	107	121	121	107	82	59	51	47	42	36
S	23	36	54	70	79	79	70	54	40	33
SW	26	29	33	36	57	86	110	124	125	111
W	27	31	34	36	37	59	98	132	153	156
NW	25	29	32	34	36	36	44	73	102	118
Hor	107	144	175	199	212	215	207	189	160	123
Zone Type D										
N	21	24	27	29	31	32	31	30	28	29
NE	90	77	58	49	48	46	44	42	39	35
E	123	124	110	85	65	60	57	53	48	43
SE	90	102	104	95	78	60	55	51	47	42
S	21	32	46	59	67	69	63	52	41	36
SW	25	28	31	34	51	74	94	106	109	100
W	28	30	33	34	35	53	84	112	130	135
NW	24	27	30	32	33	34	41	64	87	101
Hor	95	124	150	171	185	191	188	176	156	128

Source: Excerpted by permission from the *1993 ASHRAE Handbook of Fundamentals*, Chapter 26, Table 36.

108 SOLAR COOLING LOADS

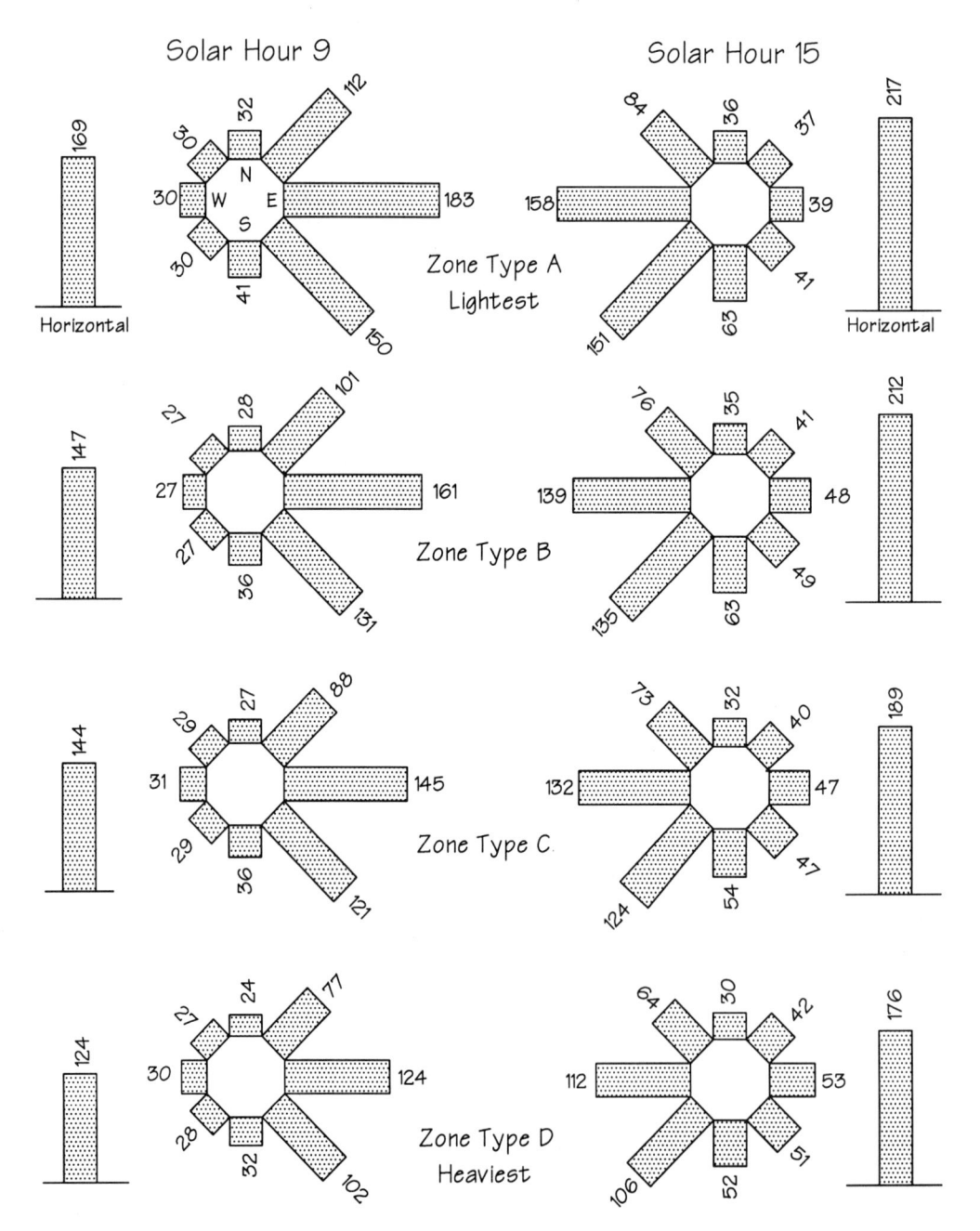

FIGURE 7.1 Graphic depiction of solar cooling loads for sunlit glass. Values in Btu/h per square foot.

TABLE 7.2 TYPICAL SHADING COEFFICIENTS FOR FENESTRATION

Glass:	1/8 in. clear double-strength sheet glass	1.00
	1/4 in. clear plate glass	0.95
	1/2 in. clear (residential) double glazing	0.88
	1 in. clear (commercial) double glazing	0.81
	1/4 in. tinted (bronze, gray or green) glass	0.60
	1 in. tinted double glazing	0.55
	Reflective glass	0.10 to 0.40
Other:	Venetian blinds	0.45 to 0.60
	Curtains	0.40 to 0.60
	Open-weave curtains	0.60 to 0.75
	Continuous overhang	0.25[a]
	Fixed vertical fins	0.30[a]
	Egg-crate shading	0.10[a]
	Mature trees	0.20 to 0.30[a]
	Young trees	0.50 to 0.65[a]

[a]Note: Actual shading for these items will depend upon geometry, orientation, latitude, or species of tree. Values shown are approximate and should only be used for rough estimates.

$$q_{\text{solar}} = A_{\text{gl}} \times \text{SC} \times \text{SCL} \quad \text{Formula 7A}$$

where

q_{solar} is the cooling load due to solar heat gain through glazing (and fenestration) in Btu/h

A_{gl} is the area of glazing in ft^2

SC is the shading coefficient of the fenestration. If two or more types of shading exist (such as tinted glass and interior and/or exterior shading), the various shading coefficients are multiplied together.

SCL is the solar cooling load factor (obtained from Table 7.1) in Btu/h · ft^2

Example 7A: Determine the cooling load due to the solar gain through the windows of "The Office" (shown in Figure 7.2) for solar hour 15 on a clear July 21st day based on the following:

A building located in Philadelphia.

Areas of glass: South 160 ft^2; West 120 ft^2.

Glazing shading coefficient: 0.81 (clear 1 in. double glazing).

Interior shading coefficient: 0.60 (venetian blinds).

Zone type information: carpeted floor covering, gypsum board partitions, and a negligible effect from interior shade.

Solution: Philadelphia is located at 40°N latitude. Therefore, the SCL values in Table 7.1 can be used. Based upon construction information, the space quali-

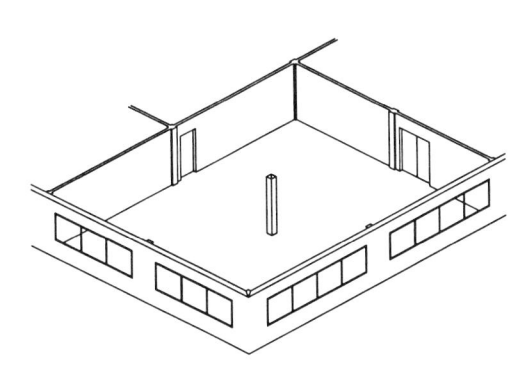

FIGURE 7.2 Floor plan of "The Office."

fies for Zone Type A (from Table 4.2). Therefore, for solar hour 15 the SCL values are (from Table 7.1):

Due to south-facing glass: 63 Btu/h · ft²
Due to west-facing glass: 158 Btu/h · ft²

Now apply Formula 7A to determine the cooling loads due to the south and west glass area. Note that the shading coefficients of the glass and interior shading are multiplied together to obtain the total shading coefficient. For south glass:

$$q_{solar} = A_{gl} \times SC \times SCL$$

(Formula 7A)

$$= 160 \times (0.81 \times 0.60) \times 63$$

$$= 4900 \text{ Btu/h}$$

For west glass:

$$q_{solar} = A_{gl} \times SC \times SCL$$

(Formula 7A)

$$= 120 \times (0.81 \times 0.60) \times 158$$

$$= 9215 \text{ Btu/h}$$

Therefore, the total cooling load due to the glazing for the conditions noted will be 14,115 Btu/h.

In computing peak cooling loads, it is good practice to assume that any interior shading devices (i.e., shades or venetian blinds) will be employed. In contrast, when solar loads are used to estimate energy use, interior shading devices should be assumed to be in place only during those hours when the glazing will be in direct sunlight.

7.2 SUN CONTROL TO REDUCE SOLAR HEAT GAIN

It is not always necessary to use tinted or reflective glazings to substantially reduce solar heat gain in commercial buildings. Fixed horizontal shading devices, such as overhangs, are particularly effective for southerly orientations while vertical projections, such as fins and deep window wall framing, work well to shade east- and west-facing glass areas.

The shadow created by horizontal projections is determined through the use of the profile analysis as defined and illustrated in Section 3.3. Profile angles can be calculated for specific conditions at any location, although that is usually not necessary.

Provided in Tables 7.3–7.5 are summer profile angles for 32°N, 40°N, and 48°N latitudes respectively. Profile angles indicated are for vertical surfaces oriented toward the directions shown for the solar hours when the sun will shine on that surface. These profile angle tables are organized to reflect the symmetry of apparent sun motion. Therefore, the profile angle for an east-facing wall at solar hour 9 will be identical to the profile angle for a west-facing wall at solar hour 15. The tables are read by using orientations on the top for the morning solar times on the left. For afternoons, solar times on the right and orientations along the bottom of the tables apply.

Examination of profile angle tables reveal that some sunlit surfaces experience large profile angles (say 45° or more) hour after hour. Obviously, the more time there is significant shading, the greater the potential benefit and cost effectiveness of shading devices.

The height of a shadow below an overhang is determined by using Formula 7B and is shown in Figure 7.3.

$$S_h = P_h \times \tan(C) \quad \text{Formula 7B}$$

where

S_h is the vertical height of shade cast by an overhang
P_h is the horizontal projection of an overhang out from a building face
C is the profile angle in degrees

TABLE 7.3 PROFILE ANGLES IN DEGREES—32°N LATITUDE APPROXIMATELY 21ST DAY OF EACH MONTH

Solar Hour	Solar Position ALT.	AZ.	E	ESE	SE	SSE	S	SSW	SW	WSW	Solar Hour
June											
7	24	103	25	29	41	71					17
8	37	97	37	41	50	70					16
9	50	89	50	52	59	72	90				15
10	62	80	63	63	67	74	85				14
11	74	61	76	74	75	78	82	88			13
12	81	0	90	87	84	82	81	82	84	87	12
			W	WSW	SW	SSW	S	SSE	SE	ESE	
July (and May)											
7	23	100	23	27	36	63					17
8	35	93	35	38	47	65					16
9	48	85	48	49	55	67	85				15
10	61	73	62	61	64	70	81				14
11	72	52	76	73	72	74	79	85			13
12	78	0	90	85	81	79	78	79	81	85	12
			W	WSW	SW	SSW	S	SSE	SE	ESE	
August (and April)											
7	19	92	19	20	27	44					17
8	31	84	32	33	38	52	80				16
9	44	74	45	44	48	57	74				15
10	56	60	59	56	57	62	71	85			14
11	65	37	74	68	66	66	70	77	87		13
12	70	0	90	82	75	71	70	71	75	82	12
			W	WSW	SW	SSW	S	SSE	SE	ESE	
September (and March)											
7	13	82	13	13	16	24	58				17
8	25	73	26	25	28	36	58				16
9	37	62	40	37	38	44	58	83			15
10	47	47	56	49	47	50	58	72			14
11	55	27	72	62	56	55	58	65	78		13
12	58	0	90	77	66	60	58	60	66	77	12
12				W	WSW	SW	SSW	S	SSE	SE	ESE

Source: Excerpted by permission from the *1993 ASHRAE Handbook of Fundamentals*, 1985, Chapter 27, Table 6.

SOLAR COOLING LOADS

TABLE 7.4 PROFILE ANGLES IN DEGREES—40°N LATITUDE APPROXIMATELY 21ST DAY OF EACH MONTH

Solar Hour	Solar Position ALT.	AZ.	E	ESE	SE	SSE	S	SSW	SW	WSW	Solar Hour
June											
7	26	100	26	30	40	66					17
8	37	91	37	40	48	64					16
9	49	80	49	50	54	65	82				15
10	60	66	62	60	61	67	77	89			14
11	69	42	76	71	69	70	74	81	89		13
12	73	0	90	84	78	75	73	75	78	84	12
			W	WSW	SW	SSW	S	SSE	SE	ESE	
July (and May)											
7	24	97	24	27	36	58					17
8	35	87	35	37	44	59	86				16
9	47	76	48	47	51	61	77				15
10	57	61	61	58	58	63	73	86			14
11	66	37	75	69	66	67	71	77	87		13
12	70	0	90	82	76	71	70	71	76	82	12
			W	WSW	SW	SSW	S	SSE	SE	ESE	
August (and April)											
7	19	89	19	20	26	41	88				17
8	30	79	31	31	35	47	72				16
9	41	67	44	41	43	51	66	90			15
10	51	51	58	52	51	55	63	77			14
11	59	29	73	64	60	59	62	69	81		13
12	62	0	90	78	69	63	62	63	69	78	12
			W	WSW	SW	SSW	S	SSE	SE	ESE	
September (and March)											
7	11	80	12	12	14	21	50				17
8	23	70	24	23	25	31	50				16
9	33	57	37	33	33	38	50	75			15
10	42	42	53	45	42	43	50	64	87		14
11	48	23	71	57	50	48	50	57	71		13
12	50	0	90	72	59	52	50	52	59	72	12
			W	WSW	SW	SSW	S	SSE	SE	ESE	

Source: Excerpted by permission from the *ASHRAE Handbook of Fundamentals*, 1985, Chapter 27, Table 7.

TABLE 7.5 PROFILE ANGLES IN DEGREES—48°N LATITUDE APPROXIMATELY 21ST DAY OF EACH MONTH

Solar Hour	Solar Position ALT.	AZ.	E	ESE	SE	SSE	S	SSW	SW	WSW	Solar Hour
7	27	96	27	30	39	61					17
8	37	85	37	38	44	58	83				16
9	47	72	48	47	50	58	74				15
10	56	55	61	56	56	60	69	81			14
11	63	31	75	67	63	63	66	73	83		13
12	65	0	90	80	72	67	65	67	72	80	12
			W	WSW	SW	SSW	S	SSE	SE	ESE	

July (and May)

7	25	93	25	27	34	54					17
8	35	82	35	35	41	53	78				16
9	44	68	46	44	47	54	69				15
10	53	51	60	54	53	57	65	78			14
11	59	29	74	65	61	60	63	70	81		13
12	62	0	90	78	69	64	62	64	69	78	12
			W	WSW	SW	SSW	S	SSE	SE	ESE	

August (and April)

7	19	87	19	20	24	38	80				17
8	29	75	29	29	32	42	64				16
9	38	61	42	38	39	45	58	82			15
10	46	45	56	48	46	48	55	69	90		14
11	51	24	72	60	53	52	54	61	74		13
12	54	0	90	74	62	56	54	56	62	74	12
			W	WSW	SW	SSW	S	SSE	SE	ESE	

September (and March)

7	10	79	10	10	12	18	42				17
8	20	67	21	20	21	26	42	88			16
9	28	53	34	29	28	32	42	66			15
10	35	38	49	39	36	36	42	55	80		14
11	40	20	68	52	43	40	42	49	63	87	13
12	42	0	90	67	52	44	42	44	52	67	12
			W	WSW	SW	SSW	S	SSE	SE	ESE	

Source: Excerpted by permission from the *ASHRAE Handbook of Fundamentals*, 1985, Chapter 27, Table 8.

114 SOLAR COOLING LOADS

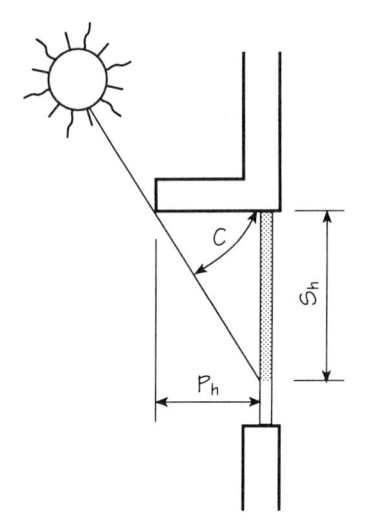

FIGURE 7.3 Shadows below horizontal projections.

Shadow width (S_w) from a vertical projection is determined by using Formula 7C and is shown in Figure 7.4.

$$S_w = P_v \times \tan(BW) \quad \text{Formula 7C}$$

where

S_w is the width of a shadow cast by a vertical projection
P_v is the horizontal projection of a vertical shading device
BW is the surface-solar azimuth: the angular difference between the angle from south which the wall is facing (angle W) and the solar azimuth angle (angle B) as shown in Figure 7.4.

Example 7B: The east-facing windows of a building are 3 ft wide and 5 ft high and are surrounded by shading devices which project out 2 ft both horizontally (P_h) and vertically (P_v). The building is located in San Diego (32.4° latitude). Determine the portion of the window that will be shaded in June during solar hour 10.

Solution: San Diego is close to 32°N latitude. Therefore, use Table 7.3 to obtain the solar angles which occur in June at solar hour 10 as follows, and as shown in Figure 7.5.

Solar altitude angle (A): 62 degrees
Solar azimuth angle (B): 80 degrees
Profile angle (C): 63 degrees

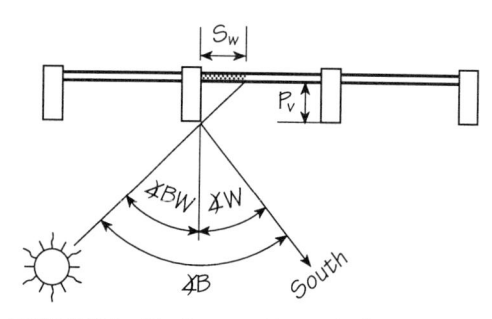

FIGURE 7.4 Shadows cast by vertical projections.

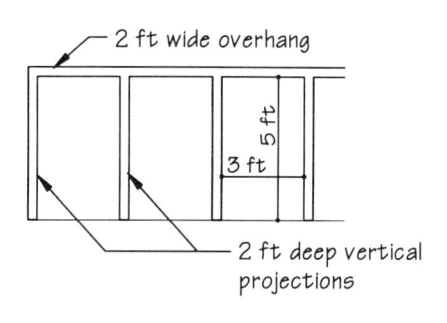

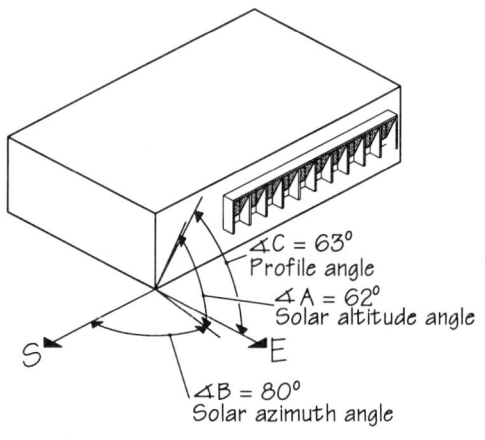

FIGURE 7.5 Solar angles for Example 7B.

Apply Formula 7B to determine the height of the shadow below the horizontal portions of the shading device. The tangent of the profile angle (C) of 63 degrees is 1.96.

$$S_h = P_h \times \tan(C)$$

(Formula 7B)

$$= 2 \times 1.96$$

= 3.92 ft below the overhanging horizontal elements

An east-facing wall faces 90 degrees (to the east of south). And the solar azimuth angle for this example is 80 degrees as noted above. Therefore, angle BW, the difference between these angles (90 − 80) is 10 degrees. Use Formula 7C to determine the width of shadow cast by the vertical elements. The tangent of 10 degrees is 0.176.

$$S_w = P_v \times \tan(BW)$$

(Formula 7C)

$$= 2 \times 0.176$$

= 0.35 ft to the north of the vertical shading elements

These shadows on the east-facing glass are illustrated in Figure 7.6. The area of unshaded glass is 2.85 ft^2, which is only 19% of the window area.

7.3 CONDUCTION OF HEAT THROUGH ROOFS, WALLS, AND GLAZING

For most buildings, by far the largest solar impact is the gain from direct radiation through glass. Nonetheless, the clear day conduction of absorbed solar heat through opaque surfaces such as walls and roofs cannot be ignored, and may represent a significant portion of the cooling load for some

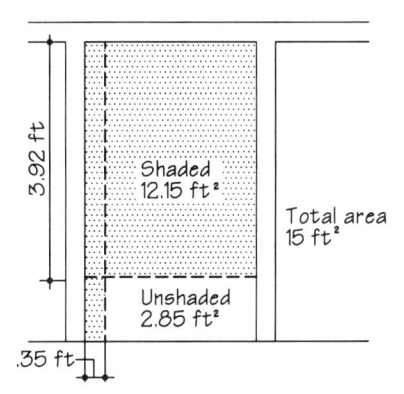

FIGURE 7.6 Shading on east-facing building facade.

building types (i.e., large and low buildings with large roof areas).

The method developed to compute conduction solar heat gain through these surfaces involves what are known as "sol–air" temperatures, since they take into account the combined impact of clear day solar radiation and outdoor air temperatures (see Figure 7.7).

The sol–air temperature approach uses a theoretical temperature known as the Cooling Load Temperature Difference (CLTD) to replace the Delta T (difference between inside and outside temperatures) used in heat transfer calculations. Using the CLTD method, the computed total heat gain is equal to the heat gain due to temperature difference plus the heat gain from solar radiation.

The mass of construction and surface color play large roles in determining CLTDs. In massive buildings with heavy construction the peak cooling loads are delayed and generally reduced. In addition, dark colors absorb and transmit more heat than light colors which reflect much of the solar radiation.

Roof Heat Gain

The intensity of sol–air heat gain through roofs over the course of a sunny day depends primarily upon:

116 SOLAR COOLING LOADS

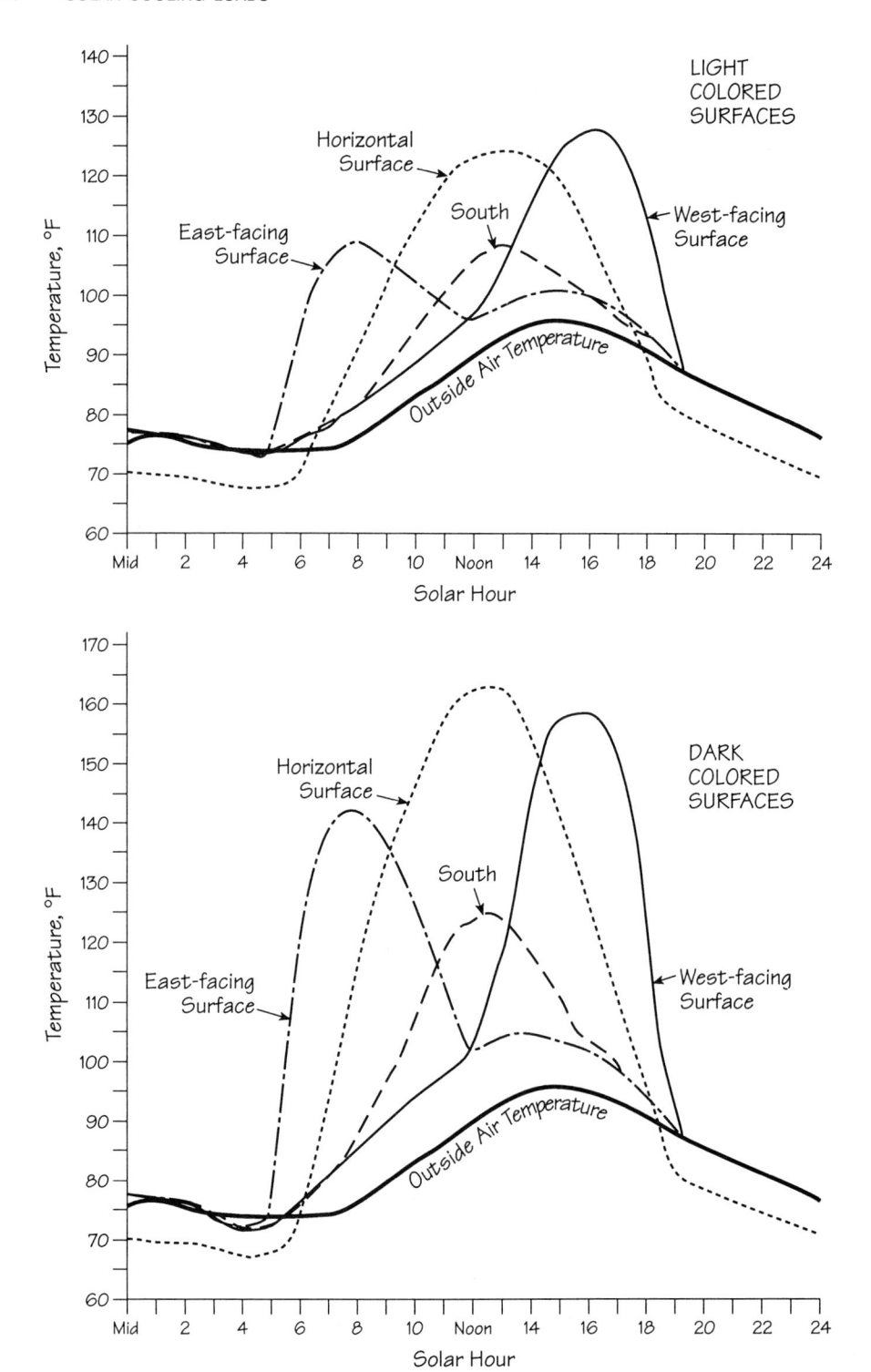

FIGURE 7.7 "Sol–air" heat gain.

The roof deck material and its massiveness

Whether the roof/ceiling system has a suspended ceiling or not

The location of thermal mass in the roofing system relative to the insulation

The R-value of the roofing system

Roof numbers are determined from Table 7.6. For example, a steel deck roof with the mass evenly placed, no suspended ceiling, and an R-value of 22 will be designated Roof Number 2.

Once a "roof number" has been determined for a particular roof assembly (from Table 7.6), CLTD values can be determined for any solar hour from Table 7.7. Values apply directly for:

40°N latitude during the month of July

An indoor design temperature of 78°F

A mean outdoor temperature of 85°F, and a dark surface color

A review of Table 7.7 shows that CLTD values are much larger, and more dynamic, for lightweight (steel deck) roof construction (e.g., roof numbers 1 and 2). Values for heavyweight (HW) concrete construction remain much more constant throughout the day (particularly assemblies with the mass inside the insulation and with suspended ceilings). Also note that the mas-

TABLE 7.6 ROOF NUMBERS FOR USE IN CLTD TABLE 7.7

Mass Location	Susp. Clg.	R-value	Wood (1 in.)	HW Conc. (2 in.)	Steel Deck	Attic/Clg. Combo.
Mass Inside the Insul.	No	0 to 10	N.A.[a]	2	N.A.	N.A.
		10 to 20	N.A.	4	N.A.	N.A.
		20 to 25	N.A.	5	N.A.	N.A.
	Yes	0 to 5	N.A.	5	N.A.	N.A.
		5 to 10	N.A.	8	N.A.	N.A.
		10 to 20	N.A.	13	N.A.	N.A.
		20 to 25	N.A.	14	N.A.	N.A.
Mass Evenly Placed	No	0 to 5	1	2	1	1
		5 to 15	2	N.A.	1	2
		15 to 20	4	N.A.	2	2
		20 to 25	4	N.A.	2	4
	Yes	0 to 5	N.A.	3	1	N.A.
		5 to 10	4	N.A.	1	N.A.
		10 to 15	5	N.A.	2	N.A.
		15 to 20	9	N.A.	2	N.A.
		20 to 25	10	N.A.	4	N.A.
Mass Outside the Insul.	No	0 to 5	N.A.	2	N.A.	N.A.
		5 to 10	N.A.	3	N.A.	N.A.
		10 to 15	N.A.	4	N.A.	N.A.
		15 to 25	N.A.	5	N.A.	N.A.
	Yes	0 to 10	N.A.	3	N.A.	N.A.
		10 to 15	N.A.	4	N.A.	N.A.
		15 to 20	N.A.	5	N.A.	N.A.

Source: Excerpted by permission from the *1993 ASHRAE Handbook of Fundamentals*, Chapter 26, Table 31.
[a]N.A.-not applicable.

118 SOLAR COOLING LOADS

TABLE 7.7 ILLUSTRATIVE FLAT ROOF CLTD VALUES 40°N LATITUDE—JULY 21ST (°F EQUIVALENT)

Roof No.	Solar Time (hour)								
	8	9	10	11	12	13	14	15	16	17
1	13	29	45	60	73	83	88	88	83	73
2	4	17	32	48	62	74	82	86	85	80
3	5	13	24	35	47	57	66	72	74	73
4	−3	0	7	17	29	42	54	65	73	77
5	2	6	12	21	31	41	51	60	66	69
8	10	12	16	21	28	35	42	48	53	56
9	4	4	7	12	19	27	36	45	53	59
10	10	9	10	12	17	23	30	37	44	50
13	16	16	17	20	24	28	33	38	42	46
14	20	19	20	22	24	28	32	36	39	42

Source: Excerpted by permission from the *1993 ASHRAE Handbook of Fundamentals*, Chapter 26, Table 30.

sive roofs actually start the business day (see column of values for solar hour 8) with higher CLTD values because of the release of heat stored the day before. Graphically illustrated in Figure 7.8 are flat roof CLTD values for several roof numbers from Table 7.7.

CLTD values are incorporated into Formula 7D, which determines sol–air solar heat gain on clear days for roofs (and also for walls and glazing).

$$q_{cond} = U \times A \times CLTD \quad \text{Formula 7D}$$

where

q_{cond} is the conductive sol–air heat gain through the roof, walls, or glazing in Btu/h

U is the component U-factor in Btu/ft^2·°F·h

A is the area of the component in ft^2

CLTD is the appropriate cooling load temperature difference (sol–air heat gain value) in °F equivalent from Tables 7.7 (roof), 7.8 (walls), or 7.9 (glass)

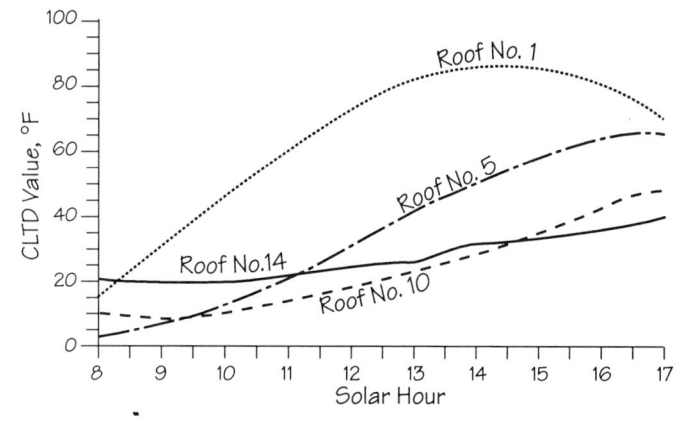

FIGURE 7.8 40°N latitude flat roof CLTD values for July 21st.

Example 7C: Compute the sol–air heat gain through the flat roof of a building in Salt Lake City for solar hour 15 in July based on the following:

- A roof area of 8000 ft^2
- A metal deck roof with 2 in. of insulation. The roof assembly has a U-factor of 0.08 ($R = 12.0$)
- The building does not have a suspended ceiling

Solution: Since Salt Lake City is located near 40°N latitude, the CLTD values in Table 7.7 apply for the month of July. Roof number 1 applies (see Table 7.6) for the construction described. Table 7.7 is then used to determine the CLTD value for roof number 1 for solar hour 15. The answer is a very significant 88°F. Formula 7D can then be used to solve for this sol–air heat gain.

$$q_{cond} = U \times A \times CLTD$$

(Formula 7D)

$$= 0.08 \times 8000 \times 88$$

$$= 56{,}320 \text{ Btu/h}$$

Opaque Wall Heat Gain

Sol–air heat gain through opaque walls is calculated using a similar procedure to that for roofs and Formula 7D. Wall CLTD values are based upon orientation, type of construction, and the R-value of the wall assembly. The *ASHRAE Handbook* describes 16 Wall Numbers based upon:

- Location of mass (i.e., inside insulation, evenly distributed with insulation, or outside insulation)
- One of 15 principal wall materials (face brick, wood, concrete, etc.)
- Secondary (exterior) construction of either: stucco and/or plaster; steel or other lightweight siding; face brick
- The R-value of the wall construction ranging from R-0 to R-27

Low Wall Numbers (i.e., 1, 2, and 3) represent lightweight wall constructions such as a metal curtain wall with varying amounts of insulation. Medium Wall Numbers (i.e., 7–10) in general represent walls with lightweight exterior siding with more massive wall materials (and insulation) beneath. And the higher Wall Numbers (i.e., 15 and 16) represent the heaviest types of wall construction such as face brick backed up by insulation and concrete or concrete block. Presented in Table 7.8 are CLTD values for the primary wall orientations for Wall Numbers 1, 5, 10, and 16 which range from the lightest (and least insulated) to the heaviest in construction.

Note the difference in values for the different Wall Numbers. Values for lightweight walls (i.e., Wall Number 1) vary greatly and have very high values for walls facing east in the morning and west in the afternoon. In contrast, CLTD values for heavyweight walls (e.g., Wall Number 16) remain fairly constant throughout the day with only minor differences due to orientation.

Opaque wall CLTD values for the four construction types (Wall Numbers) listed in Table 7.8 are illustrated in Figure 7.9. Notice how dramatically CLTD values change for lightweight wall construction (Wall Number 1).

Conduction Heat Gain Through Glass

In addition to direct solar heat gain through glazing (as discussed previously), the conductive heat gain through glazing must also be considered as an element of a building's cooling load. Cooling loads for conductive heat gain through glass are computed using Formula 7D and values from Table 7.9.

Values in Table 7.9 can be used for an inside temperature of 78°F, an outdoor maximum temperature of 93–102°F, and

TABLE 7.8 CLTD VALUES FOR SUNLIT WALLS 40°N LATITUDE—JULY 21ST (°F EQUIVALENT)

Wall Number	Wall Facing	8	9	10	11	12	13	14	15	16	17
1	N	11	11	13	17	21	25	27	29	29	28
	E	51	62	64	59	48	36	31	30	30	28
	S	4	11	21	33	43	50	52	50	44	34
	W	4	8	13	17	21	27	42	59	73	80
5	N	3	5	6	8	9	12	14	17	19	21
	E	8	17	26	33	39	40	40	38	37	35
	S	2	3	4	8	13	19	25	31	35	36
	W	4	4	5	7	9	11	15	20	28	37
10	N	5	5	5	6	7	8	10	12	14	17
	E	7	10	14	20	26	31	34	35	36	36
	S	5	4	4	5	7	11	15	20	24	28
	W	9	8	7	7	7	8	10	13	17	23
16	N	9	8	7	7	7	8	9	10	11	13
	E	11	11	12	15	19	22	26	28	30	31
	S	11	9	8	8	8	9	11	14	17	20
	W	17	15	13	12	11	11	11	12	14	17

Source: Excerpted by permission from the *1993 ASHRAE Handbook of Fundamentals*, Chapter 26, Table 32.

outdoor daily range of 16–34°F (provided the outdoor daily average temperature remains approximately 85°F). For a different inside temperature add or subtract the difference from 78°F to the value in the table. For outdoor daily average temperature less than 85°F, subtract the difference between 85°F and the daily average temperature; if greater than 85°F, add the difference.

Correction of CLTD Values

As stated before, the "basic CLTD values" included in the Tables above are based on:

- July 21st clear day conditions at 40°N latitude
- A mean outdoor temperature of 85°F and an indoor air temperature of 78°F
- A dark surface color

July is the basis of the CLTD tables since it is often the month when ambient temperature, humidity, solar heat gain, and internal heat gains produce the largest combined cooling loads. Nonetheless, many building characteristics and conditions can result in peak cooling loads during different months (as shown in Figure 8.8) and thus the reason for the CLTD correction procedure. Latitude and month modifications (LM) are provided in Table 7.10. Intermediate orientations and values for months not shown can simply be interpolated.

Corrections for latitude and month (LM), surface color (K), inside temperature other than 78°F (t_{in}), and mean outside temperature (t_{out}) other than 85°F may be made by using Formula 7E.

$$CLTD_c = [(CLTD + LM) \times K] + (78 - t_{in}) + (t_{out} - 85)$$

Formula 7E

where

$CLTD_c$ is the corrected CLTD value
CLTD is the basic value obtained for

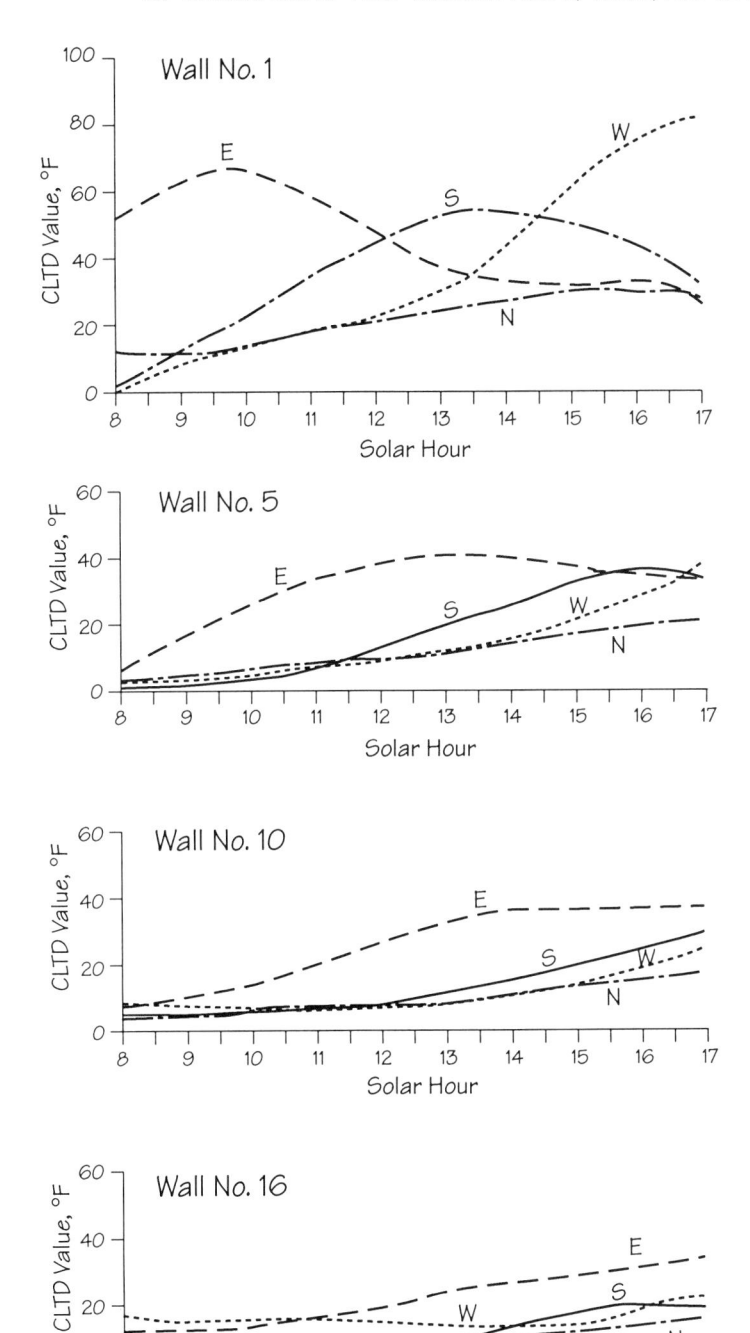

7.3 CONDUCTION OF HEAT THROUGH ROOFS, WALLS, AND GLAZING   121

FIGURE 7.9 40°N latitude opaque wall CLTD values for July 21st.

122 SOLAR COOLING LOADS

TABLE 7.9 CLTD VALUES FOR GLAZING

Solar Hour	CLTD (°F)
8	0
9	2
10	4
11	7
12	9
13	12
14	13
15	14
16	14
17	13

Source: Excerpted by permission from the *1993 ASHRAE Handbook of Fundamentals*, Chapter 26, Table 34.

	the component from Table 7.7, 7.8, or 7.9 in °F
LM	is the correction factor for other than 40°N latitude and months other than July from Table 7.10
K	is a color correction factor from Table 7.11
78	is the assumed inside air temperature (used by basic CLTD tables) in °F
t_{in}	is the actual inside air temperature in °F
t_{out}	is the actual outside air temperature in °F
85	is the assumed mean outside air temperature (used by basic CLTD tables) in °F

The CLTD correction procedure assumes that roof construction is without an attic. After all other adjustments are made, corrected CLTD values may be reduced by 25% for buildings with attics where the ceiling is insulated, and a fan is used to positively ventilate the space between the ceiling and roof.

Example 7D: Determine the sol–air heat gain through a dark-colored metal deck roof (with $1\frac{1}{2}$ in. of insulation) on a food storage facility in San Diego on a clear day for solar hour 14 in October assuming the following:

No suspended ceiling in the building
Roof area = 5000 ft^2
Roof U-factor = 0.12 Btu/ft$^2 \cdot$ °F $\cdot$ h (which is also equal to R-8 construction)
Inside temperature = 50°F

Solution: The roof assembly described is designated as Roof Number 1 in Table 7.6 since it has a metal deck roof, no ceiling and an R-value of 8. The basic CLTD value for Roof Number 1 for solar hour 14 is then found to be 88°F in Table 7.7. This CLTD value must then be modified for the following reasons.

San Diego is located at 32.4°N latitude so the appropriate adjustment for 32°N latitude listed in Table 7.10 applies and the LM correction adjustment for horizontal surfaces in October is −10.

Modifications are required for the inside temperature of 50°F, and the outside temperature in October. The normal daily maximum temperature in October in San Diego is about 74°F (obtained from NOAA annual summaries). This value can be used for solar hour 14 in the early afternoon.

TABLE 7.10 LATITUDE AND MONTH (LM) CLTD CORRECTIONS—NORTH LATITUDE WALLS AND ROOFS

	Month	N	NE/NW	E/W	SE/SW	S	HOR
	Dec	−4	−8	−4	4	13	−9
16	Jan/Nov	−4	−7	−4	4	12	−7
	Feb/Oct	−3	−5	−2	2	7	−4
°N	Mar/Sept	−3	−2	−1	0	0	−1
	Apr/Aug	−1	−1	−1	−3	−6	0
Lat	May/July	4	3	−1	−5	−7	0
	June	6	4	−1	−6	0	−7
	Dec	−5	−9	−7	3	13	−13
24	Jan/Nov	−4	−8	−6	3	13	−11
	Feb/Oct	−4	−6	−3	3	10	−7
°N	Mar/Sept	−3	−3	−1	1	4	−3
	Apr/Aug	−2	0	−1	−1	−3	0
Lat	May/July	1	2	0	−3	−6	1
	June	3	3	0	−4	−6	1
	Dec	−5	−10	−8	2	12	−17
32	Jan/Nov	−5	−9	−8	−4	9	12
	Feb/Oct	−4	−7	−4	4	11	−10
°N	Mar/Sept	−3	−4	−2	3	7	−5
	Apr/Aug	−2	−1	0	0	1	−1
Lat	May/July	1	1	0	−1	−3	1
	June	1	2	0	2	12	−15
40	Dec	−6	−10	−10	0	10	−21
	Jan/Nov	−5	−10	−9	1	11	−19
°N	Feb/Oct	−5	−8	−6	3	12	−14
	Mar/Sept	−4	−5	−3	4	10	−8
Lat	Apr/Aug	−2	−2	0	2	4	−3
	May/July	0	0	0	0	1	1
	June	1	1	1	0	−1	2
	Dec	−6	−11	−13	−3	6	−25
48	Jan/Nov	−6	−11	−11	−1	8	−24
	Feb/Oct	−5	−10	−8	1	11	−18
°N	Mar/Sept	−4	−6	−4	4	11	−11
	Apr/Aug	−3	−3	−1	4	7	−5
Lat	May/July	0	0	1	3	4	0
	June	1	2	2	2	3	2
	Dec	−7	−12	−16	−9	−3	−28
56	Jan/Nov	−6	−11	−14	−6	2	−27
	Feb/Oct	−6	−10	−10	0	9	−22
°N	Mar/Sept	−5	−7	−5	4	12	−15
	Apr/Aug	−3	−4	−1	5	9	−8
Lat	May/July	0	0	2	5	7	−2
	June	2	2	3	4	6	1

Source: Excerpted by permission from the *1989 ASHRAE Handbook of Fundamentals*, Chapter 26, Table 32.

TABLE 7.11 CLTD COLOR CORRECTION *K* VALUES

K	Description of Facade/Roof
1.0	Dark-colored roofs and walls or light-colored surfaces in industrial (dirty) areas
0.83	Permanently medium-colored walls in rural areas
0.65	Permanently light-colored walls in rural areas

Source: Adapted with permission from the *1989 ASHRAE Handbook of Fundamentals*, Chapter 26, Table 32.

The above adjustments are entered into Formula 7E. Note that a *K* value of 1.0 is obtained from Table 7.11 since it is a dark colored roof. Also note that the low temperature of 50°F maintained in the food storage facility adds to the CLTD and resultant heat gain.

$$\text{CLTD}_c = [(\text{CLTD} + \text{LM}) \times K]$$

(Formula 7E)

$$= ([88 + (-10)] \times 1) + (78 - 50) + (74 - 85)$$
$$= 78 + 28 - 11$$
$$= 95°F$$

The sol-air heat gain can now be calculated using Formula 7D, but with the corrected CLTD (CLTD_c).

$$q_{cond} = U \times A \times \text{CLTD}_c$$

(Formula 7D)

$$= 0.12 \times 5000 \times 95$$
$$= 57{,}000 \text{ Btu/h}$$

This result can then be directly used in a cooling load calculation since the massiveness of the construction was factored in when the applicable roof selection was made.

8

COOLING LOAD CALCULATIONS

The external and internal heat gains which impact on cooling loads are discussed in various portions of this book along with formulas to measure them. This chapter pulls all of this information together, to show how to compute peak cooling loads for individual spaces and entire buildings.

Central HVAC systems are sized for the time when the cooling load for a building "peaks" as a whole. Precisely when this peak hour occurs depends on many factors, as will be discussed. For starters, however, it can be said that many buildings peak during mid-afternoon summer hours (usually solar hours 14, 15, or 16) when outside air temperatures are their highest, and a building's capacity for additional heat storage is diminished.

Today, detailed cooling load calculations for complex buildings are often done with the aid of computer programs because of the many interactive and varying factors. Still, before using computer output, one should be fully conversant with the elements and factors involved and be able to spot check results to be sure that "garbage in" data entry has not led to "garbage out" results.

8.1 THERMAL "ZONING" OF A BUILDING

A "zone" is a building area under the control of a single thermostat. Ideally, all spaces within a zone have uniform heat gains, identical indoor design requirements, and similar hours of occupancy. The smallest zone is a simple room, often cooled by local (individual) cooling equipment under the control of the occupant.

Most mid- to large-size buildings have at least five zones (per floor) as shown in Figure 8.1 to account for changing solar loads by orientation.

Perimeter zones are orientation-sensitive and must be served by HVAC systems which can provide year-round conditioning. For HVAC purposes, perimeter zones are typically considered to be 12–15 ft

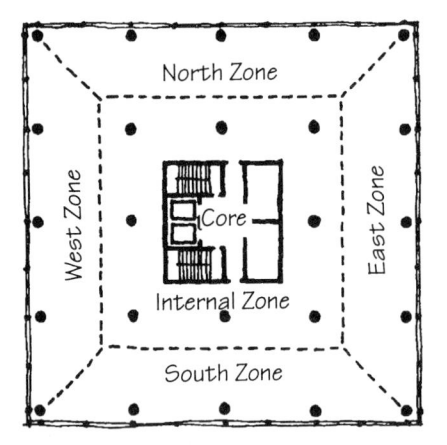

FIGURE 8.1 Basic zoning of floors by orientation.

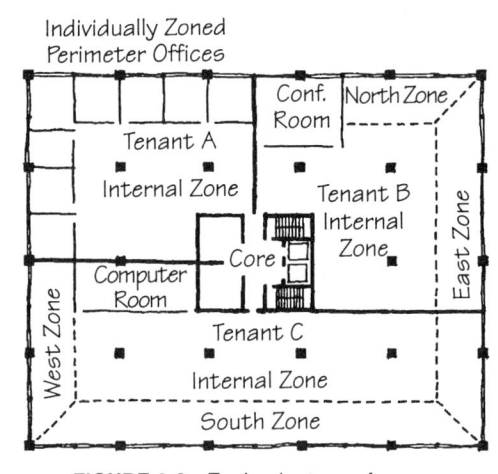

FIGURE 8.2 Zoning by type of space.

wide. Small perimeter spaces (e.g., individual offices) often are treated as small zones which are conditioned by individually controlled cooling and heating units. Larger perimeter zones are generally served by central systems which afford opportunities for improved energy efficiency through the possible use of the economizer cycle or closed-loop heat pumps.

Internal zones of large buildings generally require heat removal most of the time when a building is occupied due to the heat gain from lights, people, and office equipment. When the activity levels of areas within the interior zone are relatively constant and predictable, a very simple constant volume system can be used. When activity and occupancy within an internal zone is expected to vary greatly, a variable air volume system would more effectively provide proper space temperature control.

Many large buildings function in a more diverse and complex thermal manner and require many additional zones for spaces with intermittent use, or with more or less heat gain (e.g., conference rooms and computer rooms) as shown in Figure 8.2.

The number and location of thermal zones will determine which type(s) of cooling distribution and delivery system(s) should be used for a particular building application. For example, some multifloor buildings have perimeter areas facing a particular direction on several floors which can be zoned together and served by one air supply system. In contrast, some building types require such specialized temperature and humidity control for perimeter and interior spaces that many spaces and rooms become individual zones.

8.2 ASHRAE CALCULATION PROCEDURE

In this chapter, cooling loads will be calculated for a single room and an entire example building using the ASHRAE "CLTD/CLF Calculation Procedure." As described in Chapter 7, CLTD stands for Cooling Load Temperature Difference. CLTDs not only account for cooling loads due to outside air temperatures, but also for the impact of solar heat gain. As described in Chapter 4, CLF stands for Cooling Load Factor. CLFs are used to modify space cooling loads based upon the heat storage potential of the building structure and furnishings.

Cooling loads are calculated to determine adequate cooling capacity for reasonable design conditions. As discussed in

Chapter 2, cooling systems are not designed for the most extreme temperatures since such an approach would lead to gross oversizing.

8.3 EQUATIONS FOR EXTERNAL HEAT GAINS

Glass Solar Heat Gain

Formula 7A is used to compute cooling loads due to solar heat gain through glazed areas. Results in Btus per hour are based on component areas, the shading coefficient (SC) of the glass or fenestration, and solar cooling load factors (SCL) to account for possible heat storage of the construction and furnishings.

$$q_{solar} = A_{gl} \times SC \times SCL$$

(Formula 7A)

Conductive Heat Gain through Roofs, Exterior Walls, and Glass

Individual cooling loads for building elements are all computed by using Formula 7D. Results in Btus per hour are based on component U-factors, areas, and sol–air factors (CLTD) from the appropriate tables.

$$q_{cond} = U \times A \times CLTD$$

(Formula 7D)

Partitions, Ceilings, and Floors Conduction Heat Gain

Cooling loads can also result when partitions, ceilings, or floors of conditioned spaces are adjacent to spaces maintained at a higher temperature. Formula 8A is used.

$$q_{sep\text{-}con} = U \times A \times (t_h - t_i)$$

(Formula 8A)

where

$q_{sep\text{-}con}$ is the heat gain through separating construction (partitions, ceilings, or floors) in Btu/h

U is the U-factor of the component in units of Btu/ft$^2 \cdot °F \cdot h$

A is the area of the separating construction in ft^2

t_h is the higher temperature of an adjoining space or the outside summer design temperature in °F (as applicable)

t_i is the inside design temperature in °F

Ventilation and Infiltration Air

Sensible heat gain from ventilation and/or infiltration air is computed by using Formula 6C, based upon the air flow rate (cfm) and the outside and inside dry-bulb temperatures.

$$q_{oa\text{-}sen} = 1.10 \times cfm \times (t_o - t_i)$$

(Formula 6C)

Latent heat gain from ventilation and/or infiltration air is computed by using Formula 6B based upon the ventilation rate (in cfm) and the humidity ratio of the outside (W_o) and inside (W_i) air.

$$q_{v\text{-}lat} = 4840 \times cfm \times (W_i - W_o)$$

(Formula 6B)

Total heat gain from ventilation and/or infiltration air (sensible plus latent) is computed by using Formula 6E based on the ventilation rate (in cfm) and the enthalpy of the outside (H_o) and inside (H_i) air. See Section 19.2 for additional information on enthalpy.

$$q_{oa\text{-}tot} = 4.5 \times cfm \times (H_o - H_i)$$

(Formula 6E)

128 COOLING LOAD CALCULATIONS

8.4 EQUATIONS FOR INTERNAL HEAT GAINS

People

Sensible and latent cooling loads due to the number of people (N) are computed by using Formulas 4A and 4B based upon sensible (SHG_p) and latent (LHG_p) heat gains per person and a cooling load factor (for sensible gain only).

$$q_{p\text{-sen}} = N \times SHG_p \times CLF_p$$

(Formula 4A)

$$q_{p\text{-lat}} = N \times LHG_p$$

(Formula 4B)

Equipment and/or Appliances

Sensible heat gain from equipment and/or appliances is computed by using Formula 4C based upon equipment power density (EQ_{wsf}), floor area (A), and a cooling load factor (CLF_{eq}).

$$q_{eq} = EQ_{wsf} \times A \times 3.413 \times CLF_{eq}$$

(Formula 4C)

Lights

Heat gain from lighting systems is computed by using Formula 4D based upon lighting power density (L_{wsf}), floor area (A), and a cooling load factor. Remember that 1 W is equal to 3.413 Btu/h.

$$q_{lgt} = L_{wsf} \times A \times 3.413 \times CLF_{lgt}$$

(Formula 4D)

8.5 DETAILED COOLING LOAD CALCULATION FOR "THE OFFICE"

Example 8A: Compute the cooling load during July for solar hour 15 for "The Of-

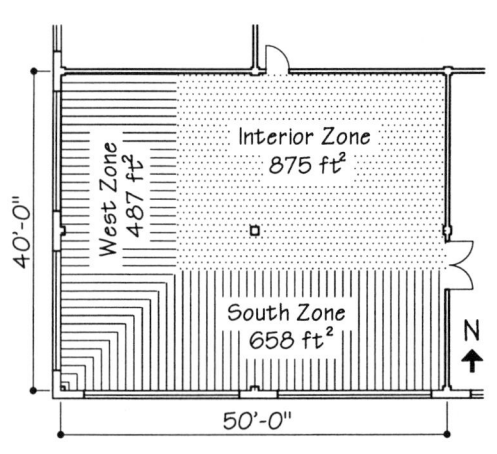

FIGURE 8.3 Floor plan of "The Office."

fice" as shown in Figure 8.3 and based on the following:

- Location in Philadelphia, PA. Outdoor design conditions of 90°F db and 74°F wb (the 2.5% summer design condition).
- Indoor design conditions of 78°F and 50% RH (65°F wb).
- Building occupancy begins at solar hour 8 and continues for 10 h. The maximum occupancy assumed for "The Office" is 20 people who are seated and doing very light work. People are assumed to be evenly distributed throughout the space.
- The space temperature is maintained constantly overnight. Therefore, the mass of the building and furnishings is available for heat storage during the day.
- Location on the top floor of a building. Assume that the building has a metal deck roof with 2.5 in. of insulation and a U-factor of 0.06 ($R_T = 16$). The office space has a hung ceiling and the roof is dark colored.
- Fairly heavyweight wall construction which satisfies the criteria for designation as Wall Number 10. The U-factor of the wall is 0.08 ($R_T = 12$).
- Exterior wall component information as follows:

8.5 DETAILED COOLING LOAD CALCULATION FOR "THE OFFICE"

Component	Area (ft^2)	U-factor	Shading Coeff.	Wall Number
South opaque wall	440	0.08	N.A.	10
South glass	160	0.50	0.60	N.A.
West opaque wall	360	0.08	N.A.	10
West glass	120	0.50	0.60	N.A.

- Zone type information: Carpeted floor covering, gypsum board partitions and a negligible effect from interior shade. According to Table 4.2 such a space qualifies as Zone Type A when cooling loads due to solar heat gain through glass are being computed. Similarly, the space qualifies as Zone Type B when cooling loads for people, equipment, and lights are being determined.
- Office equipment power density is 2.0 W/ft^2 for this area with personal computers and printers.
- Lighting power density is 2.0 W/ft^2.
- Outside air ventilation rate is 20 cfm per person, or 400 cfm (20 × 20 occupants). Assume that the space is positively pressurized so that there will be no infiltration air.
- Zoning of the space for HVAC system design purposes is as shown in the Figure.

Solution: Determine the total room and zone cooling loads due to each of the contributing factors as follows:

Solar Heat Gain Through Glazing. Solar heat gain through glazing usually produces a major portion of the cooling load. Since Philadelphia is located at 40°N latitude, use the SCL values in Table 7.1. For Zone Type A (see description above) the SCL values for solar hour 15 are:

SCL for south-facing glass: 63 Btu/h·ft^2
SCL for west-facing glass: 158 Btu/h·ft^2

The cooling loads resulting from this solar heat gain can now be computed using Formula 7A. The shading coefficient (SC) of the glass is 0.60.

South glass area:

$$q_{\text{solar}} = A_{\text{gl}} \times \text{SC} \times \text{SCL}$$

(Formula 7A)

$$= 160 \times 0.60 \times 63$$

$$= 6050 \text{ Btu/h, sensible}$$

West glass area:

$$q_{\text{solar}} = A_{\text{gl}} \times \text{SC} \times \text{SCL}$$

(Formula 7A)

$$= 120 \times 0.60 \times 158$$

$$= 11{,}380 \text{ Btu/h, sensible}$$

Roof (Sol-Air) Cooling Loads. The metal deck roof construction with a hung ceiling and a total R-value of 16 qualifies as Roof Number 2 (see Table 7.6). Now obtain the CLTD value from Table 7.7 for Roof Number 2. For solar hour 15 the CLTD value is 86°F. The U-factor was stated to be 0.06 with the entire roof area being 2000 ft^2.

$$q_{\text{cond}} = U \times A \times \text{CLTD}$$

(Formula 7D)

$$= 0.06 \times 2000 \times 86$$

$$= 10{,}320 \text{ Btu/h, sensible cooling load for the entire roof}$$

As shown in Figure 8.3, 32% of "The Office" is considered as being in the south zone, 24% in the west zone, and 44% in the interior zone. Once the entire roof cooling load is prorated by area, the cooling loads from roof heat gain for each of the three thermal zones within "The Office" are found as follows:

	Sensible Cooling Loads (Btu/h)
South zone	3290
West zone	2515
Interior zone	4515

Opaque Wall (Sol-Air) Cooling Loads. Again use Formula 7D and CLTD values for Wall Number 10 from Table 7.8. Since Philadelphia is at 40°N latitude, and the example is for July, there is no adjustment required for the CLTD value as there would be for other latitudes or months. For solar hour 15, the applicable CLTD values for Wall Number 10 are 20°F for south-facing walls and 13°F for west-facing walls.

South wall area:

$$q_{cond} = U \times A \times CLTD$$

$$\text{(Formula 7D)}$$

$$= 0.08 \times 440 \times 20$$

$$= 700 \text{ Btu/h, sensible}$$

West wall area:

$$q_{cond} = U \times A \times CLTD$$

$$\text{(Formula 7D)}$$

$$= 0.08 \times 360 \times 13$$

$$= 370 \text{ Btu/h, sensible}$$

Obviously, heat gain through well-insulated opaque walls and resulting cooling loads are relatively small.

Glass (Sol-Air) Cooling Load. To compute the sol-air (conductive) cooling load for the glazing again use Formula 7D. The CLTD value from Table 7.9 for solar hour 15 is found to be 14°F (considered to be equal for all orientations).

South glass area:

$$q_{cond} = U \times A \times CLTD$$

$$\text{(Formula 7D)}$$

$$= 0.50 \times 160 \times 14$$

$$= 1120 \text{ Btu/h, sensible}$$

West glass area:

$$q_{cond} = U \times A \times CLTD$$

$$\text{(Formula 7D)}$$

$$= 0.50 \times 120 \times 14$$

$$= 840 \text{ Btu/h, sensible}$$

Heat Gains from People. The assumed occupancy of the space is 20 persons (N). Sensible and latent heat gain per person is obtained from Table 4.1 (for people seated doing very light work). The cooling load factor (CLF) is obtained from Table 4.5 for Zone Type B and a 10-h period of occupancy. For solar hour 15 (7 h after occupancy of the space begins) the CLF is 0.93.

$$q_{p\text{-}sen} = N \times SHG_p \times CLF_p$$

$$\text{(Formula 4A)}$$

$$= 20 \times 245 \times 0.93$$

$$= 4560 \text{ Btu/h, sensible}$$

$$q_{p\text{-}lat} = N \times LHG_p$$

$$\text{(Formula 4B)}$$

$$= 20 \times 155$$

$$= 3100 \text{ Btu/h, latent}$$

The distribution of these cooling loads for the three zones is as follows:

8.5 DETAILED COOLING LOAD CALCULATION FOR "THE OFFICE"

	Sensible Load (Btu/h)	Latent Load (Btu/h)
South zone	1450	990
West zone	1110	750
Interior zone	2000	1360

Office Equipment Heat Gain. The equipment power density (EQ_{wsf}) is 2.0 W/ft² for the 2000-ft² office area where workers have personal computers and printers. Use Table 4.5 to determine a cooling load factor (CLF_{eq}) of 0.93 for Zone Type B, a total of 10 operational hours, and 7 hours after the equipment was turned on. Now use Formula 4C to compute the cooling load.

$$q_{eq} = EQ_{wsf} \times A \times 3.413 \times CLF_{eq}$$

(Formula 4C)

$$= 2.0 \times 2000 \times 3.413 \times 0.93$$

$$= 12{,}700 \text{ Btu/h, sensible}$$

	Sensible Load (Btu/h)
South zone	4050
West zone	3100
Interior zone	5550

Lighting System Heat Gain. The lighting system power density (L_{wsf}) is 2.0 W/ft² for the entire 2000-ft² office area (A). Table 4.10 is consulted to determine the cooling load factor of 0.95 for Zone Type B, lights on for 10 hours, and solar hour 15 which is 7 hours after the lights are turned on. Formula 4D can now be used to compute the cooling load.

$$q_{lgt} = L_{wsf} \times A \times 3.413 \times CLF_{lgt}$$

(Formula 4D)

$$= 2.0 \times 2000 \times 3.413 \times 0.95$$

$$= 12{,}970 \text{ Btu/h, sensible}$$

Once prorated by area, the zone cooling loads from the lighting system are as follows:

	Sensible Load (Btu/h)
South zone	4135
West zone	3160
Interior zone	5675

Ventilation Air Heat Gains. Use Formula 6C to compute sensible heat gain of ventilation air ($q_{oa\text{-}sen}$). The outside air flow rate is 400 cfm. The outside design db temperature (t_o) is 90°F with a 74°F wb design temperature. The inside design temperature (t_i) is 78°F.

$$q_{oa\text{-}sen} = 1.10 \times \text{cfm} \times (t_o - t_i)$$

(Formula 6C)

$$= 1.10 \times 400 \times (90 - 78)$$

$$= 5{,}280 \text{ Btu/h, sensible for the entire space. Zone loads are shown below.}$$

The psychrometric chart in Figure 8.4 indicates the humidity ratio conditions of the outside air and inside air. The difference is approximately 0.005 pounds of water per pound of dry air.

Now use Formula 6B to compute the latent heat gain for the 400 cfm of ventilation air ($q_{v\text{-}lat}$).

$$q_{v\text{-}lat} = 4840 \times \text{cfm} \times (W_o - W_i)$$

(Formula 6B)

$$= 4840 \times 400 \times (0.015 - 0.010)$$

$$= 9680 \text{ Btu/h, latent}$$

By zone these sensible and latent ventilation loads are distributed as follows:

132 COOLING LOAD CALCULATIONS

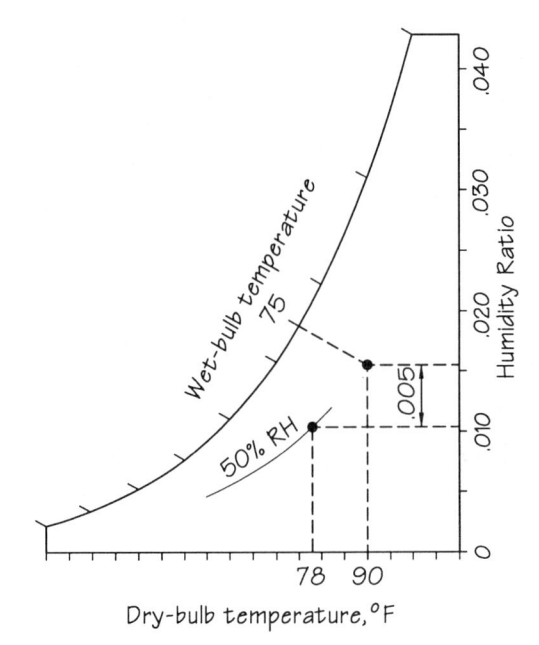

FIGURE 8.4 Office cooling load example: humidity ratios of outside and inside air.

	Sensible Load (Btu/h)	Latent Load (Btu/h)
South zone	1680	3085
West zone	1290	2355
Interior zone	2310	4240

Summation of Total Space Cooling Loads. Once summed, the individual load components add up to the following zone cooling loads:

Cooling Load Component	Sensible Cooling Loads by Zone (Btu/h) South	West	Interior
Glass (solar)	6,050	11,380	
Roof	3,290	2,515	4,515
Walls	700	370	
Glass (cond.)	1,120	840	
People	1,450	1,110	2,000
Equipment	4,050	3,100	5,550
Lights	4,135	3,160	5,675
Ventilation	1,680	1,290	2,310
Zone totals: (sensible)	22,475	23,755	20,050

Grand total sensible (all zones): 66,280 Btu/h

Cooling Load Component	Latent Cooling Loads by Zone (Btu/h) South	West	Interior
People	990	750	1,360
Ventilation	3,085	2,355	4,240
Zone totals: (latent)	4,075	3,105	5,600

Grand total latent (all zones): 12,780 Btu/h

These cooling loads for "The Office" (located in Philadelphia) in July during solar hour 15 are also illustrated in Figure 8.5. Note the relative size of each component. Examples of cooling distribution systems (fan-coil units and VAV boxes) designed to satisfy these zone loads are included in Chapter 21.

8.6 PEAK HOUR COOLING LOADS

When designing systems and selecting equipment for central cooling systems, it is necessary to determine the peak hourly cooling load for the entire building. The time of hourly peak for the entire building depends, of course, on many factors including: location, building type and orientation, amount and type of glazing, hours of operation, intensity of internal heat gains, and massiveness of construction.

Despite all of these variables, the peak cooling hour for most buildings can be estimated reasonably well. Indicated in Figure 8.6 are likely peak hours for various building characteristics and conditions.

8.7 PEAK COOLING LOADS FOR ENTIRE BUILDINGS

The example of the 40,000 ft^2 two-story building in Figure 8.7 will be used to illustrate how the totals for different hours are used to determine the peak cooling load for the building as a whole. The building is elongated along the east–west axis with major facades facing north and south. Stem-

8.7 PEAK COOLING LOADS FOR ENTIRE BUILDINGS

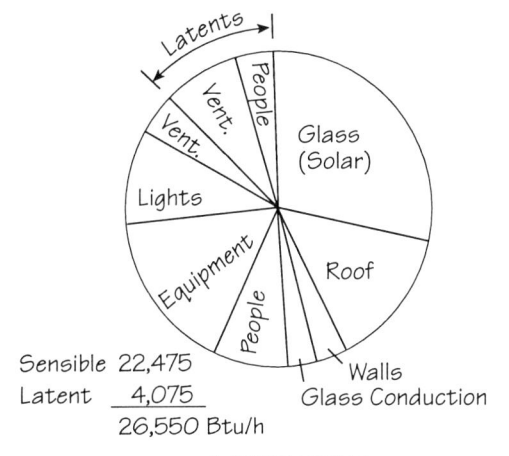

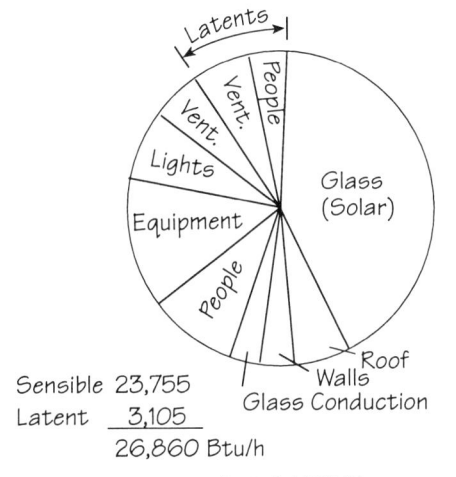

ming from its shape, the building has large interior zones, typical of many contemporary buildings.

Example 8B: Determine the hour of peak building cooling load for the building in Figure 8.7 based upon the following:

- Glazing: 33.5% of gross exterior wall, tinted double glazing (U = 0.50, SC = 0.60)
- Medium weight wall construction: U = 0.10
- Lightweight metal deck roof: U = 0.05
- Lights: 2.4 W/ft^2
- Office equipment: 1.6 W/ft^2
- Medium weight interior construction
- Ten-hour per day building occupancy by 250 people
- Indoor design conditions of 78°F db and 65°F wb
- Outdoor design conditions of 90°F db and 75°F wb
- 40°N latitude
- Ventilation rate of 20 cfm of outside air per person

Solution: A computerized spreadsheet employing formulas discussed earlier was used to perform the repetitious calculations for various hours. Originally, results were computed for the following seven hours when the peak building cooling load is likely to occur: July solar hours 13, 14, and 15; August solar hours 14 and 15; and September solar hours 14 and 15.

For the example building, the results for the various hours are very close to each other. Originally, the peak building cooling load was found to occur in September for solar hour 14. As a result, it was prudent to compute another cooling load for September during solar hour 13 to see if it was larger than for solar hour 14. As shown in the summary it was not. (Note: 1 ton of refrigeration capacity is equal to 12,000 Btu/h, as fully defined in Section 10.1).

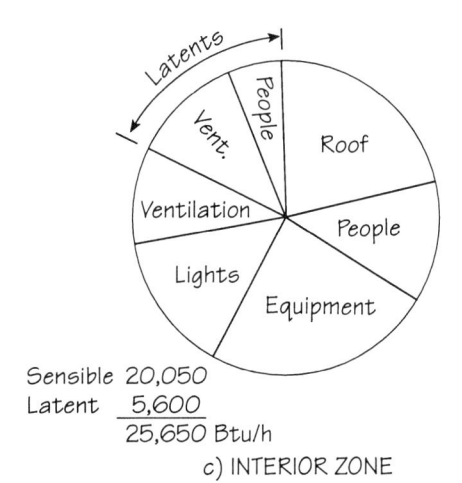

FIGURE 8.5 Cooling load for the zones of "The Office."

a) SOUTH ZONE
Sensible 22,475
Latent   4,075
          26,550 Btu/h

b) WEST ZONE
Sensible 23,755
Latent   3,105
          26,860 Btu/h

c) INTERIOR ZONE
Sensible 20,050
Latent   5,600
          25,650 Btu/h

134 COOLING LOAD CALCULATIONS

Low Square Buildings
Consistent Fenestration

Mid-afternoon
Hours 13 to 15
July or August

High Square Buildings
Consistent Fenestration

Mid- to Late Afternoon
Hours 14 to 17
July or August

Elongated Buildings
on East-West Axis
Consistent Fenestration

Early to Mid-afternoon
Hours 13 to 15
September

Elongated Buildings
on North-South Axis
Consistent Fenestration

Late Afternoon
Hours 15 to 17
June or July

Buildings with Majority
of Fenestration Facing East

Early to Mid-afternoon
Hours 9 to 11
July or August

Buildings with Large
Unshaded Skylight

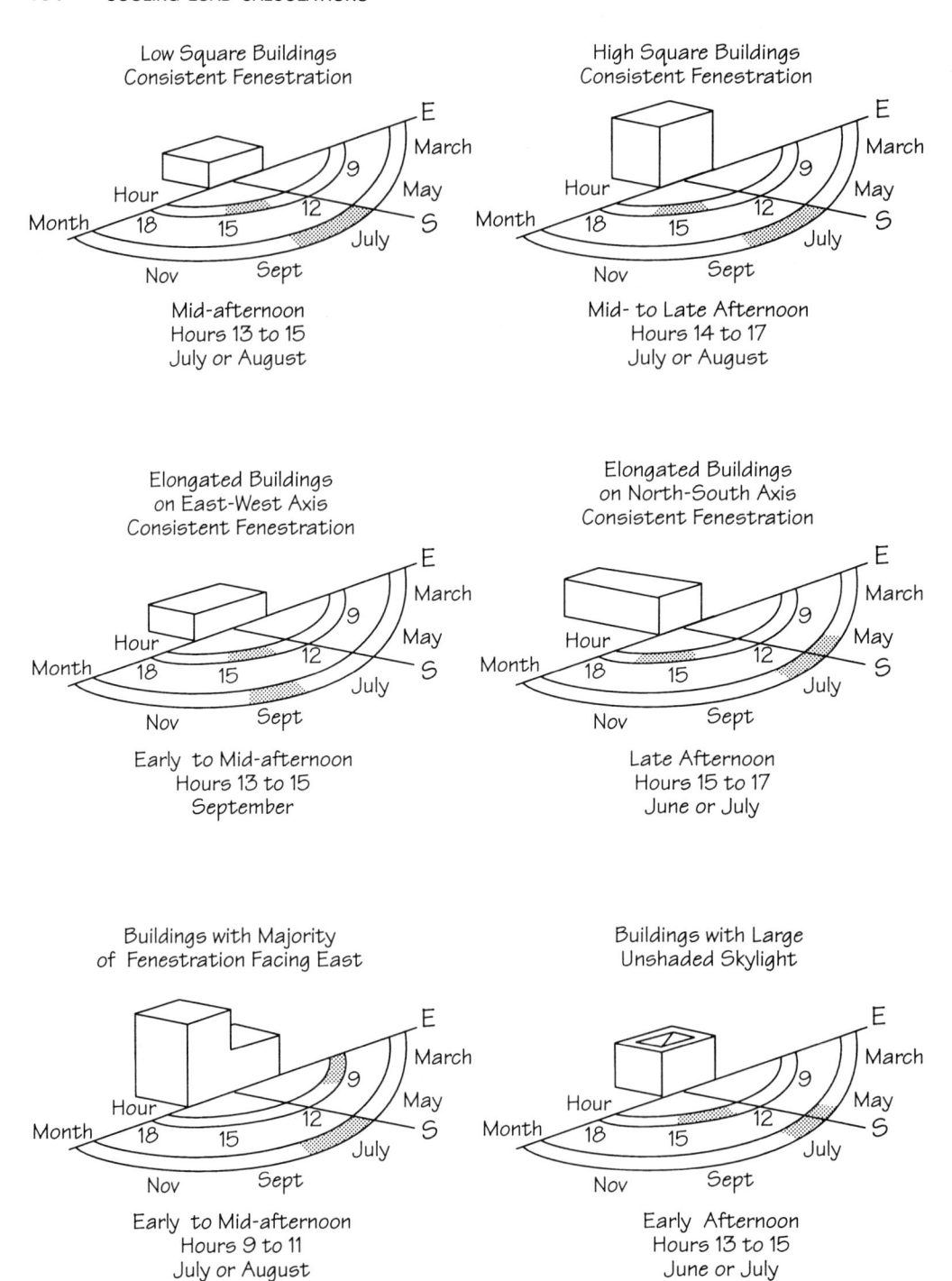

Early Afternoon
Hours 13 to 15
June or July

FIGURE 8.6 Likely peak cooling hours for entire buildings.

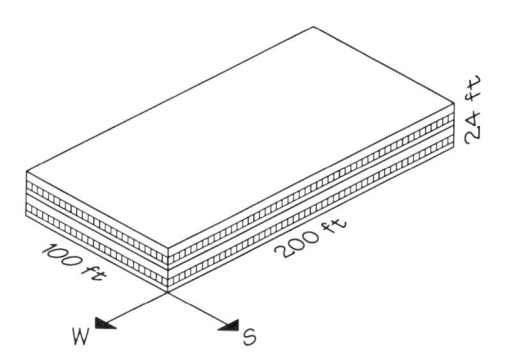

FIGURE 8.7 Two-story example building elongated along the east–west axis.

ure 8.9) so that the elongated facades now face east and west.

Solution: The wall and glass areas were adjusted on the computer spreadsheet to reflect the totals for the rotated building. The total building cooling loads were then computed for the original seven solar hours noted in Example 8B. As in the previous example, again the original peak load occurred at a solar hour (solar hour 15 in July) which was not sandwiched by other results.

SUMMARY: TOTAL ELONGATED BUILDING COOLING LOADS POSSIBLE PEAK HOURS

Month	Solar Hour	Total Bldg. Heat Gain (Btu/h)	Tons of Refrigeration Capacity
July	13	930,000	78
	14	944,000	79
	15	955,000	80
August	14	947,000	79
	15	955,000	80
September	13	950,000	79
	14	968,000	81
	15	965,000	80

Therefore, for the hours analyzed, the peak cooling load occurs during solar hour 14 in September when the total cooling load is computed to be 968,000 Btu/h (81 tons). The load components are illustrated in Figure 8.8.

As illustrated in Figure 8.8, about 19% of the total peak cooling load for the example building (when elongated on the east–west axis) comes from "orientation sensitive" sensible loads (184,000 Btu/h of the total 968,000 Btu/h).

Rotation of Example Building

Example 8C: Determine total building cooling loads for the example building when rotated 90 degrees (as shown in Fig-

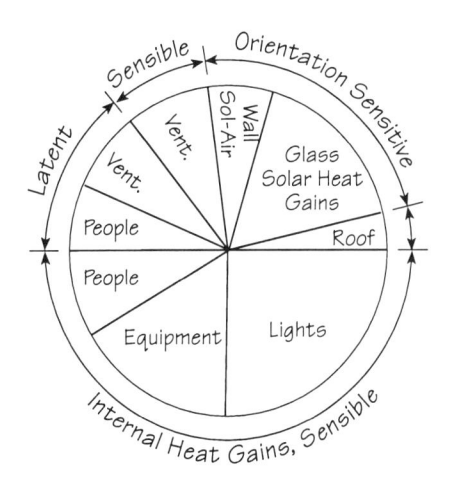

FIGURE 8.8 Components of the peak cooling load for Example 8B.

136 COOLING LOAD CALCULATIONS

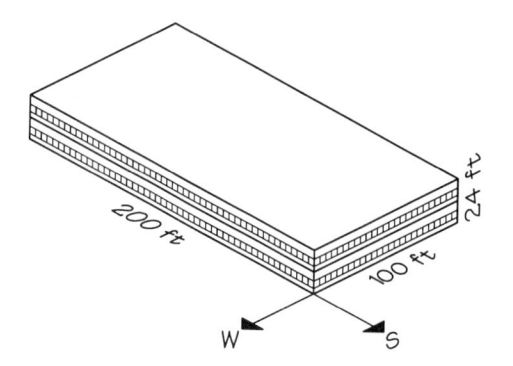

FIGURE 8.9 Rotated example building elongated on north-south axis.

Therefore, another calculation was performed to determine the peak load for solar hour 16 in July. Again it was lower.

We have seen that the peak cooling loads for the example and rotated buildings varied only slightly (2.5%) from 968,000 to 992,000 Btu/h. It should be realized that a much greater difference results from buildings with a greater proportion of floor area in perimeter zones.

8.8 SQUARE FEET PER TON

Cooling loads can also be expressed in terms of building size. For example, the highest computed load (83 tons) for the rotated building yields a result of 482 square feet per ton (sf/ton) for the 40,000 square foot building.

SUMMARY: TOTAL ROTATED BUILDING COOLING LOADS
POSSIBLE PEAK HOURS

Month	Solar Hour	Total Bldg. Heat Gain (Btu/h)	Tons of Refrigeration Capacity
July	13	924,000	77
	14	964,000	80
	15	992,000	83
	16	985,000	82
August	14	931,000	78
	15	960,000	80
September	14	937,000	78
	15	966,000	81

Therefore, the highest cooling load for the rotated building (992,000 Btu/h or 83 tons) occurs during July for solar hour 15. This is due to the late afternoon solar heat gain passing through the relatively large area of west-facing windows.

Summarized in Figure 8.10 are the total computed loads for all 16 hours computed (for the example building and the rotated building). They range from a low of 924,000 Btu/h (77 tons) for the rotated building at hour 13 solar time in July to a high of 992,000 Btu/h (83 tons) for the rotated building just two hours later at hour 15 solar time in July (month 7).

Some sources still republish "sf/ton" tables developed many years ago for different building types (office buildings, schools, apartment buildings, etc.). If you must, use published sf/ton guidelines for preliminary load estimates only. Today's buildings are very different from those of 15 or 20 years ago, having less lighting power, more computers, and other equipment which make sf/ton guidelines unreliable. In addition, sf/ton guidelines which use building type as the only variable are obviously suspect when one thinks about

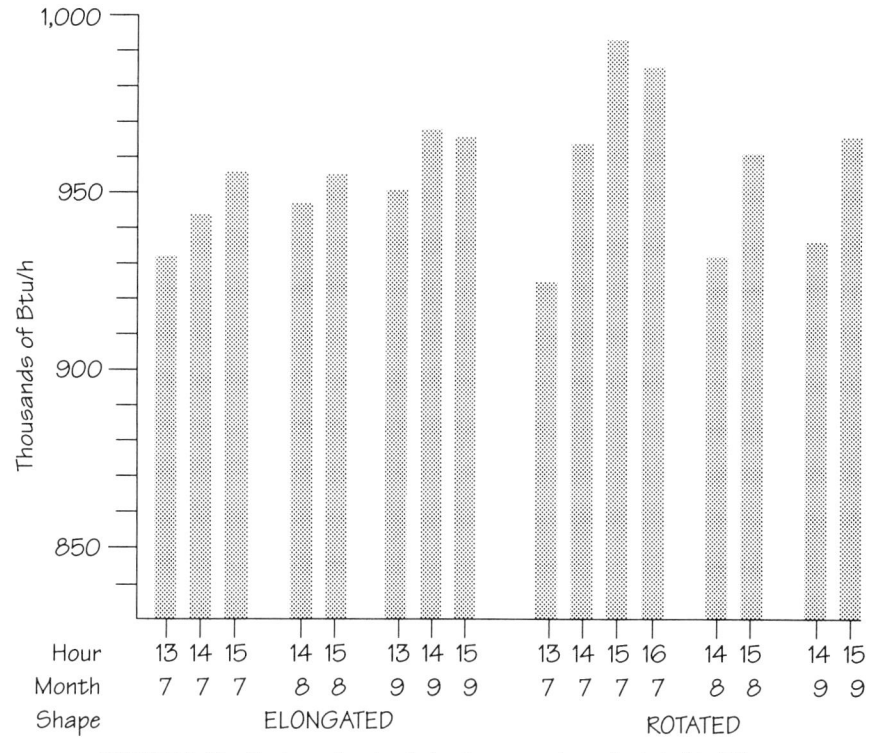

FIGURE 8.10 Peak cooling loads for the example and rotated buildings.

the variability of design conditions for different climates.

8.9 OCCUPANCY DIVERSITY

In doing a cooling load calculation, a series of judgments must be made about building occupancy and operational characteristics. For example, consider the following:

1. A mechanically cooled multipurpose university building contains a large auditorium. How likely is it that the auditorium would be fully occupied during peak conditions (probably midafternoon in July or August)? And if such a meeting were to occur, how likely is it that other spaces in the building (offices, classrooms, library, etc.) would also be used to capacity with all lights and office equipment on?

2. A headquarters building for an insurance company where many of the building occupants will be out calling on clients during the main part of the day.

3. A restaurant that serves only dinner.

For each of these building situations, significant reductions in load, equipment size, initial cost, and operational costs can be realized if reasonable occupancy and operational conditions are assumed.

Perhaps the biggest fear of HVAC designers is that the cooling equipment will be inadequately sized, resulting in discomfort and complaints by building occupants. Therefore, when appropriate, assumptions used in sizing systems should be discussed with the client for reasonableness and decisions documented.

9

AIR, DUCTS, AND FANS

Many HVAC systems deliver air for ventilation or conditioning (i.e., heating, cooling, humidity control) of spaces. Therefore, it is essential that basic information about air (and air pressure) is understood. Air is often directed to and from spaces within a building by a network of friction-creating ductwork. Discussed in this chapter are the fundamentals of duct friction and duct sizing.

Also included is information on the various types of fans used for HVAC applications. Perhaps most familiar is the axial-type propeller fan, similar to those used by small airplanes. Other fan types are other axial fans (tubeaxial and vaneaxial) and centrifugal fans. A series of fan laws based on operational properties is presented and applied.

See Chapter 21 for detailed information on air delivery systems to supply required ventilation and conditioned air.

9.1 AIR PROPERTIES

Although the properties of dry air change slightly at different temperatures, for most typical HVAC applications it is adequate to consider "standard air" (actually at 68°F) to have a density of 0.075 lb/ft^3. This means that a pound of standard air would occupy about 13.3 ft^3 (1/0.075) which is equal to a volume of 2 ft by 2 ft by 3.33 ft. The specific heat of air is 0.24 Btu/lb·°F. This means that 0.018 Btus of heat are required to raise one cubic foot of air one degree Fahrenheit (0.075 × 0.24). This value of "0.018" is known as the "heat capacity of air."

The Ability of Air to Hold Moisture

Air can hold more water vapor (moisture) when it is warmer. In fact, air at 85°F can hold three times as much water vapor as air

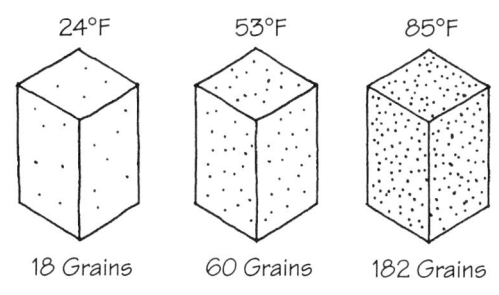

FIGURE 9.1 Maximum moisture air can hold at various temperatures.

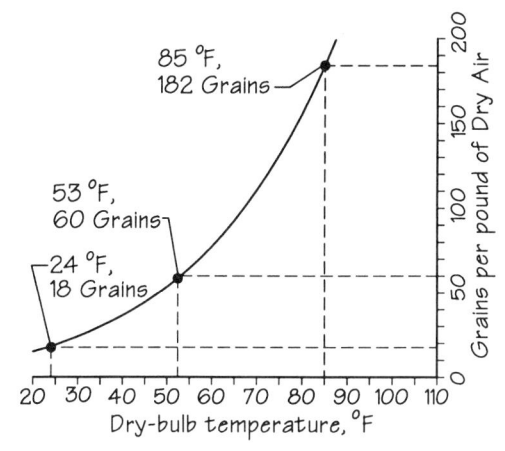

FIGURE 9.2 Plotting the ability of air to hold moisture.

at 53°F, and 10 times as much water vapor as air at 24°F.

The amount of water vapor in air is quantified by the old English term of "grain," with 7000 grains equaling one pound. Figure 9.1 indicates how many grains of moisture can be held by a pound of dry air at various dry-bulb temperatures. Note that air at 85°F can hold about 10 times the amount of water vapor as air at 24°F.

9.2 PROPERTIES OF MOIST AIR

Detailed knowledge about the ability of air to hold varying amounts of water vapor at different dry-bulb temperatures is essential for the design of HVAC systems. Plotted in Figure 9.2 are the three conditions shown in Figure 9.1 with the dry-bulb temperature on the horizontal axis, and moisture (the maximum number of grains per pound of dry air) on the vertical axis. The three points are then connected by an increasingly steep curve which is indicative of the ability of air to hold additional moisture at higher temperatures.

The figure is the beginning of a more complex "psychrometric chart" which presents many properties of air–moisture mixtures. Relative humidities are indicated by lines which curve upward towards the right, as shown in Figure 9.3. Note that the curving line representing 50% relative humidity is always midway between the horizontal base line and the curving top line of the chart as measured vertically.

Shown in Figure 9.4 are wet-bulb temperatures indicated along the top curving

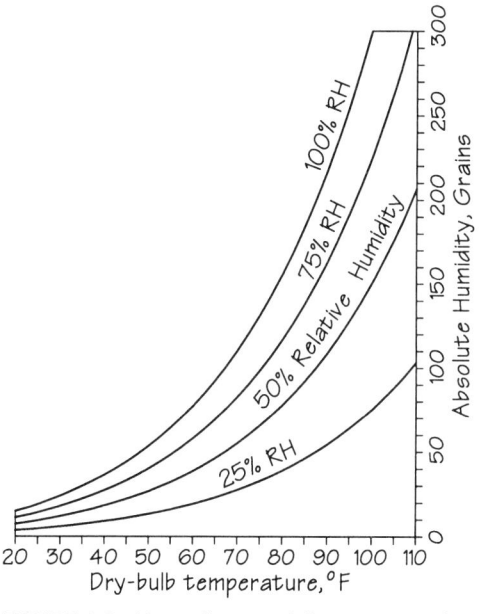

FIGURE 9.3 Upwardly curved lines representing relative humidity.

140 AIR, DUCTS, AND FANS

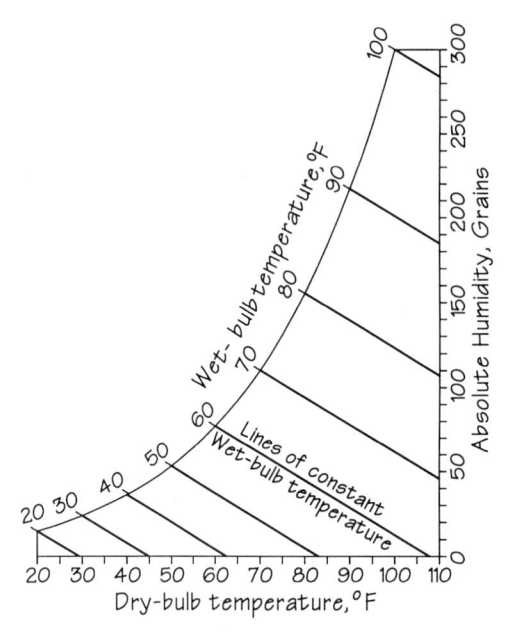

FIGURE 9.4 Wet-bulb temperature and enthalpy.

line of the chart, which is also the line of 100% relative humidity. Lines of constant wet-bulb temperature then extend downward to the right at approximately a 30-degree angle. These diagonal lines also represent lines of constant enthalpy (total heat) as explained in Chapter 19.

Presented in Figure 9.5 is a basic psychrometric chart for conditions at sea level showing the relationships of temperature and humidity. See Figure 19.2 for a complete psychrometric chart which presents additional information on enthalpy, sensible heat factors, humidity ratio and air volume per pound.

Example 9A: Determine the wet-bulb temperature of air which has a dry-bulb temperature of 90°F and 40% relative humidity.

Solution: Using the psychrometric chart in Figure 9.5, locate 90°F dry-bulb along the horizontal axis. Now move vertically along that line until the curving line representing 40% relative humidity is reached. Then move diagonally on the line of constant wet-bulb temperature to the top curve of the chart. The answer is 71°F.

Example 9B: Determine the relative humidity of air with a 75°F dry-bulb temperature which contains 92 grains of moisture per pound of dry air. Also determine the maximum amount of moisture which air can hold at this temperature, as well as the "dew point temperature" of this air.

Solution: On the psychrometric chart, find the dry-bulb temperature of 75°F on the horizontal axis, and the moisture content of 92 grains per pound of dry air on the right-hand vertical axis. Very near to where they intersect (point A), we find the curved line which indicates that this air–moisture mixture has a relative humidity of 70% (see Figure 9.6). Then, since a pound of 75°F dry air holds 92 grains when it is 70% saturated, it follows that the maximum it can hold at this temperature (100% RH) will be found by multiplying the grains by the ratio of 100 to 70, or:

92 grains/lb of dry air × 100/70
= 131.4 grains/lb of dry air

The maximum amount of moisture which can be held by 75°F air can also be determined graphically. From point A move vertically to point B along the top curve representing 100% RH. The quantity of grains, 132, can then be read on the vertical axis.

The dew point temperature of air at condition A can be found graphically by moving horizontally on the chart to the left until the condition of 100% relative humidity is reached at point C, a dew point temperature of 65°F (read on either the dry-bulb or wet-bulb temperature scales).

See Chapter 19 for more detailed "Applied Psychrometric" information.

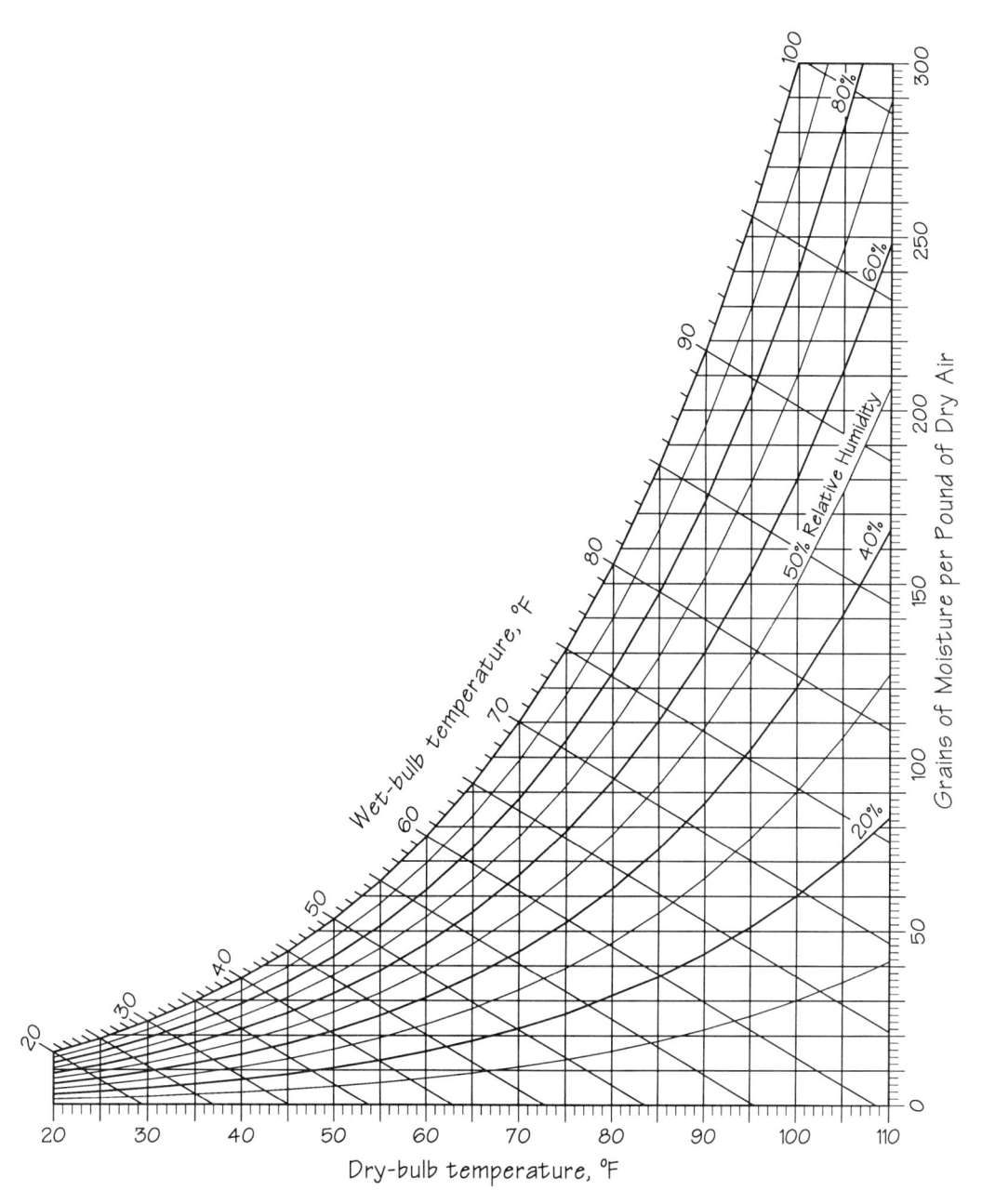

FIGURE 9.5 The psychrometric chart, which plots air–moisture mixtures.

142 AIR, DUCTS, AND FANS

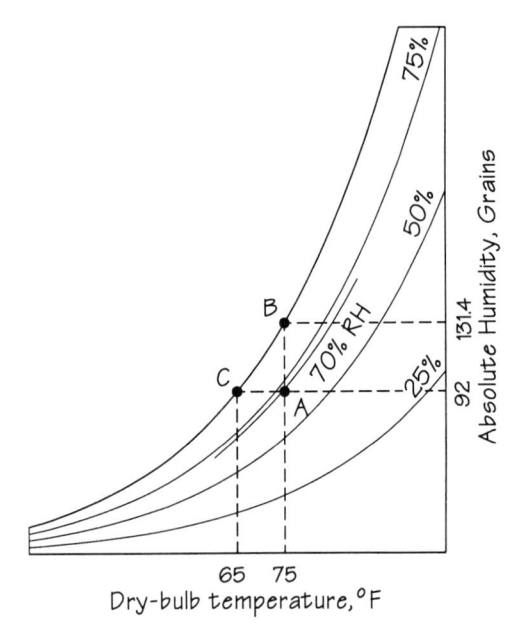

FIGURE 9.6 Grains of moisture in Example 9B.

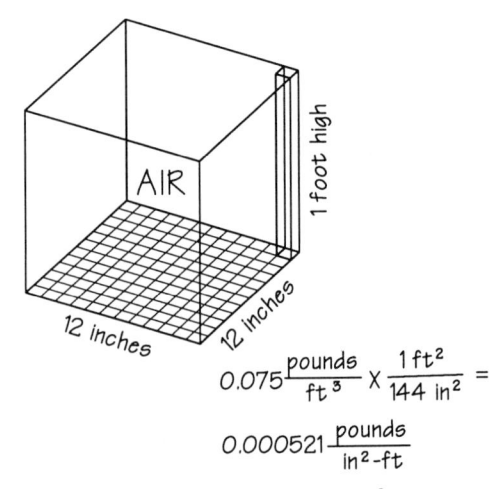

$$0.075 \frac{\text{pounds}}{\text{ft}^3} \times \frac{1 \text{ ft}^2}{144 \text{ in}^2} = 0.000521 \frac{\text{pounds}}{\text{in}^2\text{-ft}}$$

FIGURE 9.7 Pressure from 1 ft³ of air.

9.3 AIR PRESSURE

When you stand on a bathroom scale, your weight is indicated by the amount of downward pressure you have placed on the scale. But pressure was actually being placed on the scale before you even got on it—the air pressure of our atmosphere. Illustrated in Figure 9.7 is the pressure from 1 ft³ of air, 1 ft high, resulting in a pressure of 0.000521 lb/in.² per foot of height (0.075/144).

The air in our atmosphere then produces an atmospheric pressure at sea level of 14.7 lb/in.² (psi), equivalent to columns of "constant density" air 28,215 ft high, or about 5.34 mi (28,215/5280), as shown in Figure 9.8. In fact, the lower atmosphere (technically the stratosphere) extends about 30 mi above sea level with the air becoming increasingly lighter at higher altitudes.

The figure also illustrates why high altitude locations experience less atmospheric pressure such as "mile high" Denver, Colorado. Atmospheric air pressures for various altitudes in psi are indicated in Figure 9.9.

For most HVAC design applications near sea level, an air density of 0.075 lb/ft³ is routinely used as the basis of calculations and is referred to as a "constant."

Pressure Equivalents

Depending upon the HVAC application, various terms will be used throughout this book to describe pressures. Columns, or heights of each of the following, will produce a pressure equal to the pressure of our atmosphere at sea level of 14.7 psi.

Water: 33.95 ft or 407.4 in.
Mercury: 29.92 in.

The differences, of course, result from the significantly different densities of air (only 0.075 lb/ft³), water (62.4 lb/ft³), and mercury (849 lb/ft³) when all are at a temperature of 68°F.

Absolute Pressure

Here on earth some amount of atmospheric pressure (14.7 psi at sea level or less at higher elevations) is always present. In most cases, however, what is of most importance to operating HVAC systems is the

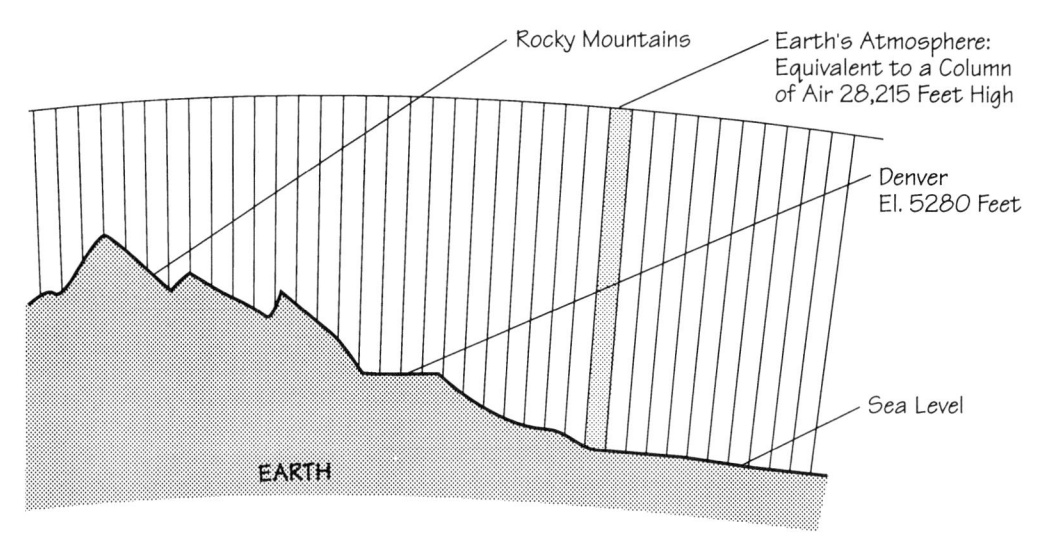

FIGURE 9.8 Air pressure from earth's atmosphere.

operational or "gauge" pressure above local atmospheric pressure. Illustrated in Figure 9.10 are the following terms: atmospheric pressure (psi_{atm}); gauge pressure (psi_g); vacuum pressure below atmospheric (psi_{vac}); and absolute pressure (psi_a).

Air pressure, though invisible, colorless, odorless, and easy to forget, is an important factor to consider in many aspects of HVAC design.

Velocity Pressure

Anyone who has ever held their hand out of the window of a moving automobile knows that moving air creates pressure. As will be seen in examples below, moving air creates what is known as "velocity pressure," which is determined using Formula 9A.

$$vp = (V/4005)^2 \quad \text{Formula 9A}$$

FIGURE 9.9 Atmospheric air pressures and air densities at altitude.

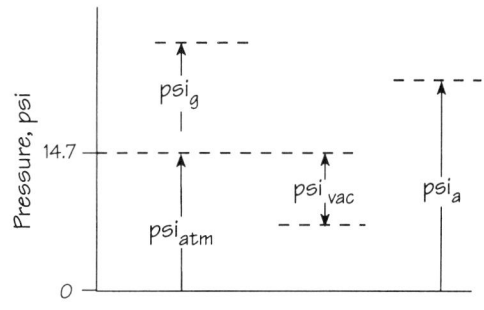

FIGURE 9.10 Pressure terminology and relationships.

where

vp   is the velocity pressure in inches of water gauge (in. wg)
V   is the air velocity in feet per minute (fpm)
4005   is a factor used for "standard air"

Observe Formula 9A and note that air with a velocity of 4005 fpm would have a velocity pressure equal to 1 in. wg while air at a velocity of about 1000 fpm has a velocity pressure of only about 0.06 in. wg.

9.4 FRICTION IN DUCTWORK

The friction produced by lengths of ductwork, fittings (i.e., transitions, elbows), and other items placed in the airstream has to be estimated in order to design duct and fan systems. Illustrated in Figure 9.11 is how ducts of different shapes compare in enclosing an area of 1 ft². As shown, circular ducts have the smallest perimeter to enclose a given area, and thus offer the least frictional resistance.

Although round ductwork clearly has an advantage in using less material, which in turn reduces friction, most ducts are rectangular to accommodate clearance considerations and also to allow for easy transitions in one dimension. Rectangular ducts have a dimensional property referred to as the "aspect ratio." A 12 in. × 6 in. rectangular duct has an aspect ratio of two since $\frac{12}{6}$ equals 2, while the aspect ratio for a 12 in. × 12 in. square duct would be 1. Ducts with aspect ratios greater than 4 are costly to fabricate, will experience a great deal of friction, and should be avoided whenever possible.

Duct sizing involves the use of charts to determine air flows, acceptable velocities, and system friction. One approach would be to have a separate chart for 6 in. × 8 in. rectangular ducts, another chart for 6 in. × 10 in. ducts, another for 6 in. × 12 in. ducts, and so on. Such an approach is, of course, plausible, but the range of possibilities is somewhat overwhelming. Therefore, the "circular equivalent method" has been developed to simplify duct sizing to a method which uses one figure and one table. Then, once you understand the basic procedures you will be able to use the "ductulator" design device (which is widely used in the industry) or to confidently access computer programs.

Circular Equivalent Method

Using the circular equivalent method, one first determines the diameter of round duct which will satisfy a given air flow requirement at an acceptable velocity. A table is then consulted for the range of rectangular duct sizes and shapes which will produce the same air flow at the same frictional losses. Selections can then be made to clear architectural barriers and structural members.

Figure 9.12 is used to size low volume,

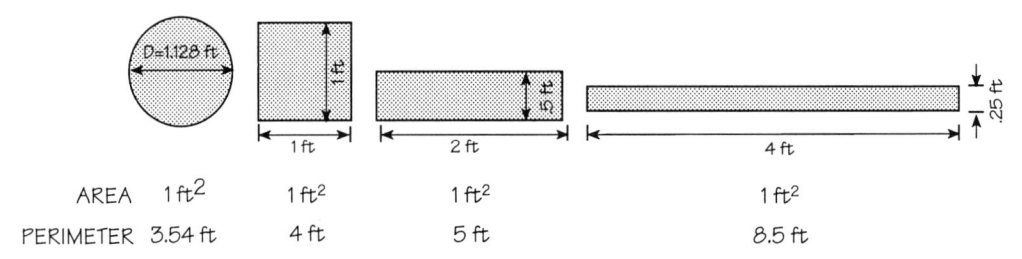

	D=1.128 ft	1 ft × 1 ft	2 ft × .5 ft	4 ft × .25 ft
AREA	1 ft²	1 ft²	1 ft²	1 ft²
PERIMETER	3.54 ft	4 ft	5 ft	8.5 ft

FIGURE 9.11 Perimeter for ducts of various shapes.

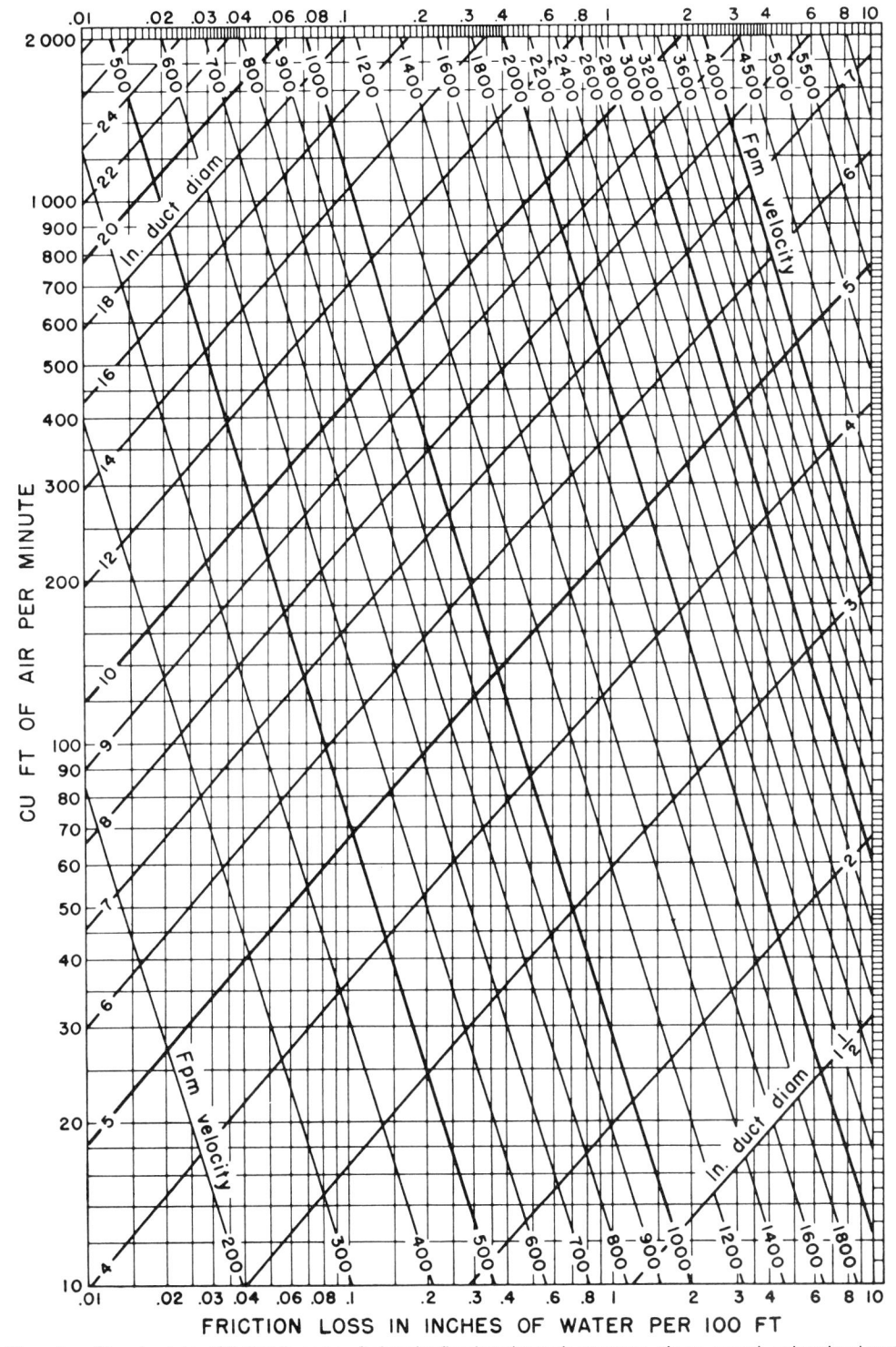

(Based on Standard Air of 0.075 lb per cu ft density flowing through average, clean, round, galvanized metal ducts having approximately 40 joints per 100 ft.) Caution: Do not extrapolate below chart.

FIGURE 9.12 Friction in round (straight) ducts. Reprinted with permission from the *1957 ASHRAE Handbook* and the *Manual for the Balancing and Adjustment of Air Distribution Systems* by the Sheet Metal and Air Conditioning Contractors Association, Inc.

straight round ductwork. It shows the relationship of air velocity, air flow, and friction loss for round ducts from 1.5–24 in. duct diameter. (Note: the figure is based upon standard air flowing through clean, round galvanized metal ducts having approximately 40 joints per 100 ft.)

Friction is very pronounced in small ducts due to the large perimeter to duct cross-sectional area ratio. For example, for a flow of 100 cfm, the frictional loss through a 6-in. diameter round duct is approximately 4 times as great as it would be if a similar flow were through an 8-in. diameter duct.

Example 9C: Determine the air flow rate through a 5-in. diameter round duct at a velocity of 500 fpm. Also determine the friction loss.

Solution: Consult Figure 9.12. Match up the diagonal 5-in. round duct line with the diagonal line in the other direction representing an air velocity of 500 fpm. From this intersection move horizontally to the left axis where the air flow is found to be 70 cfm. Now return to the applicable intersection. Move vertically to the horizontal axis where the friction loss is found to be 0.1 in. of water per 100 ft (of equivalent duct length).

A fairly complex formula (not reproduced) establishes the diameter of a circular duct which will yield the same friction and air flow capacity equivalents of rectangular ducts. Table 9.1 provides these circular equivalents for rectangular ducts with dimensions of A × B. Note that the values near the top of the table for small A dimensions are used primarily for residential ductwork which are run in exterior walls and interior partitions.

Table 9.1 and Figure 9.12 are used in concert to determine system friction loss of

TABLE 9.1 CIRCULAR DUCT EQUIVALENT DIAMETERS OF RECTANGUALR AND SQUARE DUCTS FOR EQUAL FRICTION AND CAPACITY

Dimension A of Rectangular Duct (in.)	Dimenson B of Rectangular Duct (in.)						
	6	8	10	12	14	16	18	20
3	4.6	5.2	5.7	6.2	6.6	7.0		
3½	5.0	5.7	6.3	6.8	7.2	7.7		
4	5.3	6.1	6.7	7.3	7.8	8.3		
4½	5.7	6.5	7.2	7.8	8.4	8.8		
5	6.0	6.9	7.6	8.3	8.9	9.4		
5½	6.3	7.2	8.0	8.7	9.3	9.9		
6	6.6							
8	7.6	8.7						
10	8.4	9.8	10.9					
12	9.1	10.7	12.0	13.1				
14	9.8	11.4	12.9	14.2	15.3			
16	10.4	12.2	13.7	15.1	16.4	17.5		
18	11.0	12.9	14.5	16.0	17.3	18.5	19.7	
20	11.5	13.5	15.2	16.8	18.2	19.5	20.7	21.9
30	13.6	16.1	18.3	20.2	22.0	23.7	25.2	26.6
40	15.3	18.2	20.7	22.9	25.0	27.0	28.8	30.5
50	16.8	19.9	22.7	25.2	27.6	29.8	31.8	33.7
60	18.1	21.5	24.5	27.3	29.8	32.2	34.4	36.5

Source: Excerpted with permission from the *1989 ASHRAE Handbook of Fundamentals*, Chapter 32, Table 3.

rectangular ducts. First, the circular equivalent of rectangular duct dimensions is determined from Table 9.1. Once this is known, friction loss per 100 equivalent feet of straight duct can be determined by consulting Figure 9.12.

Example 9D: Determine the friction loss of a 50-ft length of an 18 in. × 12 in. rectangular duct with air having a velocity of 800 fpm.

Solution: First use Table 9.1 to find that an 18 in. × 12 in. rectangular duct has equal friction and capacity to a 16-in. diameter circular duct. Then see in Figure 9.12 that a 16-in. diameter duct with air flowing at 800 fpm will experience a friction loss of 0.06 in. of water per 100 ft of length.

Therefore, the 50-ft length of duct in this example will experience half of that (0.06 × 0.5) or 0.03 in. of water. Calculations of this type are required to determine system pressure losses in order to provide fans which are of adequate capacity to overcome them.

9.5 OTHER FRICTION LOSSES

The previous very simple example computed friction loss for a straight length of duct. In actual systems, friction from elbows, transitions, cooling coils, and filters also contribute to total friction loss.

Air does not easily change direction, and duct systems should be designed to permit flows as straight as possible to minimize friction loss. Turns in duct systems can be made in various ways and with differing frictional results. Table 9.2 provides values for "additional equivalent length of straight duct in feet" for three types of round elbows which have a geometry where the radius of turn (R) is 1.5 times the duct diameter (D).

Example 9E: Determine the equivalent length of 8-in. diameter round ductwork

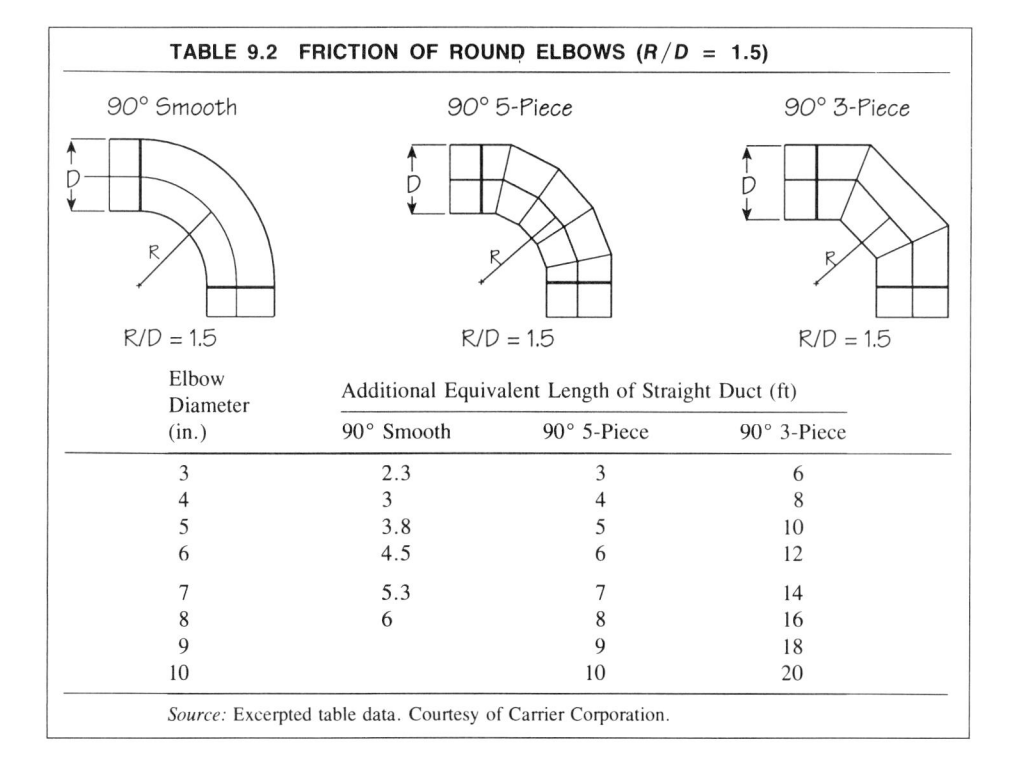

TABLE 9.2 FRICTION OF ROUND ELBOWS (R/D = 1.5)

Elbow Diameter (in.)	Additional Equivalent Length of Straight Duct (ft)		
	90° Smooth	90° 5-Piece	90° 3-Piece
3	2.3	3	6
4	3	4	8
5	3.8	5	10
6	4.5	6	12
7	5.3	7	14
8	6	8	16
9		9	18
10		10	20

Source: Excerpted table data. Courtesy of Carrier Corporation.

148 AIR, DUCTS, AND FANS

shown in Figure 9.13. The turns will be made with 3-piece elbows which turn with a geometry such that the R/D ratio is equal to 1.5.

Solution: Consult Table 9.2. Note that each 8-in. 3-piece elbow will add the equivalent of 16 ft of straight ductwork. The equivalent length of the ductwork section shown in Figure 9.13 can now be determined as follows:

Duct Section	Length or Equivalent Length
Section 1	4 ft
Elbow A	16 equivalent ft
Section 2	6 ft
Elbow B	16 equivalent ft
Section 3	6 ft
Elbow C	16 equivalent ft
Section 4	4 ft
Elbow D	16 equivalent ft
Section 5	4 ft
Total:	88 equivalent ft

In this example the actual length of straight ductwork of 24 ft provides only about 27% ($\frac{24}{88}$) of the frictional resistance provided within the section of ductwork. It should also be realized that frictional resistance could be reduced substantially to a total of only 48 equivalent ft if smooth radius elbows were used. The total pressure loss in inches of water gauge could then be determined by using Figure 9.12 if the air velocity or flow rate (in cfm) is known.

The design and geometry of duct systems can of course be much more complex and develop substantial frictional losses. Ducts in commercial buildings can have reverse bends, changes in size, dips below obstructions, and branch splits all within a matter of feet. Chapter 32 of the *1993 ASHRAE Handbook of Fundamentals* contains many pages of new and revised "Fitting Loss Coefficients" for numerous duct conditions. Finally, once you understand the basic methodology being used to compute friction loss, be sure to use some computer software for complex systems before the tortuous procedures get the best of you.

9.6 DUCT SIZING METHODS

Duct sizing is usually accomplished by one of three methods. For residential and other small systems, the "equal friction method" uses as its starting point the selection of

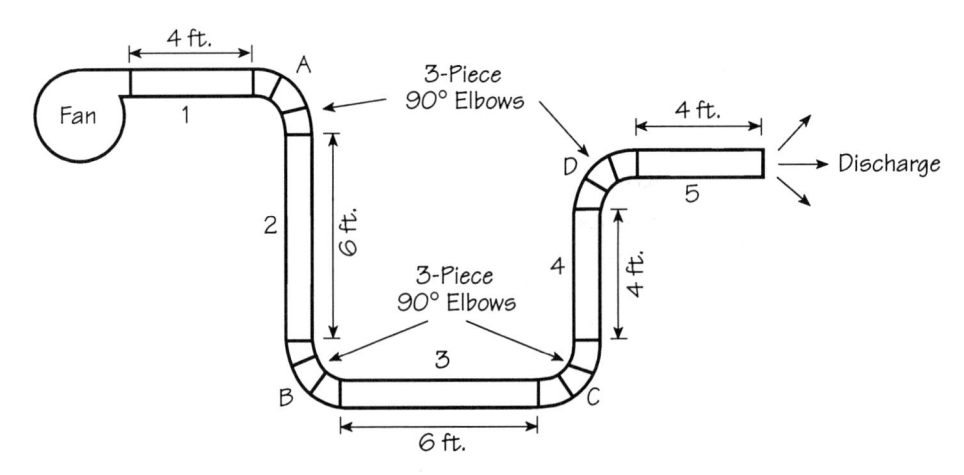

FIGURE 9.13 Friction loss in section of ductwork.

friction loss per 100 "equivalent" ft of duct. The term equivalent is used so that the friction loss that would occur for an actual length of straight ductwork can be adjusted for the additional friction loss which occurs in fittings (transitions, elbows, etc.). Using this method, a fitting of a given size and construction will be said to produce a friction loss equivalent to some length of straight duct in feet.

A friction loss of 0.1 in. of water per 100 equivalent ft of ductwork is commonly used for design purposes. Once a friction loss value has been selected, and required air flow in cfm is known, then air velocities and duct sizes can be determined.

The equal friction method is applied by first identifying the duct run which will experience the greatest friction loss (usually the longest duct run). The other duct runs are then constructed to increase their friction loss to approximately the same level through the use of "balancing dampers." In such a manner, the air flows to all outlets can be adjusted to satisfy the design requirements.

Other Methods

Other duct sizing methods are sometimes used in the design of larger buildings:

1. The "static regain method" is primarily used to design systems where air velocities are over 2000 fpm to prevent oversizing fan motors. This method accounts for the fact that velocity pressure can be converted into static pressure, and that a regain of static pressure can take place by properly sizing ductwork after a branch takeoff.
2. The "velocity reduction method" is applied by selecting a starting fan discharge velocity in feet per minute and then estimating reductions in velocity which are likely to occur downstream. It is a method that requires a great deal of experience and even then should only be used on fairly simple and predictable duct distribution systems.

9.7 HVAC FANS

The purpose of a fan in an HVAC system is to propel air at an adequate velocity (after overcoming frictional resistance) to provide the required quantity of air for ventilation and/or to heat or cool. HVAC fans generally fall into two categories—axial and centrifugal.

Axial Fans

With axial fans, air flow is in a straight line into and through the fan. Various types of axial fans are shown in Figure 9.14 with direct connection to drive motors. Belt drives can also be used and are very common on simple propeller fans. A propeller fan is a type of axial fan which is generally used to propel air without an enclosure in open air. This is known as a "free field" or "free delivery" operation. When a propeller-type fan blade (impeller) is placed within a ducted system, it is known as a tubeaxial or vaneaxial fan.

Propeller Fans

With propeller fans, air is pushed by a rotating impeller having two or more blades. The housing can be a circular ring, orifice plate, or venturi. Propeller fan performance is best when the housing is close to the blade tips. The maximum efficiency of propeller fans is reached near free delivery.

Applications: Propeller fans are used for low-pressure, high-volume applications such as air circulation within a space or unducted ventilation through a wall or roof.

150 AIR, DUCTS, AND FANS

a) Propeller Fan

b) Tubeaxial Fan

c) Vaneaxial Fan

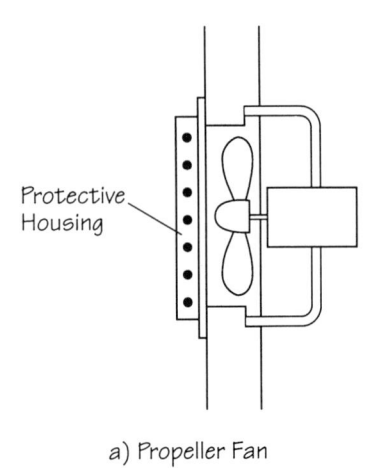

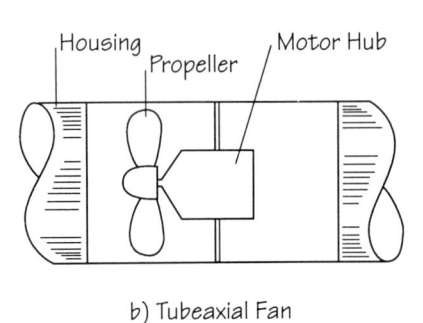

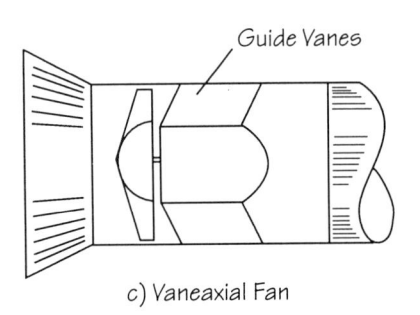

FIGURE 9.14 Types of axial fans.

Tubeaxial Fans

Tubeaxial fans have larger hubs than propeller fans (up to 50% of the fan outlet diameter). They usually have reduced tip clearance, and operate at higher speeds to overcome more pressure resistance than propeller fans. The fan wheel usually has from four to eight blades which can be of flat cross section or an air-foil design. Under a moderately low flow rate, tubeaxial (and vaneaxial) fans perform poorly due to aerodynamic stall. Neither type should be operated under those conditions.

Applications: Tubeaxial fans are used in low- and medium-pressure systems where downstream air distribution is not critical, such as in industrial exhaust systems (drying ovens, paint spray booths, fume hood exhausts, etc.).

Vaneaxial Fans

Vaneaxial fans are a special type of axial fan similar to tubeaxial, but with fixed guide vanes either mounted on the hub behind the blades or upstream in the housing. Large hubs are usually more than 50% of the fan outlet diameter. They are more compact and expensive than tubeaxial fans.

Vaneaxial fans can overcome additional pressure with greater efficiency. They operate quietly and usually have 12 or more adjustable airfoil-type blades. Exercise caution in selecting these fans (see tubeaxial discussion above), as they also operate poorly at moderately low flow rates.

Applications: Vaneaxial fans are used for low-, medium-, and high-pressure systems where straight flow (yielding good downstream air distribution) is required.

Centrifugal Fans

In centrifugal fans, air enters from one or both sides at the center (the "eye") and is propelled outward to the scroll-shaped housing where it discharges through a rectangular outlet. Centrifugal fans are driven by belt drives and direct connections to motors. A belt-driven centrifugal fan and a variety of blade types used for various HVAC applications are shown in Figure 9.15.

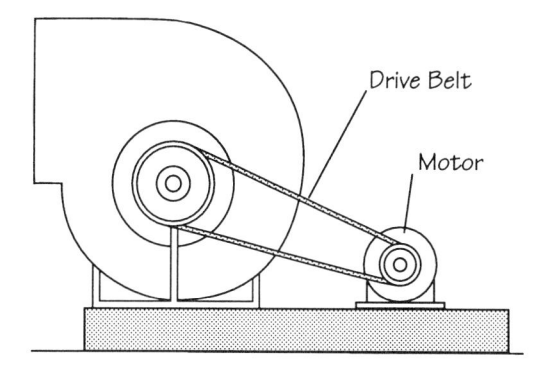

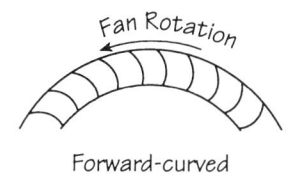

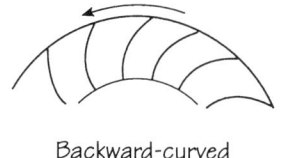

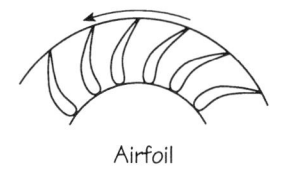

FIGURE 9.15 Centrifugal fan blade types.

Forward-Curved Fans

Forward-curved fans are lightweight and low-cost in construction. The impeller has 24–64 shallow blades with both the tip and heel curved forward. Air is propelled at a speed greater than the fan tip speed. Forward-curved fan wheels are the smallest of all centrifugal type fans. They run slower than backward-inclined fan wheels since they actually push the air rather than inducing it through the housing.

Applications: Forward-curved fans are used in low-pressure (less than 5 in. wg), small capacity systems such as residential warm air furnaces, room air conditioners, and small packaged and rooftop equipment.

Backward-Curved Fans

Backward-curved fans typically have impellers with 10–16 blades curved backward (or inclined) from the direction of fan rotation. They are only slightly less efficient than forward-curved airfoil fans, and are mounted in similar scroll-type housing. They operate more quietly than forward-curved fans and are inherently nonoverloading of motors.

Applications: Backward-curved fans are used on large low-, medium-, or high-pressure systems where high efficiency is desired.

Airfoil Fans

Airfoil fans have the highest efficiency of all centrifugal type fans. The impeller typically has 10–16 airfoil-shaped blades curved away from the direction of fan rotation. The impeller is housed within a scroll designed to efficiently convert velocity pressure to static pressure. Construction tolerances between wheel and air inlet must be very close to achieve maximum efficiency.

Applications: Air-foil fans are used for applications similar to those of backward-curved fans where operational efficiency is paramount and the additional cost is acceptable.

9.8 FAN PERFORMANCE

Fan performance is illustrated on "fan curves" which plot their ability to deliver air on the horizontal axis versus static pressure (in inches of water gauge) on the vertical axis. When fans operate without any static pressure, this condition is known as "free delivery." "Shutoff static pressure" refers to a condition when friction totally prevents the flow of air. Shown in Figure 9.16 are typical system curves for forward- and backward-curved fan blades.

Stable Operation

Fan operation is referred to as "stable" if a slight change in static pressure results in only a minor change in air flow. Fan instability (known as "surging") is manifested in pulsations which may occur when a forward-curved fan is operated at a small percentage of its free delivery flow rate, or, even more likely, when two or more forward-curved fans are connected in parallel.

Fans should be selected so that their point of operation on the fan curve results in a high static efficiency and stable operation. In most cases, selection will be based on operation which will deliver approximately 60% of the free delivery capacity of the fan. However, the exact performance of a given fan and its operational limitations should always be obtained from the fan manufacturer.

Volume Control

For proper system operation, the fan capacity must be matched to the system pressure drop characteristics. This is commonly accomplished by:

- Variable frequency drive (VFD) controls which use either alternating current (AC) or direct current (DC) motors. This type of control is common for variable air volume (VAV) systems.
- Dampers on forward-curved fans (either inlet vane or discharge). In reducing flow, this approach saves a moderate amount of power by operating the motor at a higher efficiency. When the static pressure is increased, the air volume is decreased and the work of the motor is also decreased.
- Variable pitch motor sheaves to change the fan rpm. Significant power savings result, in accordance with the fan laws, as described below.

Brake Horsepower (bhp)

Brake horsepower (bhp) is the total horsepower required by the fan motor to overcome turbulence and fan construction friction (including bearing losses, belt losses, duct system friction losses) plus the horsepower consumed to actually move the air.

9.9 FAN LAWS

Once the specific operating conditions of a particular fan are known, other operating conditions can be computed based upon application of mathematical "laws" which relate the following terms:

cfm   is the air flow rate in cubic feet per minute (cfm)
P is static pressure in inches of water gauge (in. wg)
D is the diameter of the fan wheel in inches
N is fan speed in revolutions per minute (rpm)
bhp   is the brake horsepower required by the fan motor

There are various fan laws which relate the above terms, three of which pertain to fan speed, and are of primary importance to HVAC design.

Fan laws are basically ratios. They are used to compare four terms at a time (such as two each for cfm and N). When three of the four values are known, the fourth and final value can be determined.

Fan Law 1. The air flow rate in cfm is

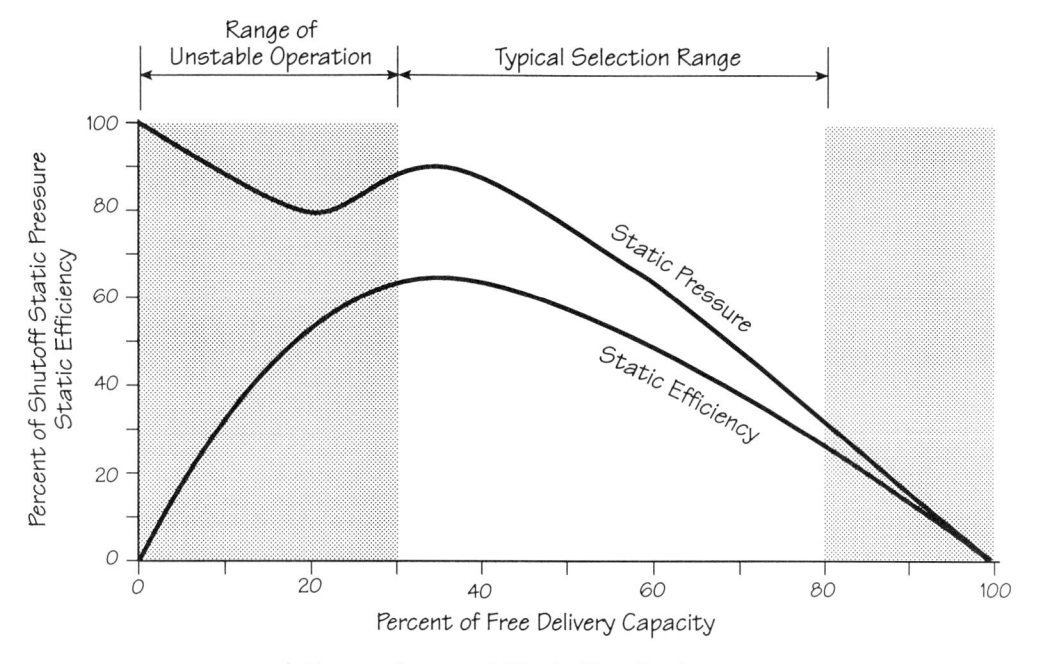

a) Forward-curved Blade Fan Performance

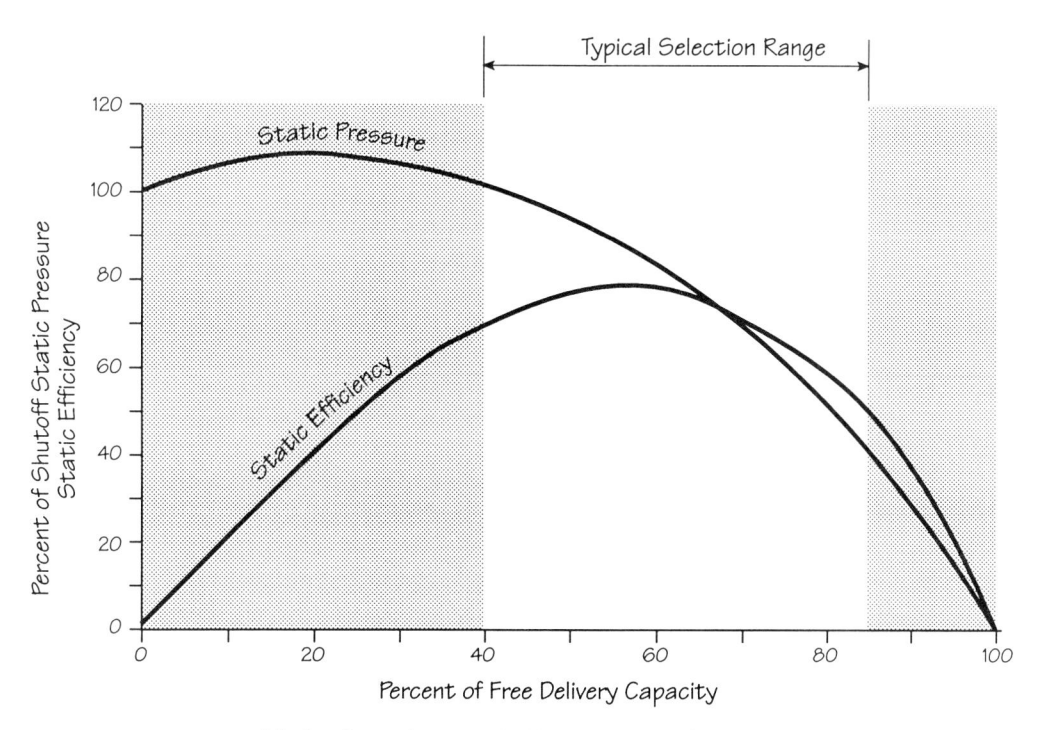

b) Backward-curved Blade Fan Performance

FIGURE 9.16 Typical system curves.

154   AIR, DUCTS, AND FANS

directly proportional to fan speed (N) as expressed in Formula 9B. Thus if fan speed doubles, the air flow rate also doubles. A similar directly proportional relationship also exists for air flow rate and the diameter (D) of the fan wheel.

$$\frac{cfm_1}{cfm_2} = \frac{N_1}{N_2} \qquad \text{Formula 9B}$$

Depending upon which term needs to be found, the fan laws are often reworked. For instance Fan Law 1 can also be:

$$cfm_1 = \frac{N_1}{N_2} \times cfm_2$$

or

$$cfm_2 = \frac{N_2}{N_1} \times cfm_1$$

or

$$N_1 = \frac{cfm_1}{cfm_2} \times N_2$$

or

$$N_2 = \frac{cfm_2}{cfm_1} \times N_1$$

or

$$\frac{cfm_1 \times N_2}{cfm_2 \times N_1} = 1$$

Fan Law 2. Static pressure (P) varies directly with the square of the air flow (cfm) as expressed in Formula 9C. Similar relationships also hold that friction head varies directly with the square of fan speed (N) as well as the square of the fan wheel diameter (D).

$$\frac{P_1}{P_2} = \left(\frac{cfm_1}{cfm_2}\right)^2 \qquad \text{Formula 9C}$$

Fan Law 3. Brake horsepower (bhp) varies directly with the cube of the air flow (cfm) as expressed by Formula 9D. Similar relationships also exist between bhp with the cubes of fan speed (N) as well as the fan wheel diameter (D).

$$\frac{bhp_1}{bhp_2} = \left(\frac{cfm_1}{cfm_2}\right)^3 \qquad \text{Formula 9D}$$

The following detailed example will illustrate how the fan laws are applied in fan selection.

Example 9F: Assume fan operational conditions as follows:

Air flow rate (cfm_1): 6000 cfm
Static pressure (P_1): 4 in. wg
Fan speed (N_1): 2000 rpm
A brake horsepower (bhp_1): 6

Determine the new air flow rate (cfm_2), static pressure (P_2), and brake horsepower (bhp_2) if the fan speed (N_2) is revised to 1500 rpm.

Solution: First, use Fan Law 1 (Formula 9B) to determine the flow rate (cfm_2) at the slower fan speed.

$$cfm_2 = cfm_1 \times \frac{N_2}{N_1} \qquad \text{(Formula 9B)}$$

$$= 6000 \times \frac{1500}{2000}$$

$$= 4500 \text{ cfm}$$

Then, Fan Law 2 (Formula 9C) is used to determine the static pressure at the slower fan speed.

$$P_2 = P_1 \times \left(\frac{N_2}{N_1}\right)^2 \qquad \text{(Formula 9C)}$$

$$= 4 \times \left(\frac{1500}{2000}\right)^2$$

$$= 2.25 \text{ in. wg}$$

Finally, the required horsepower of the motor (bhp_1) to drive the fan at the reduced cfm can be found by using Fan Law 3 (Formula 9D).

$$\text{bhp}_2 = \text{bhp}_1 \times \left(\frac{\text{cfm}_2}{\text{cfm}_1}\right)^3$$

(Formula 9D)

$$= 6 \times \left(\frac{1500}{2000}\right)^3$$
$$= 6 \times 0.422$$
$$= 2.53$$

Therefore, a 3-horsepower motor is required.

Example 9F has shown through the application of Fan Law 3 how relatively small reductions in air flow rates in cfm can substantially reduce motor and energy requirements. As a result, architectural and HVAC design which reduces loads and required flows can be energy efficient while reducing operational costs.

10

WATER, PIPES, AND PUMPS

Water is used in many HVAC systems to distribute or store energy for heating or cooling purposes. As such, a thorough understanding of the properties and pressures of water is required.

Pipes are used in HVAC systems to distribute and deliver hot water, chilled water, condenser water, condensate, and boiler feedwater. To circulate the water, pumps are required. A hot water heating system is considered "closed" since the job of the pump is to merely overcome friction encountered in the piping system; the height of water column within the piping system going up is balanced by the height of water column in the piping coming down. In contrast, in an "open" system the pump must be sized to also handle the unbalanced height of water column (such as the water to and from a cooling tower).

Piping systems impose a frictional resistance to flow based upon pipe size, length, valves, the type and number of elbows and other fittings, and the speed of the fluid carried within. All of these factors need to be considered in determining the total head (frictional resistance) which must be overcome by a pump in terms of feet of water.

Pump performance is depicted by "pump curves" which indicate the capacity in gallons per minute (gpm) depending on the total head encountered. A series of "pump laws" (also known as affinity pump laws) relate the factors of flow rate, pressure head, and other factors which must be considered in selecting a pump. For example, one pump law states that motor horsepower varies by the cube of a rate change in flow rate.

10.1 WATER PROPERTIES AND PRESSURE

Water is much, much denser than air. Humans can withstand the downward pressure of over 5 mi of air in our atmosphere, but the tolerance of our bodies to pressure from water ceases at a depth of only about 150

10.1 WATER PROPERTIES AND PRESSURE

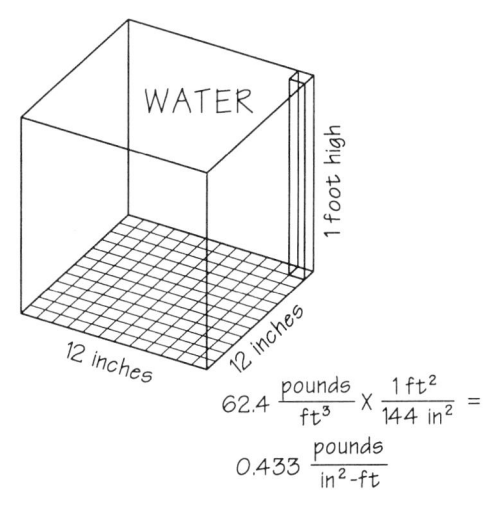

FIGURE 10.1 Water pressure per foot and per inch.

$$62.4 \frac{\text{pounds}}{\text{ft}^3} \times \frac{1 \text{ ft}^2}{144 \text{ in}^2} = 0.433 \frac{\text{pounds}}{\text{in}^2\text{-ft}}$$

Boiling Point Temperature

It is easy to think of 212°F as the boiling point of water. Well, it is—but only at an atmospheric pressure (sea level) of 14.7 psia. Water will also "boil," or change state from a liquid to vapor, at only 150°F if the pressure is subatmospheric at 4 psia, or at the high temperature of 350°F if the pressure is about 140 psia.

Local altitude (and resulting pressure) also affects the temperature at which water will boil. For example, the boiling point of water will be only 202.7°F at an altitude of 5000 ft, such as Denver, where the atmospheric pressure is only 12.2 psi (see Section 9.3). Also see Figure 14.1 for additional information about the relationship of boiling point and pressure for water.

ft. As illustrated in Figure 10.1, water with a density of 62.4 lb/ft³ exerts a pressure of 0.433 psi per foot of height or 0.0361 psi per inch of height. Such a water pressure is 832 times the pressure from air (62.4/0.075).

Inches and Feet of Water

As will be seen in other chapters, the pressure equivalent of water is used extensively. For example, duct systems with a frictional resistance of 0.144 psi are usually referred to instead as having a pressure loss of "4 inches of water," or more fully, 4 inches of water gauge (4 in. wg) as follows:

$$\frac{0.144 \text{ psi}}{0.0361 \text{ psi/in. of water}} = 4 \text{ in. wg}$$

In piping systems that carry water which is much denser than air, the frictional resistances are much larger. As a result, the term commonly used to express pressure head (frictional resistance) in water piping systems is "feet of water."

States of Water

Indicated in Figure 10.2 are the various states of water (at standard atmospheric pressure), including "superheated" steam which is hotter than 212°F.

The importance of understanding the pressure-sensitive nature of liquids and vapors cannot be overstressed. As will be explained in Section 20.2, the vapor compression cycle used by most mechanical air conditioning systems works by having a "refrigerant" change state from a liquid to a gas.

"Ton" of Refrigeration Effect

Note in Figure 10.2 that it takes 144 Btus to change or melt one pound of solid water (ice) to a liquid. This value is the latent heat of fusion of water. A term commonly used in the sizing of air conditioning systems is that a "ton" of air conditioning can remove 12,000 Btu/h of heat. The basis of this term can be seen as follows, with the 2000 lb representing a ton of ice.

$$\frac{2000 \text{ lb} \times 144 \text{ Btu/lb}}{24 \text{ h}} = 12,000 \text{ Btu/h}$$

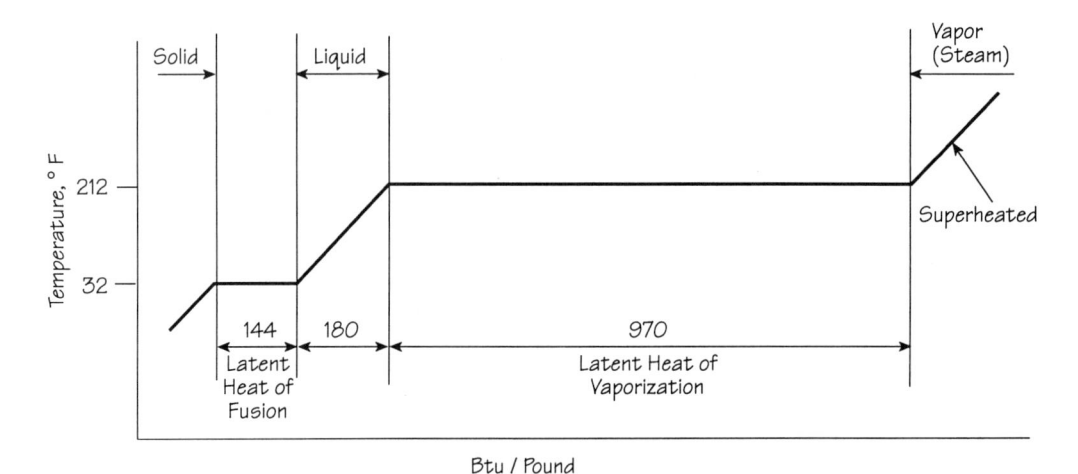

FIGURE 10.2 States of water at sea level.

10.2 USING WATER TO STORE AND DELIVER HEAT

Water has the capacity to deliver (or store) a very large amount of heat per unit volume (3467 times more than air—see Table 12.1). Therefore, it is no wonder that hot water is used in so many HVAC systems.

Water weighs 8.33 lb/gal (at 68°F) and has a specific heat of 1.0 Btu/lb·°F. These values are considered sufficiently accurate for use in low-temperature HVAC calculations. Formula 10A is used to compute the heat required to raise the temperature of a quantity of water.

$$Q = 8.33 \times G \times (t_2 - t_1)$$

Formula 10A

where

Q is the amount of heat required in Btus
8.33 is the weight in pounds of 1 gal of water multiplied by the specific heat of water (1.0 Btu/lb·°F)
G is the number of gallons of water to be heated
t_2 is the water temperature produced in °F
t_1 is the initial water temperature in °F

Example 10A: Consider a tank with 100 gal of water at an initial temperature of 60°F. Determine the quantity of heat required to raise the water temperature in the tank to 130°F.

Solution: Use Formula 10A.

$$Q = 8.33 \times G \times (t_2 - t_1)$$

(Formula 10A)

$$= 8.33 \times 100 \times 70$$

$$= 58,310 \text{ Btus}$$

Such an amount of heat is significant, and could serve a variety of HVAC needs. Also, consider that a 100-gal tank has a volume of about 13.3 ft^3, which could be provided by a 2-ft diameter tank which is 4.25 ft high (about the volume of two 55-gal drums). In comparison, if air were to hold the same quantity of heat (58,310 Btu) with a 70°F temperature increase, the volume would have to be 46,280 ft^3, equivalent to a 60-ft diameter tank over 16 ft high.

Also keep in mind the significant weight of water for its impact on structural design. For example, it will normally be difficult (or impossible) to add rooftop storage tanks on a building not originally designed to support the weight.

10.3 PIPE DATA

Heavy wall Type K copper piping (tubing) is used for underground piping and severe service applications. The most widely used copper piping for HVAC applications, Type L, is of moderate strength. Slightly thinner-wall Type M copper is sometimes used for heating system piping which will not be subject to external damage.

Schedule 40 steel pipe is of moderate wall thickness, and is commonly used in steam heating systems (and sometimes in hot water systems if large pipe sizes are required). Schedule 80 steel pipe has thicker walls, and is generally employed for applications such as high-pressure steam lines.

Table 10.1 provides useful pipe data for copper and steel pipes.

Notice that actual inside diameters are larger than the nominal pipe sizes. For example, 1/2 in. Schedule 40 pipe has an actual inside diameter of 0.622 in. and an actual outside diameter of 0.840 in.

Also see how dramatically the volume of water increases for larger pipe size. For example, 1/2 in. Type L copper tubing can hold only 0.012 gal of water per linear foot whereas the 2 in. Type L copper tubing holds 0.161 gal of water per linear foot, or more than 13 times as much.

Weights per foot of piping when filled with water are used to determine spacing of pipe hangers and sizing of support rods. The National Fire Protection Association (NFPA) recommends that pipe hangers be designed to handle the weight of the pipe filled with water and a man who has fallen off a ladder. Table 10.2 provides rules for typical hanger spacing.

Plastic Pipe

Nonmetallic plastic piping is increasingly being used in plumbing systems and various cold and hot water supply piping applications (where allowed by the Building Code). Advantages of plastic piping include its light weight, corrosion resistance, and ease of joining. Disadvantages include limited use in hot water piping systems and very high thermal expansion (5–20 times as great as steel depending upon plastic type).

The types of plastic pipe most commonly used include:

- PVC (polyvinyl chloride): Commonly used for cold water, drain, waste and vent piping, condensate lines, and some underground piping. Not recommended for hot water piping. PVC is easily joined by socket joints and solvent cementing.
- CPVC (chlorinated polyvinyl chloride): Similar to PVC except it can withstand temperatures up to 140°F.
- PB (polybutylene): A very flexible plastic which can withstand temperatures to 210°F. Can be used for hot and cold plumbing water piping. Not recommended for hot water heating system piping.
- PP (polypropylene): Used for chemical waste lines since it is inert to many chemicals.
- ABS (acrylonitrile butadiene styrene): A high-strength material with good weather and impact resistance. Used for compressed air systems.
- RTRP (reinforced thermosetting resin plastic): The only plastic piping material recommended for use in hot water heating systems (200°F) as well as other less demanding piping applications.

TABLE 10.1 COPPER AND STEEL PIPE DATA

Material	Nom.	Diameter (in.) In.	Diameter (in.) Out.	Inside Cr-Sec. Area (in.^{2})	Outer Circum. (in.)	Weight Per Lin. Ft. of Pipe and Water (lb)	Gal. of Water per Lin. Ft.
Copper:							
Type L	$\frac{3}{8}$	0.430	0.500	0.145	1.57	0.26	0.008
Type L	$\frac{1}{2}$	0.545	0.625	0.233	1.96	0.39	0.012
Type L	$\frac{3}{4}$	0.785	0.875	0.484	2.75	0.67	0.025
Type L	1	1.025	1.125	0.825	3.53	1.01	0.043
Type L	$1\frac{1}{4}$	1.265	1.375	1.257	4.32	1.43	0.065
Type L	$1\frac{1}{2}$	1.505	1.625	1.779	5.10	1.91	0.093
Type L	2	1.985	2.125	3.093	6.67	3.09	0.161
Type L	$2\frac{1}{2}$	2.465	2.625	4.770	8.24	4.55	0.248
Type L	3	2.945	3.125	6.808	9.81	6.29	0.354
Type L	$3\frac{1}{2}$	3.425	3.625	9.214	11.39	8.29	0.479
Type L	4	3.905	4.125	11.971	12.95	10.58	0.622
Steel:							
Sch. 40	$\frac{1}{4}$	0.364	0.540	0.104	1.69	0.475	0.005
Sch. 40	$\frac{1}{2}$	0.622	0.840	0.304	2.65	0.992	0.016
Sch. 40	$\frac{3}{4}$	0.824	1.050	0.533	3.29	1.372	0.028
Sch. 40	1	1.049	1.315	0.864	4.13	2.055	0.045
Sch. 40	$1\frac{1}{4}$	1.380	1.660	1.495	5.21	2.929	0.077
Sch. 40	$1\frac{1}{2}$	1.610	1.900	2.036	5.96	3.602	0.106
Sch. 40	2	2.067	2.375	3.356	7.46	5.114	0.174
Sch. 40	$2\frac{1}{2}$	2.469	2.875	4.788	9.03	7.873	0.248
Sch. 40	3	3.068	3.500	7.393	10.96	10.781	0.383
Sch. 40	$3\frac{1}{2}$	3.548	4.000	9.888	12.56	13.397	0.513
Sch. 40	4	4.026	4.500	12.730	14.13	16.316	0.660
Sch. 80	$\frac{1}{2}$	0.546	0.840	0.234	2.65	1.189	0.012
Sch. 80	$\frac{3}{4}$	0.742	1.050	0.433	3.29	1.686	0.026
Sch. 80	1	0.957	1.315	0.719	4.13	2.483	0.037
Sch. 80	$1\frac{1}{4}$	1.278	1.660	1.283	5.21	3.551	0.067
Sch. 80	$1\frac{1}{2}$	1.500	1.900	1.767	5.96	4.396	0.092
Sch. 80	2	1.939	2.375	2.954	7.46	6.302	0.154
Sch. 80	$2\frac{1}{2}$	2.323	2.875	4.238	9.03	9.491	0.220
Sch. 80	3	2.900	3.500	6.605	10.96	13.122	0.344
Sch. 80	$3\frac{1}{2}$	3.364	4.000	8.890	12.56	16.225	0.458
Sch. 80	4	3.826	4.500	11.496	14.13	19.953	0.597

TABLE 10.2 SUGGESTED HANGER SPACING AND (A36 Steel) ROD SIZE FOR STRAIGHT HORIZONTAL RUNS

Nominal Pipe Size (in.)	Standard (Water)	Steel Pipe[a] (Steam)	Copper Tube (Water)	Rod Size (in.)
$\frac{1}{2}$	7	8	5	$\frac{1}{4}$
$\frac{3}{4}$	7	9	5	$\frac{1}{4}$
1	7	9	6	$\frac{1}{4}$
$1\frac{1}{2}$	9	12	8	$\frac{3}{8}$
2	10	13	8	$\frac{3}{8}$
$2\frac{1}{2}$	11	14	9	$\frac{3}{8}$
3	12	15	10	$\frac{3}{8}$
4	14	17	12	$\frac{1}{2}$

Source: Excerpted with permission from the *1992 ASHRAE Handbook HVAC Systems and Equipment*, Chapter 42, Table 6.

[a] Spacing does not apply where span calculations are made or where concentrated loads are placed between supports such as flanges, valves, and specialties.

10.4 FRICTION IN PIPING

Copper tubing is by far the most common HVAC system piping material for small pipes ($1\frac{1}{2}$ in. or less). Provided in Figure 10.3 is a chart for friction or "head" loss through various types of copper tubing in feet of water per 100 ft of piping equivalent. The chart is based upon circulating water at 60°F, as is the norm for pipe friction charts (so that they can also be used for cold water conditions). In actuality when hotter water circulates it has a lower viscosity (the property of a fluid which resists any force tending to produce flow), and thus will flow with less friction. In spite of that, the difference is slight in low-temperature hot water applications and usually ignored as a design consideration.

One group of diagonal lines on Figure 10.3 indicates a range of tubing (pipe) sizes. The other diagonal lines indicate velocities of water flow in feet per second (fps). Piping for closed piping systems (such as a hot water heating system) should be designed to limit friction to less than 3 ft of water (head loss) per 100 ft and velocity to less than 10 fps. Design velocities in piping for open systems which contain air (such as cold water supply piping) are lower (4–8 fps) to minimize noise.

For a given flow in gpm, various pipe sizes can be used; as pipe sizes get smaller, water velocities increase and friction losses become greater.

Similar charts are available for friction loss in steel and plastic piping (primarily for low-temperature applications). They show that only very slight friction differences occur because of the different piping materials, and the differences are almost unnoticeable in pipes larger than 3 in. in diameter.

Example 10B: Select the Type L copper tubing size which can best meet a required hot-water heating system flow rate of 10 gpm for a small building.

Solution: Figure 10.3 indicates that for a flow of 10 gpm, 1 in. Type L copper tubing would produce an acceptable velocity equal to 4 fps but a relatively high friction loss of about 7 ft of water per 100 ft of equivalent piping length. In contrast, $1\frac{1}{4}$ in. Type L copper tubing would produce a ve-

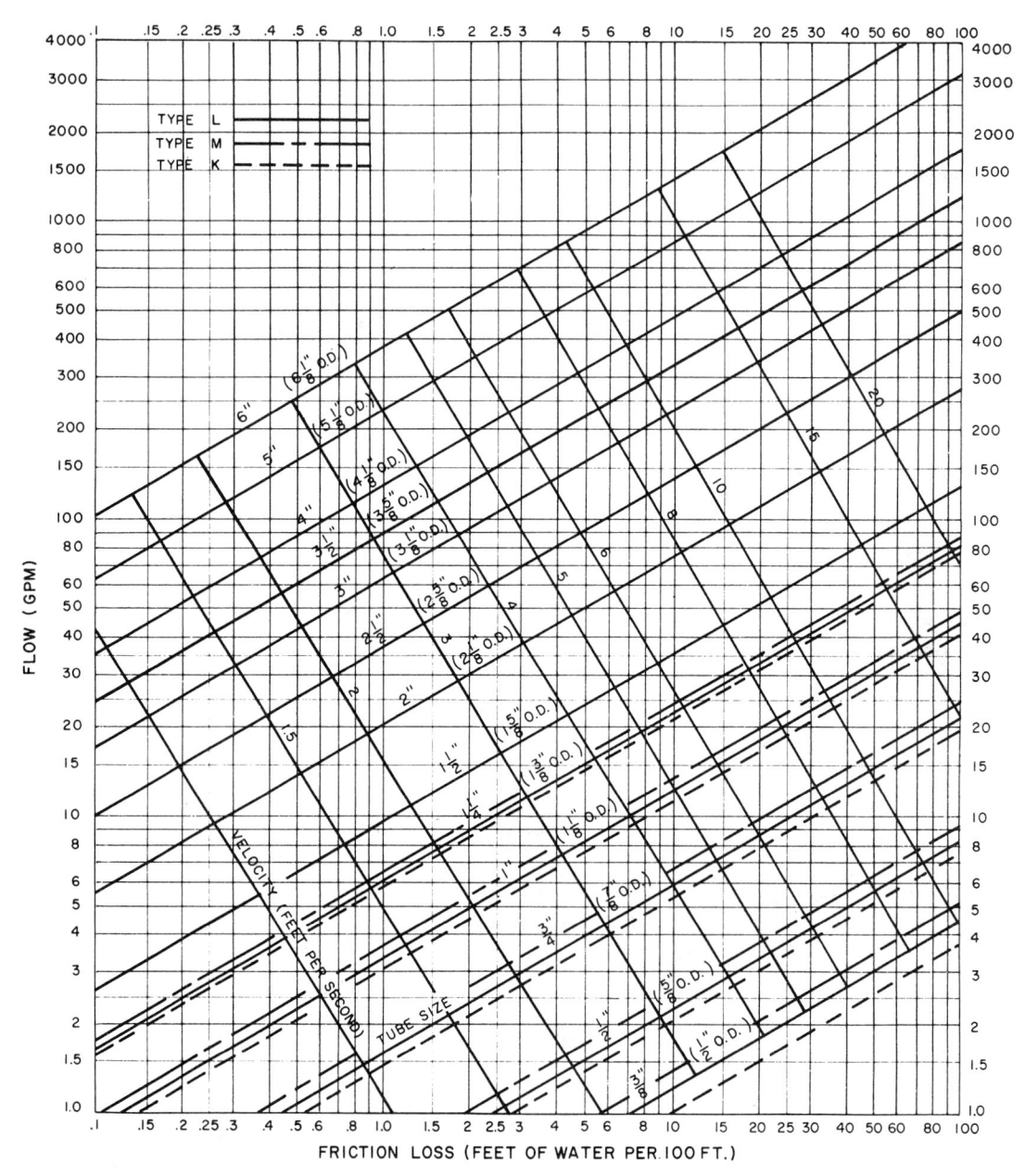

FIGURE 10.3 Copper tubing friction loss. Reprinted courtesy of The Carrier Corporation.

locity of about 2.7 fps, and a reduced friction loss of only about 2.5 ft of water per 100 ft of equivalent piping length.

Pipe sizing always requires a weighing of the importance of various factors including noise, possible pipe erosion in heavily used piping systems with high velocities, and initial and operational costs. For most applications the $1\frac{1}{4}$ in. pipe which requires a smaller pump would be selected.

Friction Loss in Fittings

Friction loss occurs not just in lengths of pipe, but also in elbows and other fittings including valves. Pipe friction in hydronic piping systems typically results in a frictional loss of between 1 and 6 ft of water head loss per 100 ft of "piping equivalent." Equivalent pipe length (EPL) refers to the actual length of the piping plus an additional length to account for friction in fittings (elbows, valves, radiators, etc.). A commonly used design "rule of thumb" is simply to double the actual length of piping to obtain an approximate EPL.

When greater precision is required to determine the EPL, use the ASHRAE "Elbow Equivalent Method" of pipe sizing as summarized in Table 10.3.

Example 10C: Determine the EPL for a 100-ft length of 1 in. hot water heating system supply piping which has twenty (20) 90° elbows in it. The water velocity is 6 fps.

Solution: Table 10.3 indicates that for a water velocity of 6 fps, each 1-in. elbow will impose a frictional resistance equal to 3 ft of piping of that size. Therefore, 20 elbows will impose a frictional resistance equal to 60 ft of piping (20 × 3) and the total equivalent pipe length would be 160 ft (100 + 60).

For more complex systems, determine the equivalent pipe length by consulting the *ASHRAE Handbook of Fundamentals* for complete tables for elbows, couplings, valves, and other friction-producing fittings.

10.5 SYSTEM HEAD LOSS

Once the EPL of a system is determined, it is then possible to compute frictional system head loss (SHL) for the entire piping system including all fittings using Formula 10B.

$$SHL = HL_{100ft} \times EPL \quad \text{Formula 10B}$$

where

SHL   is the system head loss due to friction in feet of water

HL_{100ft} is the head loss in feet of water per 100 equivalent feet of pipe from Figure 10.3

EPL   is the equivalent pipe length of the system in feet

TABLE 10.3 EQUIVALENT LENGTH IN FEET OF PIPE FOR 90° ELBOWS

Fluid Velocity (fps)	Pipe Size (in.)					
	$\frac{1}{2}$	$\frac{3}{4}$	1	$1\frac{1}{4}$	$1\frac{1}{2}$	2
2	1.4	1.9	2.5	3.3	3.9	5.1
4	1.5	2.1	2.8	3.7	4.4	5.6
6	1.7	2.3	3.0	4.0	4.7	6.0
8	1.7	2.4	3.1	4.2	4.9	6.3
10	1.8	2.5	3.2	4.3	5.1	6.5

164 WATER, PIPES, AND PUMPS

Example 10D: Determine the system head loss (SHL) for a hot water heating system which has a water flow rate of 10 gal/min per minute through 1 in. copper tubing. The piping system has an equivalent pipe length (EPL$_{ft}$) of 300 ft (2 × 150 ft actual pipe length).

Solution: From Figure 10.3, a flow of 10 gpm through 1-in. copper pipe will result in a velocity of about 4 fps, and a head loss of about 7 ft/100 ft (HL$_{100ft}$). Use Formula 10B for this closed piping system.

$$SHL = HL_{100ft} \times EPL$$

(Formula 10B)

$$= \frac{7.0 \text{ ft}}{100 \text{ ft}} \times 300 \text{ ft}$$

$$= 21 \text{ ft of water}$$

This resulting system head loss (caused by friction) is one factor used in the selection of a pump as discussed below.

10.6 CENTRIFUGAL PUMPS

Most HVAC applications utilize centrifugal pumps in which the fluid to be pumped (usually water) enters a casing in the center of the pump and is propelled outward by an impeller which rotates like a pinwheel. Most of these centrifugal pumps are the "volute type" (spiral- or scroll-shaped form) which discharges the fluid perpendicularly to the pump shaft as shown in Figure 10.4.

Pumps are classified by their installation arrangement (see Figure 10.5). "Circulators" are low-flow, low-head, fractional horsepower pumps which are mounted directly on and supported "in-line" by the piping. Circulator pumps are used for hydronic heating systems and domestic hot water recirculation. Most large pumps are

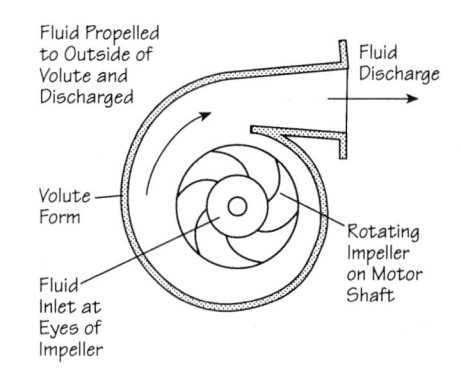

FIGURE 10.4 Volute-type centrifugal pump.

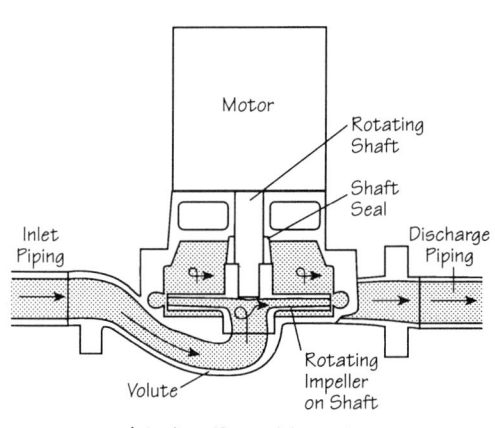

a) In-line Pump Mounting

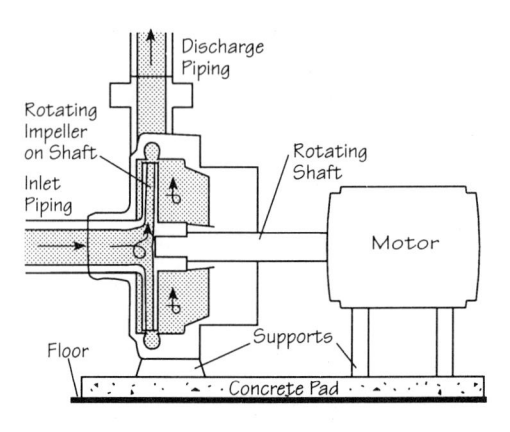

b) Base-mounted Pump

FIGURE 10.5 Pump classifications.

"base-mounted" to properly support the pump and to remove the strain of the pump and motor weight from the piping system.

Other Pump Types

Positive displacement pumps are used for select HVAC applications such as pumping of high viscosity fluids (e.g., fuel oil) from storage tanks. Gear and screw-type pumps are commonly used.

Net Positive Suction Head (NPSH)

NPSH is a term related to pumps in some HVAC and industrial systems with low suction pressures. A problem can exist with centrifugal pumps when operated at a lower pressure in the impeller eye. This pressure difference can result in the flashing (boiling) of some circulating liquid to vaporous bubbles. These bubbles then collapse (known as cavitation) which results in noise, pump impeller corrosion, and ultimately pump failure. NPSH is generally not a concern in closed pump applications. But it is a concern in open systems (e.g., condensate return, boiler feed, and condenser water).

10.7 PUMP CURVES

Throughout this book many items and equipment selection alternatives are presented in tabular form. Such is not the case for pumps since graphic presentation of pump performance is much more revealing. Pump performance is illustrated by "pump curves" which plot their flow rate in gallons per minute (gpm) versus the total head in feet of water. Pump curves are prepared by manufacturers from test data.

Pump curve "A" in Figure 10.6 is usually considered to be too flat since there can be very large changes in flow for relatively small changes in head. For example, a small

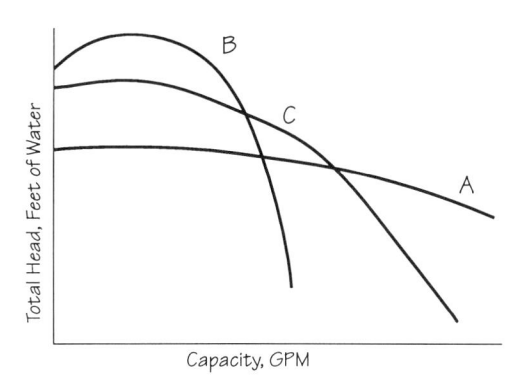

FIGURE 10.6 Pump capacity curves.

change in total head may change the flow rate from 100 to 300 gpm. Such an increased flow may result in unnecessary noise, water use, or be responsible for overloading the pump motor as discussed below.

In contrast, pump curve "B" is very steep, which may be a possible advantage in establishing known flows in a system though this advantage can be overstated. Greater attention should probably be given to flow rates at individual terminal devices rather than the flow rate at the central pump itself. Therefore, pumps with an intermediate curve such as "C" are usually most desirable for HVAC applications. Readers should note that pump curves included herein are illustrative only. Consult manufacturers' published data for actual pump performance information.

10.8 SYSTEM CURVES

Closed Systems

The system curve, as shown in Figure 10.7, illustrates how total head varies as a function of the gpm flow rate. The pump will operate at the point corresponding to the intersection of the pump curve with the system curve.

In Figure 10.7, Point A (the operational

166 WATER, PIPES, AND PUMPS

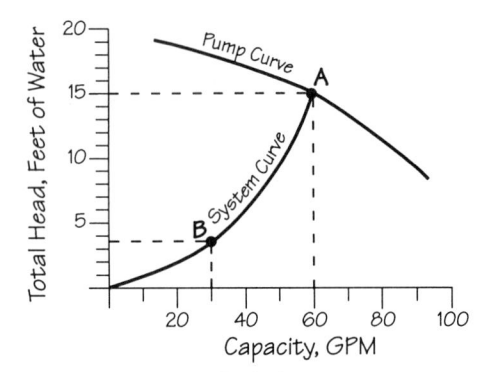

FIGURE 10.7 Typical system curve.

condition on the pump curve) indicates a flow of 60 gpm with a total head of 15 ft. The shape of system curves is governed by the mathematical relationship that total head will change as the square of the change in gpm flow (see Formula 10F). Therefore, Point B which represents one-half of system flow (30 gpm) will correlate to a total head which will be 1/4 of the total head for full flow. This will be equal to 3.75 ft of water (1/4 × 15).

Most pumps perform somewhat differently than their "design" conditions since the worst case friction assumptions are often not realized. Illustrated in Figure 10.8 is the pump curve for a system designed for a flow rate of 80 gpm and a pressure head of 35 ft of water (Point A). In operation, the actual flow rate is 100 gpm with an actual pressure head of 25 ft of water. This condition is represented by Point B at the top of the actual system curve, which must, of course, also be on the pump curve.

Open Systems

Pumps in open systems do more than overcome friction in the pipes and fittings. Illustrated in Figure 10.9 is the open piping system of a cooling tower. As indicated, the static head which a pump must overcome is based upon pump location relative to open system components and resulting static suction and discharge heads. In this case, the static head is 10 ft.

The system curve for an open system must overcome the total head (TH), which is determined by using Formula 10C.

$$TH = SHL + SH \quad \text{Formula 10C}$$

where

TH   is the total head of a piping system in feet of water

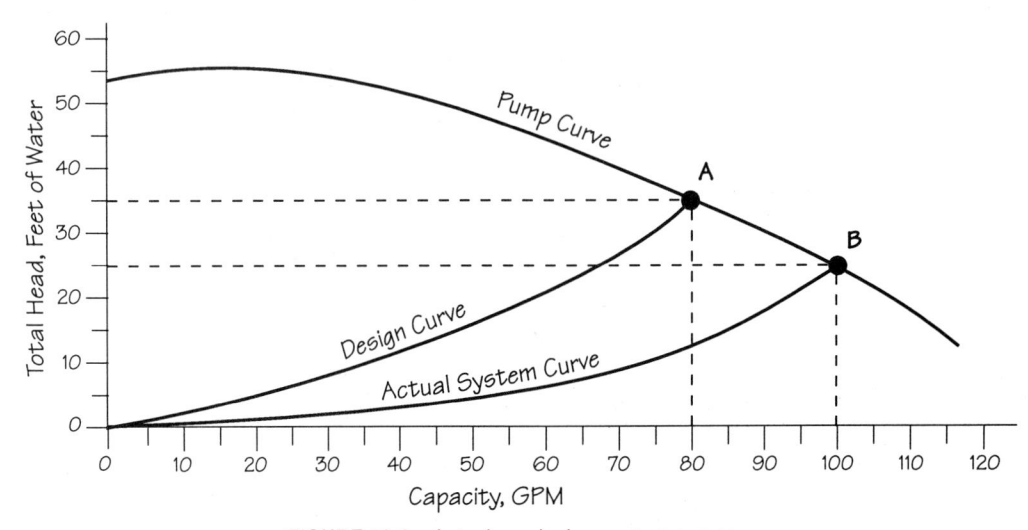

FIGURE 10.8 Actual vs. design system curves.

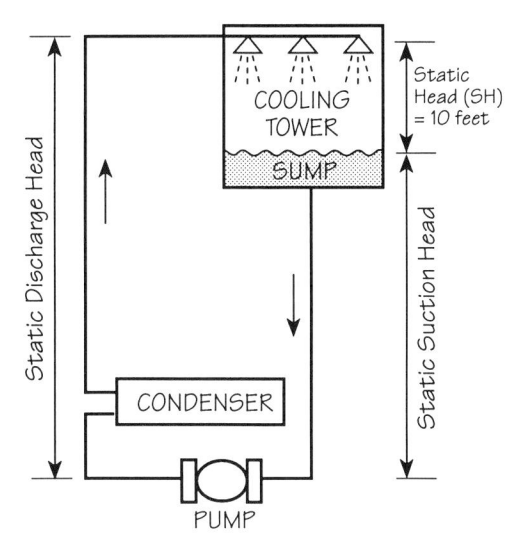

FIGURE 10.9 Cooling tower piping—an "open system."

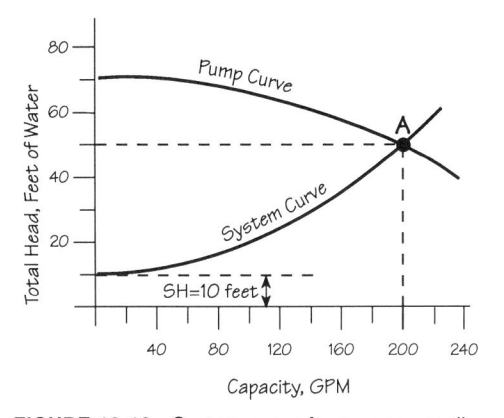

FIGURE 10.10 System curve for an open cooling tower circuit.

SHL   is the system head loss in feet of water (see Formula 10B)
SH   is the static head of an open system in feet of water

Example 10E: The piping for an open cooling tower circuit similar to that depicted in Figure 10.9 has a design system head loss (SHL) of 40 ft of water and a required flow rate of 200 gpm. The static head (SH) between the cooling tower nozzles and the sump is 10 ft. Determine the design operating point for this system.

Solution: First determine the total head (TH) in feet of water by using Formula 10C.

$$TH = SHL + SH$$
(Formula 10C)
$$= 40 + 10$$
$$= 50 \text{ ft of water}$$

Therefore, the design operating point is a flow rate of 200 gpm and a total head of 50 ft of water.

Presented in Figure 10.10, is the system curve for this open system with Point A representing the operating point. The curve begins at 0 gpm flow and a total head of 10 ft (the constant static head). No flow will occur unless the pump develops at least a 10-ft head.

10.9 PUMP PERFORMANCE

Efficiency

Similar to paddling a canoe, the efficiency of a pump is maximized when very little turbulence is produced by the impeller and volute design. Manufacturers' charts show the mechanical efficiency of a particular pump housing based upon impeller size, total head in feet of water, and capacity in gpm as illustrated in Figure 10.11. Note that efficiencies for this pump are indicated by curved lines and range from 45 to 68%.

The various impellers which can be used in a particular pump housing to produce varying performance are sometimes known as a "family." It is normally good practice to select a pump housing and an impeller near the middle of the available range. With such an approach, if operating conditions warrant, a larger or smaller impeller can be easily installed.

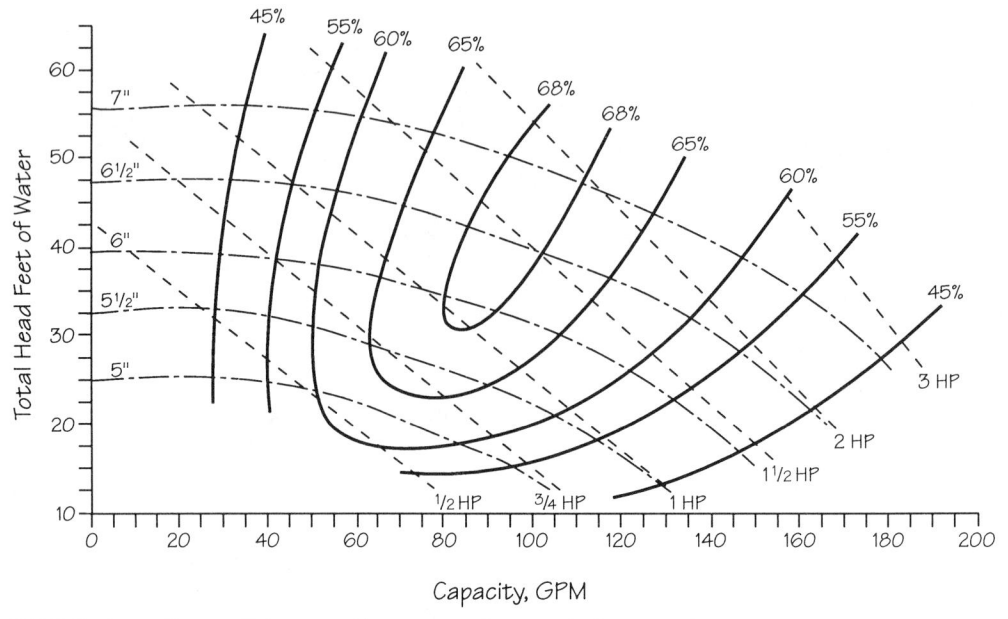

FIGURE 10.11 Pump efficiency and horsepower for a particular pump housing and various diameter impellers.

Pump Motors

Manufacturers' pump diagrams also indicate required motor horsepower to achieve the pump performance shown. The downward sloping dashed lines in Figure 10.11 indicate the brake horsepower (bhp) motor requirement (see below) in standard motor horsepower increments.

As discussed earlier, actual operational pump flow rates can be hard to accurately predict during design. Pump motor selection based upon the estimated design system curve is known as "to the point" selection. There is a danger in selecting pump motors for "to the point" design conditions since such a selection does not allow for a shift in the operating point to the right which may result in an overloaded motor.

Example 10F: The design conditions for a pump are a flow rate of 70 gpm and a total head of 28 ft of water (Point A in Figure 10.12). Select a pump motor for these conditions.

Solution: Point A (the design condition) coincides with the dashed diagonal line representing a 3/4 hp (horsepower) motor which would be selected if "to the point" design is employed. But this motor will overload if the flow increases on the pump curve to the right! Therefore, select the 1-hp motor which will be nonoverloading for the entire pump curve. Although the motor is larger, its operation at partial load will only minimally increase motor power requirements.

Brake Horsepower (bhp)

The brake horsepower is the total power required by the pump to move the water plus bearing losses, pipe and fitting friction losses, and pump construction friction and to overcome turbulence. The brake horsepower (bhp) for a pump is determined by using Formula 10D.

$$\text{bhp} = \frac{\text{gpm} \times \text{SHL}}{3960 \times M_{\text{eff}}} \quad \text{Formula 10D}$$

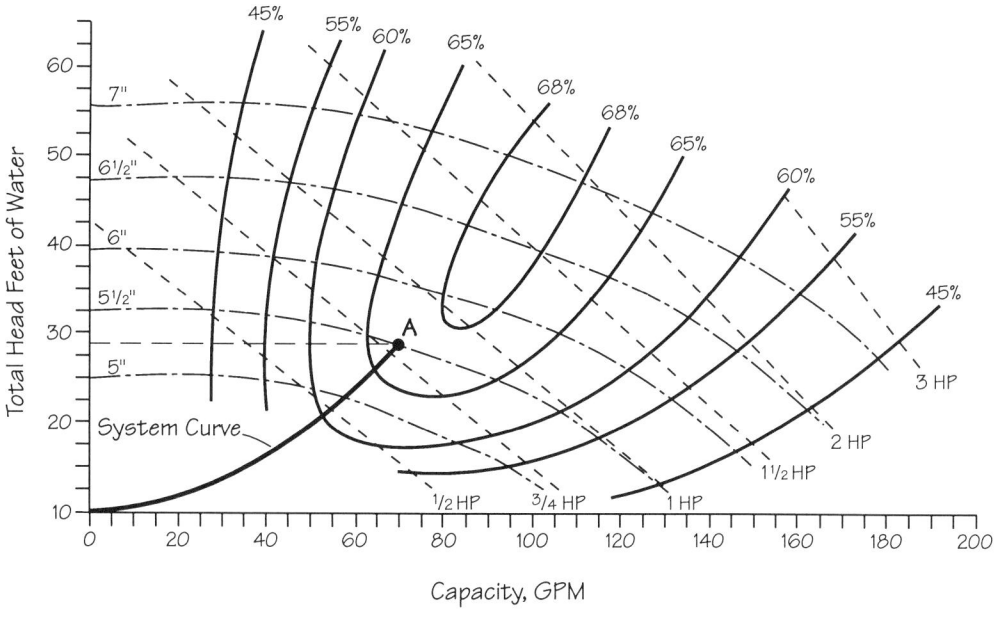

FIGURE 10.12 Pump—Example 10E.

where

bhp   is the required brake horsepower for a pump motor
gpm   is the water flow rate in gallons per minute
SHL   is the system head loss in feet of water
3960 is a value which results from dividing one horsepower (33,000 foot pounds per minute) by the density of water of 8.33 lb/gal
M_{eff} is the mechanical efficiency of the pump

Another term sometimes used is water horsepower (whp) which is equal to the brake horsepower (bhp) multiplied by the mechanical efficiency (M_{eff}).

Example 10G: Determine the brake horsepower of a pump to handle a flow of 100 gpm in a closed system with a system head loss (SHL) of 30 ft of water. Assume a mechanical efficiency (M_{eff}) for the pump of 0.65 (65%). (Note: Efficiency for a particular pump will vary greatly, typically from about 40 to 80%, depending on pump size and actual system flows and head loss.)

Solution: Determine the required bhp using Formula 10D.

$$bhp = \frac{gpm \times SHL}{3960 \times M_{eff}} \quad \text{(Formula 10D)}$$

$$= \frac{100 \times 30}{3960 \times 0.65}$$

$$= 1.2$$

Compare this result with the pump and motor information provided in Figure 10.11. For this example, a $1\frac{1}{2}$ horsepower motor would be used. Note that such a motor would not overload since the $1\frac{1}{2}$ hp line in Figure 10.11 is above the 6 in. impeller pump curve which satisfies the design conditions at all points of the pump curve.

170 WATER, PIPES, AND PUMPS

Pump Speed

Pump motors for most open-system HVAC applications (e.g., cooling towers) have a rated operational speed of 1750 revolutions per minute (rpm), whereas many small circulators in hot water heating systems operate at higher speeds and are typically water-lubricated. Large pumps with a rated operational speed of 3450 rpm transmit more noise into the system and are thus usually limited to condensate return and industrial applications. Two-speed motors are also available with speeds such as 1750/1150 rpm, 1750/850 rpm, and 1150/850 rpm.

Pumps that can operate at variable speed can be used for increased efficiency for a wide range of conditions. Speed can be varied in several ways, including use of variable frequency, direct current, eddy-current drives, and hydraulic drives.

10.10 PUMP LAWS

Once the specific operating conditions of a pump are known, other operating conditions can be computed based upon application of mathematical "laws" which are similar to the fan laws presented in Section 9.9.

The following terms will be used to explain the pump laws:

gpm   is the liquid flow rate in gallons per minute
H is the pump head in feet of water column
D is the diameter of the pump impeller in inches
N is the pump speed in revolutions per minute (rpm)
bhp   is the brake horsepower required by the pump

Pump Law 1. The liquid flow rate in gpm is directly proportional to pump speed (N) as expressed in Formula 10E. Thus if pump speed doubles, the flow rate also doubles.

$$\frac{\text{gpm}_1}{\text{gpm}_2} = \frac{N_1}{N_2} \qquad \text{Formula 10E}$$

A similar directly proportional relationship also exists for gpm flow rate and impeller diameter (D).

Pump Law 2. The total head (H) developed varies directly with the square of the pump speed (N) as expressed in Formula 10F.

$$\frac{H_1}{H_2} = \left(\frac{N_1}{N_2}\right)^2 \qquad \text{Formula 10F}$$

Similar relationships also hold that developed head varies directly with the fluid flow in gpm as well as the square of impeller diameter (D).

Pump Law 3. Brake horsepower (bhp) varies directly with the cube of pump speed (N) as expressed by Formula 10G.

$$\frac{\text{bhp}_1}{\text{bhp}_2} = \left(\frac{N_1}{N_2}\right)^3 \qquad \text{Formula 10G}$$

Similar relationships also exist between bhp with the cubes of gpm fluid flow as well as impeller diameter (D).

Example 10H: The design conditions for a hot water heating system are established as a flow rate of 20 gpm, and a total head of 10 ft of water. Use the pump laws to compute the operating conditions for flow rates which are 75%, 50%, and 25% of the design flow rate, and plot the results on a pump curve diagram.

Solution: Use Formula 10F (Pump Law 2).

10.10 PUMP LAWS

$$\frac{H_1}{H_2} = \left(\frac{\text{gpm}_1}{\text{gpm}_2}\right)^2 \quad \text{(Formula 10F)}$$

$$H_2 = H_1 \times \left(\frac{\text{gpm}_2}{\text{gpm}_1}\right)^2$$

where gpm_1 is the design flow rate of 20 gpm and H_1 is the total head of 10 ft of water. Then gpm_2 is set at 75% of gpm_1 or 15, 50% of gpm_1 or 10, and 25% of gpm_1 or 5 to determine the corresponding total heads in feet of water.

To determine the head at 75% flow rate:

$$H_2 = H_1 \times \left(\frac{\text{gpm}_2}{\text{gpm}_1}\right)^2$$

$$= 10 \times \left(\frac{15}{20}\right)^2$$

$$= 5.6 \text{ ft of water}$$

In a similar manner the total head at the 50% flow rate is determined to be 2.5 ft of water, and the total head at the 25% flow rate is determined to be 0.625 ft of water. These results are plotted as the system curve on Figure 10.13. Point A represents the design point on the pump curve, while Points B, C and D represent 75%, 50% and 25% of the design flow rate, respectively. Lower flow rates achieved by throttling can be used by hot water heating systems when mild outdoor conditions exist.

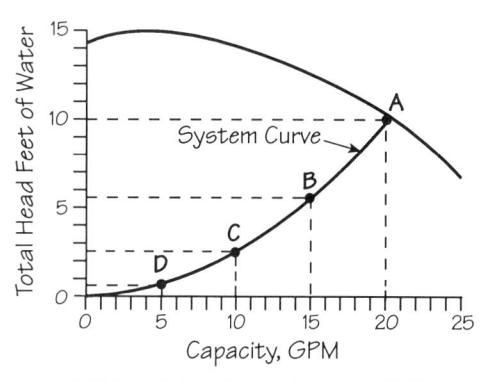

FIGURE 10.13 Pump—Example 10H.

Example 10I: Select an impeller size for the pump housing performance curves shown in Figure 10.14 for a system that has a high flow rate of 300 gpm, but a relatively low total head (30 ft of water). The selected pump will be operated at a speed of 1150 rpm (N_1) rather than the normal rated speed of 1750 rpm.

Solution: The first step is to restate the desired design conditions of 300 gpm, a total head of 30 ft, and a pump speed of 1150 rpm in terms of the 1750 rpm used as the basis of published manufacturer pump curves. Use Pump Law 1 (Formula 10E) to determine the gpm flow rate at the lower speed.

$$\frac{\text{gpm}_1}{\text{gpm}_2} = \frac{N_1}{N_2} \quad \text{(Formula 10E)}$$

$$\text{gpm}_2 = \frac{N_2}{N_1} \times \text{gpm}_1$$

$$= \frac{1750}{1150} \times 300$$

$$= 457 \text{ gpm}$$

Then, Pump Law 2 (Formula 10F) is used to determine the total head (H) at a pump speed of 1750 rpm (which is equivalent to the head of 30 ft at the pump speed of 1150 rpm).

$$\frac{H_1}{H_2} = \left(\frac{N_1}{N_2}\right)^2 \quad \text{(Formula 10F)}$$

$$H_2 = \left(\frac{N_2}{N_1}\right)^2 \times H_1$$

$$= \left(\frac{1750}{1150}\right)^2 \times 30$$

$$= 70$$

172 WATER, PIPES, AND PUMPS

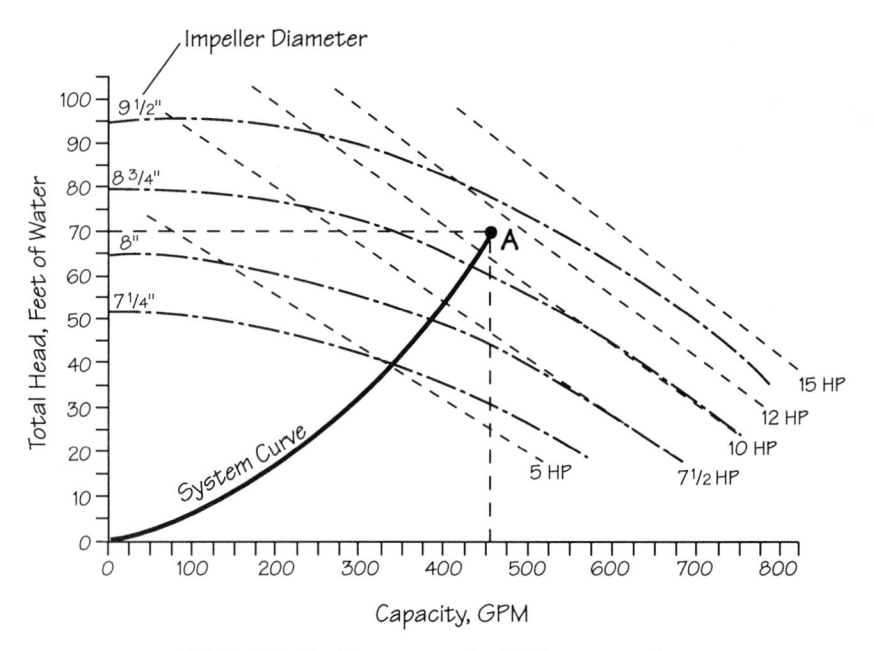

FIGURE 10.14 Pump curves for 1750 rpm operation.

The design condition (Point A) has now been redefined to select a pump which can deliver 457 gpm at a total head of 70 ft of water (at 1750 rpm). Examine this condition on Figure 10.14 which shows a family of 1750 rpm universal pump curves for a given pump housing. The figure indicates that the pump housing will meet the design conditions with an impeller of between $8\frac{3}{4}$ in. and $9\frac{1}{2}$ in. By interpolation a 9.1 in. impeller is required which manufacturers can provide by trimming a $9\frac{1}{2}$ in. impeller on a lathe.

Finally, the required horsepower of the motor (bhp$_1$) to drive the pump at the reduced speed of 1150 rpm (N_1) can be found by using a variation of Pump Law 3 (Formula 10G). By graphic interpolation on Figure 10.14, Point A indicates the need for a motor of approximately 10.8 hp (bhp$_2$) if operation is at 1750 rpm (N_2). Note that the computation is a compound one which also takes into account the trimmed impeller size.

$$\frac{bhp_1}{bhp_2} = \left(\frac{N_1}{N_2}\right)^3 \times \left(\frac{D_1}{D_2}\right)^3$$

(Formula 10G)

$$bhp_1 = bhp_2 \times \left(\frac{N_1}{N_2}\right)^3 \times \left(\frac{D_1}{D_2}\right)^3$$

$$= 10.8 \times \left(\frac{1150}{1750}\right)^3 \times \left(\frac{9.1}{9.5}\right)^3$$

$$= 10.8 \times 0.28 \times 0.88$$

$$= 2.66 \text{ bhp}$$

This result, which suggests use of a 3 horsepower motor, is based on "to the point" computations and use of the pump laws. As discussed earlier, use of a 5-bhp motor may be prudent to provide assurance against possible overloading if the system flow exceeds the design value. For most commercial applications, the difference of a larger size motor is minimal in comparison to the total cost of a job and gives a great deal of security.

11

SELECTION OF HVAC SYSTEMS

The HVAC system selection for houses and a wide range of commercial building types will be covered in this chapter. After digesting this material, you should be ready to choose appropriate systems options for various building types in different locations.

The first step in efficient HVAC design should be to minimize heating and cooling loads. Primary targets in residential-type buildings are improvements to the envelope such as more energy-efficient windows, greater levels of insulation in walls and roofs, and reduced infiltration. Primary targets for most commercial buildings are the use of more efficient lighting systems, improved sun control for glazing, and to a lesser extent envelope improvements.

HVAC system selection and design is influenced by many factors. In most cases the primary factors will be building type, climate, initial cost, and fuel availability. Secondary factors include space requirements, ease of control and zoning, reliability and maintainability, operational cost, response time, noise and vibration, fire safety, and aesthetics.

11.1 PRIMARY SELECTION FACTORS

Often HVAC system selections will depend upon consideration of the primary selection factors indicated in Figure 11.1 and discussed below.

Building Type

Buildings fall into two broad categories: residences (houses, multifamily housing, dormitories, etc.) and nonresidential buildings of various types.

Systems for nonresidential buildings often depend upon floor plan layout, space size, ceiling heights, construction clearances, and exterior wall design (size and type of windows, etc.). Building and space type will also determine whether loads are likely to change quickly (e.g., conference

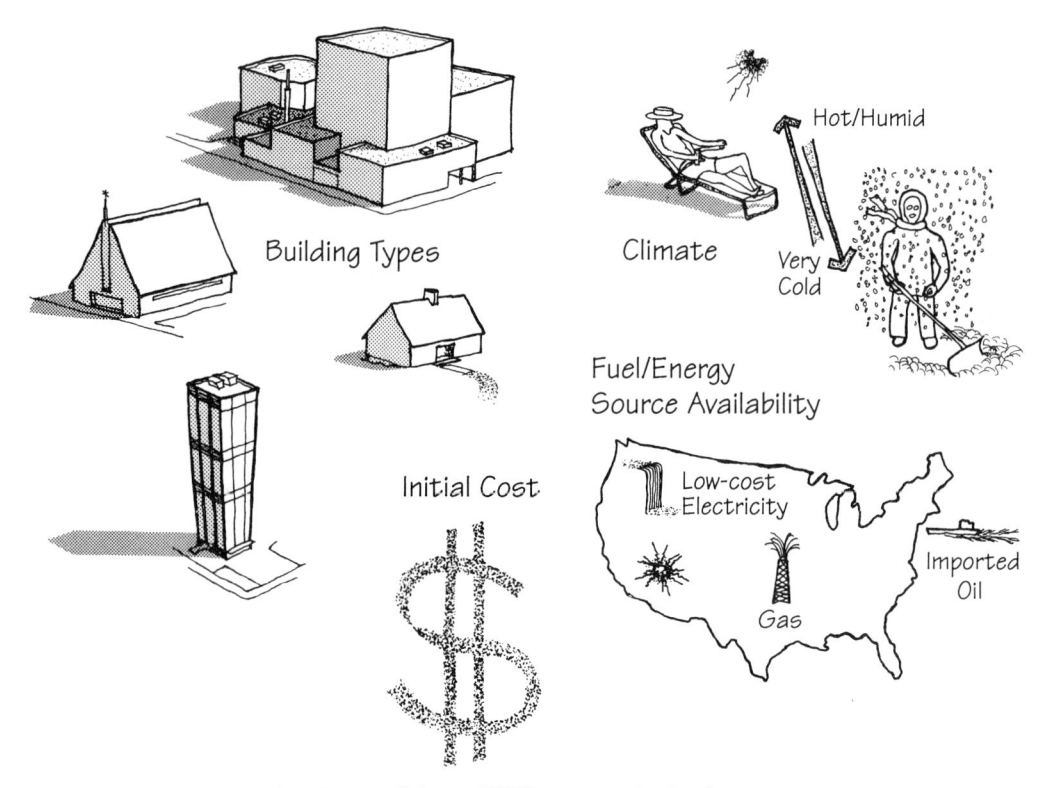

FIGURE 11.1 Primary HVAC system selection factors.

rooms, religious buildings, and airline terminals), or if there is a need for many units with individual control (e.g., hotels, motels, and condominiums).

Climate

Mechanical heating will be required by almost all buildings. The amount required and the annual cost for heat will influence the type and efficiency of heating equipment chosen.

Mechanical cooling is increasingly desired in residences throughout the United States, even those in northern climates, and is usually required in commercial buildings. The temperature and humidity of a local climate will also impact upon the benefits derived from an economizer cycle.

Mechanical ventilation is required in most large buildings to bring outside air to occupants who do not have access to operable windows. Mechanical ventilation is also becoming increasingly popular in energy-efficient homes for improved indoor air quality.

Initial Cost

Most new homeowners are aware of the importance of energy efficiency. This market factor, together with code requirements and utility company promotions, will generally result in the use of equipment which is efficient. Investment in more efficient (and expensive) equipment and improved building envelope components can be analyzed through a cost/benefit "simple payback" calculation (see Section 24.2).

In larger buildings, initial cost often dictates system selections to suit building type as discussed below. Most developers will build to meet, but not exceed, code requirements. In contrast, owners of new buildings

11-18-16

which they plan to occupy for many years will want to explore more efficient systems, even at a higher capital cost.

Fuel/Energy Source Availability

Throughout most of the country, gas (natural or propane) is readily available. This has not always been the case, and may not be in the future. During the 1970s, restrictions on new natural gas hookups resulted in increased use of all-electric HVAC systems in some locations. Use of fuel oil remains popular in the Northeast but is rare elsewhere.

In all but the remotest locations, electricity is readily available. Steam is available from utilities in many "older" cities which can be used for the heating and even cooling of large buildings through use of an absorption cycle. Finally, solar energy can provide useful heating (particularly of houses and small buildings) and daylighting for most building types.

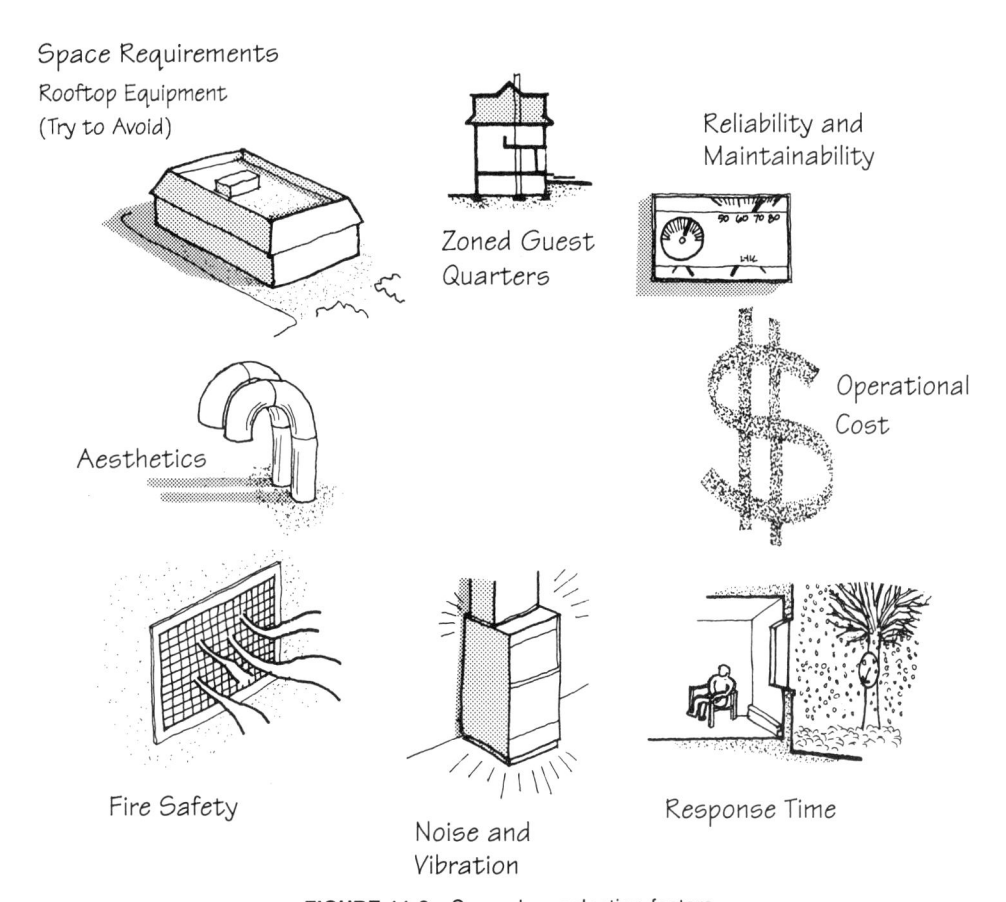

11.2 SECONDARY SELECTION FACTORS

Secondary selection factors will often determine the attributes of a particular system after basic system type has already been chosen based upon the primary selection factors discussed above. Secondary selection factors include those indicated in Figure 11.2. They too often require hard choices and a balancing of priorities.

FIGURE 11.2 Secondary selection factors.

Space Requirements

This issue pertains mostly to various types of larger buildings. For some building types such as low-rise offices and retail buildings, the desire to preserve rentable space leads to widespread use of rooftop HVAC equipment. In multistory commercial buildings, systems that preserve usable space and minimize floor-to-floor heights will be favored.

Systems using hot and cold water are often preferred for conditioning of perimeter spaces because the pipe runs are much more easily accommodated than large volume air ducts for the same loads.

Ease of Control and Zoning

Buildings usually contain many spaces which have different thermal needs. In houses, it is fairly common to have a wing or group of spaces which are only occasionally used (guest bedrooms, basements, etc.). While those spaces are unoccupied, zoning can maintain the temperature below a habitable temperature (perhaps 50°F). Such zoning can be easily accomplished with electric resistance heat or a hot water heating system while being more difficult to accomplish with a warm air system. In larger buildings, zoning requirements and system implications vary greatly by building type as discussed below.

Reliability and Maintainability

For most residential applications, systems that have few moving parts and require little maintenance are preferred. Some systems such as electric resistance heat are so simple that they should last indefinitely. Other "systems" such as a passive solar house design which also incorporates movable insulation requires an "active" owner to put the insulation in place at night. Luckily, advances in glazing performance, including low-E glass, have now all but eliminated the need for movable insulation.

In larger buildings, system reliability and perceived safety is paramount. Concern for indoor air quality has led to increased ventilation levels and air distribution systems with constant (rather than variable) flow rates.

In the design of cooling equipment for many larger buildings, a choice must be made between types of heat rejection equipment. The primary choices are one or more air-cooled condensers which require little maintenance (but are electrically intensive) or in the case of fairly large buildings, a cooling tower (which demands careful attention and maintenance).

Operational Cost

Energy costs, particularly for electricity, vary greatly across the country and have a profound effect on system selection. For example, in areas with low electrical kilowatt hour charges such as the Pacific Northwest, all-electric heating/cooling systems such as heat pumps are attractive. In contrast, large buildings in areas with high peak electrical demand charges should minimize peak demand by using ice storage, daylighting, gas- or steam-fired cooling, and so on.

Selection of heating and cooling equipment with efficient part-load performance characteristics can significantly lower operational costs. And in multitenant buildings, use of dedicated HVAC systems for tenant spaces encourages operational economy because the tenant pays the fuel cost.

Response Time

In houses, this is a design concern that pertains mostly in cold climates during reoccupancy after shutdown (e.g., a weekend or vacation house). Some system types, such as a warm air heating system, can quickly restore comfortable indoor air temperatures. By contrast, an electric resistance heating system has a very slow response time. And in many "weekend" and vacation houses, getting a fire going in the fireplace often bridges the gap until the mechanical heating system gets up to speed. Still, for those who can't wait, control systems can be programmed to prepare space temperatures for an anticipated time of arrival or even be accessed and programmed by telephone.

Space temperatures of commercial buildings are typically set back to 50 or 55°F when unoccupied at night and on weekends during the heating season. These buildings typically go through a warm-up cycle before occupancy begins each morning. The period required for warm up depends on the outside temperature, heating system type, insulating quality of the building envelope, and the massiveness of construction. Old buildings with load bearing masonry walls may require a warm-up cycle of several hours.

Noise and Vibration

The HVAC systems that employ fans and compressors tend to produce the most noticeable (and perhaps disturbing) noise. Rarely is noise a significant factor in HVAC system selection for houses. Yet in many types of larger buildings, noise and vibration require careful design attention.

Fire Safety

The type of HVAC system employed in a building can have a profound effect upon the fire safety of a building. Vertical shaft spaces need to be fire-safed and ducts that penetrate fire-rated partitions require fire dampers.

Creation of positive air pressure by HVAC fans has various fire-related applications. A fire on an upper floor of a building can often be contained by pressurizing floors above and below and exhausting the fire floor. Also many codes require activation of dedicated fans for the pressurization of fire stairs to prevent smoke entry during a fire. This allows occupants to escape from the building while breathing fresh air. Use of building fan systems to aid in fire safety should be under the direction of the fire department from a lobby control panel. Most codes require HVAC systems of 2000 cfm and larger to shut down in the event of a fire so as not to provide a supply of oxygen to the fire.

Aesthetics

Selection can be made for compatibility with the building or to improve aesthetics. For example, many designers historically disliked having to locate devices below windows. Now, however, employment of high-performance glazings can reduce or eliminate the need for perimeter heaters.

11.3 SYSTEMS FOR RESIDENCES

Design Strategies by Climate

Some examples of how climate affects HVAC system selections for houses and other residences are noted in Figure 11.3. General guidelines for these climate types (heating degree-days, base 65) are provided below.

Cold Climates (over 7500 heating degree-days). Start with proper orientation for passive solar heating in clear regions. "Superinsulate" the envelope to very high R-values to minimize heat loss. The mechanical heating system will normally include a gas- or oil-fired furnace or boiler depending upon fuel availability and cost. Also consider the possibility of radiant heating to complement any thermal mass installed as part of a passive solar design. In these climates, electric heat pumps will only be a good choice if local electric rates are very low. Consider mechanical

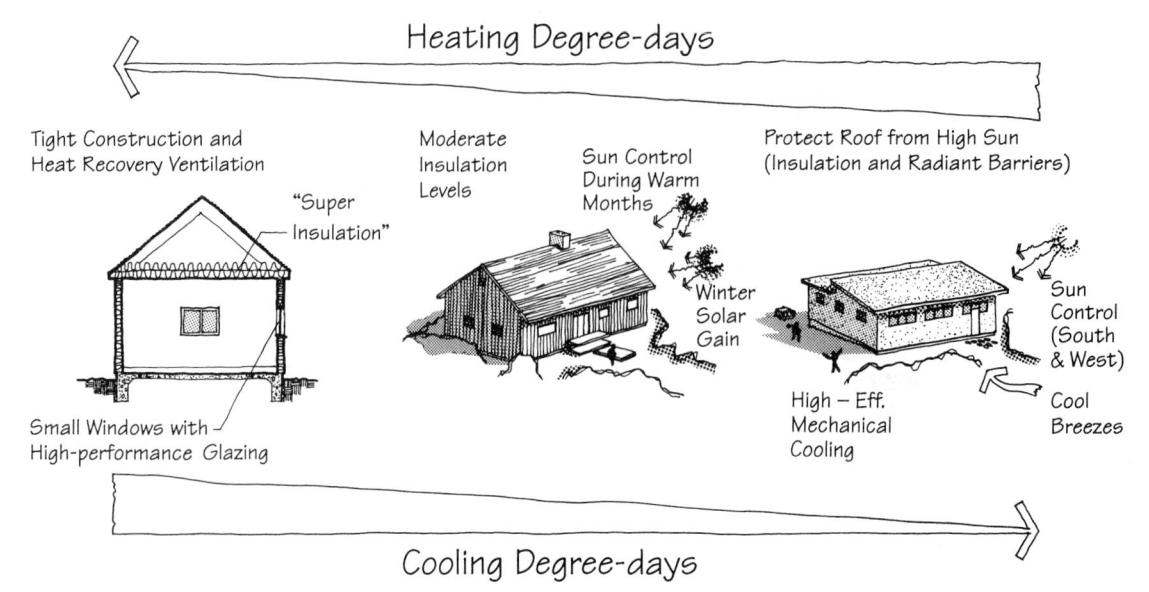

FIGURE 11.3 HVAC systems for new houses.

ventilation (perhaps with heat recovery) for assured indoor air quality.

Temperate Climates (2500–7500 heating degree-days and at least 600 cooling degree-days). These climates have significant periods when heating or cooling is required over the course of the year. Proper orientation and design can allow the sun to provide significant solar heat during the winter.

Most new houses are now designed with centralized ducted systems to provide both heating and cooling. Heating is provided either by a furnace, boiler (which generates hot water or steam for use in a coil within an air handler and perhaps also baseboard heaters or radiant floors), or heat pump (usually air-to-air), depending upon local fuel availability and cost. Cooling is provided by a cooling coil and either a condensing unit or heat pump. Consider mechanical ventilation for assured indoor air quality (perhaps without heat recovery due to lessened cost effectiveness in a temperate climate).

Warm Climates (less than 2500 heating degree-days). Heat removal (air conditioning) is the primary HVAC process required. Design houses to benefit from natural ventilation to the greatest extent possible. Provide sun control for south- and west-facing windows. Well-insulated building envelopes (particularly roofs) will reduce conducted solar heat gains. Consider providing a radiant barrier as part of the roof/ceiling design to minimize roof heat gain. In dry climates, such as in parts of the Southwest, consider evaporative cooling.

Mechanical air conditioners or heat pumps should be very efficient with a seasonal energy efficiency ratio (SEER) of 10 or more (see Section 16.3). The small amount of space heating needed will normally be provided by a heat pump, electric baseboard, or an electric resistance duct heater.

Replacement Heating Systems. When home heating equipment needs to be replaced, several questions should be asked before automatically replacing the unit with a new unit of equal heating capacity. Have there already been conservation improvements such as installation of attic insulation, new windows, or caulking to reduce infiltration? If not, might this be the time to make such improvements? If conservation has been (or will be) improved then the new heating appliance should be sized accordingly.

The replacement market is an area where very high-efficiency condensing furnaces (see Section 13.4) have some of their best applications. Replacement equipment in old, large houses which experience a great deal of heat loss can generally improve efficiency 20% or more and save many dollars per year. New condensing furnaces also make good replacements for houses with expensive to operate heat pumps installed in the 1970s when embargoes were placed on new gas services. Since houses with heat pumps were built without chimneys, the condensing appliances which need only a plastic pipe flue offer an attractive alternative when the original heat pumps reach the end of their useful operating lives.

Large Residential Buildings

Large residential buildings include multifamily housing, dormitories, and portions of some nursing homes. The trend in multifamily housing is for individually controlled and metered HVAC systems including mechanical ventilation for interior kitchen and bath areas. Apartment heat transfer (loss and gain) per square foot of floor area is proportionately much less than for a house since there are many shared walls, floors, and ceilings (see Figure 11.4).

In recent years, manufacturers have brought to market more heat producing units with low (50,000 Btu/h and less) capacities. This is in response to the lower heating loads (sometimes known as "microloads") which occur in units of multifamily buildings and the desire to have in-

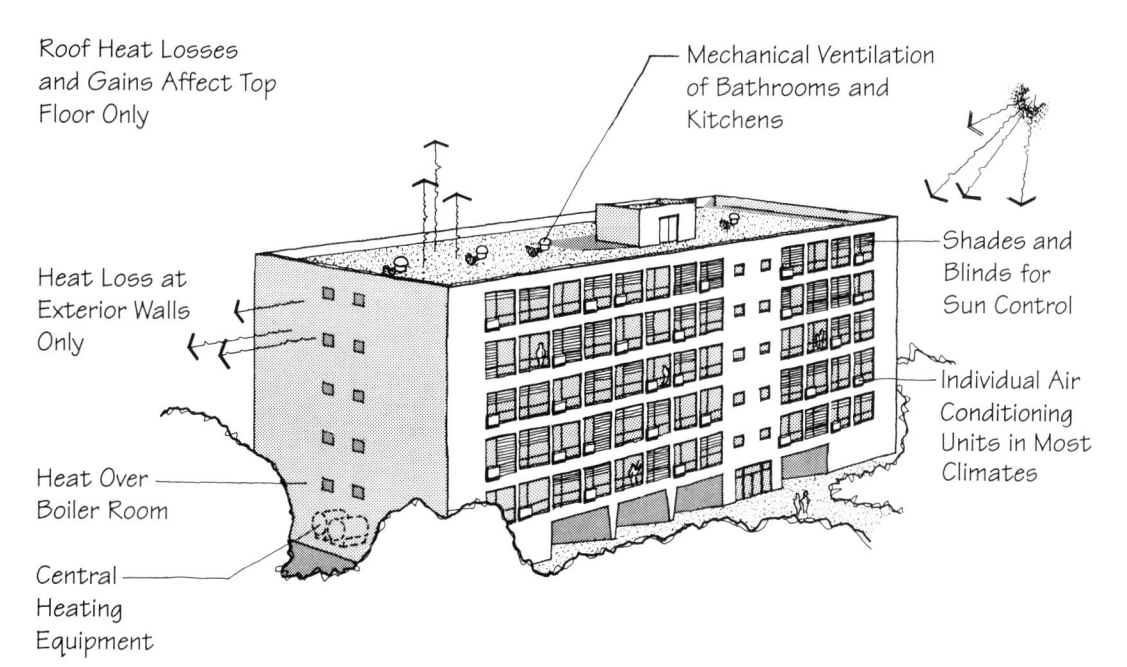

FIGURE 11.4 Large residential buildings.

dividual heating systems (see discussion on wall-mounted boilers, Section 15.5). These low-capacity units also are used in new houses which are modest in size and very energy efficient.

11.4 SYSTEMS FOR NONRESIDENTIAL BUILDINGS

In most locations, the electrical needs of large commercial buildings present a challenge for electric utilities. The cost and difficulties caused in planning for new power plants have now led most utilities to aggressively push energy conservation measures known as Demand Side Management or "DSM." Through DSM, kilowatt hour consumption of electricity is reduced and, more importantly, the kilowatt (KW) demand is lowered (see Section 24.8). Lighting is a primary target. Many utilities also encourage use of high-efficiency equipment and peak demand reducing strategies such as ice storage.

All but very small commercial buildings in northern climates require mechanical systems for ventilation and heat removal. Since mechanical ventilation is required for most spaces (particularly internal spaces), HVAC systems typically incorporate ductwork systems. Proper selection of equipment and integration of heating, cooling, and ventilation delivery systems depends largely on building type.

Retail Buildings

These (usually low-rise) buildings often employ low-cost rooftop HVAC equipment in order to preserve floor space (see Figure 11.5). Unfortunately, once units are placed on a roof they are more apt to be forgotten (and not maintained) until the day they are not working. All-electric units will generally be used for buildings in warm climates unless the building requires gas service for other purposes (such as a restaurant).

High-Rise Office Buildings

Space is at a premium in urban high-rise buildings, both vertically and horizontally (see Figure 11.6). Therefore, high air velocities are normally employed to reduce duct sizing. Systems for building interiors primarily address cooling and ventilation requirements with some form of air distribution system (either constant volume or variable air volume) typically used. Perim-

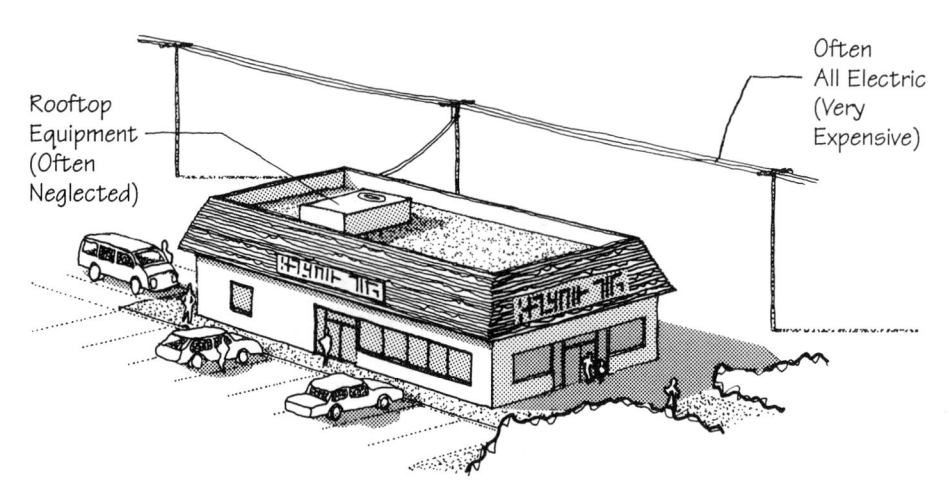

FIGURE 11.5 Retail buildings.

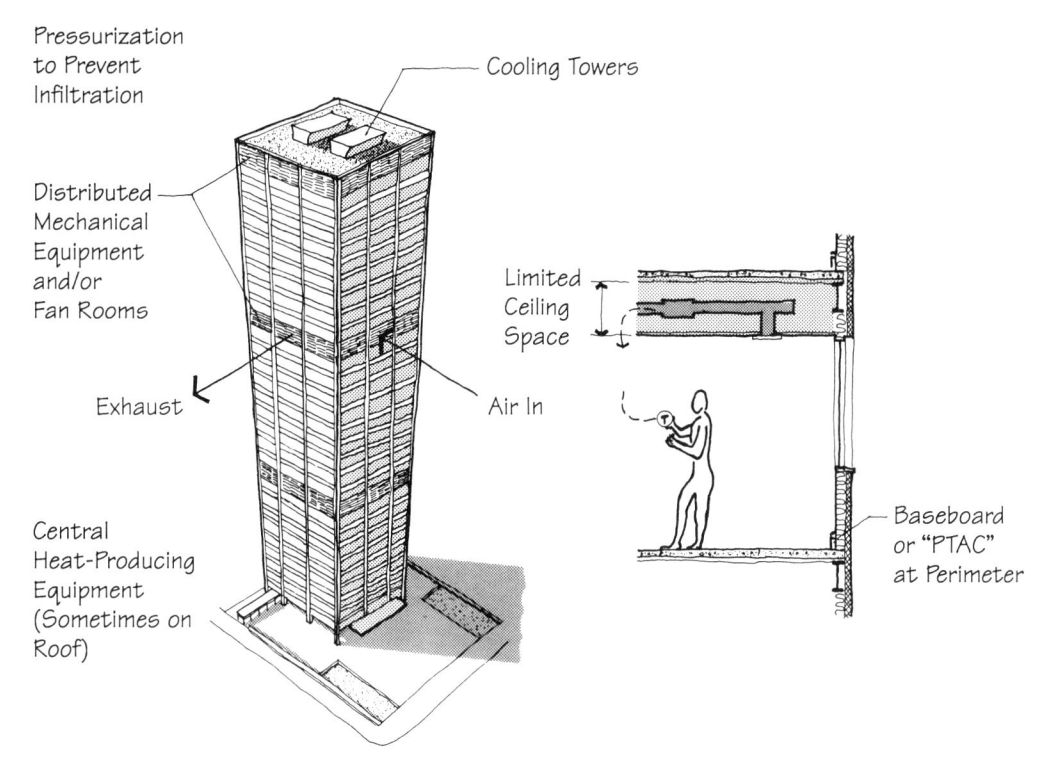

FIGURE 11.6 High-rise office buildings.

eter zones of these buildings typically have cooling loads that fluctuate widely and usually have a need for heating as well. Various approaches are taken to heat and cool perimeter spaces depending mostly upon climate and the desire for individual control. Where local control is desired, fan-coil units are often used since only small pipes need to be brought to the perimeter rather than large ductwork. Air supply systems can also be used in perimeter zones [i.e., some type of variable air volume (VAV) system] often with baseboard heating.

These buildings are now often designed to allow for individual tenant control of HVAC equipment and/or dedicated air-handling units. Very large urban buildings often have one or more vertically spaced mechanical equipment floors dedicated to HVAC equipment and one or more rooftop cooling tower(s). In these buildings, the HVAC floors often play a role in the aesthetics of the building's facade.

Low-Rise Office Buildings

For basic air delivery, these buildings typically use systems as described previously for high-rise buildings. However, the smaller scale and reduced need to minimize hung ceiling heights allow for easier extension of air delivery systems to handle loads (particularly cooling) in perimeter spaces. Therefore, in cold climates these buildings typically have baseboard heaters (usually hot water or electric where rates are low) below windows. The need for perimeter heaters is reduced when high-performance window glazing is used (see Figure 11.7).

Many buildings of four stories or less have rooftop penthouses or screened areas for HVAC equipment. Air-cooled condens-

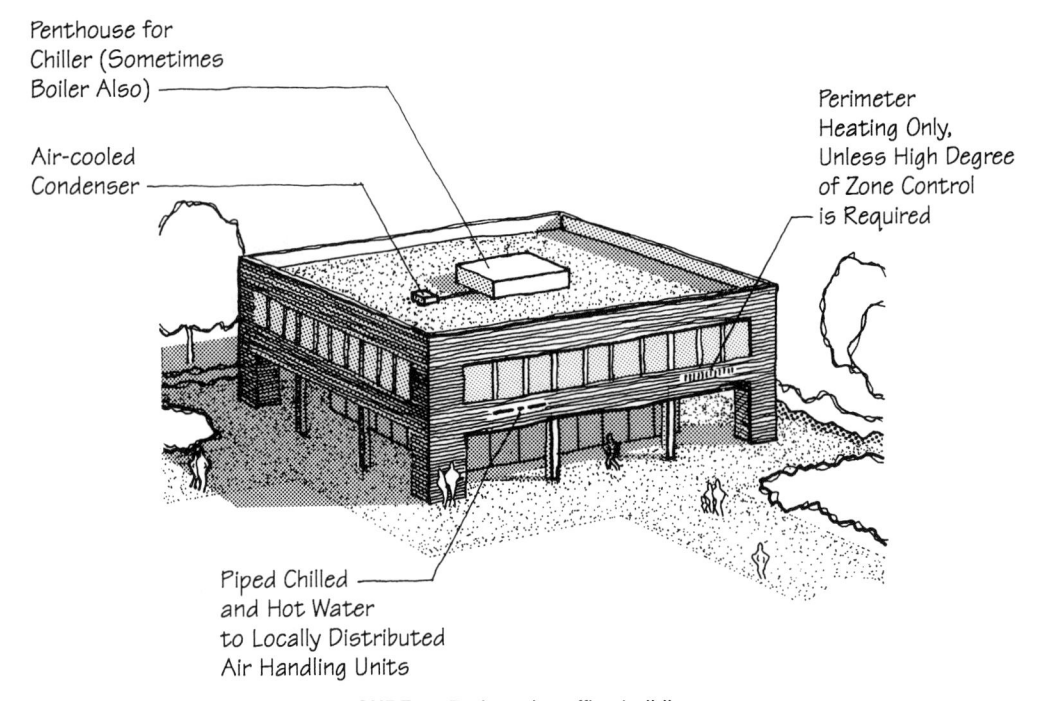

FIGURE 11.7 Low-rise office buildings.

ers are generally used for cooling system heat rejection to minimize maintenance. Systems should be designed to allow for efficient partial building occupancy as discussed above.

Computer Rooms

Main-frame computer equipment generates a tremendous amount of heat which must be cooled year round by dedicated package units which circulate thoroughly filtered air. Major units of computer equipment are often cooled via a floor plenum (12 in. high or more) beneath the raised floor through which the extensive wiring is routed. Computer rooms also require close humidity control based upon equipment requirements. As such, they generally are enclosed by construction which incorporates a vapor barrier.

Hotels and Motels

Occupants want and need individual control if these sporadically occupied spaces are to be conditioned efficiently. These spaces routinely employ "through the wall" packaged terminal air conditioners (PTAC), with building size, local climate, and utility cost influencing the method of heat delivery. Large hotels (see Figure 11.8) in cold climates will generally use piped hot water while buildings in warmer climates might have electric resistance elements. Some hotels and motels also have packaged terminal heat pumps (PTHP) but consideration should be given to year-round compressor noise and wear.

Schools

School building HVAC systems vary greatly. Occupancy density in classroom-

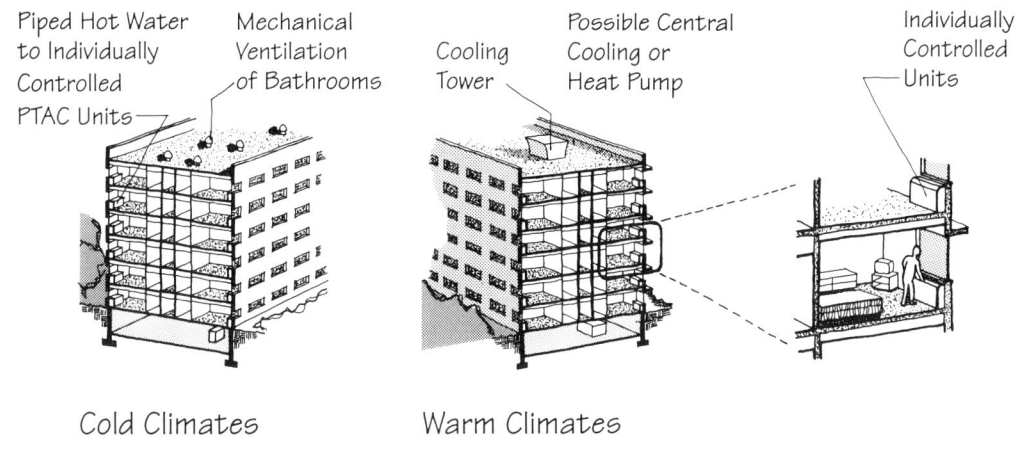

FIGURE 11.8 Hotels and motels.

type buildings requires large quantities of ventilation air. Historically it has been common to employ "unit ventilators" with local outside air intakes (see Section 15.10) to satisfy the need for ventilation as well as heating and sometimes cooling (in southern climates and schools anticipating heavy summer use). Many newer schools typically have central systems to provide for necessary ventilation and space tempering.

Many spaces in school buildings need HVAC and control systems which can quickly respond to changing populations (e.g., sudden occupancy of the auditorium). In larger schools, there is a need for a large quantity of hot water for gymnasium showers and the cafeteria. Often this can be produced by some type of heat recovery system or solar collectors (see Figure 11.9).

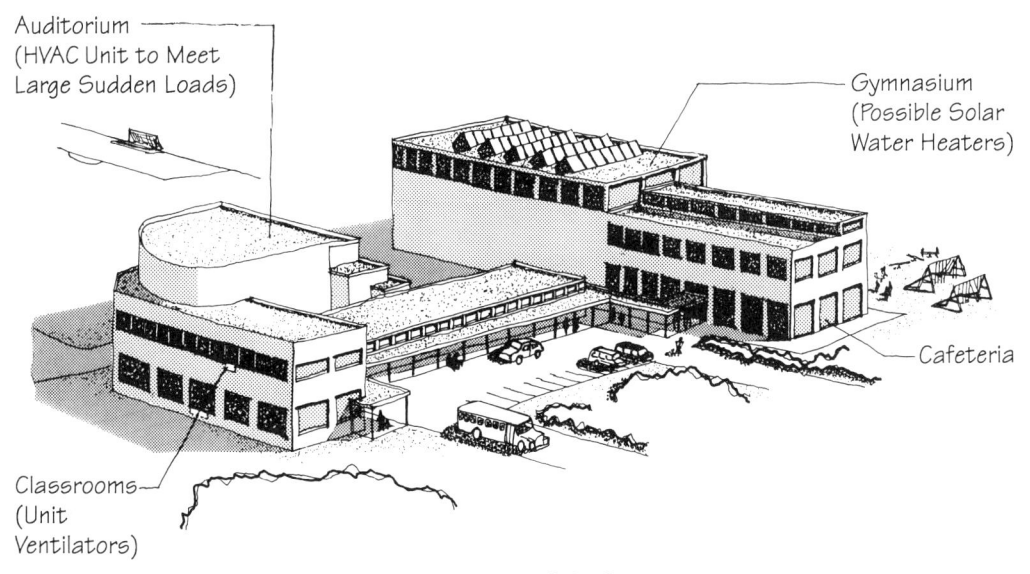

FIGURE 11.9 Schools.

Hospitals

The complexity of hospitals makes them a demanding building type for which to provide HVAC systems (see Figure 11.10). Many spaces are continuously occupied to some degree and most lights (except in patient rooms) burn continuously. Hours of operation, space function, and required temperature and humidity conditions translate into the need for many separate units (patient care, operating, ICU, computer, etc.). Many of these spaces require highly filtered and/or a high percentage (to 100%) of outside air. This suggests the need for heat recovery systems which must be designed to avoid cross-contamination.

Hospitals need steam for sterilization purposes. When they are located in areas served by utility company steam, it usually is also used for heating, cooling (through use of a steam turbine or the absorption refrigeration cycle), and domestic hot water which is required in large quantities. A backup electrical generator sized to maintain essential services, the HVAC systems, and a portion of other electrical loads will be required.

Airline Terminals

Airline terminals (see Figure 11.11) are among the most demanding building types for good HVAC design when one considers the great fluctuations in occupancy density and the need for large quantities of outside ventilation air from areas subject to jet fumes and vehicles (buses, cabs, and cars).

Often these buildings have large glass areas with heat provided by perimeter baseboard radiation. Most of these buildings are continuously occupied to some degree with lights constantly on. The variability of the people load (remember people are like little furnaces) requires a very well-controlled air distribution system.

Religious Buildings

Today many religious buildings (churches, synagogues, mosques, etc.) have sections which have very different profiles of occupancy. For example, the sanctuary may be used only during several specific periods of worship and for special occasions, whereas another room or area may be intensively used for child care, exercise classes, and

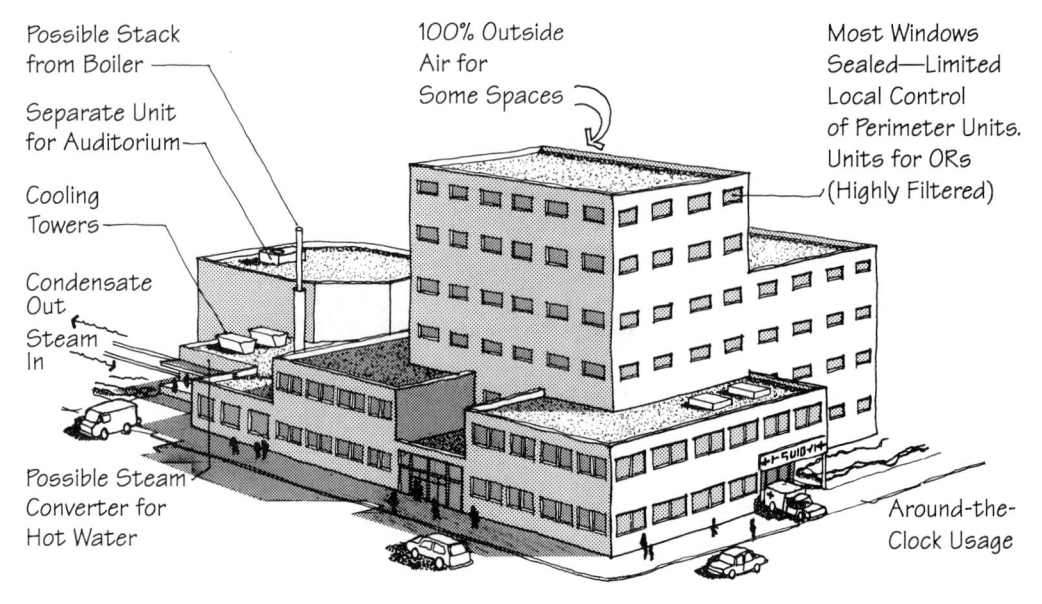

FIGURE 11.10 Hospitals.

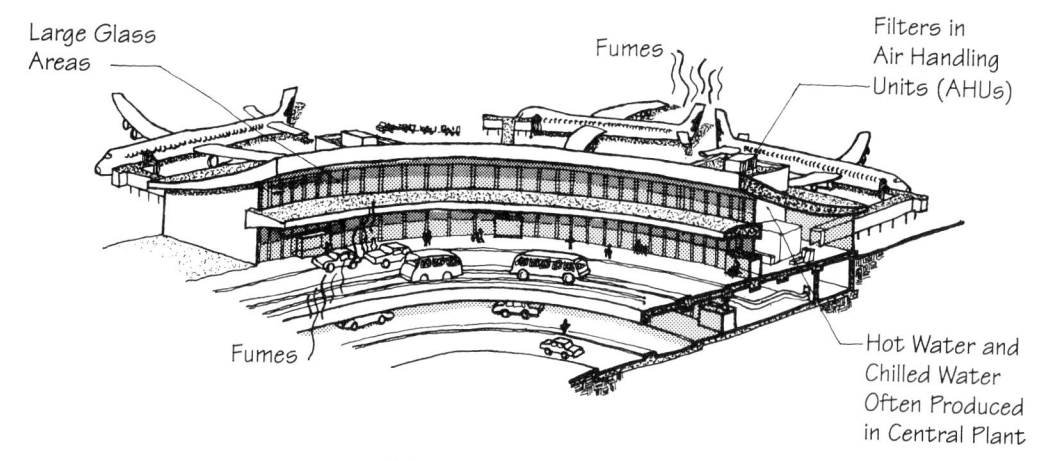

FIGURE 11.11 Airline terminals.

other group activities. For churches in relatively cold climates, modular boilers can respond nicely to large and rapid changes in heating loads. Significant cooling needs for large groups normally occur during off-peak weekend periods (see Figure 11.12).

Libraries

Libraries (see Figure 11.13) normally have large high-ceilinged spaces with large plenum spaces above the hung ceiling as well. Libraries must of course be quiet. Therefore, particular attention should be given to reduction of cooling loads (lighting and solar heat gains) so that the reduced loads can be satisfied by relatively low-velocity air distribution, usually with a VAV system. All HVAC equipment should be installed to minimize transmission of noise and vibration, with ductwork designed to prevent transmission of noise between adjoining spaces (e.g., study rooms). Low-noise air diffusers should be used (see Section 21.5). Careful control of humidity is also required for the preservation of old books.

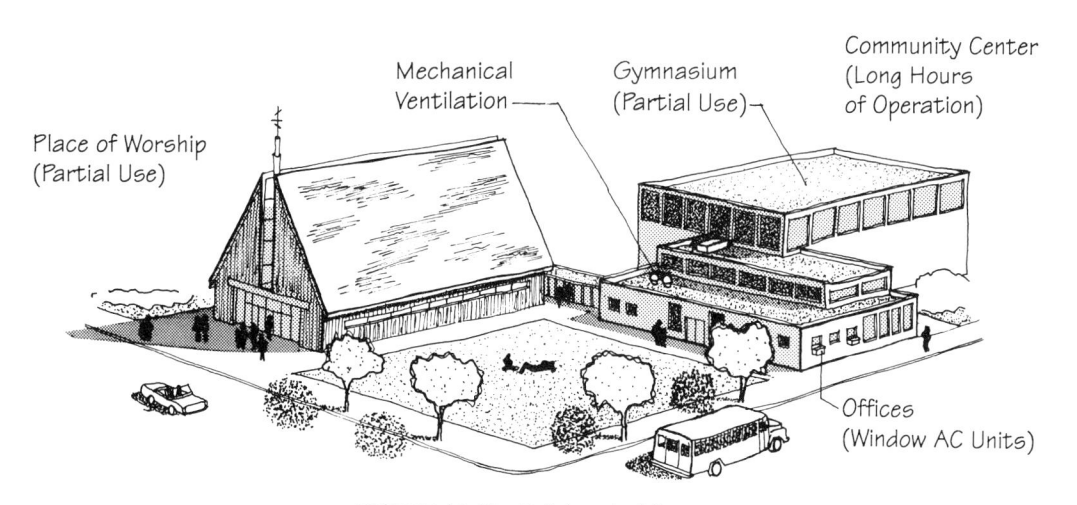

FIGURE 11.12 Religious buildings.

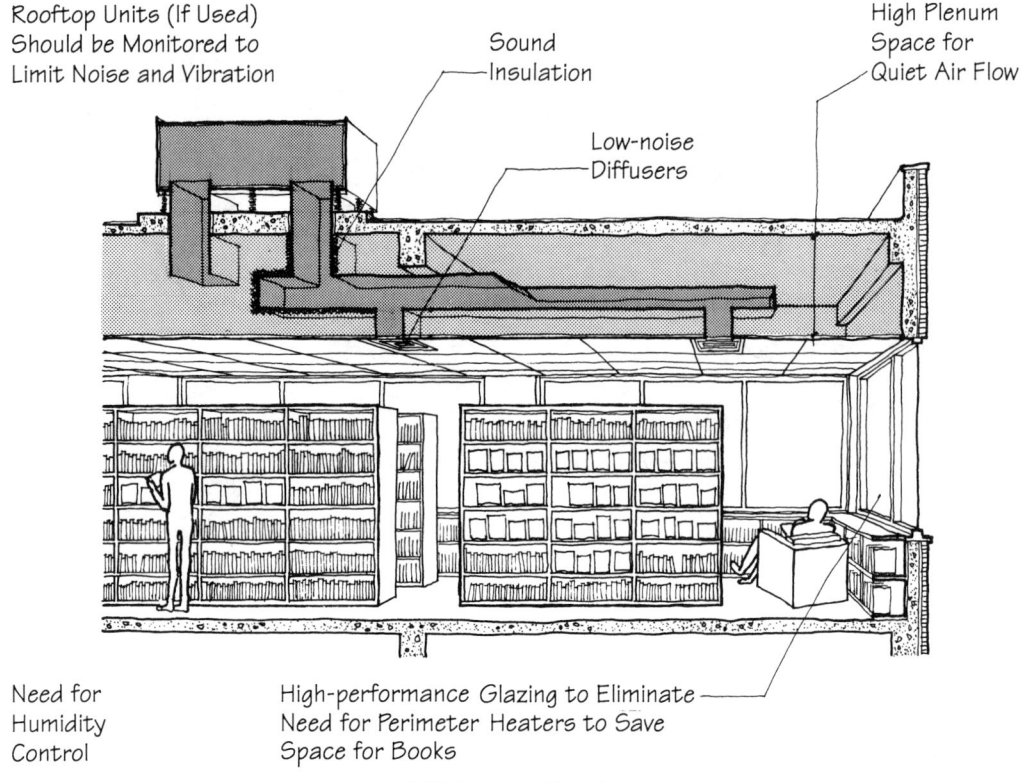

FIGURE 11.13 Libraries.

Labels on figure:
- Rooftop Units (If Used) Should be Monitored to Limit Noise and Vibration
- Sound Insulation
- High Plenum Space for Quiet Air Flow
- Low-noise Diffusers
- Need for Humidity Control
- High-performance Glazing to Eliminate Need for Perimeter Heaters to Save Space for Books

Manufacturing Plants

Here HVAC system design is driven by type of use—from modest bench work to complex manufacturing. Most manufacturing plants are one-story buildings containing high, large spaces. Possible summer overheating or the use of mechanical cooling can be minimized by employing a well-insulated roofing system.

Ventilation rates are often high when dealing with chemicals, smoke, or dust to control air quality. Despite high ventilation rates and the need for local exhaust, in some cases the process heat coming from the manufacturing process will still be adequate to heat the space unless the outside temperature becomes very low. This heat may also be useful (and worth recovering) if the production area is adjoined by large office areas and the facility is located in a generally cold climate (see Figure 11.14).

Buildings where manufacturing processes do not produce a significant amount of heat are often heated by unit heaters or ducted air handlers if the space is also mechanically cooled. Office areas in manufacturing plants will normally be individually conditioned by unitary equipment.

Warehouses

Warehouse storage areas (see Figure 11.15) usually have no windows. Most warehouses are heated (to 60°F or so) by ceiling-mounted unit heaters or radiant heaters. Usually there is a small office area which is likely to be conditioned either by an individual room PTAC (packaged terminal air

FIGURE 11.14 Manufacturing plants.

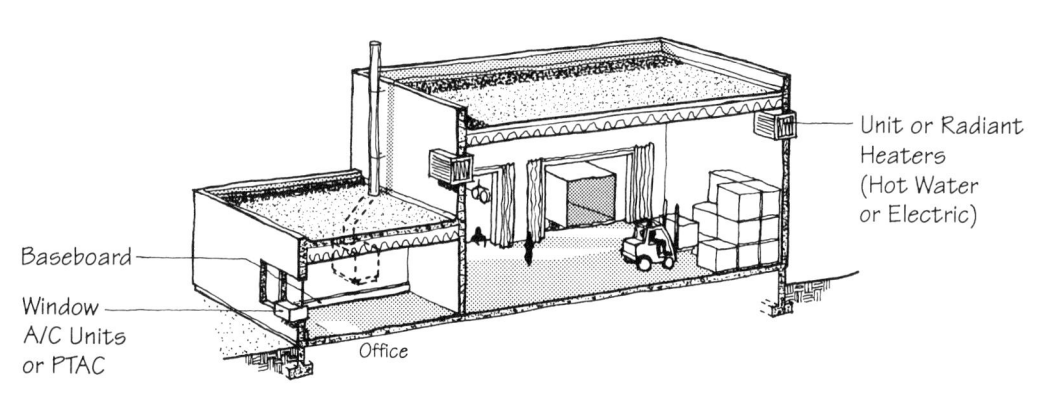

FIGURE 11.15 Warehouses.

conditioner) or a unitary system. In cold climates, doors leading to loading dock areas often have air curtains or plastic curtains to block wind and limit heat loss. A well-insulated roofing system will minimize indoor temperatures in the summer.

The use of ceiling mounted "casablanca" propeller fans can help redistribute hot stratified air back to the floor (where workers may be) in the winter time, and in the summer provide an evaporative cooling effect.

Climate-Controlled Warehouses

Warehouses used to store food or other items with precise requirements for temperature and humidity (refrigeration or freezing) require HVAC systems which are very dependable. These buildings need modular and/or backed-up heating and cooling systems so that a unit failure will not result in a loss of or damage to building contents. These buildings also generally have their own back-up electrical generators.

12

SOLAR HEATING SYSTEMS

The HVAC needs of residential buildings closely follow climate. When comfortable conditions exist outside, the need for a heating system of any type is nonexistent or slight. Nonetheless, to maintain comfortable conditions when it gets cold outside, the collection of "free" solar heat can be very beneficial, particularly in clear, cold climates.

Solar heating systems fall into several basic categories. "Passive" solar heating systems use only natural processes to collect, store, and distribute heat. In contrast, "active" solar systems employ pumps and/or fans to move collected heat. One is advised, though, to also consider "hybrid" systems which are inherently passive but use small fans or pumps to enhance or improve performance.

As discussed at the beginning of Chapter 3, passive solar design is not new, having been used by Aristotle and his competitors. What is relatively new, however, is a simple quantitative manual method to estimate solar benefits.

The application of solar design to houses is beneficial and cost effective in many locations. Solar heat gain can also be used in some types of commercial buildings, such as buildings with atriums and daylit circulation spaces. In such buildings, the solar heating benefit often is best achieved if the mechanical HVAC system is designed to utilize this heat, such as in a closed-loop heat pump. Also realize that many large buildings have heat removal problems even during cold winter periods because of large internal heat gains from lights, people, and equipment. As a result, designers usually attempt to limit direct solar heat gain from most commercial spaces.

12.1 ACTIVE SOLAR HEATING SYSTEMS

One of the first reactions to the "energy crisis" of 1973 was the rush to demonstrate that solar energy could supplant fossil fuels in heating houses and commercial buildings. Initially, the focus of activity was on the use of "active" solar collectors to produce hot air or hot water for space heating purposes. By the late 1970s, though, the use of active solar collectors for space heating purposes was all but abandoned in favor of passive heating systems.

Factors that contributed to the decline of interest in active solar systems for space heating included the high initial cost for collectors and their installation, a better understanding of operational costs for pumps and fans, maintenance costs and problems for systems only used during the winter months, aesthetics, and the end of federal and local tax credits. See Section 22.7 for information on active solar systems for production of domestic hot water.

12.2 PASSIVE SOLAR HEATING SYSTEMS

From the earliest times, humans have understood that facing the sun themselves, and having their dwellings also face the sun, provides additional comfort in cold weather. Significant storage of solar heat, however, only became possible once south-facing openings were sealed by transparent glazings.

Efforts to harness solar heat have evolved over the last 2500 years—a fascinating story, beyond the scope or purpose of this volume. The topic of contemporary passive solar heating systems and their performance is explored in detail below.

Direct Gain

The simplest, and most common solar design approach, is based on "direct gain" with glazing (double-glazing or low-E) oriented south to allow for penetration of low winter sun rays, as illustrated in Figure 12.1a. Solar performance can be enhanced if night-time heat loss through the glass is significantly reduced by placing "mov-

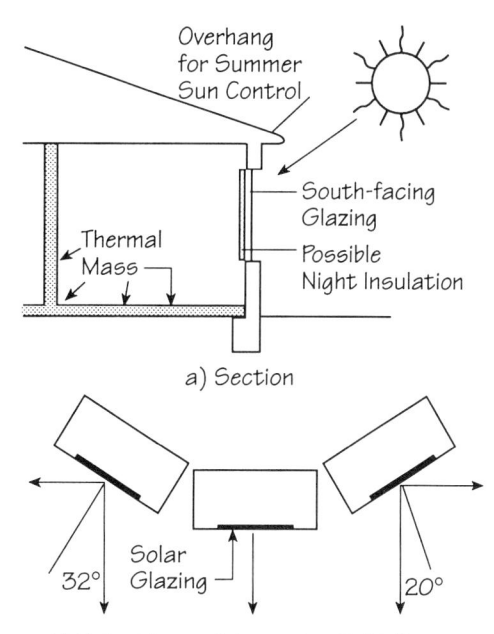

FIGURE 12.1 Direct gain passive solar heating.

able" insulation over the glazing as indicated. An often used rule-of-thumb is that solar glazing should lie between 20 degrees east and 32 degrees west of true south, as shown in Figure 12.2b. With such orientations, solar performance will be reduced only slightly below that for true south.

Still, it is important to remember that the chance for swing-season and summer overheating increases as the glass orientation varies toward east or west from solar south.

As a rule of thumb, the heat storage capacity of conventional construction (gypsum wallboard, etc.) is adequate for direct gain houses having a modest amount of south-facing glass area (of up to about 7% of the house's total heated floor area). These houses are known as "suntempered."

To improve solar performance, additional glass area is needed as well as the presence of sunlit materials which have the ability to absorb and store heat ("thermal mass"). Thermal mass is also required in high-performance solar houses to prevent overheating of the indoor air, as discussed below.

Provided in Table 12.1 are heat capacities of air, water, and various materials used in building construction which may serve as heat-storing thermal mass.

Notice how the heat capacity of various materials compares to that of air. Yes, a given volume of water can hold over 3000 times the heat of an equal volume of air when experiencing a similar temperature rise.

Nevertheless, a word of caution is in order. Table 12.1 might appear to suggest that a material such as white oak provides better heat storage than materials such as brick and concrete, which are commonly used for thermal mass in passive solar buildings. This is not true once "thermal conductivity," which is the rate of heat storage, is factored in, since hardwoods have a thermal conductivity which is only one-fourth that of brick.

Thermal Storage Walls

The importance of thermal mass to solar heating systems has led to the evolution of other design approaches. The concept of placing massive vertical walls several inches behind a glazed air space was first developed in France. These are named "Trombe" or "Trombe-Michel" walls after their designers, engineer Felix Trombe and architect Jacques Michel.

Variations on this design include venting, to facilitate a convective heat flow through vents, or solid wall construction, to delay heat delivery until later in the day or evening. Another design approach, known as a "Water Wall," employs a large volume of water as the thermal storage medium. These are shown in Figure 12.2.

Sunspaces

The design of a glazed south-facing sunspace can take many forms including:

Use of sloped or overhead glazing

The inclusion of movable shading devices or insulation

Incorporation of thermal mass

Using the space for growing plants and vegetables. In this case the design is normally called a solar greenhouse and attention must be given to the maintenance of warm temperatures

Trombe walls are sometimes criticized because their massive opaque construction is unattractive and expensive. In this regard, sunspaces can be thought of as wide trombe walls that you can live in.

Illustrated in Figure 12.3 is a variety of

TABLE 12.1 HEAT CAPACITIES OF MATERIALS USED IN CONSTRUCTION AND HVAC APPLICATIONS

Material	Specific Heat (Bt/lb·°F)	Density (lb/ft³)	Heat Capacity (Btu/°F·ft³)	Ratio of Heat Capacity of Material to Air
Air	0.24	0.075	0.018	1:1
Gypsum board	0.26	50	13.0	722:1
Brick	0.30	112	22.4	1244:1
Concrete (stone)	0.16	144	23.0	1277:1
White oak	0.57	47	26.8	1489:1
Water	1.00	62.4	62.4	3467:1

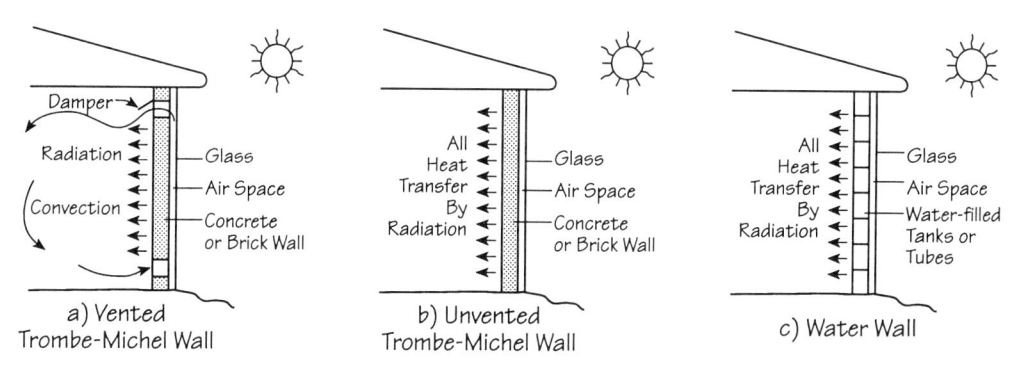

FIGURE 12.2 Types of thermal storage walls.

sunspace designs for which passive solar performance can be estimated using the LCR method described below.

12.3 PASSIVE SOLAR PERFORMANCE

Simplified evaluation of passive solar performance is most reliably accomplished using methods developed and refined by Dr. J. Douglas Balcomb and his associates at Los Alamos National Laboratory beginning in the late 1970s and continuing now at the National Renewable Energy Laboratory (NREL). Known as the "Load Collector Ratio" method, this procedure allows users to quickly estimate the savings achieved by a house with solar design features when compared to an otherwise identical "reference house" without such features. This work was initially published by the American Solar Energy Society, Inc. as the *Passive Solar Design Handbook, Volume 3* and later republished by ASHRAE as *Passive Solar Heating Analysis*. The most current reference is *Passive Solar Design Strategies: Guidelines for Home Builders* and accompanying BuilderGuide computer software available from the Passive Solar Industries Council (PSIC) in Washington, DC.

What is a Load Collector Ratio (LCR)? It is the heating "load" divided by the area of solar "collector" (vertical south-facing glass), hence the name.

More specifically, the "load" is known as the Building Load Coefficient (BLC). The BLC is the heat loss of a solar house (in units of Btus per degree-day) without including the heat loss of any solar (south-facing) glazing. But doesn't south-facing glazing also lose heat? Of course it does—read on!

A good way to think about south-facing glazing is that it functions like a two-way valve. Heat is lost, since all glass loses a lot of heat compared to well-insulated walls. However, heat is also gained through south-facing glazing in the form of solar radiation, which in almost all climates (except for very cloudy locations) will exceed the heat lost through the glass during the heating season.

"Reference House"

The net solar benefit will vary greatly because of local climate. It is for just this reason that the area of south glass is not considered in computing the Building Load Coefficient (BLC). Instead, LCR tables establish the net solar benefit for various designs when located in different climates.

This key assumption regarding the LCR methodology is illustrated in Figure 12.4. Note that the collector area in the "refer-

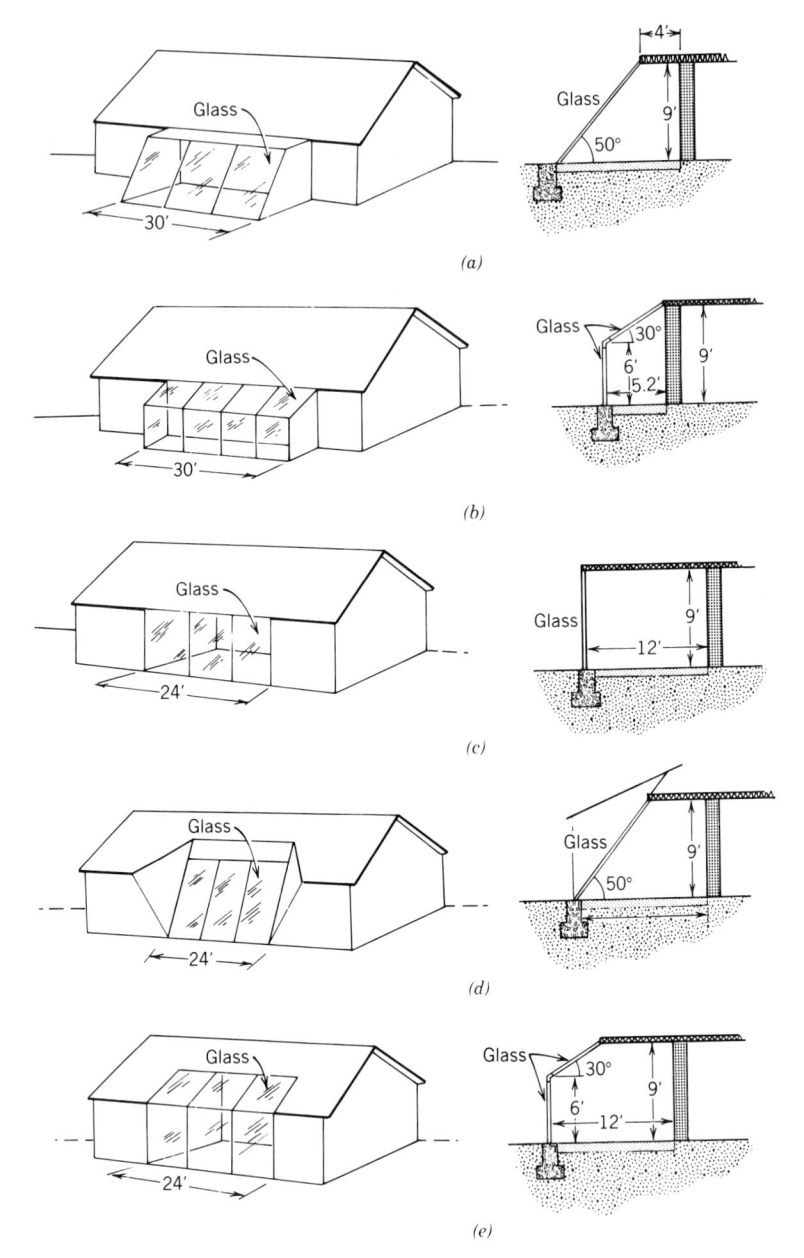

FIGURE 12.3 Sunspace design variations. (Reprinted by permission from J. D. Balcomb et al., Passive Solar Heating Analysis published by ASHRAE, 1989)

ence house" is considered to be energy-neutral, and does not impact upon the BLC heat loss calculation in any way. For the solar house, the collector area (A_p) is used to compute the Load Collector Ratio.

Correction for Nonvertical Glazing

The LCR method bases performance on area of vertical glazing. When an expanse of south-facing glazing is not vertical (such

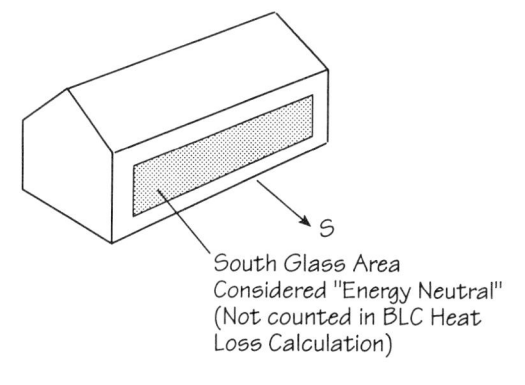

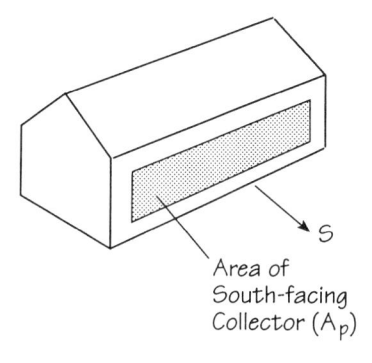

12.3 PASSIVE SOLAR PERFORMANCE

L   is the length of collector glazing in feet

Application of Formula 12A to determine the total A_p for tilted glass surfaces is illustrated in Figure 12.5.

The projection of tilted glazing areas to their vertical equivalent is an arbitrary convention employed by the LCR method. In fact, it should be realized that south-facing glazing, which is tilted back slightly from vertical, may yield slightly improved heating performance due to more direct penetration by low winter sun angles.

Building Load Coefficient (BLC)

The total heat loss for a solar reference house which is known as the "BLC" is determined by using Formula 12B.

$$BLC = [q_{tot} - (A_p \times U_c)] \times 24$$

Formula 12B

where

BLC   is the building load coefficient in Btu/°F·day

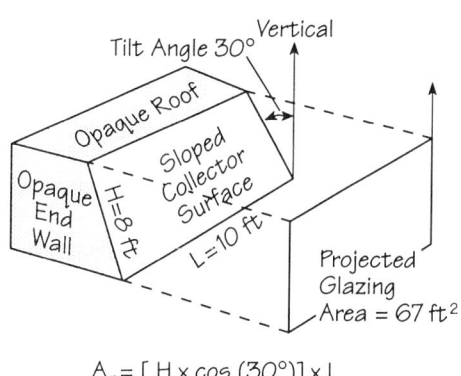

South Glass Area Considered "Energy Neutral" (Not counted in BLC Heat Loss Calculation)

REFERENCE HOUSE

Area of South-facing Collector (A_p)

SOLAR HOUSE

FIGURE 12.4 "Energy neutral" south-facing collector area.

as in many sunspaces), adjustment can be made by using Formula 12A.

$$A_p = [H \times \cos(TILT)] \times L$$

Formula 12A

where

A_p is the area equivalent of sloped collector glazing projected to the vertical in ft²

H   is the actual height of the sloped collector glazing in feet

TILT   is the tilt angle of sloped collector glazing from vertical in degrees

Tilt Angle 30°
Vertical
Opaque Roof
Opaque End Wall
Sloped Collector Surface
H = 8 ft
L = 10 ft
Projected Glazing Area = 67 ft²

$A_p = [H \times \cos(30°)] \times L$
     $= [8 \text{ ft} \times .833] \times 10 \text{ ft}$
     $= 67 \text{ ft}^2$

FIGURE 12.5 Projected area (A_p) for nonvertical glazing.

q_{tot} is the total heat loss of the house in Btu/°F·h (see Formula 5I)
A_p is the area of south-facing collector glazing in ft², projected to vertical (if necessary using Formula 12A)
U_c is the U-factor of the south-facing collector glazing
24 is the conversion of 24 hours in 1 day

LCR Methodology

The LCR methodology employs Formula 12C. It is a simple ratio of a house's heat load compared to the area of collector available to meet part of that heating load.

$$\text{LCR} = \frac{\text{BLC}}{A_p} \qquad \text{Formula 12C}$$

where

LCR is the load collector ratio
BLC is the building load coefficient in Btu/°F·day
A_p is the collector area in ft² (projected to vertical if necessary using Formula 12A)

Computed values of LCR are used to determine solar savings in LCR tables which are provided for over 200 U.S. and Canadian locations. The LCR table for each location contains values for 94 different designs which take advantage of direct gain (DG), incorporate trombe walls (TW), or water walls (WW) for thermal storage, or have an attached greenhouse-like sunspace (SS).

Direct Gain Reference Designs

As noted above, the most common solar design is direct gain by south-facing windows. The LCR tables provide performance data for nine reference direct gain (DG) design types for all of the cities.
The nine direct gain (DG) designs are referenced by letters (A, B, C) which indicate the amount of thermal mass. The three "DG-A" designs all have the same amount of thermal storage capacity, 30 Btu/°F·ft²g ("ft²g" is per square foot of south glazing, as illustrated in the example below). The three "DG-B" designs have more thermal storage capacity (45 Btu/°F·ft²g), while the three "DG-C" designs are the best performers since they have the most thermal storage capacity (60 Btu/°F·ft²g).

The numbers following DG-A, DG-B, and DG-C indicate the type of glazing. The "1" in DG-A1 refers to designs with conventional double glazing. The designation "2" refers to triple glazing and can also be used to estimate the performance of high-performance (low-E) glazing which is now commonly used. Finally, "3" refers to conventional double glazing, which is covered at night by movable insulation (R-9). Table 12.2 describes the features of each of the nine direct gain reference designs.

Thermal Storage Capacity

To select the correct reference design (e.g., DG-C2), use Formula 12D to compute the amount of thermal storage capacity.

$$\text{TM}_{cap} = \frac{S_{tm} \times D_{tm} \times V_{tm}}{\text{ft}^2 g}$$

Formula 12D

where

TM_{cap} is the thermal storage capacity of the thermal mass in units of Btu/°F per ft² of collector area
S_{tm} is the specific heat of the thermal mass material in Btu/lb·°F
D_{tm} is the density of the thermal mass material in pounds per ft³
V_{tm} is the volume of the thermal mass (obtained by multiplying the area directly illuminated by

12.3 PASSIVE SOLAR PERFORMANCE

TABLE 12.2 DIRECT GAIN REFERENCE DESIGNS

Desig.	Thermal[a] Storage Capacity (Btu/ft^{2}g·°F)	Mass[b] Thickness (in.)	Mass-to-Glazing Area Ratio	No. of Glazings	Night Insulation
A1	30	2	6	2	No
A2	30	2	6	3	No
A3	30	2	6	2	Yes
B1	45	6	3	2	No
B2	45	6	3	3	No
B3	45	6	3	2	Yes
C1	60	4	6	2	No
C2	60	4	6	3	No
C3	60	4	6	2	Yes

[a] Per unit of projected area.
[b] Assuming volumetric heat capacity of 30 Btu/ft^3·°F.

sunlight on a winter day by the thickness of the mass)

ft^{2}g is the area of solar glazing (equal to A_p, the projected collector area) in ft^2

Example 12A: Determine the correct direct gain reference design designation for a house with the following characteristics:

Exposed floor area of 360 ft^2 which is 6 in. thick

The specific heat (S_{tm}) of the concrete floor is 0.20 Btu/lb·°F and the density (D_{tm}) is 150 lb/ft^3

An area of south-facing vertical glazing (ft^{2}g) of 100 ft^2

Solution: The quantity of thermal storage capacity (TM$_{cap}$) is calculated using Formula 12D as follows:

$$TM_{cap} = \frac{S_{tm} \times D_{tm} \times V_{tm}}{ft^2g}$$

(Formula 2D)

$$= \frac{0.20 \times 150 \times (360 \times 0.5)}{100}$$

$$= 54 \text{ Btu/°F·ft}^2g$$

The result is very close to the 60 Btu/°F·ft^2 in the "DG-C" reference designs. Therefore, use that group, with final designation depending upon glazing type.

Table 12.3 provides performance data for the nine Direct Gain reference designs for Louisville, Kentucky. The numbers within the field of the table are LCRs (load collector ratios). For example, on the first line of the table for the DG-A1 reference design type you will find the LCR value of 68 in the first column. To understand its significance assume that a particular solar house meets this reference design and has an LCR of 68 (determined by Formula 12C). Note at the top of the table that the SSF or "solar savings fraction" for a house with such an LCR of 68 will be .10 (or 10%) when compared to the reference house which had no south glazing.

Solar Savings Fraction (SSF)

The computed LCR in most cases will fall between the two values that bracket it and require interpolation using Formula 12E to determine the solar savings fraction.

$$SSF_{int} = SSF_A + [.10 \times (C/B)]$$

Formula 12E

196 SOLAR HEATING SYSTEMS

TABLE 12.3 LCR TABLE DIRECT GAIN DESIGNS—LOUSVILLE, KENTUCKY

Reference Design Type	Solar Savings Fractions (SSF)								
	.10	.20	.30	.40	.50	.60	.70	.80	.90
DG-A1	68								
DG-A2	104	43	24	13	7				
DG-A3	145	65	39	26	19	13	9	5	
DG-B1	70	21							
DG-B2	107	46	26	16	10	6			
DG-B3	149	68	41	28	20	15	11	8	4
DG-C1	98	36	18	6					
DG-C2	132	58	33	22	15	10	6	2	
DG-C3	173	80	48	33	24	18	14	10	6

where

SSF_{int} is the interpolated solar savings fraction

SSF_A is the solar savings fraction (i.e., .10, .20, etc.) corresponding to the LCR value which is higher than the one computed for a specific design

B is the numerical difference between the two tabular LCR values that frame (are above and below) a computed LCR

C is the difference between the computed LCR and the higher of the tabular LCR values that frame it

Heat Saved by the Passive Solar Heating System

Formula 12F is used to estimate the amount of heat ($Q_{sol\text{-}yr}$) saved by the passive solar heating system.

$$Q_{sol\text{-}yr} = SSF \times BLC \times HDD_x$$

Formula 12F

where

$Q_{sol\text{-}yr}$ is the quantity of heat provided by the solar design in Btus per year

SSF is the solar savings fraction
BLC is the building load coefficient (see definition above)
HDD_x is the number of heating degree-days per year below base temperature X

Heating degree-day base temperatures (HDD_x) are similar to balance point temperatures (see Section 4.5). In this context, however, they do not include any solar heat gain. For the purposes of the LCR method, the base temperature is the outside air temperature at which heat from the sun and/or a mechanical heating system is needed to maintain the desired indoor set-point temperature.

Auxiliary Heat Required

Formula 12G is used to estimate the amount of auxiliary heat ($Q_{aux\text{-}yr}$) required annually from a mechanical heating system.

$$Q_{aux\text{-}yr} = (1 - SSF) \times BLC \times HDD_x$$

Formula 12G

where

$Q_{aux\text{-}yr}$ is the quantity of auxiliary heat required in Btus per year.

Example 12B: A small house to be built in Louisville has a computed heat loss not including the south glass of 300 Btu/°F·h; a substantial amount of thermal storage capacity to qualify for the "C" grouping; and 100 ft² of south-facing low-E, high-performance glazing.

Determine the Solar Savings Fraction (SSF), the quantity of heat saved by the solar heating system, and the quantity of auxiliary heat which will still be required. Note that the house has a computed heating base temperature of 60°F.

Solution: Apply Formula 12C to obtain the Load Collector Ratio. For this method, the total house heat loss (BLC) must be in units of Btu/°F·day, which may require multiplication of total heat loss (if known in units of Btu/°F·h) by the conversion of 24 hours in 1 day.

$$\text{LCR} = \frac{\text{BLC}}{A_p} \quad \text{(Formula 12C)}$$

$$= \frac{(300 \times 24)}{100}$$

$$= 72$$

Now identify the applicable reference design. Since the thermal mass qualifies for the "C" grouping and the south-facing glass is "low-E," the DG-C2 design is applicable. Looking on that line in Table 12.3, one finds that an LCR of 132 would yield a SSF (solar savings fraction) of .10 (10%); whereas an LCR of 58 is required for a solar savings of .20 (20%).

Use the interpolation Formula 12E, because the LCR of 72 falls between the two listed LCR values which frame it. The LCR of 132 corresponds to an SSF$_A$ of .10; and the LCR of 58 which corresponds to an SSF of .20. Therefore, when using Formula 12E "B" in this case is equal to 74 (132 − 58) while "C" is equal to 60 (132 − 72).

$$\text{SSF}_{\text{int}} = \text{SSF}_A + [.10 \times (C/B)]$$

$$\text{(Formula 12E)}$$

$$= .10 + [.10 \times (60/74)]$$

$$= .10 + .08 = .18 \text{ or } 18\%$$

To determine the energy savings due to the solar heating system, apply Formula 12F for this house in Louisville. The quantity of heating degree-days to use is 3560 (from Table 2.3) since the heating base temperature of the house is stated to be 60°F.

$$Q_{\text{sol-yr}} = \text{SSF} \times \text{BLC} \times \text{HDD}_x$$

$$\text{(Formula 12F)}$$

$$= .18 \times 7200 \times 3560$$

$$= 4{,}613{,}760 \text{ Btus per year}$$

Then use Formula 12G to determine auxiliary heating energy requirements.

$$Q_{\text{aux-yr}} = (1 - \text{SSF}) \times \text{BLC} \times \text{HDD}_x$$

$$\text{(Formula 12G)}$$

$$= (1 - .18) \times 7200 \times 3560$$

$$= 21{,}018{,}240 \text{ Btus per year}$$

By the way, never forget that while passive solar houses save energy over the course of the year, an auxiliary heating system must be sized to satisfy the entire space heating load of a house *including* heat loss through the solar collection area, to provide heat at night and on cloudy/overcast days.

The material contained herein is an introduction to computing passive solar performance. For further study, see the *Passive Solar Design Handbook, Vol. III* which provides the complete methodology including various refinements for designs which do not meet the specific parameters of the detailed reference designs.

Reverse LCR tables are also published with LCR parameters (i.e., 100, 70, and 50) at the head of the table and SSF values within the field. A limited number of these reverse tables are published in *Mechanical and Electrical Equipment for Buildings, 8th Ed.*, by Stein and Reynolds.

When using the LCR tables, try not to let the tremendous amount of information about other designs in other cities get in the way of your finding the correct line in the correct table for your design. Remember that the solar performance (SSF) of a house design improves as the LCR is lowered. The LCR is lowered by either reducing the heat loss of the rest of the house (lowering BLC), or by increasing the collector area (A_p), or both.

12.4 PERFORMANCE OF DG-A2 HOUSE DESIGNS

Table 12.4 contains a listing of the LCR values which would be required by houses fitting the DG-A2 description, for a selection of North American cities (having at least 3500 HDD_{65}). Also listed are the latitudes and average outdoor January air temperatures (Jan T_o) for other design purposes.

The DG-A2 design was chosen for illustration since, with low-E glazing (having high solar transmission), and a moderate amount of thermal mass for heat storage, it is a popular solar design choice in many locations. It should be realized, though, that enhanced solar performance with low-E glazing will result from the use of additional thermal mass to qualify for the DG-B2 or DG-C2 designations.

The values within the field of Table 12.4 are LCR values which yield the solar savings fractions (.10, .20, .30, etc.) indicated at the top of each column when a house design has the indicated LCR. SSF values of .70 or greater are not shown since they can only possibly be achieved in optimum climates with great design difficulty, and even then indoor overheating is likely.

Example 12C: A house design meets the criteria for DG-A2 designation. The Building Load Coefficient (BLC) is 4440 Btu/°F·d, and there is 120 ft^2 of vertical south-facing low-E glazing (with high solar transmission). This yields a Load Collector Ratio (LCR) of 37 (4440/120). Determine the Solar Savings Fraction (SSF) for this house if it were built in various North American cities.

Solution: Refer to Table 12.4 for SSFs for houses meeting the DG-A2 designation. As indicated, a house with an LCR of 37 would yield an SSF of .10 if built in Binghamton, New York (a very cloudy climate). This means that this solar design will provide only 10% of the heating energy needed by the house each year and the additional cost for any extra south glazing and/or thermal mass may not be justified.

Further inspection of Table 12.4 reveals that for the same LCR of 37, the SSF will be about .20 for that house design in New York City or Des Moines, Iowa.

An SSF of about .30 would be achieved if the same house were built in Roanoke, Virginia or Casper, Wyoming.

Finally, an SSF of about .40 would be the result if the house were built in Reno, Nevada and an SSF of almost .50 if it were built in Prescott, Arizona.

Although house design is important, the clearness of the climate is critical to achieving high SSFs, as indicated in this example.

12.5 INDOOR AIR TEMPERATURE

It is good practice to design passive solar houses so that indoor air temperatures will fall within the comfort range during January on a clear day. Use Formula 12H together with Figure 12.6.

$$t_{\text{i-JAN}} = t_{\text{o-JAN}} + \text{TD}_{\text{sol}} + \text{TD}_{\text{int}}$$

Formula 12H

TABLE 12.4 LCR VALUES REQUIRED TO ACHIEVE INDICATED SSFs "DG-A2" SOLAR DESIGN

City	Lat.	Jan T_0	.10	.20	.30	.40	.50	.60	.70
Albany, NY	42.7	21.5	56	19					
Albuquerque, NM	35.0	35.2	227	107	67	46	32	23	16
Asheville, NC	35.4	37.9	158	71	42	28	18	12	7
Baltimore, MD	39.2	33.4	110	48	28	17	10	4	
Billings, MT	45.8	21.9	108	46	25	15	7		
Binghamton, NY	42.2	22.0	37						
Bismarck, ND	46.8	8.2	71	27	12				
Boise, ID	43.6	29.0	152	66	37	22	13	6	
Boston, MA	42.4	29.2	85	34	18	9			
Buffalo, NY	42.9	23.7	41						
Caribou, ME	46.9	10.7	52	17					
Casper, WY	42.9	23.2	138	63	37	24	16	10	5
Chicago, IL	41.8	24.3	79	31	15				
Cincinnati, OH	39.1	31.1	84	33	17	8			
Cleveland, OH	41.4	26.9	56	16					
Concord, NH	43.2	20.6	57	20					
Denver, CO	39.7	29.9	172	80	49	33	23	15	10
Des Moines, IO	41.5	19.4	90	36	19	10			
Detroit, MI	42.4	25.5	62	22					
Dodge City, KN	37.8	30.8	174	77	47	31	21	14	9
Duluth, MN	46.8	8.5	44						
Eagle, CO	39.6	18.0	119	53	31	20	12	7	
Edmonton, AL	53.6	6.6	78	29	11				
Fargo, ND	46.9	5.9	54	16					
Halifax, NS	44.6	25.9	80	32	16	7			
Hartford, CT	41.9	24.8	63	24	10				
Indianapolis, IN	39.7	27.9	76	29	14				
Intern. Falls, MN	48.6	1.9	40						
Kansas City, MO	39.3	27.1	117	50	28	17	10	4	
Louisville, KY	38.2	33.3	104	43	24	13	7		
Madison, WI	43.1	16.8	68	25	11				
Minneapolis, MN	44.9	12.2	59	20					
Mt. Shasta, CA	41.3	33.6	161	69	40	25	16	10	5
Newark, NJ	40.7	31.4	99	43	24	14	7		
New York, NY	40.8	32.2	86	36	19	10			
North Omaha, NE	41.4	20.2	99	41	22	12	6		
Philadelphia, PA	39.9	32.3	102	44	25	14	8		
Pittsburgh, PA	40.5	28.1	60	20					
Portland, OR	45.6	38.1	135	53	27	14			
Prescott, AZ	34.6	37.1	237	110	69	47	34	24	17
Providence, RI	41.7	28.4	82	33	18	9			
Rapid City, SD	44.0	21.9	109	47	26	15	8		
Reno, NV	39.5	31.9	208	95	58	39	27	18	12
Roanoke, VA	37.3	36.4	140	62	37	24	15	9	5
Salt Lake City, UT	40.8	28.0	157	69	40	26	16	10	5
Seattle, WA	47.4	38.2	127	50	24	11			
Springfield, IL	39.8	26.7	100	41	22	12	5		
St. Louis, MO	38.7	31.3	126	53	30	18	11	5	
Syracuse, NY	43.1	24.0	45						
Toronto, ON	43.7	25.2	78	31	15	6			
Vancouver, BC	49.3	37.2	129	52	25	12			
Washington, DC	38.9	32.1	100	43	24	14	7		
Wilmington, DE	39.7	32.0	104	45	26	15	9		

200 SOLAR HEATING SYSTEMS

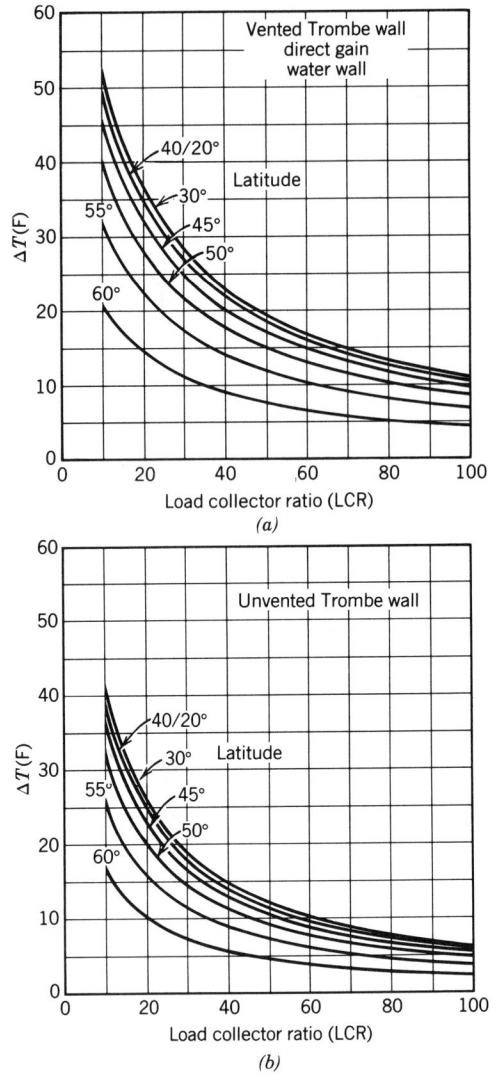

FIGURE 12.6 Indoor/outdoor temperature difference: clear January days (TD$_{sol}$). *Source:* Excerpted from the *Passive Solar Design Handbook*, Volume Two, January 1980.

where

- $t_{i\text{-JAN}}$ is the average indoor air temperature (°F) in January after a succession of clear days
- $t_{o\text{-JAN}}$ is the average outside air temperature (°F) in January
- TD$_{sol}$ is the expected temperature difference on a clear January day between the average inside air temperature and average outside air temperature. Determined from Figure 12.6 based upon design type, latitude and the LCR (load collector ratio)
- TD$_{int}$ is the heat gain from internal sources (people, lights, appliances, etc.). Normally about 5 to 7°F for residential applications.

Example 12D: Determine whether the house design described in Example 12C will produce excessive indoor air temperatures in January for any of the cities referenced in the solution. Assume that the heat gain from internal sources (TD$_{int}$) is 5°F.

Solution: Obtain the latitude and average January outdoor air temperature ($T_{o\text{-JAN}}$) for the referenced cities from Table 12.4.

Use city latitudes to determine TD$_{sol}$ based upon an LCR of 37. Use Figure 12.6a for the direct gain design type.

Then use Formula 12H to determine the average indoor air temperature in January ($T_{i\text{-JAN}}$) after a succession of clear days. For Prescott, AZ, where the average January outdoor air temperature is 37.1, and TD$_{sol}$ is found to be approximately 23°F from Figure 12.6a. The average indoor air temperature is then determined as follows:

$$t_{i\text{-JAN}} = t_{o\text{-JAN}} + \text{TD}_{sol} + \text{TD}_{int}$$

(Formula 12H)

$$= 37.1 + 23 + 5$$
$$= 65.1°F$$

With results for all of the referenced cities as follows:

City	Lat.	$T_{o\text{-JAN}}$		TD_{sol}		TD_{int}		$T_{i\text{-JAN}}$
Binghamton	42.2	22.0	+	22	+	5	=	49.0
New York	40.8	32.2	+	23	+	5	=	60.2
Des Moines	41.5	19.4	+	22	+	5	=	46.4
Roanoke	37.3	36.4	+	23	+	5	=	64.4
Casper	42.9	23.2	+	22	+	5	=	50.2
Reno	39.5	31.9	+	23	+	5	=	59.9
Prescott	34.6	37.1	+	23	+	5	=	65.1

Clearly, January overheating is not a problem for this design in any of these cities. In fact, some auxiliary heat will be required in some of the locations even on sunny days to maintain a comfortable indoor temperature. Further, Figures 12.6a and 12.6b indicate that large indoor–outdoor temperature differences (TD_{sol}) require very aggressive solar designs which attain LCRs of 30 or less.

Temperature Swing

In direct gain solar houses, the indoor temperature swing depends upon the indoor–outdoor temperature difference (TD_{sol}), and the ratio of mass surface to solar glazing area. Factors from Table 12.5 are then used in Formula 12I, which clearly shows that the addition of thermal mass will significantly reduce the swing in indoor air temperature.

$$t_{sw} = TSF \times TD_{sol} \quad \text{Formula 12I}$$

where

t_{sw} is the indoor air temperature swing on a clear January day in °F

TSF   is the temperature swing factor from Table 12.5

Once a temperature swing has been found, the maximum indoor air temperature can be found by using Formula 12J.

$$t_{max} = t_{i\text{-JAN}} + (t_{sw}/2) \quad \text{Formula 12J}$$

where

t_{max} is the maximum indoor air temperature on a clear January day in °F

$t_{i\text{-JAN}}$ is the average indoor air temperature on a clear January day in °F determined by Formula 12H

t_{sw} is the indoor air temperature swing on a clear January day in °F determined by Formula 12I

TABLE 12.5 DIRECT GAIN SOLAR HOUSES TEMPERATURE SWING FACTORS

Ratio: Mass Surface/Glazing Area	Temperature Swing Factors (TSF)
1.5	1.11
3.0	0.74
9.0	0.37

Source: Excerpted from the *Passive Solar Design Handbook,* Volume Two, January 1980.

In a similar manner, the minimum indoor air temperature (t_{min}) can be found by using Formula 12K.

$$t_{min} = t_{i\text{-JAN}} - (t_{sw}/2) \quad \text{Formula 12K}$$

Example 12E: Determine the January indoor air temperatures (minimum and maximum) for the solar house in Example 12D when located in Prescott, AZ. Assume that the mass area to glass area ratio is 3.0 (applicable to the direct gain "B" group of reference designs).

Solution: In Example 12D, the following were determined for the house when located in Prescott.

Indoor–outdoor temperature difference (TD_{sol}): 23°F Average
January indoor temperature ($T_{i\text{-JAN}}$): 65.1°F

The temperature swing (t_{sw}) is then found by using Formula 12I. Obtained from Table 12.5 for a mass area to glass area ratio of 3.0 is a TSF value of 0.74.

$$t_{sw} = TSF \times TD_{sol} \quad \text{(Formula 12I)}$$
$$= 0.74 \times 23$$
$$= 17$$

The maximum and minimum indoor air temperature in January can now be determined by using Formulas 12J and 12K.

$$t_{max} = t_{i\text{-JAN}} + (t_{sw}/2) \quad \text{(Formula 12J)}$$
$$= 65.1 + (17/2)$$
$$= 73.6°F$$

$$t_{min} = t_{i\text{-JAN}} - (t_{sw}/2) \quad \text{(Formula 12K)}$$
$$= 65.1 - (17/2)$$
$$= 56.6°F$$

The maximum temperature which will occur during late afternoon hours will be quite comfortable for any house occupants during that period. The minimum temperature (which will occur during overnight hours) is a bit low and may require the use of some form of auxiliary heat.

12.6 PRODUCTIVITY OF GLAZING

The amount of solar heat which solar glazing admits depends on its location and the clearness of the sky during the heating season. However, this glazed area also loses a large amount of heat compared to the insulated wall it replaces. The net benefit is the yield of useful solar heat, the difference between the solar gain and the heat loss. This is the "solar productivity" of the glass in units of Btus per square foot of glazing area.

The LCR tables can also be used to approximate the productivity of solar glazing for a particular city.

Example 12F: Determine the approximate solar productivity of south glazing for a house design meeting the DG-A2 designation which is built in Denver, Colorado. Assume a Building Load Coefficient (BLC) of 10,000 Btu/°F·d which is representative of a fairly well-insulated typical size house.

Solution: First obtain the LCR data from Table 12.4 for Denver for the DG-A2 design designation.

	\multicolumn{7}{c}{Solar Savings Fractions}
City	.10	.20	.30	.40	.50	.60	.70
Denver	172	80	49	33	23	15	10

To start, assume a SSF of .10 which corresponds to an LCR of 172. Use Formula 12C to determine the required area of vertical south-facing glass (A_p).

$$\text{LCR} = \frac{\text{BLC}}{A_p} \quad \text{(Formula 12C)}$$

$$A_p = \frac{\text{BLC}}{\text{LCR}}$$

$$= \frac{10,000}{172}$$

$$= 58 \text{ ft}^2 \text{ of south glass is required to achieve an SSF of .10}$$

This result can be used to determine the productivity of the south glazing using Formula 12F to compute the amount of heat saved by adding south glass to achieve an SSF of .10 (assuming base 65 heating degree-days).

$$Q_{\text{sol-yr}} = \text{SSF} \times \text{BLC} \times \text{HDD}_x$$
$$\text{(Formula 12F)}$$
$$= .10 \times 10,000 \times 6016$$
$$= 6,016,000 \text{ Btu}$$

From this we can calculate the productivity of each square foot of south glazing as follows:

$$6,016,000 \text{ Btu}/58 \text{ ft}^2 = 103,700 \text{ Btus}/\text{ft}^2$$

Other DG-A2 designs in Denver require ever-increasing amounts of glazing to achieve higher SSFs. Table 12.6 lists the area of south glazing needed, the additional glass area compared to the previous increment, and the solar productivity for each .10 increment of SSF developed by using the same method outlined above.

Note how the incremental south glazing productivities decrease sharply as the SSF increases, evidence of the "law of diminishing returns," as also illustrated in Figure 12.7. Therefore, adding south glazing to achieve a higher SSF must be carefully evaluated.

Shown in Table 12.7 is the approximate solar productivity (net yield of Btus per square foot of south glazing collector area) for the cities listed. These values are for the DG-A2 design, based on calculations similar to those used in the example above.

The yearly Btu savings shown in Table 12.7 are of course approximate and should be treated as such. Still, they provide an indication of what is achievable in various cities, and allow the reader to begin to understand the energy savings which can be

TABLE 12.6 SOUTH GLASS PRODUCTIVITY DENVER, COLORADO—DG-A2 SOLAR DESIGN

SSF Improvement	South Glass Area (ft²)	Additional ft² of South Glazing Req.	Total Btu Savings (Millions)	Incremental Glazing Productivity Net (Btu/ft²)
From 0 to .10	58	58	6.02	103,700
From .10 to .20	125	67	12.03	89,800
From .20 to .30	204	79	18.05	76,200
From .30 to .40	303	99	24.06	60,800
From .40 to .50	435	132	30.08	45,600
From .50 to .60	667	232	36.10	25,900
From .60 to .70	1000	333	42.11	18,100

TABLE 12.7 APPROXIMATE YEARLY BTU SAVINGS (SOLAR PRODUCTIVITY)—SOUTH HIGH-PERFORMANCE (Low-E or Triple) GLAZING DG-A2 DIRECT GAIN REFERENCE DESIGN

City	Btu Savings per ft^2 for South-Facing Glass Area to Improve SSF From			
	0 to .10	.10 to .20	.20 to .30	.30 to .40
Albany, NY	38,600	19,800		
Albuquerque, NM	97,400	86,900	77,000	63,000
Asheville, NC	66,900	54,600	43,600	35,600
Baltimore, MD	52,000	40,300	31,800	20,500
Billings, MT	78,500	58,200	39,800	27,200
Binghamton, NY	27,000			
Bismarck, ND	64,200	39,400	19,500	
Boise, ID	88,700	68,000	49,100	31,700
Boston, MA	47,800	31,900	21,500	10,100
Buffalo, NY	28,400			
Caribou, ME	50,100	24,300		
Casper, WY	104,300	87,600	67,700	51,600
Chicago, IL	48,400	31,300	17,800	
Cincinnati, OH	42,600	27,600	17,800	7,600
Cleveland, OH	34,500	13,800		
Concord, NH	42,000	22,700		
Denver, CO	103,500	90,000	76,100	36,200
Des Moines, IO	60,400	40,300	27,000	14,200
Detroit, MI	38,600	21,200		
Dodge City, KN	87,800	69,700	60,900	46,000
Duluth, MN	39,000			
Eagle, CO	100,300	80,500	62,900	47,500
Edmonton, AL	83,000	49,100	18,900	
Fargo, ND	50,100	21,100		
Halifax, NS	57,700	38,500	23,100	9,000
Hartford, CT	40,000	24,600	10,900	
Indianapolis, IN	42,400	26,200	15,100	
Intern. Falls, MN	42,200			
Kansas City, MO	62,700	46,800	34,100	23,200
Louisville, KY	48,300	34,100	25,200	13,200
Madison, WI	52,600	30,600	15,200	
Minneapolis, MN	48,100	24,700		
Mt. Shasta, CA	94,800	71,100	56,100	39,300
Newark, NJ	49,800	38,300	27,300	16,900
New York, NY	41,700	30,000	19,500	10,200
North Omaha, NE	65,400	46,200	31,300	17,400
Philadelphia, PA	49,600	37,600	28,200	15,500
Pittsburgh, PA	35,600	17,800		
Portland, OR	64,700	41,800	26,400	13,900
Prescott, AZ	105,600	91,500	82,500	65,700
Providence, RI	49,000	33,000	23,600	10,800
Rapid City, SD	79,800	60,500	42,600	26,000
Reno, NV	125,300	105,300	89,700	71,700
Roanoke, VA	60,300	47,900	39,500	29,400
Salt Lake City, UT	93,900	73,700	56,900	44,400
Seattle, WA	65,900	42,800	23,900	10,500
Springfield, IL	55,600	38,600	26,400	14,700
St. Louis, MO	59,900	43,500	32,900	21,400
Syracuse, NY	30,100			
Toronto, ON	53,300	35,100	19,800	6,800
Vancouver, BC	72,100	48,700	26,900	12,900
Washington, DC	50,100	38,000	27,200	16,800
Wilmington, DE	51,400	39,200	30,400	17,500
53 City Average:	60,800	N.A.	N.A.	N.A.

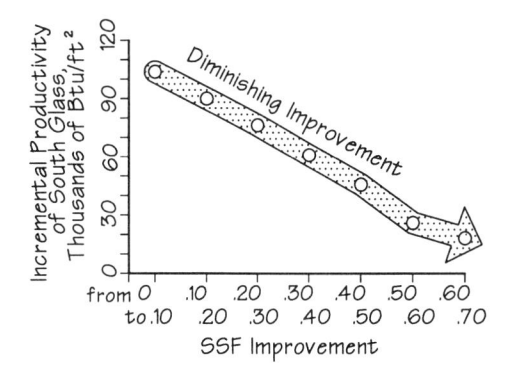

FIGURE 12.7 Productivity of south-facing glazing.

gained by this popular design (using high performance south-facing glazing and some thermal mass).

A similar analysis can of course be done for any of the other reference solar designs.

12.7 PASSIVE SOLAR HEATING OF "THE HOUSE"

Example 12G: Determine the solar heating performance of "The House" introduced in Section 5.14 if located in St. Louis, Missouri. Assume the following regarding the thermal mass:

- 450 ft² effective area of 6 in. thick thermal mass (brick over concrete slab). (Note: only two-thirds of the entire areas of the Dining Room, Living Room, and Bedroom #1 are counted to allow for furnishings.) This yields a volume (V_{tm}) of 225 ft³ of effective thermal mass.
- The specific heat (S_{tm}) of the thermal mass is 0.20 Btu/lb·°F.
- The density (D_{tm}) of the thermal mass is 120 lb/ft³.

Figure 12.8 shows a floor plan and cross section of "The House." Three rooms (the Dining Room, Living Room, and Bedroom #1) are oriented to the south.

Solution

Step 1: Compute the Building Load Coefficient (BLC) in units of Btus per degree-day by using Formula 12B. The house's total heat loss for average winter conditions (q_{tot}) is 08.3 Btu/°F·h as determined in Section 5.14. The area of south collector glazing (A_p) is 195 ft², and the U-factor of this glazing (U_c) is 0.35 Btu/ft²·°F·h.

$$\text{BLC} = [q_{tot} - (A_p \times U_c)] \times 24$$

(Formula 12B)

$$= [408.3 - (195 \times .35)] \times 24$$
$$= 340.0 \times 24$$
$$= 8160 \text{ Btu/°F·d}$$

Step 2: Compute the load collector ratio using Formula 12C.

$$\text{LCR} = \frac{\text{BLC}}{A_p} \quad \text{(Formula 12C)}$$

$$= \frac{8160}{195}$$

$$= 42$$

Step 3: To determine which direct gain reference design applies, compute the thermal storage capacity (TM_{cap}) using Formula 12D and values given in the description above.

$$TM_{cap} = \frac{S_{tm} \times D_{tm} \times V_{tm}}{\text{ft}^2 g}$$

(Formula 12D)

$$= \frac{0.20 \times 120 \times 225}{195}$$

$$= 27.7 \text{ Btu/°F per ft}^2$$

This thermal storage capacity result is close to that required by the "DG-A"

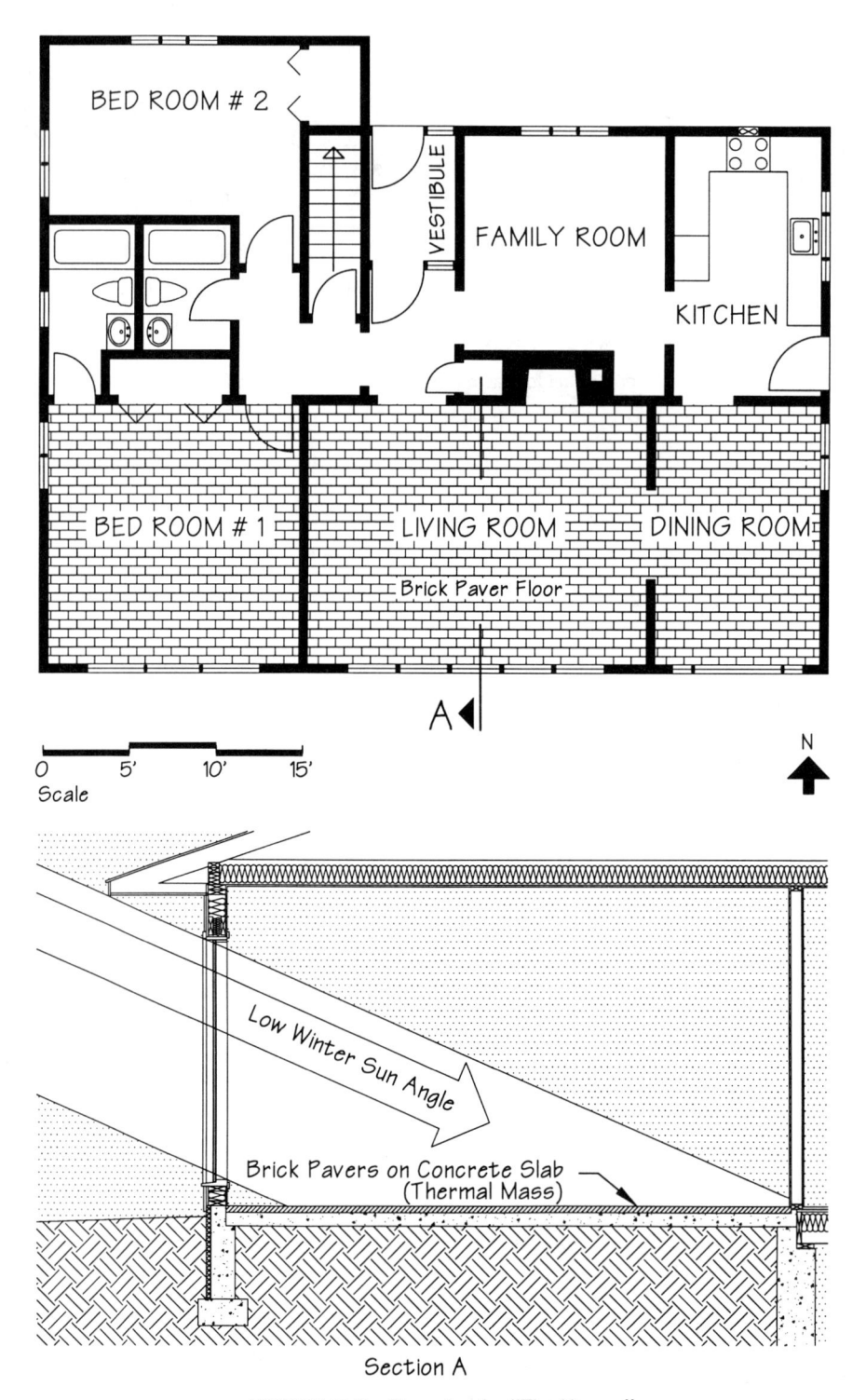

FIGURE 12.8 Floor plan for "The House."

reference design of 30. Therefore, this house design qualifies as a "DG-A2" design due to the amount of thermal mass and its high performance low-E glazing.

Step 4: Use Table 12.4 to determine the solar savings fraction for this house. Here is an excerpt of the information for the DG-A2 design in St. Louis.

City	Lat.	Jan t_o	.10	.20	Solar Savings Fractions .30	.40	.50	.60
St. Louis, MO	38.7	31.3	126	53	30	18	11	5

Since the computed LCR of 42 for the house in this example lies between the SSF values for .20 (LCR = 53) and .30 (LCR = 30). Formula 12E is then used to interpolate.

$$SSF_{int} = SSF_A + [.10 \times (C/B)]$$

(Formula 12E)

$$= .20 + [.10 \times ((53 - 42)/(53 - 30))]$$

$$= .20 + [.10 \times (11/23)]$$

$$= .20 + [.10 \times .48]$$

$$= .25 \ (25\%)$$

Step 5: Now that the interpolated solar savings fraction of 0.25 has been determined, use Formula 12F to compute the annual heating energy saved by the solar design (Q_{sol-yr}). The heating base temperature of the house has been computed to be approximately 65°F. Based upon this base temperature a total of 4750 heating degree-days is obtained from Table 2.3.

$$Q_{sol-yr} = SSF \times BLC \times HDD_x$$

(Formula 12F)

$$= 0.25 \times 8160 \times 4750$$

$$= 9,690,000 \ Btu/year$$

Such energy savings would result in operational savings of about $121 per year if natural gas were used costing $1.00 a therm (100 ft^3 of gas) and the Annual Fuel Utilization Efficiency (AFUE) is 80%. (See Section 24.1 for additional information on operational costs.)

Step 6: Finally, use Formula 12G to compute the annual quantity of auxiliary (nonsolar) heat which the house will still require.

$$Q_{aux-yr} = (1 - SSF) \times BLC \times HDD_x$$

(Formula 12G)

$$= (1 - 0.25) \times 8160 \times 4750$$

$$= 29,070,000 \ Btu/year$$

Such a quantity of auxiliary heat provided by natural gas (and using the same assumptions noted in Step 5) will cost approximately $363 per year.

Example 12G shows that passive solar design can yield significant energy savings, even in a city such as St. Louis where solar productivity is about average for major North American cities, as shown in Table 12.7. Moreover, solar savings can be particularly cost effective if they result primarily from proper house and room orientation. Well-insulated, tightly sealed passive solar houses in climates with high solar productivity can be designed to produce SSFs of .50 or greater.

Always remember, though, that an auxiliary heating system sized for the entire heat loss of the house including the south glass area must be provided to maintain comfort during extended periods of cold, cloudy winter weather.

13

WARM AIR HEATING SYSTEMS

Warm air heating systems are very popular in residences because of low cost, and because the same duct system which delivers warm air in the winter can also be used to deliver cool air during the summer. In fact, recent Census Bureau statistics show that in 1990 over 60% of new homes nationally were equipped with some form of central warm air heating system.

Warm air flow rates are widely referred to in terms of "cfm." Remember, that cfm means cubic feet of air per minute (ft^3/m), and to avoid errors the proper units should be written down when doing any type of calculation.

13.1 BASIC SYSTEM DESCRIPTION

Depicted in Figure 13.1 are the basic components of a warm air heating system.

In such a system, a centrifugal fan (usually referred to as a "blower") forces return air through the furnace where it is filtered and warmed by the heat exchanger. The air continues out of the furnace (often over a cooling coil used during the summer), and perhaps encounters an in-stream humidifier or electronic air cleaner for greatly improved indoor air quality.

The warmed supply air then travels through the duct distribution system before entering rooms and spaces to be heated through supply air outlets. The supply air outlets should be placed so that the warm supply air mixes with room air to create comfortable room conditions. The furnace blower then sucks return air through return air outlets and ducts back to the furnace for reheating and recirculation.

System Advantages

As stated above, the principal advantage of warm air heating systems is the ability to use system components for cooling in the summer simply by adding a cooling coil to the ducted airstream. Another system advantage is quick response to calls for heat (or cooling).

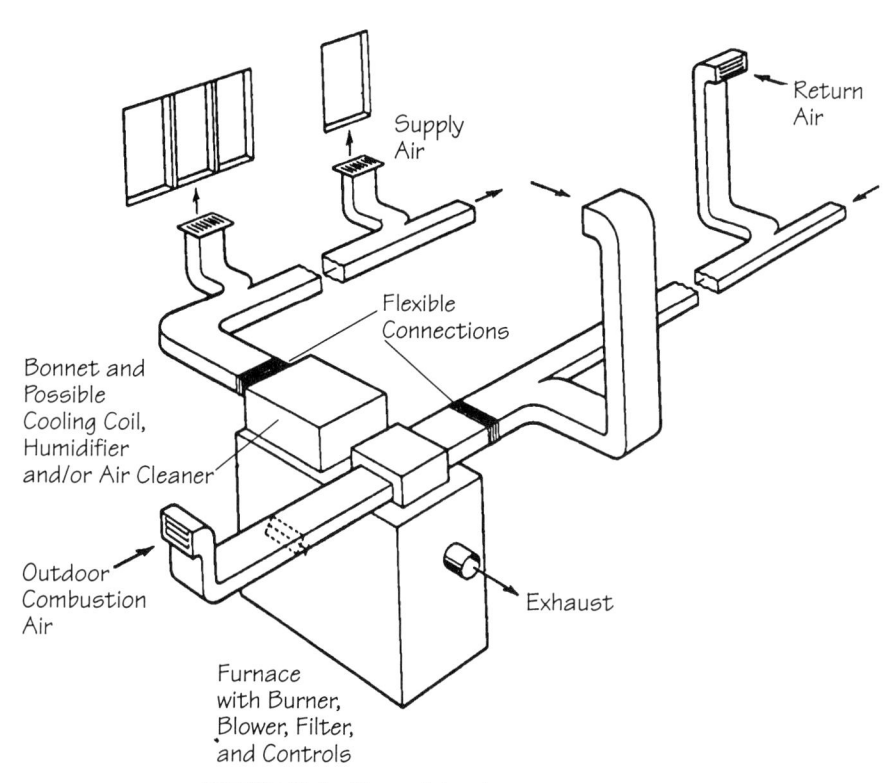

FIGURE 13.1 Warm air heating system components.

Warm air duct systems can also easily be fitted with a humidifier to prevent winter indoor air dryness or an electrostatic air cleaner to improve indoor air quality and all but eliminate dust from circulating and settling. Other than these very important and compelling advantages, hot air systems require more space, operate with more noise, and use more energy (the fan) than many other types of heating systems.

13.2 EARLY WARM AIR FURNACES

Early warm (or hot) air heating systems were simple coal-fired "pipeless furnaces" as shown in Figure 13.2a, where circulation was by "gravity": warm air rising from the unit and cool air dropping to be reheated. Such systems had serious shortcomings including manual control (i.e., how much coal to shovel in the furnace plus adjustment of dampers), uneven temperatures throughout the house, health threats of coal gas from incomplete combustion, and the potential fire hazard from a large open flame. Improvements to warm air gravity furnaces included supply and return ducts to provide for more evenly distributed heat as shown in Figure 13.2b.

13.3 MODERN FURNACE DESIGN

Modern furnace design has overcome the shortcomings noted above, and gravity systems are rare (e.g., used only for simple one-room structures). Today's furnaces generally use centrifugal fans which must be of adequate size to deliver required air quantities after overcoming system friction losses in the filter, furnace passageways,

210 WARM AIR HEATING SYSTEMS

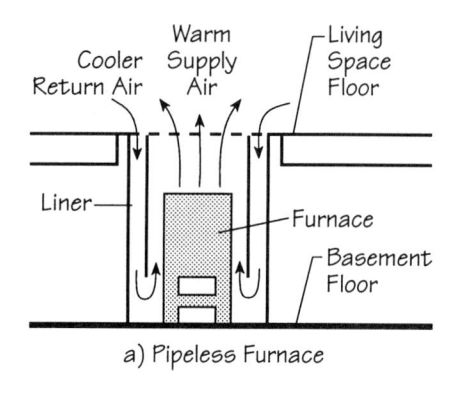

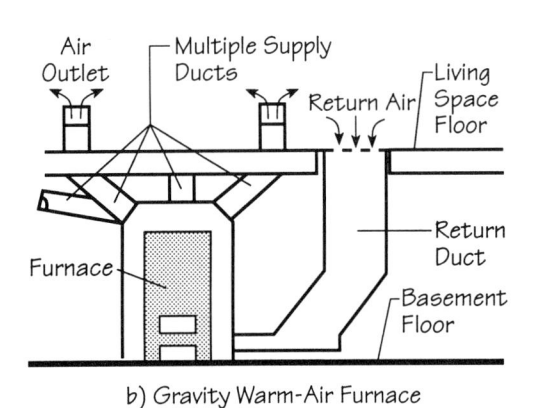

FIGURE 13.2 Early warm air furnaces.

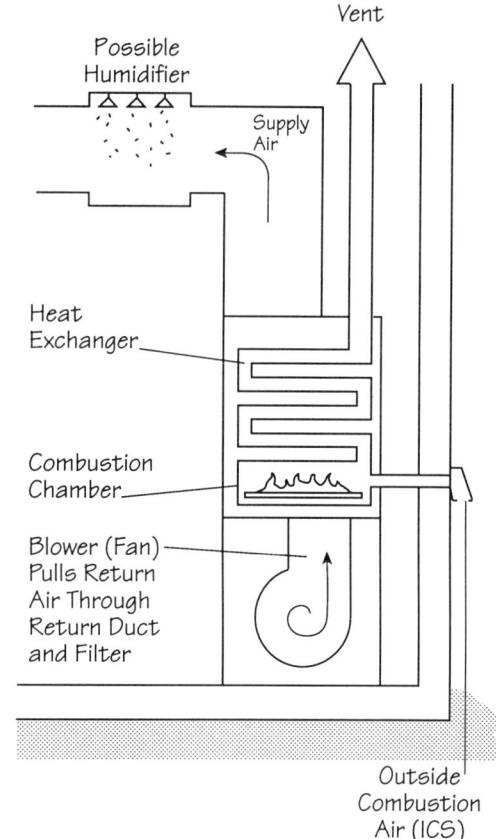

FIGURE 13.3 Schematic of modern warm air furnace (noncondensing).

cooling coil (if any), and ducts. The basic components of a modern mid-efficiency (noncondensing) furnace are shown schematically in Figure 13.3. When combustion air is supplied directly to the combustion chamber, as shown, it is known as an isolated combustion system (ICS) or "sealed combustion."

Furnace models are made to allow for upward, downward, or lateral distribution of heated supply air as shown in Figure 13.4. Basic house layout and architectural constraints such as a low basement ceiling may necessitate the use of a wide cabinet "low boy" unit. By far and away the most typical units are the "upflow" type.

The Blower

Blowers on furnaces can be operated at various speeds to provide varied air flows under different pressure conditions. The term "external static pressure" refers to system pressure losses which occur outside the furnace. Heating-only systems in houses normally encounter external static pressures of between 0.1 and 0.25 in. of water gauge (in. wg). In contrast, systems that contain a cooling coil in the airstream normally encounter external static pressures in the range of 0.4 to 0.6 in. wg. Blower control logic can be set in the factory, or adjusted in the field, to utilize the appropriate fan speeds

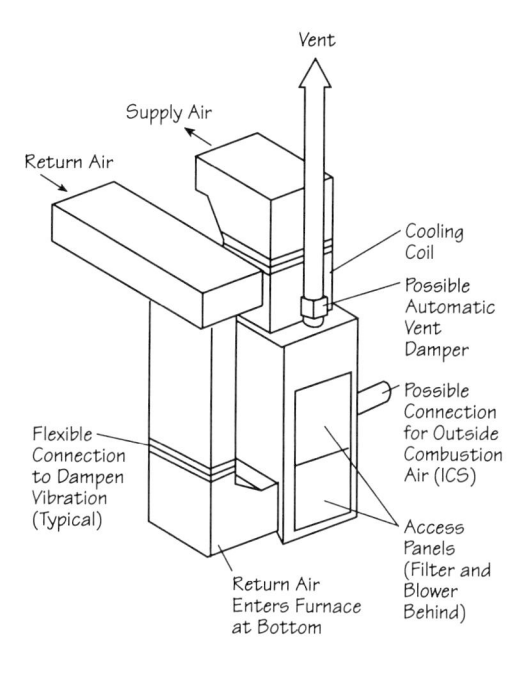

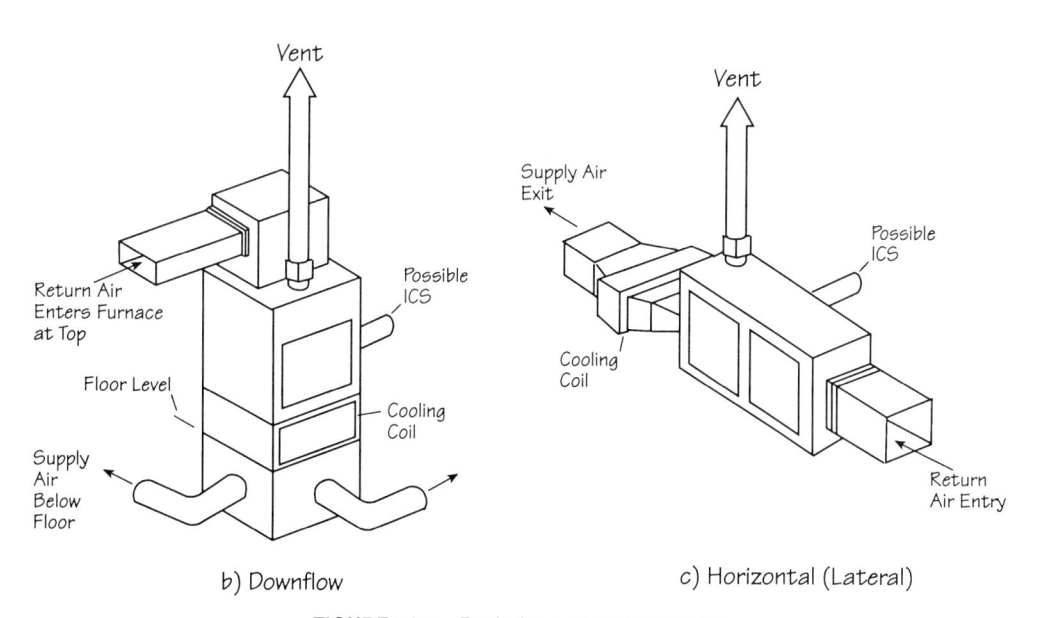

FIGURE 13.4 Basic furnace arrangements.

WARM AIR HEATING SYSTEMS

to satisfy required heating and cooling air flows. See Chapter 9 for a discussion of friction in air delivery systems.

Supply Air Temperature

Air typically returns to a furnace at about 65°F. The amount of heat then added to the supply air varies by manufacturer, unit size, and speed of blower. Small furnaces with heating outputs of less than 50,000 Btu/h may have a temperature rise of 20-50°F (yielding supply air which is 85-115°F). Larger capacity furnaces generally have a higher temperature rise of 50-80°F, producing supply air of 115-145°F, temperatures which may be uncomfortable near air supply outlets.

Another comfort problem can result from sporadic operation of a warm air heating system. If the blower is off for a prolonged period, air in the duct system cools down (especially if ducts are in an unheated basement or crawl space). Then when the thermostat calls for heat, the first air blown out is uncomfortably cold (like an automobile heating system).

Control

Warm air heating systems generally are "one-zone" systems controlled by a single thermostat. On a call for heat, warm air systems respond more quickly than hot water or steam systems since they are not constructed of massive, heat-absorbing elements. Improved comfort through less air stratification can also be achieved through continuous operation of the system fan, even when additional heat is not required.

Low Winter Humidity

Cold outside air cannot hold much moisture, and when it is heated, very low relative humidities (RH) can result. Figure 13.5 shows how air at 30°F and 50% RH when heated to 70°F will have a RH of only about 10%, which is too dry and well below the

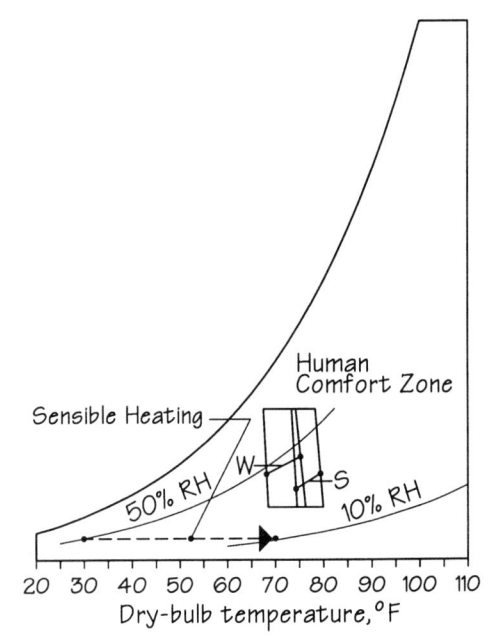

FIGURE 13.5 Heating of cold and dry outside air in winter.

human comfort zone. Therefore, humidifiers are added to increase the relative humidity of the airstream. Most humidifiers require cleaning and maintenance to prevent the growth of potentially unhealthy microorganisms and to prolong what often is a short useful life.

The quality of circulated air in warm air systems is maintained by installing a filter in the airstream ahead of the blower and heat exchanger. Many units employ disposable filters which should be changed often. Increasingly, systems now use permanent filters which can be washed (again often) and reused. For a very high level of air quality, including removal of smoke, electronic air cleaners can be used. These create an electric field which charges dust particles which are then collected on a plate with the opposite polarity.

Efficiency

In accordance with Federal Law, conventional furnaces manufactured after January

1, 1992 must have a tested AFUE (Annual Fuel Utilization Efficiency) of at least 78%. These units are referred to as "mid-efficiency" appliances which typically have AFUEs of between 78 and 84%. See Chapter 17 for information on efficiency and venting of combustion heating appliances.

13.4 VERY HIGH-EFFICIENCY FURNACES

Condensing Appliances

Condensing furnaces derive their high efficiency by recovering the heat from water vapor which traditionally has gone up the chimney as a high temperature (300–600°F) flue gas.

This is accomplished by lowering the temperature of the combustion gases below 212°F through contact with additional heat exchange area within the furnace. In the process the water vapor will condense back into a liquid and release about 1000 Btu for each pound of water (just like radiators in a steam heating system). As a result, the water from the combustion process exits as a low-temperature liquid (about 150°F) with only a fraction of the volume of the "high" temperature water vapor in the flue gas (0.0006 or less).

Condensing appliances do not require traditional chimneys. Exhaust of the remaining gases (carbon dioxide, nitrogen, and excess air) can be by small diameter (2–4 in.) corrosion-resistant PVC taken through a nearby side wall. With most of the heat recovered, the liquid condensate is drained to the sewage system (in some cases through a limestone filter to neutralize its slight acidity). The basic components of a condensing furnace are shown in Figure 13.6.

Some caution is advised in using condensing appliances since the technology is relatively new, and it does carry an initial cost premium. Nonetheless, benefits can be substantial, particularly in the case of replacement units in houses with a large heating load or in new buildings where the construction of a chimney may be avoided.

Condensing technology is applied in furnaces far better than in boilers for most applications. This is because a condensing appliance works by significantly lowering the temperature of the combustion gases within the heat exchanger. Such a reduction in heat exchanger temperature is incompatible with boiler return water from baseboard radiation (or other terminal devices) which usually is 160°F or more. This is substantially higher than furnace return air, which is about 65°F. One potential application of condensing boilers is in radiant systems where the supply water temperature required for circulation is much lower.

Pulse Combustion

"Pulse" combustion furnaces use a different technological approach to combustion, to obtain efficiencies of over 90%. They work by having gas and air enter a small combustion chamber where they are mixed and ignited in a manner very similar to that of an automobile internal combustion engine.

The explosive combustion process is initiated by a spark plug. The first explosion closes the fuel/air valves and forces the heated combustion gases out of the chamber and through the heat exchanger. This results in a pressure drop within the chamber, causing the fuel/air valves to reopen. The heat in the chamber remaining from the first explosion then is enough to ignite the second charge, and the spark plug is not needed. Therefore, the combustion process becomes a self-generating "chain-reaction" with 60 explosions per second. The products of combustion for pulse heating appliances are liquid water and low temperature gases, which are disposed of in a similar manner to other condensing equipment.

Pulse technology was brought to the marketplace in the late 1970s and has

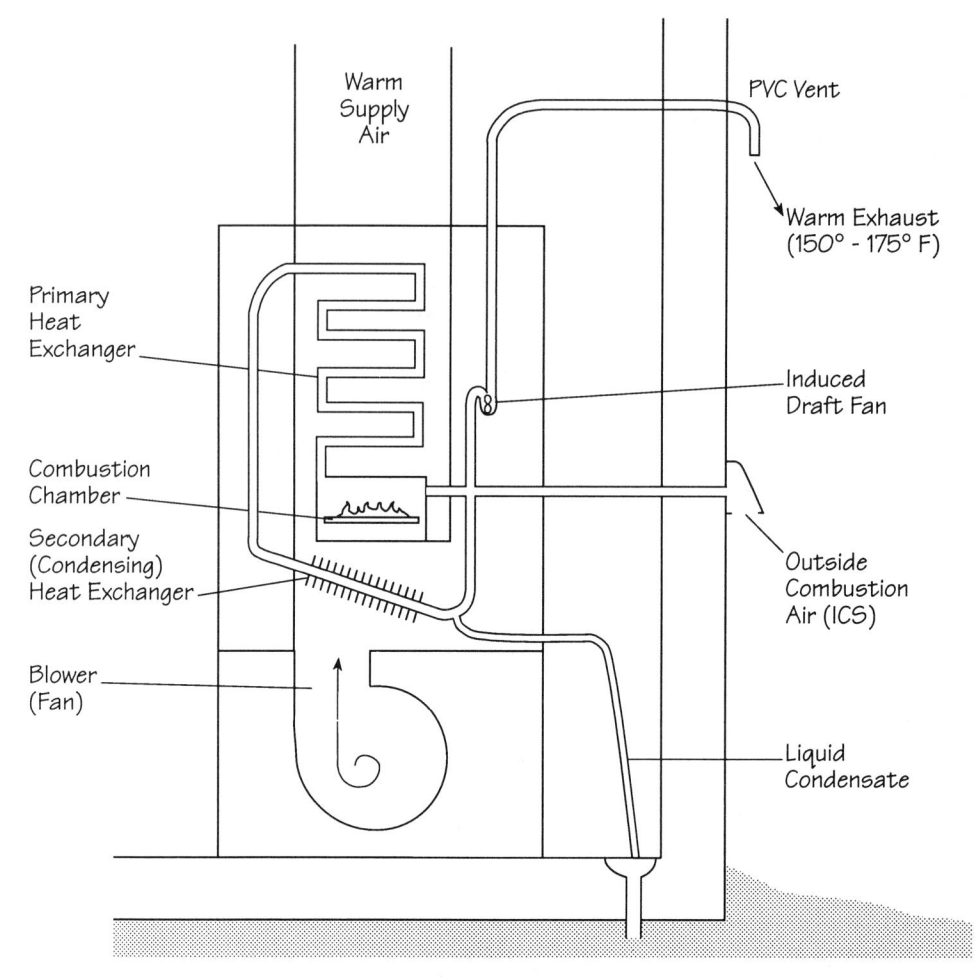

FIGURE 13.6 Condensing warm air furnace, which yields improved operational efficiency.

proven to be reliable. Still, noise can be a problem, since the pulse combustion process produces a humming vibration which requires isolation from occupied spaces.

13.5 FURNACE SIZING

Residential furnaces are available for heating output capacities ranging from about 30,000 to 200,000 Btu/h with many new small-capacity units being brought to the market in recent years at the low end of the range to accommodate town houses, condominiums, and energy-conserving houses.

Sizing is based upon the quantity of Btus per hour which the furnace must provide to offset building heat loss under winter design conditions. Sizing also involves use of an oversizing factor (OSF) to assure adequate capacity. Adding about 10–15% is typical if a system is being engineered closely. Greater oversizing of up to 25% (the limit in ASHRAE Standard 90.1) may be appropriate in systems for commercial buildings if night-time temperatures are to be set back significantly, to assure the capacity for quick pick-up of space temperatures the following day. Formula 13A is used to determine the required output of useful heat by the furnace in Btus per hour.

$$\text{Furn}_{out} = q_{tot} \times (t_i - t_o) \times \text{OSF}$$

Formula 13A

where

Furn$_{out}$	is the required furnace heat output in Btu/h
q_{tot}	is the total winter design heat loss in Btu/°F·h
t_i	is the inside winter design temperature in °F
t_o	is the outside winter design temperature in °F
OSF	is the oversizing factor (Note: a factor of 1.10 represents an increase in size of 0.10 or 10%)

Example 13A: The design heat loss of a house to be built in Philadelphia has been calculated to be 620 Btu/°F·h (q_{tot}). Determine the required furnace heat output (F_{out}). Assume an indoor design temperature (t_i) of 68°F, and an oversizing factor (OSF) of 1.10 (an increase in capacity of 10%).

Solution: The winter design dry-bulb temperature for Philadelphia is 14°F (97.5% value from Table 2.1). First calculate the heat output required to meet the design condition by using Formula 13A.

$$\text{Furn}_{out} = q_{tot} \times (t_i - t_o) \times \text{OSF}$$

(Formula 13A)

$$= 620 \times (68 - 14) \times 1.10$$

$$= 36,830 \text{ Btu/h}$$

After consulting manufacturers' literature, one may find that units which will deliver outputs of 30,000, 45,000, or 60,000 Btu/h are available from a particular manufacturer. This presents a problem. The 30,000 Btu/h unit is too small, since it is less than the design heat loss of 33,480 Btu/h (620 × 54) even before the oversizing factor is applied. However, if the unit is too large it will cycle (go on and off and on again, etc.) too often and be inefficient.

Calculate the oversizing factor for the 45,000 Btu/h output unit. It would be 1.34 (45,000/33,480), much too large. A 37,500 Btu/h output furnace by another manufacturer is finally selected. This unit yields an acceptable oversizing factor of 1.12 (37,500/33,480).

13.6 RESIDENTIAL AIR DISTRIBUTION OPTIONS

The basic approach to air distribution from the furnace depends upon the construction features of the house and predominant need for either heating (in cold or temperate climates) or cooling (in warm or hot climates).

Houses with Basements

The extended plenum system as shown in Figure 13.7 is typically used in houses with basements to limit the locations where headroom problems might exist. With this approach, a large main duct extends from the supply outlet of the furnace in one or two directions. Branches run between floor joists, then deliver heated supply air to each perimeter air outlet.

Houses Without Basements

Two basic choices are available for slab-on-grade and crawl space houses, depending upon severity of climate and house geometry. Depicted in Figure 13.8 is a loop perimeter system well suited to slab-on-grade construction since the distribution system will help to warm the floor slab at the perimeter. Such a system needs to have a high level of insulation at the slab edge and below the slab near the perimeter to limit heat loss and prevent condensation.

Shown in Figure 13.9 is a simple radial perimeter distribution system which may require less ductwork than the loop perimeter system for some house configurations.

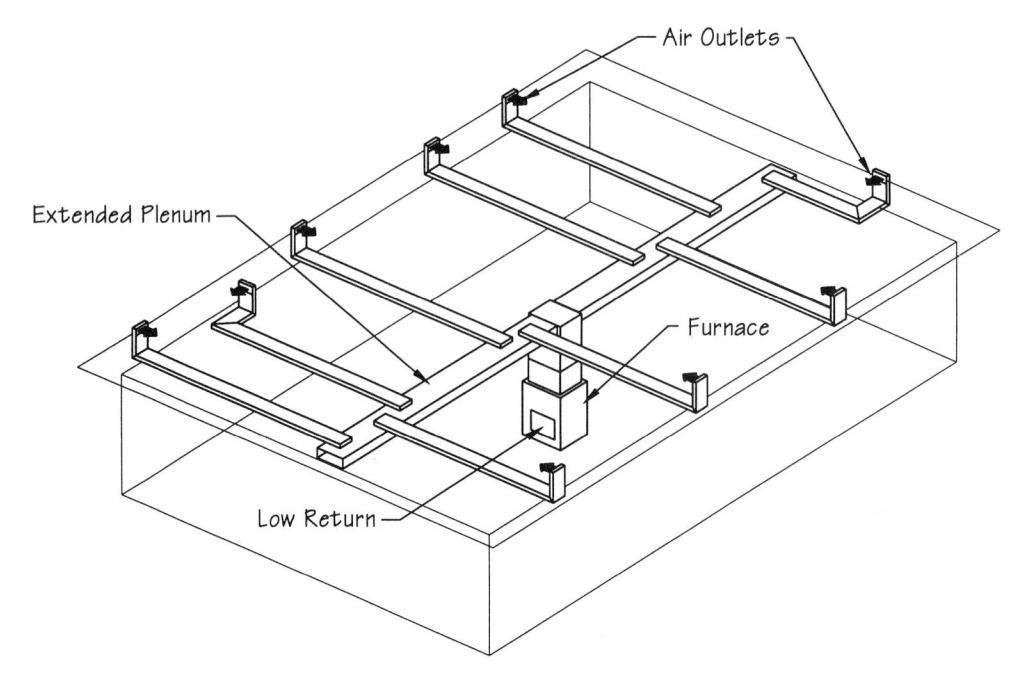

FIGURE 13.7 Extended plenum duct distribution method.

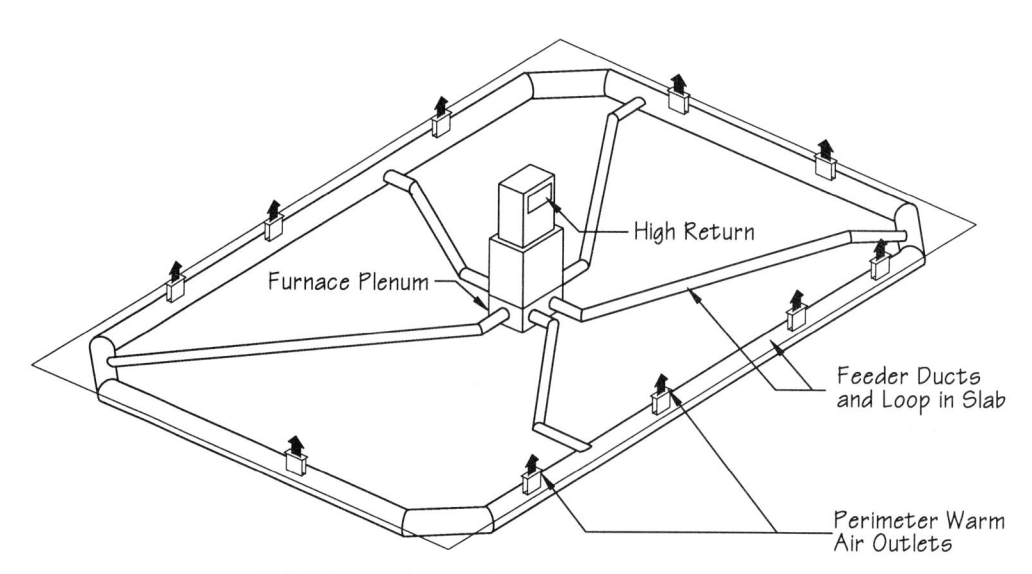

FIGURE 13.8 Loop perimeter air distribution system.

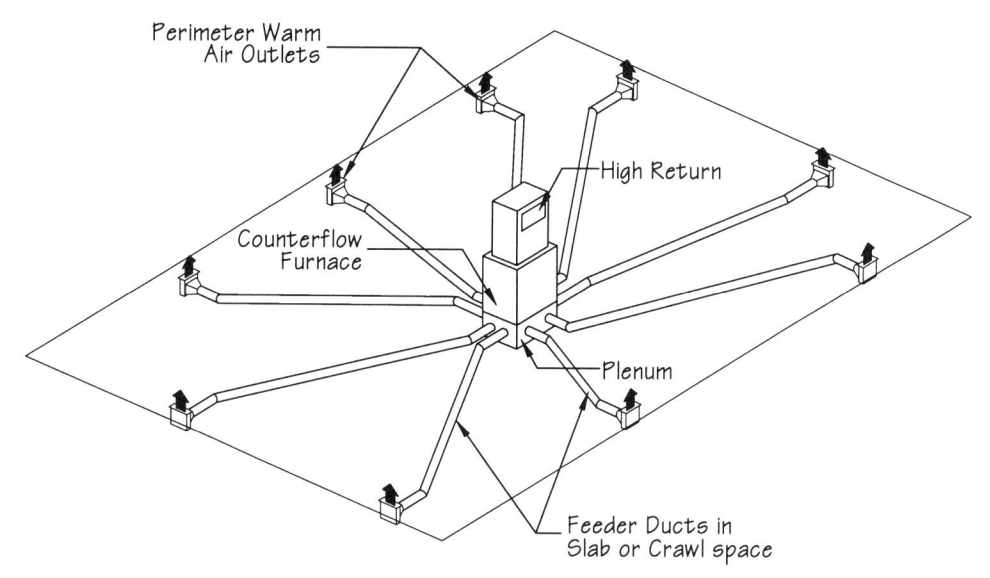

FIGURE 13.9 Radial perimeter air distribution system.

13.7 AMOUNT OF WARM AIR REQUIRED

The quantity of delivered air needed depends upon the design heat loss of the space (or entire zone) and the temperature of supply air compared to the air temperature being maintained in the space.

Use Formula 13B to determine the required flow rate of warm air.

$$\text{cfm} = \frac{q_{\text{tot}}}{1.08 \times (t_s - t_i)} \quad \text{Formula 13B}$$

where

- cfm is the air flow requirement in cubic feet per minute
- q_{tot} is the total quantity of heat needed by the space or zone being supplied in Btu/h
- 1.08 is an air constant (the heat capacity of air, 0.018 Btu/°F·ft³ × 60 m/1 h) in Btu·m/°F·ft³·h
- t_s is the supply air temperature in °F
- t_i is the inside design temperature in °F

Supply air ductwork can be expected to lose 10–15% of air and heat through leakage. When supply ducts are within the heated volume of the building, this loss is inconsequential. Ducts that pass through unheated spaces such as basements and crawl spaces should have their joints taped to minimize losses and supply air flow should be increased by 5–10% to compensate.

Example 13B: A house has a design heat loss of 35,000 Btu/h (q_{tot}). Determine the required air flow if the supply air will have a temperature (t_s) of 110°F, and the house has an inside design temperature (t_i) of 70°F.

Solution: Use Formula 13B to determine the air flow.

$$\text{cfm} = \frac{q_{tot}}{1.08 \times (t_s - t_i)}$$

(Formula 13B)

$$= \frac{35,000}{1.08 \times (110 - 70)}$$

$$= \frac{35,000}{43.2}$$

$$= 810 \text{ ft}^3/\text{m}$$

13.8 SIZING DUCTS

Ducts to supply heated air can be easily sized once the required air flow in cfm is known and the velocity of the heated supply air is chosen. For residences, warm air velocities of approximately 750–1000 feet per minute (fpm) are generally recommended for main ducts, and about 500 fpm for branch ducts. Use Formula 13C to approximate required duct area and see Chapter 9 for additional duct sizing procedures.

$$A_{duct} = \frac{\text{cfm} \times 144}{V_{fpm}} \times \text{FA} \quad \text{Formula 13C}$$

where

A_{duct} is the required duct area in in.^{2}
cfm is the air flow rate in ft^3/m
FA is the frictional allowance as follows:
  1.0 for round or nearly square ducts
  1.10 for rectangular ducts where the ratio of width to depth is about 1 : 3
  1.25 for thin rectangular ducts where the ratio of width to depth is about 1 : 5
144 is the conversion of 144 in.^{2} in 1 ft^2
V_{fpm} is the air velocity in feet per minute

Once required duct area has been determined, the actual shape and dimensions can be established.

Example 13C: Determine the required size of the rectangular ductwork in Example 13B for an air flow of 810 cfm. Assume a velocity of 800 fpm.

Solution: Use Formula 13C to determine the required duct area.

$$A_{duct} = \frac{\text{cfm} \times 144}{V_{fpm}} \times \text{FA}$$

(Formula 13C)

$$= \frac{810 \times 144}{800} \times 1.10$$

$$= 160 \text{ in.}^2$$

Therefore, the required air flow could be met by a rectangular supply duct which is 20 in. wide and 8 in. deep.

Supply Ducts

It is important to understand the factors that determine duct size as illustrated in the examples above. However, in the layout of small duct systems for houses it is often useful to determine approximate duct sizes as a preliminary design guideline. Provided in Table 13.1 is the average capacity of residential supply ducts with acceptable air velocities.

Return Air Ducts and Plenums

The approximate size of return ducts and plenums can be determined from Table 13.2. Note how the cfm capacity of returns is much less than the cfm capacity of equally sized supply ducts (as shown in Table 13.1). This is because of the reduced air velocity of the room air which the blower induces into the returns.

TABLE 13.1 AVERAGE cfm CAPACITY OF TYPICAL RESIDENTIAL SUPPLY DUCTS

Duct Type	Duct Size (in.)	Nominal Cross-Section Area (in.)	Nominal cfm Capacity
Round or oval branch duct or (rectangular equivalent)	4	13	40
	5 ($3\frac{1}{4} \times 8$)	20	70
	6 ($3\frac{1}{4} \times 10$)	28	120
	7 ($3\frac{1}{4} \times 14$)	38	170
	8	50	250
Rectangular trunk duct	8×8	64	300
	12×8	96	500
	16×8	128	750
	20×8	160	1000
	24×8	192	1250
	28×8	224	1500

Source: Reprinted with permission from Ronald K. Yingling, Donald F. Luebs, and Ralph J. Johnson, *Residential Duct Systems: Selection and Design of Ducted HVAC Systems,* © 1981 Home Builder Press, National Association of Home Builders, 1201 15th Street, NW, Washington, DC 20005, (800) 223-2665, p. 62.

TABLE 13.2 AVERAGE cfm CAPACITY OF TYPICAL RESIDENTIAL RETURN DUCTS AND PLENUMS

Nominal Size (in.)	Approximate Capacity (cfm)
Floor joist plenum:	
8 in. deep, 16 in. oc	300
10 in. deep, 16 in. oc	420
6×8	110
8×8	170
10×8	220
12×8	280
14×8	250
16×8	400
18×8	460
20×8	510
22×8	590
24×8	650
26×8	710
28×8	770
30×8	840

Source: Extracted from *Summer Air Conditioning* by Konzo, Carroll, and Bareither as published by The Industrial Press.

13.9 DUCT INSULATION

The function of ducts in warm air systems is to transport heated air to its intended destination. Where supply ducts pass through unheated basements, garages, and crawl spaces, insulation is needed to minimize heat loss. Use Formula 13D to determine the minimum R-value of insulation required by many energy codes and standards. Fiberglass duct wrap blankets ($1\frac{1}{2}$ or 2 in. thick) with vapor barrier facing (and taped joints) are typically used.

$$R_i = \frac{TD}{15} \qquad \text{Formula 13D}$$

where

R_i is the minimum R-value of the insulation
TD   is the design temperature difference between the air in the duct and the space through which it travels in °F
15   is an empirical factor

Example 13D: A duct with 130°F warm air passes through a garage with a winter design temperature of 40°F. Determine the minimum R-value of duct insulation.

Solution: Use Formula 13D.

$$R_i = \frac{TD}{15} \qquad \text{(Formula 13D)}$$

$$= \frac{130 - 40}{15}$$

$= 6$, the minimum R-value for the duct insulation

13.10 AIR OUTLETS

Warm supply air travels through a system of ductwork to outlets placed either in the floor, wall, ceiling, or simply in the side or at the end of a duct located within a space. After entering and mixing with room air, air then is induced to enter return inlets by the furnace blower. The following types of air outlets and returns are used:

Grilles: Simple louvered or perforated covers which fit on the side or end of a supply or return duct, or as an entry path to a return air plenum. Most commonly, grilles are rectangular in shape with louvered vanes which may be adjustable to redirect air flow.

Registers: Similar to grilles, they also have a damper to control the volume of air flow.

Diffusers: Ceiling outlets used in air systems which provide cooling, or heating and cooling. Diffuser design includes deflectors designed to mix supply air with large quantities of room air without creating a draft. And in the case of cool air, diffusers prevent dense cool air from simply dropping from the ceiling and causing uncomfortable conditions.

The function of supply air outlets is to deliver warm or cool air (known as "primary air") to a space. The inflow of primary air induces some room air (known as "secondary air") to join in a flow pattern which results in mixing. The amount of secondary air induced depends upon many factors such as outlet type, room geometry, outlet velocity, and air volume. Secondary air values of 10–20 times the amount of primary air are typical. Therefore, a supply air outlet which delivers 60 cfm of primary air typically induces circulation of between 600 and 1200 cfm of secondary air.

Illustrated in Figure 13.10 are two important terms used to describe outlet performance:

Throw: the distance that an outlet will propel supply air (which exits at a ve-

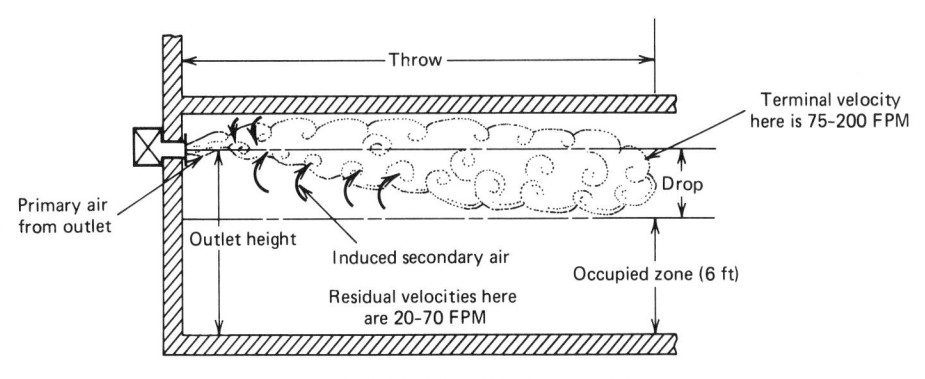

FIGURE 13.10 Air outlet throw and drop.

locity of 300–800 fpm) before the velocity of the air decreases to about 50 fpm.

Drop: the vertical distance that cool primary air has fallen at the end of the "throw" when projected horizontally.

Also important to performance is spread.

Spread: the pattern of air flow in a room once it exits an air outlet. Spread depends upon outlet design, with the use of angled vanes making it possible to distribute primary air over a wider area but with less throw. See Figure 13.11 which also indicates the suction of cool air known as "aspiration."

Air Outlet Locations

Warm air supply outlets have normally been located below windows to counter the dropping cool air. Floor registers, such as those illustrated in Figure 13.11, are often used in heating-only systems. Distribution of air during the heating season works well since the warm air is thrown up past windows to combat drafts. Then the less dense heated air will continue to rise and pass across the ceiling before flowing to return outlets on opposite walls.

However, for dual service systems used for both heating and cooling, such an approach will be less than ideal during the cooling season when dense cool air exits from floor registers. The horizontal position of floor registers also make them prone to collect dirt and dust and even loose articles.

For climates where the need for cooling dominates, the use of high wall outlets is a good solution. As shown in Figure 13.12, the total cool air pattern is well mixed. The potential problem of stratification when heat is needed is also shown. This can be minimized by using adjustable deflection vanes and/or ceiling fans in spaces over about 12 ft high.

Air distribution from ceiling outlets in residential warm air heating systems is rare. When used (i.e., with an attic-mounted horizontal furnace) the design concerns are generally similar to high wall outlets. Additional information on air outlets is contained in Section 21.5.

Sizing Air Outlets and Returns

The capacity of air outlets or returns depends upon their "free area" and air velocity. Most outlets have a free area which is approximately two-thirds of the nominal area. Therefore, a 10 × 6 register with a nominal overall area of 60 in.^{2} would have

222 WARM AIR HEATING SYSTEMS

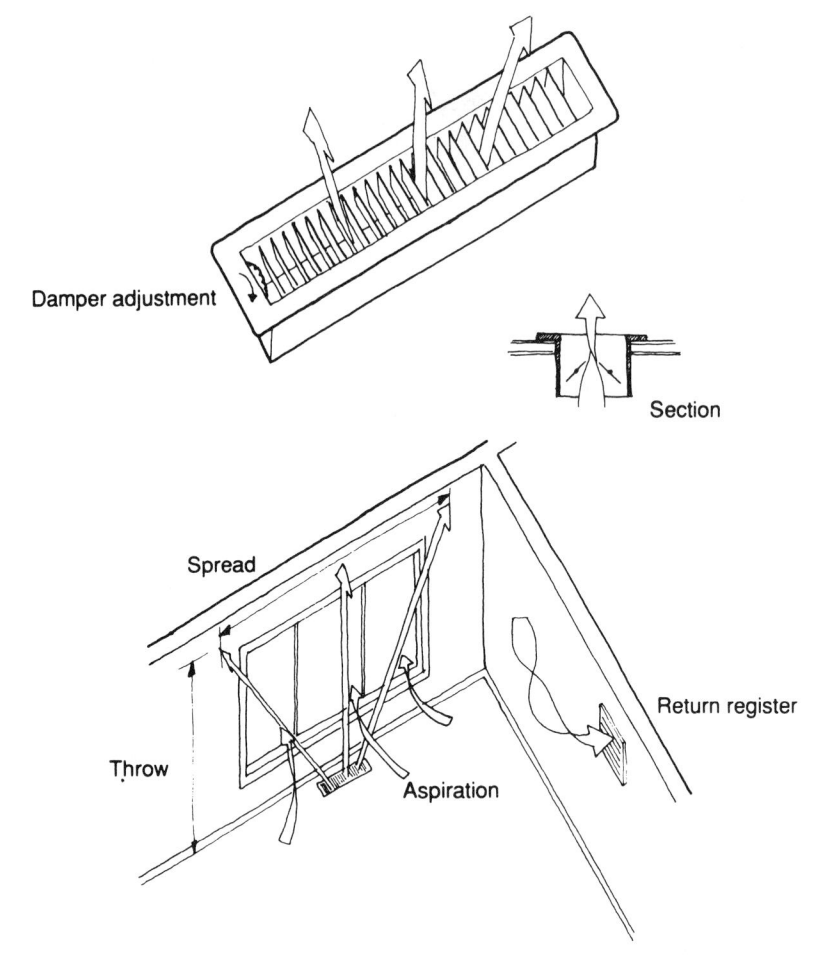

FIGURE 13.11 Spread of air from a floor outlet.

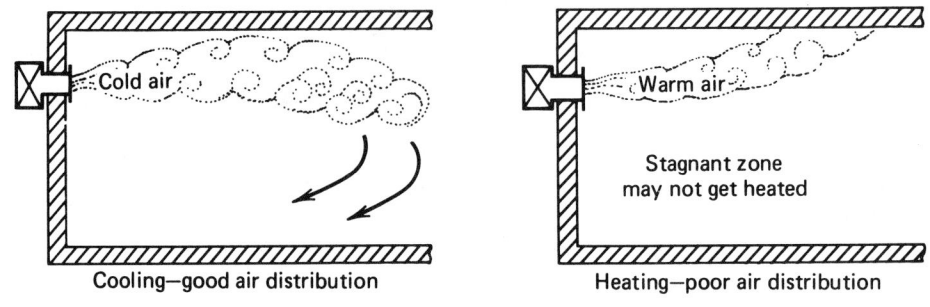

FIGURE 13.12 High wall air outlets.

TABLE 13.3 TYPICAL CAPACITIES OF SUPPLY AIR OUTLETS

Nominal Register Sizes (in.)	Air-flow rate (cfm) Wall Registers	Air-flow rate (cfm) Floor Registers
$10 \times 2\frac{1}{4}$	NA	50 to 70
10×4	80 to 120	100 to 140
10×6	130 to 190	NA
10×8	180 to 250	NA
$12 \times 2\frac{1}{4}$	NA	60 to 90
12×4	100 to 140	120 to 170
12×6	160 to 230	NA
12×8	220 to 310	NA
12×10	280 to 400	NA
$14 \times 2\frac{1}{4}$	NA	70 to 110
14×4	80 to 120	140 to 200
14×6	190 to 260	NA
14×8	260 to 370	NA
14×10	330 to 460	NA

Source: Extracted with permission from *Summer Air Conditioning* by Konzo, Carroll, and Bareither as published by The Industrial Press.

a free area of about 40 in.² for air to actually flow through.

As a preliminary design guideline, the capacity and nominal sizes of typical residential supply registers (with the range reflecting air speeds from 500 to 750 fpm) are provided in Table 13.3. Consult manufacturers' catalog information for actual values for a particular outlet design.

Provided in Table 13.4 is typical return intake capacity based upon a velocity of 500 fpm.

13.11 WARM AIR HEATING OF "THE HOUSE"

Example 13E: Design a warm air heating system for "The House."

TABLE 13.4 SIZES AND CAPACITIES OF RETURN INTAKES

Nominal Size (in.)	Approximate Capacity (cfm)
10×4	80
10×6	130
10×8	180
12×4	100
12×6	160
12×8	220
12×10	280

Source: Extracted with permission from *Summer Air Conditioning* by Konzo, Carroll, and Bareither as published by The Industrial Press.

Solution: As detailed in Section 5.14, "The House" has room-by-room design heat losses as shown in the following summary.

"THE HOUSE" ROOM-BY-ROOM HEAT LOSS FOR WINTER DESIGN CONDITIONS

Room	Room Design Heat Loss (Btu/h)
Dining room	5480
Living room	8100
Bedroom #1	6440
Ext. bathroom	1340
Bedroom #2	4680
Vest./hall	1460
Family room	2620
Kitchen	3030
	q_{tot} = 33,150 Btu/h

Furnace Sizing

Since the design heat loss which incorporates the difference between inside and outside design temperatures is already known (q_{tot}) in units of Btu/h, a variation of Formula 13A can be used to determine the required furnace output in Btus per hour (Btu/h). Assume an oversizing factor of 1.10.

$$\text{Furn}_{out} = q_{tot} \times \text{OSF}$$

(Formula 13A variation)

$$= 33,150 \times 1.10$$

$$= 36,465 \text{ Btu/h, the required furnace output}$$

Supply Air Flow

Determine the required supply air flow in cfm for each room of the house by using Formula 13B as illustrated for the Dining Room.

Dining Room Supply Air. The Dining Room has a design heat loss of 5480 Btu/h (q_{tot}). Base the design on supply air with a temperature (t_s) of 100°F and an inside design temperature (t_i) of 70°F.

$$\text{cfm} = \frac{q_{tot}}{1.08 \times (t_s - t_i)}$$

(Formula 13B)

$$= \frac{5480}{1.08 \times (100 - 70)}$$

$$= 170 \text{ ft}^3/\text{m}$$

This required air flow can then be provided by air outlets (usually two or more for good distribution in large rooms). In a similar manner, the supply air requirements of all rooms can be computed and "rounded" as shown below. Also indicated are the nominal sizes of wall registers and return air intakes and the number required (indicated in parentheses) based upon Tables 13.3 and 13.4. Sizing of supply air outlets is based upon an air speed of 500 fpm.

ROOM-BY-ROOM SUPPLY AIR AND OUTLET REQUIREMENTS

Room	Required cfm	Supply Wall Registers	Return Air Intakes
Dining room	170	12 × 4 (2)	10 × 6 (2)
Living room	250	12 × 4 (2)	10 × 6 (2)
Bedroom #1	200	12 × 4 (2)	10 × 6 (2)
Exterior bathroom	40	10 × 4	No return
Bedroom #2	150	12 × 4 (2)	10 × 4 (2)
Vest./hall	50	No supply[a]	No return
Family room	80	10 × 4	10 × 4 (2)
Kitchen	100	12 × 4	No return
Total:	1040		

[a]Note: The supply air requirement of the vestibule/hall has been added to the adjoining Family Room.

13.11 WARM AIR HEATING OF "THE HOUSE"

It is common to supply air to kitchens and bathrooms but not to return it. Instead potentially humid and odorous air from these spaces is exhausted to the outside by vents, range hoods, or operable windows.

Supply Duct Layout

For "The House," having a partial basement, the extended plenum duct distribution method combined with a perimeter loop will be used. The furnace will be centrally located as shown in Figure 13.13. The main supply duct to the left (east) of the furnace will serve the Vestibule, Family Room, Kitchen, Dining Room and Living Room and deliver 520 cfm. The main supply duct to the right (west) of the furnace will serve Bedroom #1, the exterior Bathroom, and Bedroom #2 and also deliver 520 cfm.

Supply Duct Sizing

Air velocities to be used will be 800 feet per minute (fpm) for main supply ducts and

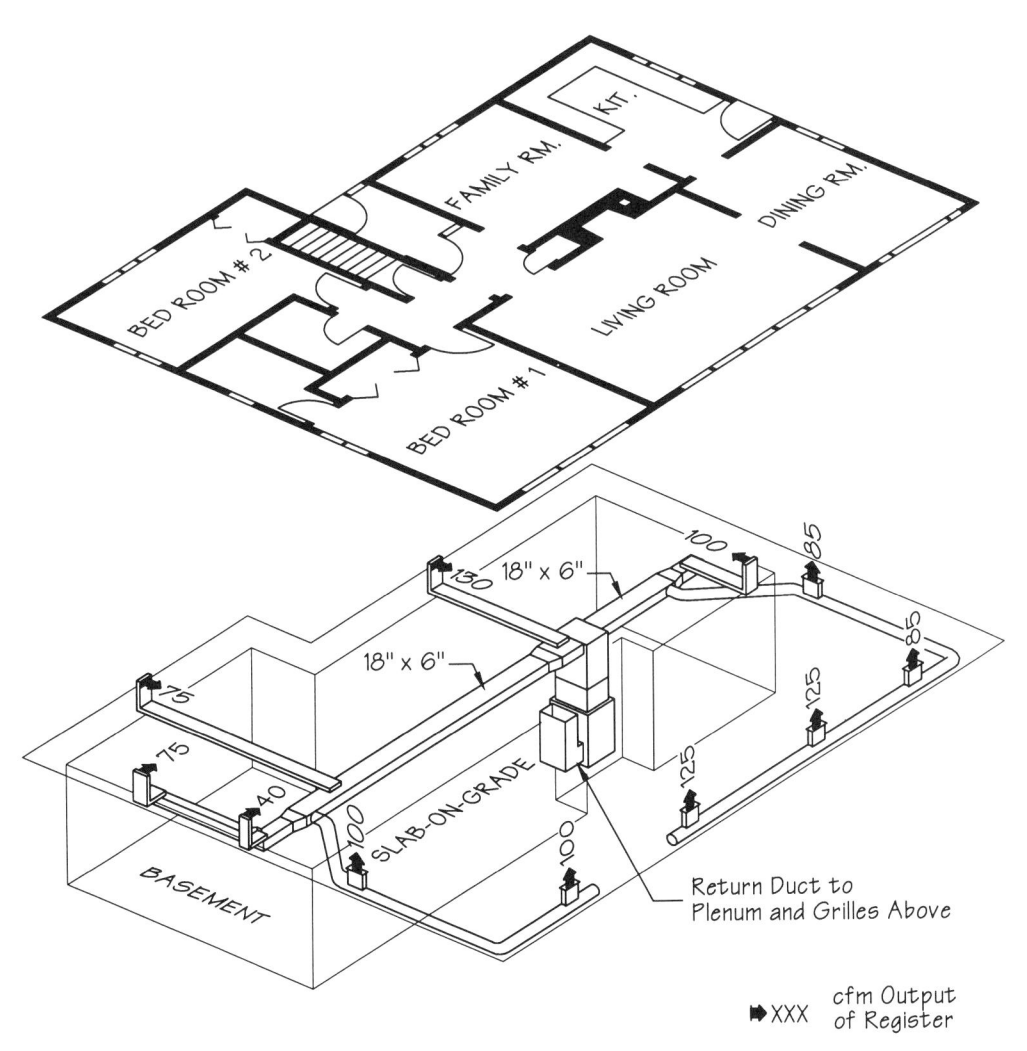

FIGURE 13.13 Possible duct layout for warm air supply to "The House."

500 fpm for branches. Formula 13C is used to determine the required size of supply ducts. Assume that rectangular ducts will be used which have a friction allowance (FA) of 1.1.

To illustrate the process, the required size of the two supply mains is computed as follows for the required flow (of 520 cfm) and an air velocity of 800 fpm.

$$A_{duct} = \frac{\text{cfm} \times 144}{V_{fpm}} \times \text{FA}$$

(Formula 13C)

$$= \frac{520 \times 144}{800} \times 1.10$$

$$= 103 \text{ in.}^2$$

Therefore, the two main supply ducts going to each side of the furnace can be 18 in. × 6 in. (108 in.^{2}) at the beginning. These supply ducts could then be reduced in size after the takeoff of the supply branches. Common practice though is to maintain the size of extended plenum ducts of less than about 24 ft. The branch supply ducts are sized in a similar manner using a velocity of 500 fpm.

Yearly Energy Use and Operational Cost

The yearly energy use (of 46.5 million Btus), and approximate operational cost (of $580), to heat "The House" with a warm air furnace is detailed in Section 5.14.

14

STEAM HEATING SYSTEMS

Lessons learned from the mighty steam engines of the 1700s quickly led to development of steam heating systems which produced steam in "boilers" which function like giant tea kettles. The first systems used one pipe to supply steam to radiators and return condensate to the boiler. These early one-pipe systems had clear limitations. Large pipes were needed to allow steam and returning water to flow in opposite directions, and they often were noisy!

In this chapter the fundamentals of steam, and steam heating and distribution systems will be stressed, enabling the reader to understand the operation and functioning of the many existing systems. Through such an understanding, systems can often be "redesigned" to make them operate more quietly and efficiently, and with better control and temperature balance.

Currently, steam heating systems are still used in some new buildings and spaces such as warehouses, garages, small commercial buildings, and apartment buildings. Steam is also sometimes used in tall buildings where the height of water column produced by a closed hot water system would produce very high pressure in the boiler. Instead, steam can be fed to steam-to-water heat exchangers (known as "converters") on upper floors where the steam condenses and transfers its heat to hot water. This hot water is then circulated through a local hot water heating distribution system.

14.1 STEAM PROPERTIES

Once water is heated within a boiler to its boiling point of 212°F (at a standard sea level pressure of 14.7 psi), water becomes steam if about another 1000 Btus of heat is added per pound of water (the "latent heat of vaporization"). This steam rises naturally through a network of supply piping and steam risers, delivering large quantities of heat to occupied spaces. For good steam heating system operation, it is essential that this be "dry" steam (without entrained liquid water).

228 STEAM HEATING SYSTEMS

Table 14.1 shows properties of dry saturated steam for a range of pressures at sea level. Gauge pressures above atmospheric (14.7 psia) are expressed in psig. Therefore, if a system at sea level has an absolute pressure of 19.7 psia, it is said to have a pressure of 5.0 psig (19.7–14.7). By industry convention, gauge pressures below atmospheric are referred to in inches of mercury vacuum (in. Hg vac) with 1 in. of mercury equaling 0.491 psi.

Saturation temperature is the temperature at which liquid water boils and becomes a vapor (steam) at a given pressure. Quite simply, the higher the pressure, the higher the boiling point as shown in Figure 14.1.

Heat of evaporation is equivalent to the quantity of heat (in Btus) released at a radiator as a pound of steam condenses back into liquid water. As indicated in Table 14.1, the heat of evaporation does not change much at different pressures (e.g., 1002 Btu/lb for 20 in. Hg vac and 946 Btu/lb for 15 psig or 29.7 psia). The value of 960 Btu/lb is utilized in ASHRAE tables for the sizing of low-pressure systems operating at slightly above atmospheric pressure.

Operation at vacuum pressures below atmospheric can also be useful. In particular, "variable vacuum" steam systems can adjust the heat output as outside temperature conditions change. As shown in Table 14.1, it takes much less steam to fill a system when it is operated at a substantial vacuum since the specific volume of the steam increases dramatically.

14.2 STEAM BOILERS

Low-pressure boilers with working pressures of up to 15 psig are used in steam heating systems for buildings. High pressure systems are used only in very large installations and, by code, usually require a full-time operating engineer. Common low-pressure steam boilers are constructed of cast iron or steel, as discussed below.

TABLE 14.1 PROPERTIES OF SATURATED STEAM

Pressure Gauge	Absolute (psia)	Saturation Temperature (°F)	Heat of Evaporation (Btu/lb)	Specific Steam Volume (ft^3/lb)
28 in. Hg vac	0.9	101	1037	341
20 in. Hg vac	4.9	161	1002	75
10 in. Hg vac	9.8	192	983	39
4 in. Hg vac	12.7	205	975	31
2 in. Hg vac	13.7	209	972	29
0 psig (atmos.)	14.7	212	970	27
1 psig	15.7	216	968	25
2 psig	16.7	218	966	24
3 psig	17.7	222	964	22
4 psig	18.7	225	962	21
5 psig	19.7	227	960	20
10 psig	24.7	240	952	16
15 psig	29.7	250	946	14
50 psig	64.7	281	924	8
100 psig	114.7	338	881	4

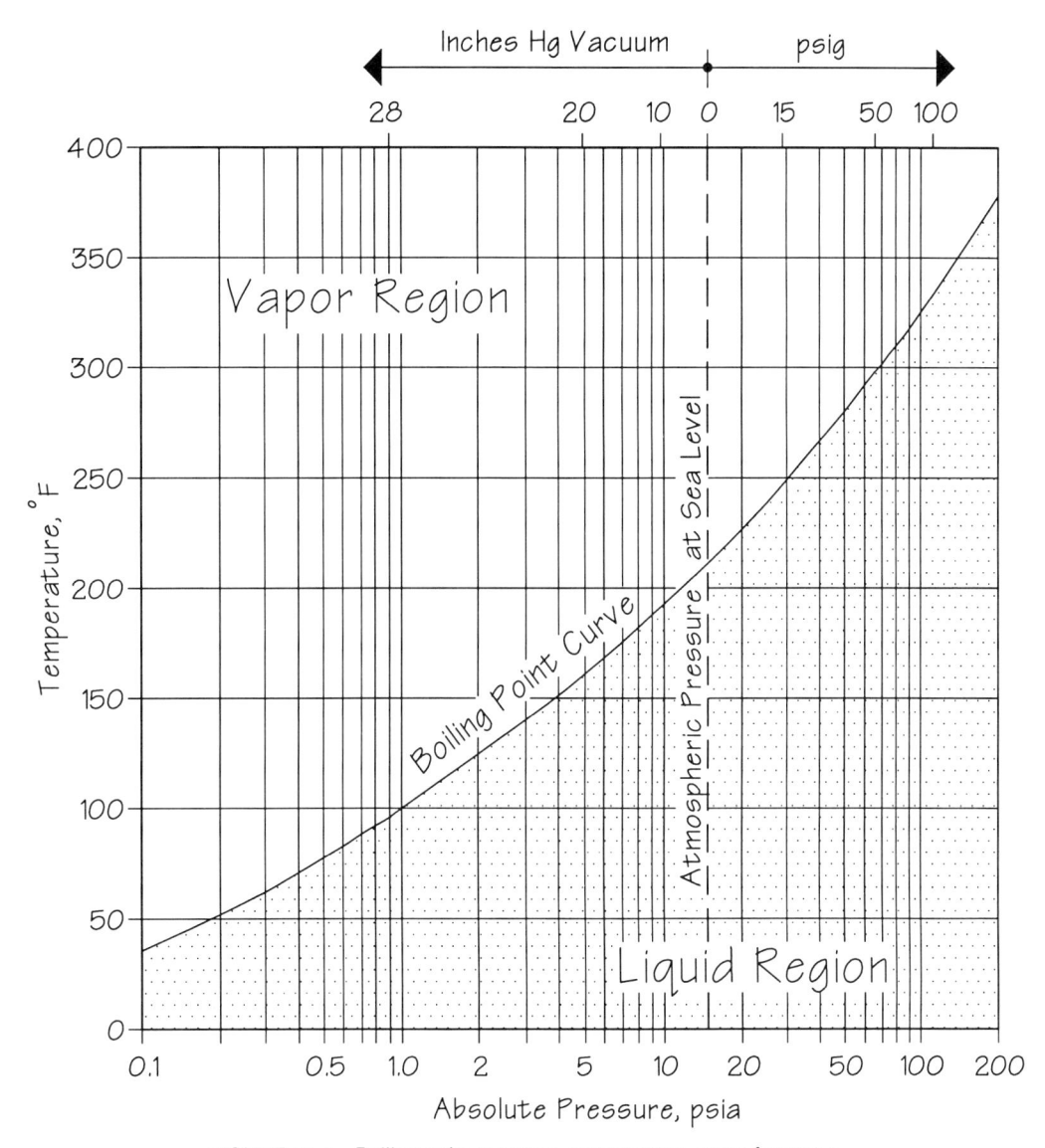

FIGURE 14.1 Boiling point pressure–temperature curve for water.

Cast-Iron Sectional Boilers

Cast-iron boilers are very similar in appearance to cast-iron radiators. They consist of sections which are connected and assembled together so as to be watertight and pressure resistant. And just like simple radiators, the number of sections used determines the capacity to deliver heat in Btu/h.

The key internal components of a typical cast-iron sectional steam boiler are shown in Figure 14.2. Note the indication of an optional tankless coil to produce domestic hot water.

Cast-iron sectional boilers are available for a very wide range of gross heating out-

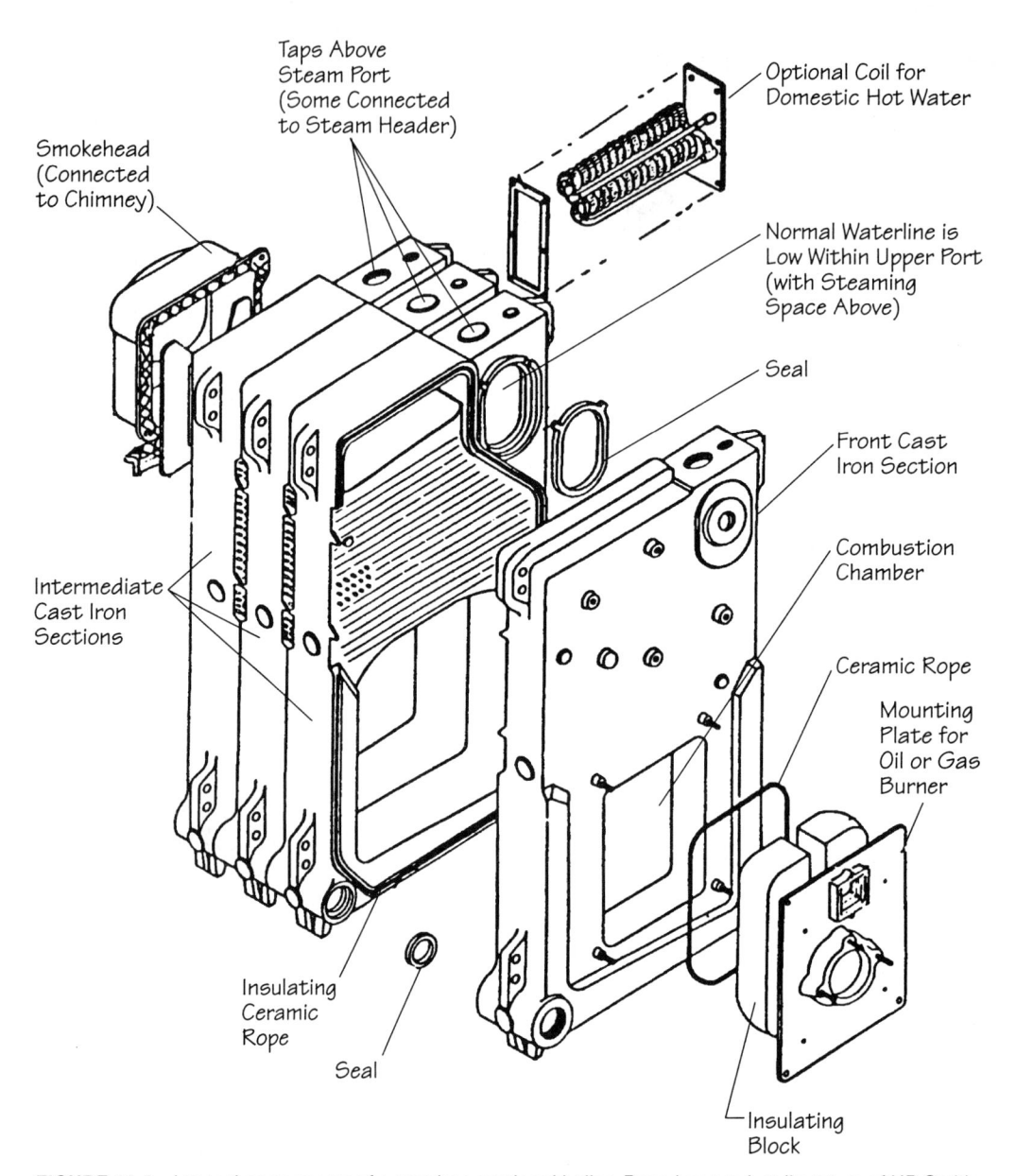

FIGURE 14.2 Internal components of a cast-iron sectional boiler. Based on product literature of HB Smith.

puts from about 30,000 Btu/h for small residences to more than 10,000,000 Btu/h for large commercial boilers.

Steel Boilers

Steel steam heating boilers are usually classified as firetube or watertube. "Firetube" boilers pass the combustion gases formed in the firebox through tubes surrounded by water. In contrast, "watertube" boilers circulate the water through tubes which are surrounded by the hot combustion gases. Scotch (or Scotch Marine) boilers are large capacity cylindrical steel firetube boilers, as illustrated in Figure 14.3.

Note how the combustion gases flow through the length of the boiler to the boiler's furnace and transfer heat to the surrounding water. This journey through the furnace passage is known as the "first pass." At the back of the boiler, the combustion gases enter a chamber, reverse direction and flow back through the firetubes constituting the "second pass," or exposure to heat transfer. Most modern Scotch boilers then reverse the direction of combustion gases yet again for a "third pass" (as shown) while some boilers even have a "fourth pass" before the gases are vented to the outside.

Another type of firetube boiler (similar in general construction features and heat transfer technique to the Scotch boiler) is known as a firebox boiler. Firebox boilers are squarish in construction with a circular top over the steam space.

Steam Boiler Operation

In the boiler, water is first heated to approximately 212°F and it begins to boil. Then as the fire continues, the steam temperature and pressure gradually rise as steam is being produced. Once steaming has occurred, burners in small residential-scale boilers simply cycle on and off. Larger boilers (e.g., serving a small apartment building) usually operate more efficiently by having low–high–low (L-H-L) burner operation or full burner modulation.

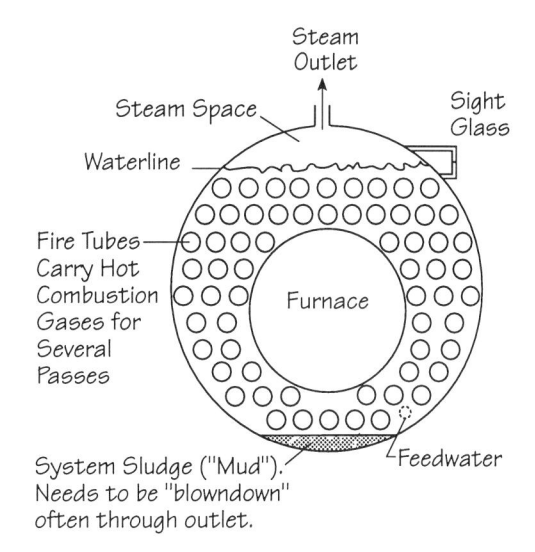

Steam Quality

Steam quality refers to the quantity of liquid water suspended within the saturated

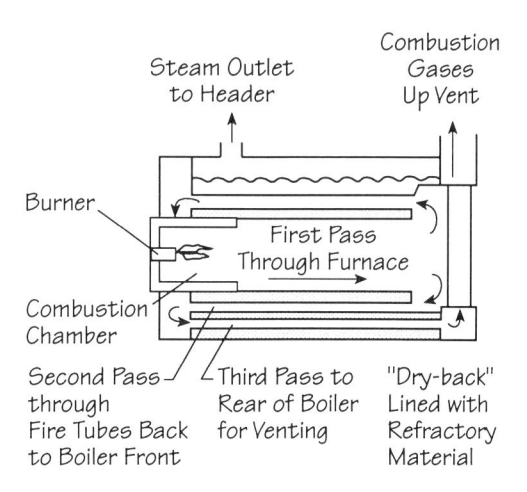

FIGURE 14.3 Steel firetube "Scotch Marine" boilers.

vapor. Zero percent steam is pure liquid water. One hundred (100) percent steam quality is all steam (water in the form of a gas), and considered totally dry. Steam of at least 98% quality is the generally accepted standard and considered dry steam in heating systems. Anything less is considered overly "wet," and may lead to noisy water hammer and poor steam distribution.

Years ago, boilers were advertised as making dry steam "at the boiler outlet." This was generally true because the steam space inside the boiler in these old, large boilers was large enough to allow the steam to shake off water before it left the boiler. Many of today's "modern" boilers are not as good at producing dry steam since their steam space is usually a lot smaller than that of their predecessors for an identical capacity.

The waterline in the boiler should be maintained at the lowest recommended safe height in order to maximize the volume where the steam can separate itself from the water inside the boiler, and maximize the area of steaming water surface. The waterline is of particular importance in cylindrical boilers. A rise of the waterline will decrease both the steaming surface area and the volume of space available for steam separation, as shown in Figure 14.4.

When a boiler is steaming, some boiler water is suspended momentarily above the boiler waterline where it can be sucked into the steam pipe if it isn't separated from the steam. Therefore, any water that leaves the boiler as wet steam must be removed in the piping adjacent to the boiler. This piping is known as the "steam header."

Water Quality

Impurities in boiler water are common and can be damaging for steam boilers. Boiler water should be treated to remove materials such as sand, algae, oil, and corrosion products that can cause "fouling," which restricts circulation and hampers heat transfer. Treatment can involve filtration and use of chemical additives in boiler water.

Dissolved mineral salts form scale deposits which cause corrosion and reduce the efficiency of heat transfer surfaces. At their low point, boilers collect rust, sediment, and water impurities (known as "mud") which must be removed from ("blown off") the boiler on an as-needed basis to prevent buildup and restriction of water passages.

Poor quality water can also result in "foaming" when impurities create an oily scum on the water surface. This can lead to "carry over" of water into the supply piping system, resulting in possible water hammer and damage to radiator valves and

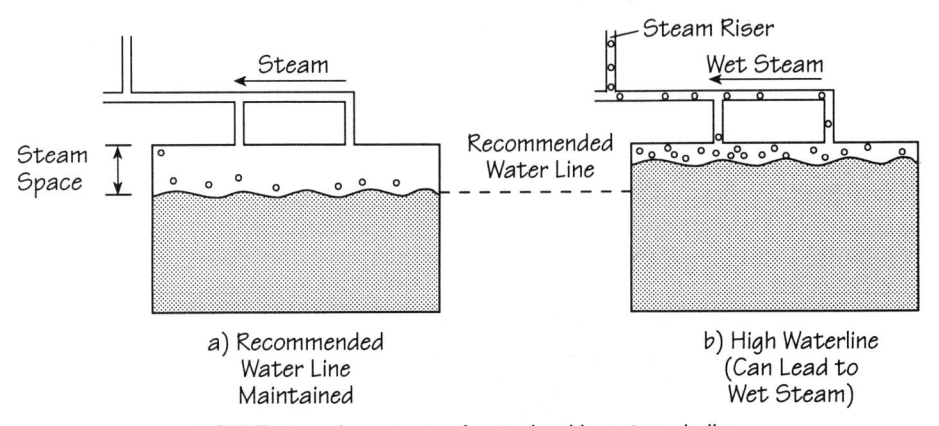

FIGURE 14.4 Importance of water level in a steam boiler.

steam traps. Water hammer also loosens rust in pipes and radiators which is eventually flushed back to the boiler where steaming is impaired.

Water in steam heating system boilers must be regularly monitored (and treated when necessary) to minimize corrosion and scaling. Alkaline water treatments (such as caustic soda, soda ash, and sodium silicate) are used to reduce corrosion by raising the pH to approximately 8–10. Scale inhibitors are then used since scale formation is promoted by a more basic pH.

Acid Gases and Condensation Damage

The combustion of fuel oil generates vaporous sulfuric acid while the burning of gas generates vaporous hydrochloric acid. The combustion of each also results in the production of water vapor as a by-product. If this water then condenses on boiler surfaces or in the vent (flue), damaging corrosion can result as discussed in Section 17.5.

14.3 STEAM BOILER SIZING

The following terms, which apply to steam heating system design and components, should be well understood.

Boiler Horsepower (Hp)—Equal to the evaporation of 34.5 pounds of water (at 212°F) to steam at the same temperature. Also equal to 33,475 Btu/h. Therefore, a 40-Hp boiler has a gross output of 1,339,000 Btu/h (40 × 33,475).

Equivalent Direct Radiation (EDR)—An amount of heating surface in ft^2 which will give off 240 Btus of heat per hour. Therefore, a radiator with an output of 5 EDR will deliver 1200 Btu/h of heat (5 × 240). Although heat outputs are consistently expressed in Btu/h in this text, it is not uncommon on renovation projects to have to consult old manufacturer's literature to determine the rated output of old radiators in EDR.

Pound of Steam—Equal to 960 Btus (the amount of heat given off when one pound of steam condenses and releases its heat of vaporization at slightly above atmospheric pressure). Also note that each pound of steam can be thought of as being equal to 4 EDR (960/240).

Presented in Table 14.2 is performance data for a typical listing of cast iron sectional boilers. Gross and net boiler outputs are listed in thousands of Btus per hour (MBH).

TABLE 14.2 CAST-IRON SECTIONAL BOILERS

Model-Sections	Boiler Horsepower	I=B=R Ratings[a] Gross Output (MBH)	Net Output (MBH)
A-4	24	798	600
B-5	31	1,055	792
C-6	40	1,338	1,008
D-7	47	1,566	1,200
E-8	55	1,855	1,440

[a]Note: I=B=R refers to the ratings of The Hydronics Institute, which formerly was the Institute of Boiler and Radiation Manufacturers.

Steam boiler sizing involves more than simply satisfying the sum of the heating requirements of the terminal devices (radiators). Steam systems must also be sized for a significant "piping and pickup" factor which accounts for the extra "start-up" heat needed to heat the heavy network of steel pipes and terminal devices (radiators). As can be seen in Table 14.2, a comparison of gross to net I=B=R ratings shows that piping and pickup factors ranging from 1.333 (i.e., 798/600 = 1.333 for Model A-4) to 1.288 (i.e., 1855/1440 = 1.288 for Model E-8) are applied.

Design experience suggests that the piping and pickup factors incorporated into ratings tables for steam systems are often inadequate and use of a value of 1.5 or even 1.6 is generally suggested. This of course makes the net rating unusable. Instead, sizing is accomplished by using Formula 14A and consulting the gross output values in equipment performance tables.

$$\text{StBoil}_{out} = q_{rads} \times P\&P \quad \text{Formula 14A}$$

where

StBoil_{out}	is the gross output required by a steam boiler in Btu/h
q_{rads}	is the rated heating output for all of the radiators or other terminal devices in Btu/h
P&P	is the piping and pickup factor

Example 14A: The radiators of a steam heating system have a total rated heating output (q_{rads}) of 1,000,000 Btu/h. Select a boiler from Table 14.2 which satisfies this load. Assume a piping and pickup factor (P&P) of 1.5.

Solution: Use Formula 14A to determine the gross output of steam boiler required.

$$\text{StBoil}_{out} = q_{rads} \times P\&P$$
$$\text{(Formula 14A)}$$
$$= 1,000,000 \times 1.5$$
$$= 1,500,000 \text{ Btu/h}$$

Such a required gross output can be satisfied by Model D-7 which has a gross output of 1,566,000 Btu/h. Note that the net rating (which assumes a different piping and pickup factor) is ignored in this sizing method.

Many replacement boilers (and some new boilers) must also be sized to satisfy the domestic hot water load (see Chapter 22). Exercise particular caution in sizing steam boilers with tankless coils to produce DHW. With such a system, a large demand for DHW can result in the loss of steam production and impeded distribution of heat.

Whatever the total design load on a boiler (actual heating load, piping and pickup factor, and possible domestic hot water load) do not take a chance on undersizing a steam boiler. True, there is some lost efficiency in an oversized boiler. More importantly, though, an undersized boiler will often result in an unbalanced system and poor temperature control.

14.4 BOILER-RELATED PIPING

Shown in Figure 14.5 are the basic elements of a steam heating boiler and related piping.

Steam Header Design

A steam system boiler that produces wet steam needs to rely on the piping above it (known as the "steam header") to separate entrained water. Steam exits a boiler through one or more outlets at the top depending on the boiler size and design. The

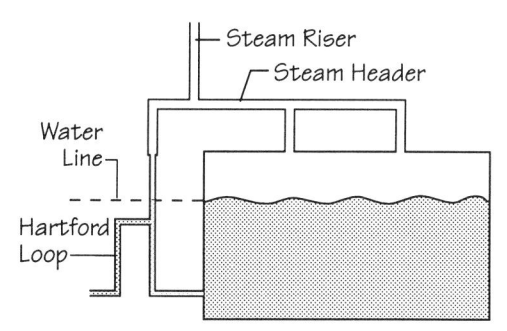

FIGURE 14.5 Steam heating boiler and related piping.

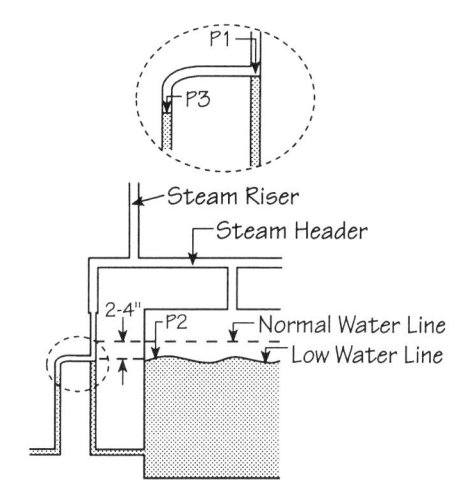

FIGURE 14.6 Hartford loop to prevent low-water firing of boiler.

designer should tap into all available boiler outlets to limit the distance steam must travel over the steaming water surface. This will reduce "priming," the term given for overly turbulent boiling which can produce very wet steam.

The outlet(s) from the boiler connect into a horizontal steam header, arranged to get all of the steam going in one direction, and to prevent entrained moisture (wet steam) from rising further in the system. The velocity and momentum of the entrained water cause it to pass by the vertical system riser (where the dry steam rises into the system). Then, the entrained moisture continues to the end of the header and down through the equalizer, which is the beginning of the Hartford Loop described below. Never try to save a fitting by having a riser come out of a boiler and continue straight up since it will carry entrained moisture with it!

Hartford Loop

The "Hartford Loop," so named because of its relevance to the insurance industry, was originally developed to prevent low-water firing of coal fired boilers (see Figure 14.6).

The piping to the Hartford Loop is known as the "equalizer." It causes pressures in the boiler and the equalizer to be the same. Therefore, P1 will be equal to P2 even when pressure in the return line, P3, is less. When the water level reaches the bottom of the top bend of the loop (typically 2–4 in. below the normal waterline), pressure cannot force any more water from the boiler and a minimum water level is maintained.

Although modern low-water cutoff controls now do the same job, incorporation of a Hartford loop is still common design practice, expected by most building officials as an additional safety measure. Note the section of piping at the top of the Hartford Loop. Historically this connection has been made with a very short pipe to minimize possible noise. Notwithstanding, at least one steam system troubleshooter is now designing with a wide pipe connection since he deliberately wants the system to be extremely noisy if a low-water cutoff has failed.

System Friction and Return Main Location

"Pressure drop" refers to the loss of steam pressure due to friction in the pipes and fittings. As shown in Section 10.1, a column of water produces a pressure of 0.433 psi for each foot (12 in.) of height and a pressure of 1 psi is produced by a water column of 27.7 in. Therefore, a small steam system with a system pressure drop of 1/8 psi would produce a water column of 3.5 in. whereas larger systems with a system pressure drop of 1/2 psi would produce a water column of about 14 in.

The term "static head" refers to the friction which condensate encounters in the wet return piping below the boiler waterline. A static head equal to 4 in. of water column is typically assumed in the design of low-pressure systems.

In gravity return steam systems, the return main piping must be located high enough above the boiler waterline to allow condensate to "back up" before it is allowed back into the boiler. This is known in the field as Dimension A as shown in Figure 14.7.

For small systems where piping is sized for a pressure drop of 1/8 psi, Dimension A should be at least 14 in. based upon a system pressure drop of 3.5 in., a static head of 4 in., and a factor of safety. For larger systems with piping sized for a pressure drop of 1/2 psi, Dimension A should be at least 28 in. above the waterline based upon a system pressure drop of 14 in., a static head of 4 in. and a reasonable factor of safety.

Return piping is referred to as "dry" when it is located above the waterline and "wet" when it is located below the waterline.

14.5 SYSTEM PRESSURE AND STEAM VELOCITY

Low-pressure Operation

Low-pressure steam boilers are normally operated at the lowest pressure needed to overcome the system pressure drop. This pressure drop is due to frictional contact of steam with the longest run of pipes and fittings in the steam distribution system. It is

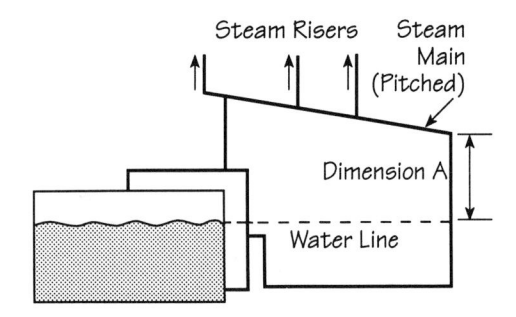

FIGURE 14.7 Dimension A to allow for condensate back-up in return mains.

good practice to limit system pressure drop to one-half or less of the initial gauge pressure. Therefore, if a system is operated at an initial pressure of 1 psig, the system pressure must be limited to one-half of that value or 1/2 psi.

Typical design pressure drops per 100 ft (of "equivalent pipe length") commonly used for pipe sizing of low-pressure steam systems are shown in Table 14.3. Note that "$oz/in.^2$" refers to ounces of water pressure per square inch, and that 16 oz is equal to 1 lb.

On small (house scale) low-pressure systems delivering less than 200,000 Btu/h, a total design pressure drop of 2 $oz/in.^2$ is typical. Larger low-pressure steam systems may have a total design pressure loss of 1/2 psi or more depending upon the required heating capacity and length of supply piping.

To determine an approximate "equivalent pipe length" it is common practice to determine the actual length of the longest run of steam supply piping and double it to account for the additional friction caused by elbows, valves, and other fittings (see Section 10.4 for additional discussion and formulas). When greater precision is required, consult the detailed fitting-by-fitting procedure contained in the *1993 ASHRAE Handbook of Fundamentals*.

Formula 14B uses equivalent pipe length to determine the design pressure drop for a system.

TABLE 14.3 PRESSURE DROPS USED FOR SIZING (LOW PRESSURE) STEAM PIPE

Initial Steam Pressure, psig	Pressure Drop Per 100 Ft
Vacuum return	2–4 oz/in.^{2}
0 psig	0.5 oz/in.^{2}
1 psig	2 oz/in.^{2}
2 psig	2 oz/in.^{2}
5 psig	4 oz/in.^{2}
10 psig	8 oz/in.^{2}
15 psig	16 oz/in.^{2}

Source: Excerpted with permission from the *1993 ASHRAE Handbook of Fundamentals*, Chapter 33, Table 11.

$$\text{SPD} = \text{PD}_{100\text{ft}} \times \text{EPL} \quad \text{Formula 14B}$$

where

- SPD is the system pressure drop in oz/in.^{2}
- PD$_{100\text{ft}}$ is the pressure drop in oz/in.^{2} per 100 ft of equivalent pipe length
- EPL is the equivalent pipe length of the system in feet

Example 14B: Determine the design pressure drop for a system where the longest supply pipe run is 300 ft for an initial gauge pressure at the boiler of 2 psig.

Solution: As indicated in Table 14.3, a pressure drop of 2 oz/in.^{2} per 100 ft of equivalent pipe length is commonly used for systems with an initial steam pressure of 2 psig. The equivalent length of pipe used for design purposes will be 600 (300 ft × 2).

Use Formula 14B to determine the design pressure drop for this system.

$$\text{SPD} = \text{PD}_{100\text{ft}} \times \text{EPL}$$

(Formula 14B)

$$= \frac{2 \text{ oz/in.}^2}{100 \text{ ft}} \times 600 \text{ ft}$$

$$= 12 \text{ oz/in.}^2$$

The result of 12 oz/in.^{2} (or 3/4 psi) is less than one-half the initial steam pressure of 2 psig, and is acceptable.

Steam Velocity

A steam velocity in supply piping of between about 100 and 200 ft/sec (6,000 to 12,000 fpm) will result in quiet operation. Steam velocity in feet per second can be calculated using Formula 14C.

$$\text{StV}_{\text{fps}} = \frac{\text{lb}_h \times V_{\text{sp}}}{25 \times A} \quad \text{(Formula 14C)}$$

where

- StV$_{\text{fps}}$ is the steam velocity in feet per second
- lb$_h$ is the steam flow in pounds per hour
- V_{sp} is the specific volume in cubic feet of 1 lb of steam at the system pressure (from Table 14.1)
- 25 is a conversion constant (0.00694 ft^2/in.^{2} × 3600 sec/h)
- A is the internal cross-sectional area of the steam supply piping

Example 14C: Compute the steam velocity for a 1$\frac{1}{2}$ in. Schedule 40 pipe ($A = 2.04$ in.^{2}) being considered for a riser in a two-pipe system assuming a steam flow of 336 pounds per hour (lb$_h$) and a steam pressure of 1 psig.

Solution: In Table 14.1, a pressure of 1 psig has a specific steam volume (V_{sp}) of 25 ft^3/lb. Employ Formula 14C.

$$\text{StV}_{fps} = \frac{\text{lb}_h \times V_{sp}}{25 \times A} \quad \text{(Formula 14C)}$$

$$= \frac{336 \times 25}{25 \times 2.04}$$

$$= 165 \text{ feet per second with } 1\tfrac{1}{2} \text{ in. pipe}$$

The result is a velocity near the 200 fps upper threshold. Perhaps a larger pipe should be used. Determine the velocity if 2 in. pipe having an internal pipe diameter of 3.36 in.^{2} were used instead.

$$\text{StV}_{fps} = \frac{\text{lb}_h \times V_{sp}}{25 \times A} \quad \text{(Formula 14C)}$$

$$= \frac{336 \times 25}{25 \times 3.36}$$

$$= 100 \text{ feet per second with 2 in. pipe}$$

This velocity is much better. It is important to see how velocity in a steam system can be controlled through pipe sizing.

14.6 ONE-PIPE SYSTEMS

One-pipe gravity return systems are the most basic and least expensive steam systems. In such a system, steam rises from the boiler through the header piping into the main steam riser. The riser then feeds the steam to a piping distribution system which leads to terminal units (generally radiators). There the steam condenses and releases the latent heat of vaporization as it becomes a liquid. The liquid water (known as condensate) then flows by gravity back to the boiler through the same radiator connections and vertical piping system, but in the opposite direction of the fresh steam flowing up the pipes.

One-pipe systems as shown in Figure 14.8 are rarely employed in new systems but will be encountered in many existing buildings. As will be seen below, most pipes are larger in one-pipe systems since they must allow for the flow of both steam and condensate without forcing the water back up the pipes and creating excessive noise.

Air Venting of One-Pipe Systems

In one-pipe steam systems, the entire system of pipes and radiators is filled with air when steam is not being produced. When heat is called for, this air must be totally and evenly expelled from all elements of the system if steam is to provide heat in an even and balanced way.

There are two types (and general locations) of vents: vents on radiators and vents on steam mains and risers. Radiator vents should be sized to vent out only the air which is in the radiator itself and the local piping for that radiator. These vents should be as small as possible to promote gradual radiator warm-up so the thermostat will have enough time to control the system.

Vents on mains and risers, however, need to have a much larger venting capacity since much larger pipe and air volumes are involved. Large capacity vents should be installed near the end of the steam mains and at the top of the steam risers. Based upon the volume of air in each part of the system, vents should be sized to remove the air from the steam mains in one minute, and all air from the steam risers in an additional minute when the system is operated just above atmospheric pressure. This will result in balanced distribution, with steam entering all parts of the system at approximately the same time. This approach is sometimes referred to as "master venting," as shown in Figure 14.9.

Master venting is frequently the remedy used to correct unbalanced existing steam systems. However, an existing steam system must be producing dry steam and be

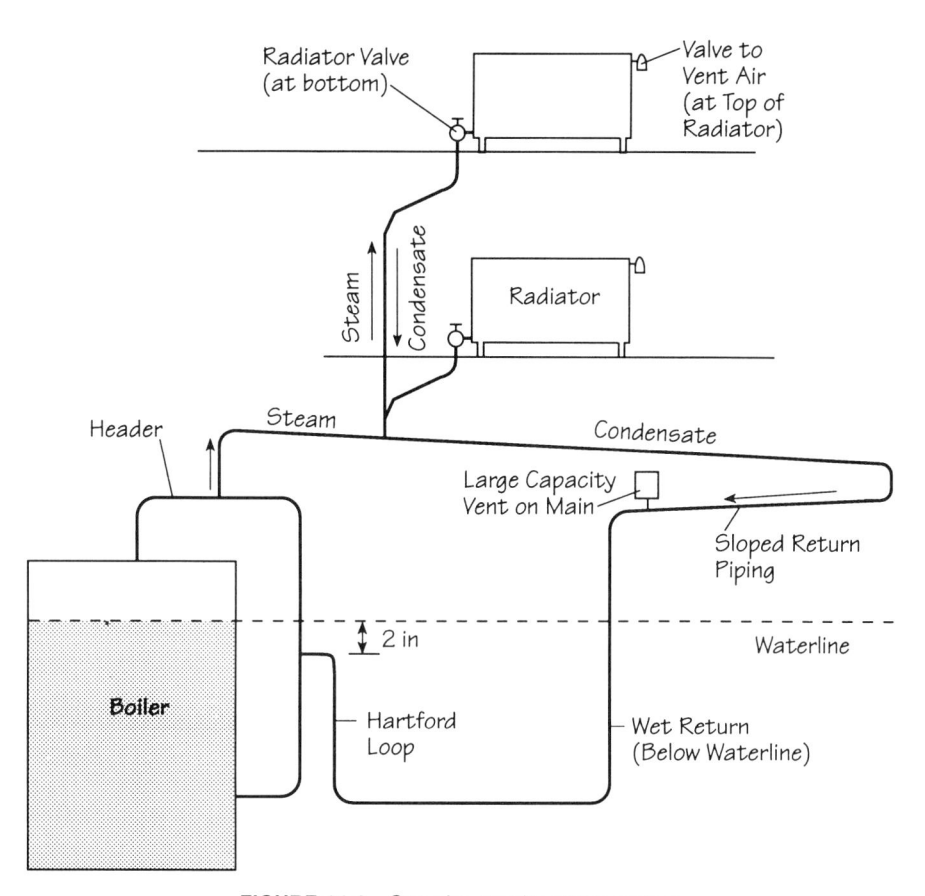

FIGURE 14.8 One-pipe gravity steam system.

FIGURE 14.9 Master vents to expel air from piping network.

totally free of water hammer before master venting is installed. Otherwise large and potentially damaging quantities of hot water are likely to emerge from the large vents.

14.7 TWO-PIPE SYSTEMS

Two-pipe gravity steam heating systems have separate piping systems for the supply of steam and the return of condensate (see Figure 14.10). Two-pipe steam systems installed after World War I also include radiator valve orifices and/or "steam traps" to prevent steam from getting into the return piping.

Many two-pipe systems can use gravity alone to return the condensate. Yet in systems where much of the return piping is be-

240 STEAM HEATING SYSTEMS

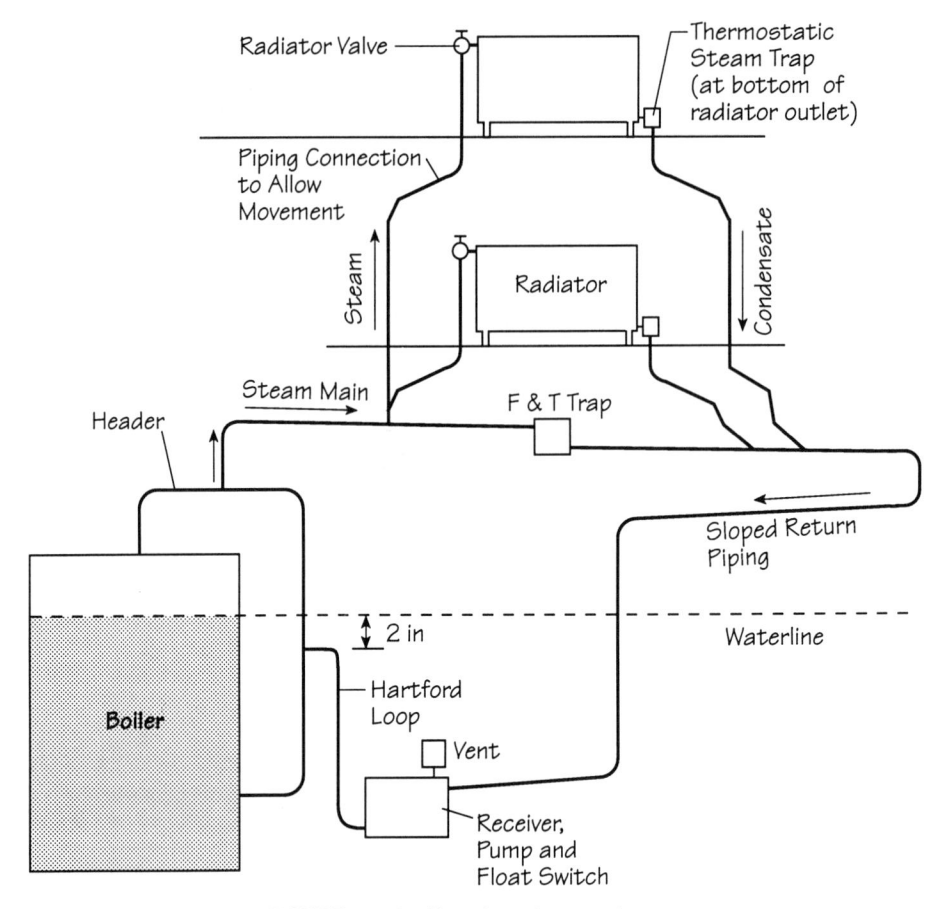

FIGURE 14.10 Two-pipe steam system.

low or only slightly above the height of the boiler, a condensate pump located at the low point of the return piping and near the boiler is required.

Steam Traps

Steam traps are required on all two-pipe steam systems to isolate or trap the supplied steam within the supply piping network and radiators. The traps do allow returning condensate (liquid water) to pass through as well as air so that it can be vented from the system. Two-pipe steam heating systems generally employ the following two types of traps shown in Figure 14.11 and described below.

Thermostatic traps: Are used on radiators, with 1/2 in. traps being typical. Activation is based on the differing temperatures of steam and condensate. One type uses a water leg of condensate behind a bellows which closes the outlet valve when steam is present. Other thermostatic traps use bimetallic elements which expand differently in response to a temperature change, and close the trap valve when steam saturation temperatures are being approached.

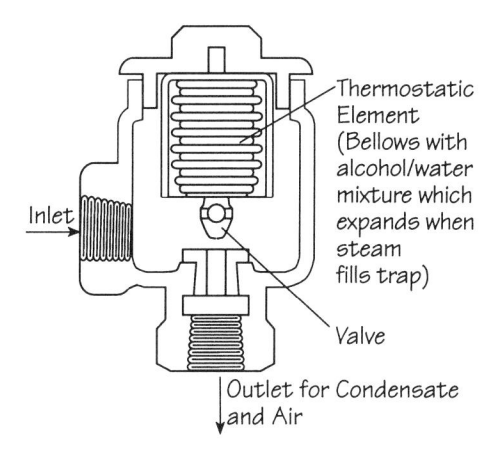

and condensate. The thermostatic element operates as described above.

Traps must be sized properly, based upon the amount of condensate to be handled (lb/h), the pressure differential between inlet and outlet at the trap, and a safety allowance to prevent condensate from backing up during start-up or warm-up cycles. When F & T traps are used, they should be equipped with strainers and drain valves to allow for clean-out of "mud" (system corrosion). Manufacturer ratings of steam traps already carry a considerable oversizing allowance. Do not oversize traps because they may fail prematurely through erosion.

Venting of Two-Pipe Systems

Air is vented from two-pipe systems in two basic ways. Old existing systems without steam traps are locally vented by a manual vent at each radiator. Newer two-pipe systems employ steam traps located at the outlet of each radiator. Air and condensate pass through the steam trap and enter the return line. The air is then vented at the condensate receiving tank.

Vacuum Systems

"Variable vacuum systems" are employed in large buildings (such as apartment buildings) to deliver a relatively constant flow of steam at a temperature and vacuum (below atmospheric pressure) to control heat output based upon outside temperature. This is possible, since the saturation temperature of steam is well below 212°F at pressures below atmospheric as shown in Table 14.1 (e.g., the saturation temperature of steam is 161°F for a vacuum pressure of 20 in. Hg). Therefore, in mild weather these systems will operate at a high vacuum, and during the coldest weather they may operate at pressures above atmospheric.

a) Thermostatic Trap

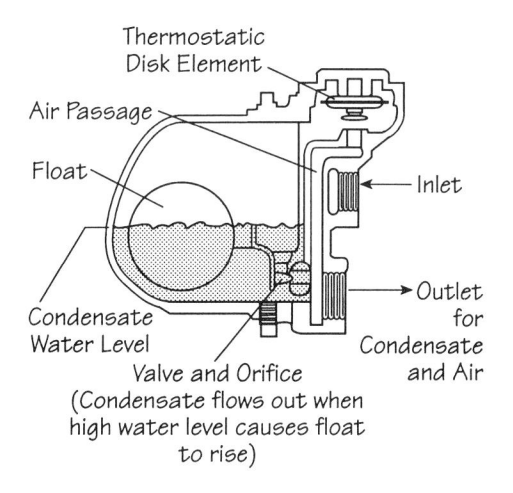

b) Float and Thermostat (F & T) Trap

FIGURE 14.11 Common heating system steam traps.

Float and thermostat traps: Known as "F & T" traps, they are typically used for draining low-pressure steam mains, and must be fitted with a thermostatic air vent to release air from the system. Float trap operation is based upon density (buoyancy) differences between steam

Variable vacuum systems can be difficult to maintain because piping can develop small leaks that make a vacuum hard to maintain except by using large (high horsepower) vacuum pumps.

14.8 PIPE SIZING

Tables 14.4 and 14.5 are used to size Schedule 40 pipe for various elements of low-pressure steam heating systems in a manner which will produce acceptable steam velocities.

Sizing Pipe when Condensate Opposes Steam Flow

Indicated in Table 14.4 is the capacity in pounds per hour (lb/h) of various pipe sizes in low-pressure steam systems for use in one-pipe systems and sections of two-pipe systems where some condensate may oppose steam flow (such as in risers or runouts that do not contain downward pipe extensions known as "drips").

Capacities shown assume an average pressure of 1 psig. Values also assume that horizontal runouts to risers and radiators are pitched at least 1/2 in. per foot.

Observe how dramatically capacities increase as pipe size increases. For example, a 3/4-in. supply riser in a one-pipe system has a capacity of 6 lb/h, whereas a 1-in. supply riser has a capacity of 11 lb/h (nearly double), and $1\frac{1}{4}$-in. pipe has a capacity of 20 lb/h (nearly double again).

Example 14D: Determine the pipe size for a supply riser in a one-pipe low-pressure steam system which supplies 100,000 Btu/h of radiation.

Solution: First convert the heat requirement of 100,000 Btu/h to pounds of steam per hour.

$$\text{Pounds of steam per hour (lb/h)} = \frac{100{,}000 \text{ Btu/h}}{960 \text{ Btu/h}}$$

$$= 104.2 \text{ lb/h}$$

As indicated in Table 14.4, a $2\frac{1}{2}$-in. riser could supply 116 lb/h, and is therefore selected.

TABLE 14.4 STEAM PIPE CAPACITIES FOR LOW-PRESSURE SYSTEMS IN POUNDS PER HOUR (lb/h)

Nominal Pipe Size (in.)	Two-Pipe Systems Sections where Condensate Flows Against Steam Vertical[a]	Horizontal	One-Pipe Systems Supply[b] Risers Up-Feed	Radiator Valves & Vertical Connections	Radiator & Riser Runouts
3/4	8	7	6	—	7
1	14	14	11	7	7
1¼	31	27	20	16	16
1½	48	42	38	23	16
2	97	93	72	42	23
2½	159	132	116	—	42
3	282	200	200	—	65
4	511	425	380	—	186
6	1800	1400	—	—	—

Source: Excerpted with permission from *1993 ASHRAE Handbook of Fundamentals*, Chapter 33, Table 15.
[a] Do not use for pressure drops of less than $\frac{1}{16}$ psi per 100 ft of equivalent run. Use Table 14.5 instead.
[b] Do not use for pressure drops of less than $\frac{1}{24}$ psi per 100 ft of equivalent run. Use Table 14.5 instead.

TABLE 14.5 FLOW RATE OF STEAM IN SCHEDULE 40 PIPE EXPRESSED IN POUNDS PER HOUR (lb/h)

Nominal Pipe Size (in.)	Pressure Drop—Psi Per 100 Feet of Equivalent Length						
	$\frac{1}{16}$	$\frac{1}{8}$	$\frac{1}{4}$	$\frac{1}{2}$	$\frac{3}{4}$	1	2
$\frac{3}{4}$	9	14	20	29	36	42	60
1	17	26	37	54	68	81	114
$1\frac{1}{4}$	36	53	78	111	140	162	232
$1\frac{1}{2}$	56	84	120	174	218	246	360
2	108	162	234	336	420	480	710
$2\frac{1}{2}$	174	258	378	540	680	780	1150
3	318	465	660	960	1190	1380	1950
$3\frac{1}{2}$	462	670	990	1410	1740	2000	2950
4	640	950	1410	1980	2450	2880	4200
5	1200	1680	2440	3570	4380	5100	7500
6	1920	2820	3960	5700	7200	8400	11900
8	3900	5570	8100	11400	14500	16500	24000

Source: Excerpted with permission from *1993 ASHRAE Handbook of Fundamentals*, Chapter 33, Table 14.

Sizing Supply Piping in Two-Pipe Systems

Table 14.5 is used for the sizing of pipes where the flow of condensate does not inhibit the flow of steam, for initial boiler pressures from 1 to 6 psig. The flow rates shown are for the indicated pressure drop in psi per 100 ft of (equivalent) length, and not the total pressure drop for the supply piping.

Example 14E: A supply riser in a two-pipe low-pressure steam system must be able to supply 100,000 Btu/h (104.2 lb/h) of radiation. Determine which size pipe should be selected if the pressure drop is 1/4 psi per 100 ft of (equivalent) length.

Solution: A capacity of more than 104.2 lb/h is required. As indicated in Table 14.5, a $1\frac{1}{2}$-in. riser could supply 120 lb/h. This is more than adequate. A comparison of this pipe requirement for a two-pipe system to the $2\frac{1}{2}$ in. riser needed for a similar capacity one-pipe system (in Example 14D) is presented in Figure 14.12.

Sizing Return Piping in Two-Pipe Systems

Use Table 14.6 for the sizing of various types of return piping for two pressure drops (1/8 psi and 1/4 psi) per 100 ft of equivalent length. For systems with other pressure drops see the complete ASHRAE Table.

In Table 14.6, the columns for "Vac" or vacuum returns apply to systems using a combination of a steam control valve and a vacuum pump controlled by a differential pressure controller to guarantee circulation.

14.9 GUIDELINES FOR LOW-PRESSURE STEAM PIPING

Supply Piping:

1. Pitch pipes at least 1/4 in. per 10 ft uniformly in the direction of steam and condensate flow where steam and condensate flow in the same direction. Where steam and condensate flow in the opposite direction the pitch should be much greater, typically 1/4 in. per foot.

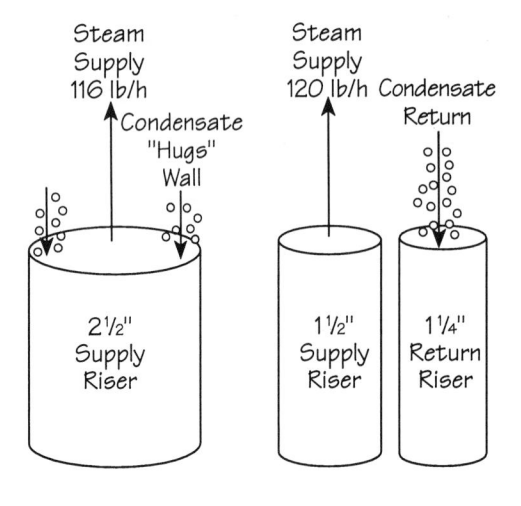

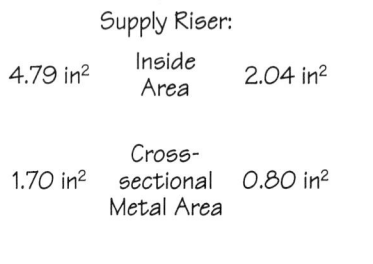

244 STEAM HEATING SYSTEMS

```
Steam              Steam
Supply             Supply
116 lb/h           120 lb/h  Condensate
   Condensate                Return
   "Hugs"
   Wall

  2½"              1½"       1¼"
  Supply           Supply    Return
  Riser            Riser     Riser
```

Supply Riser:

4.79 in²   Inside Area   2.04 in²

1.70 in²   Cross-sectional Metal Area   0.80 in²

ONE-PIPE SYSTEM   TWO-PIPE SYSTEM

FIGURE 14.12 Riser pipe size comparison.

2. Insulate pipes to prevent premature condensation and potential scalding from pipe contact.

3. Design so that condensate will be removed from supply piping to prevent water hammer and to improve steam quality (dry steam is wanted).

4. When horizontal piping changes in size, use eccentric reducers so that the bottom elevation remains constant to prevent water from being trapped. This will yield smooth condensate flow which prevents water hammer.

5. In one-pipe systems, be sure that piping arrangements (including bends around obstructions) are such that returning condensate cannot block the passage of rising steam. Also provide "drips" (downward pipe extensions) at the bottom of steam risers connected to the downstream return piping. If not so equipped, any falling condensate in the supply riser will be sure to encounter rising steam and result in poor system operation and water hammer.

6. Always remember that the piping in a steam heating system is subject to a great deal of expansion, contraction, and movement, which must be accommodated by joints and connections. Steam risers two or more floors in height should be connected as shown in Figure 14.13 to allow for bending or twisting movement as the riser moves up and down.

7. In apartment buildings and similar building types, consider piping layouts which minimize the need to enter occupied units to service or balance the system.

Return Piping:

1. To ensure prompt return of condensate, pitch return pipes up to 1/2 in. per 10 ft whenever possible.

2. Insulate piping to preserve heat and prevent possible contact injury. Do not insulate returns upstream of steam traps.

3. Include provisions for draining or "blowing off" the return piping to remove the residue of corrosion "mud" from within piping. Also consider oversizing wet return piping since it will contain the mud until it is removed.

14.10 HEATERS (RADIATORS)

Heat is usually delivered to spaces by radiators which are cast-iron sectional units or cabinet convectors with gravity air circulation over cast-iron or steel-finned elements. In years gone by when steam heat was much more common, a variety of radiator types as shown in Figure 14.14 were used.

Radiators are rated in Btu/h to allow for

TABLE 14.6 RETURN MAIN AND RISER CAPACITIES FOR LOW-PRESSURE SYSTEMS—FLOW RATE OF CONDENSATE IN SCHEDULE 40 PIPE EXPRESSED IN POUNDS PER HOUR (lb/h)

	Nominal Pipe Size (in.)	$\frac{1}{8}$ psi or 2 oz. Wet	$\frac{1}{8}$ psi or 2 oz. Dry	$\frac{1}{8}$ psi or 2 oz. Vac	$\frac{1}{4}$ psi or 4 oz. Wet	$\frac{1}{4}$ psi or 4 oz. Dry	$\frac{1}{4}$ psi or 4 oz. Vac
R	$\frac{3}{4}$	—	—	142	—	—	200
E	1	250	103	249	350	115	350
T	$1\frac{1}{4}$	425	217	426	600	241	600
U	$1\frac{1}{2}$	675	340	674	950	378	950
R	2	1400	740	1420	2000	825	2000
N	$2\frac{1}{2}$	2350	1230	2380	3350	1360	3350
	3	3750	2250	3800	5350	2500	5350
M	$3\frac{1}{2}$	5500	3230	5680	8000	3580	8000
A	4	7750	4830	7810	11,000	5380	11,000
I	5	...	...	13,700	...	...	19,400
N	6	...	...	22,000	...	...	31,000
	$\frac{3}{4}$	...	...	249	...	48	350
	1	...	113	426	...	113	600
	$1\frac{1}{4}$	...	248	674	...	248	950
R	$1\frac{1}{2}$	...	375	1420	...	375	2000
I	2	...	750	2380	...	750	3350
S	$2\frac{1}{2}$	...	...	3800	...	...	5350
E	3	...	...	5680	...	...	8000
R	$3\frac{1}{2}$	...	...	7810	...	...	11,000
	4	...	...	13,700	...	...	19,400
	5	...	...	22,000	...	...	31,000

Source: Excerpted with permission from the *1993 ASHRAE Handbook of Fundamentals*, Chapter 33, Table 16.

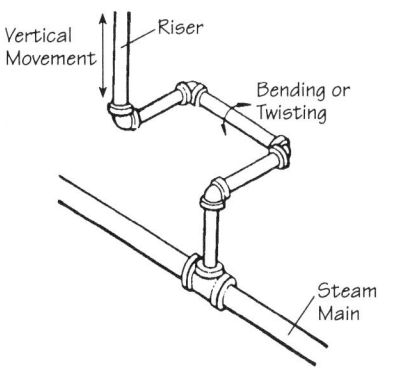

FIGURE 14.13 Piping connections to allow for expansion and movement. Reprinted with permission from the *Handbook of Air Conditioning Heating and Ventilating, Third Edition*, Stamper and Koral, The Industrial Press.

simple correlation with heat loss calculations. In addition, the heat output of steam radiators is still sometimes referred to in terms of square feet of equivalent direct radiation (EDR), where 1 ft^2 of EDR is equal to 240 Btu/h (at a steam pressure of 1 psig). Guides of the Hydronics Institute and other sources should be consulted to determine the heat output (EDR) of various types of older cast-iron radiators which may be encountered in renovation projects.

Today the only cast-iron sectional radiators commonly made are of the small-tube variety, 4 or 6 inches per section, and with radiator heights of 19 and 25 in. Provided in Table 14.7 is the approximate heat out-

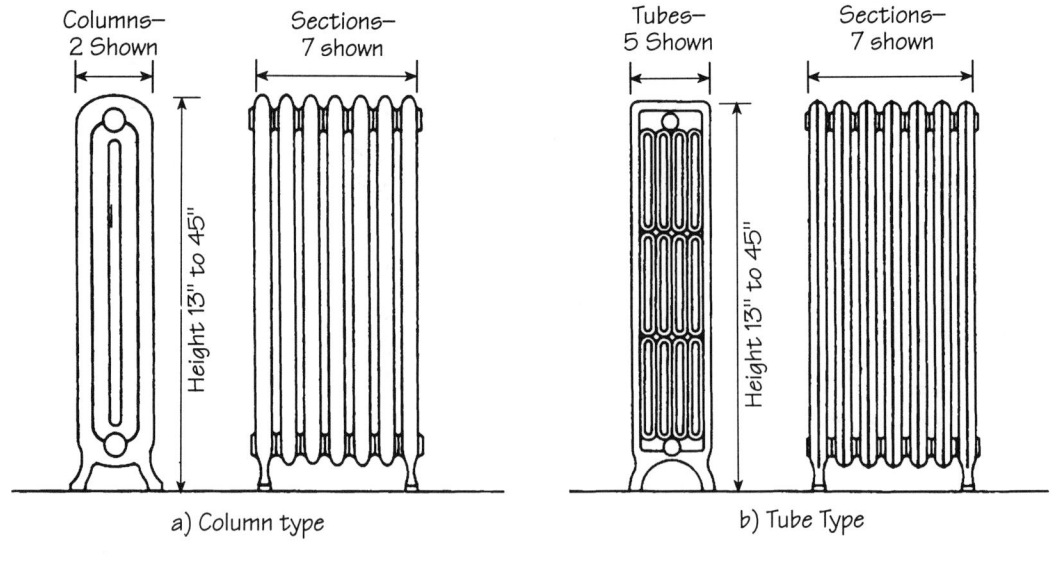

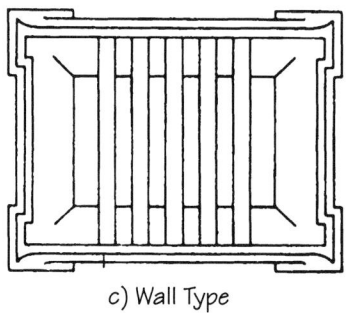

FIGURE 14.14 Radiator types found in older existing buildings.

a) Column type

b) Tube Type

c) Wall Type

Columns—2 Shown
Sections—7 shown
Height 13" to 45"

Tubes—5 Shown
Sections—7 shown
Height 13" to 45"

TABLE 14.7 SMALL-TUBE CAST-IRON RADIATORS

Height	Number of Tubes	Number of Sections	Length (in.)	Heat Output (Btu/h)
19	4	4	7	1540
		6	$10\frac{1}{2}$	2300
		8	14	3070
		10	$17\frac{1}{2}$	3840
		12	21	4600
		14	$24\frac{1}{2}$	5380
25	4	4	7	1920
		6	$10\frac{1}{2}$	2880
		8	14	3840
		10	$17\frac{1}{2}$	4800
		12	21	5760
		14	$24\frac{1}{2}$	6720
25	6	4	7	2880
		6	$10\frac{1}{2}$	4320
		8	14	5760
		10	$17\frac{1}{2}$	7200
		12	21	8640
		14	$24\frac{1}{2}$	10,080

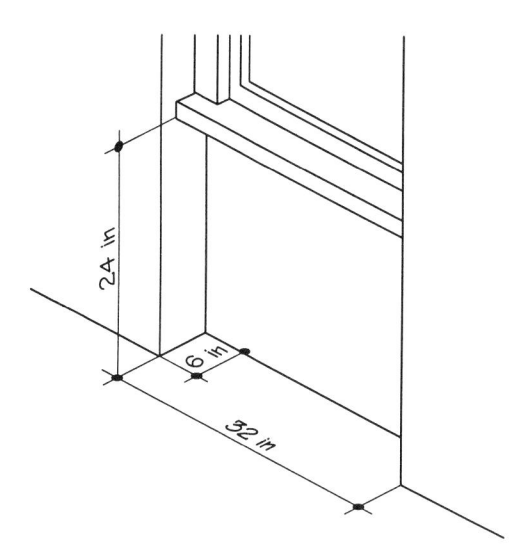

FIGURE 14.15 Window recess for Example 14F.

put of small-tube cast-iron radiators currently manufactured up to about 2 ft in length. During World War II radiators were standardized. The standard spacing of sections was set at $1\frac{3}{4}$ in., the standard leg height at $2\frac{1}{2}$ in., and unit width of about $4\frac{1}{2}$

in. for 4 tube radiators and about 7 in. for 6 tube radiators.

Example 4F: The design heat loss of a room in a converted factory with bearing wall construction has been computed to be 15,000 Btu/h. Select radiators to meet this heat loss for placement below three recessed windows. The height to the bottom of the sill is 24 in. Window recesses are 6 in. deep and 32 in. wide as shown in Figure 14.15.

Solution: Radiators should be selected to fit into the recesses below the windows without projecting outward. Since the sill is only 24 in. high, 19-in high radiators (which have 4 tubes) will have to be used.

The room heat loss of 15,000 Btu/h can be met by three equally sized radiators. Therefore, each unit must deliver a heat output of at least 5000 Btu/h. As shown in Table 14.7, 19 in. tall radiators with 14 sections have a heat output of 5380 Btu/h each. Since they are $24\frac{1}{2}$ in. long they will

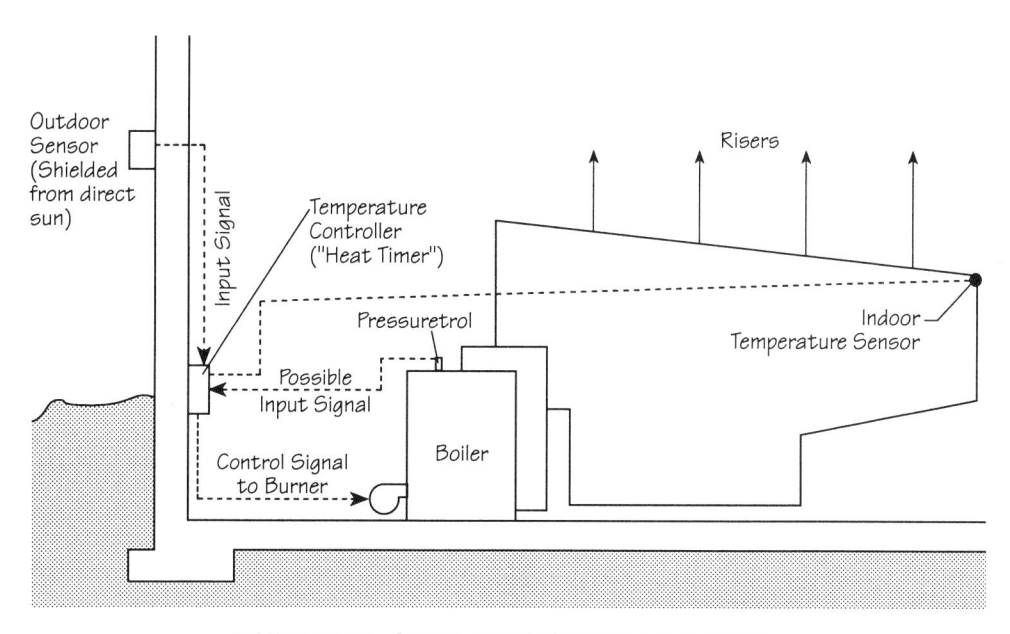

FIGURE 14.16 Central control of steam heating system.

248 STEAM HEATING SYSTEMS

easily fit within each window recess, allowing room for associated end piping.

Other Types of Heaters

Radiators are not the only type of heaters that can be used in a steam heating system. Steam unit heaters are commonly used to provide heat to large areas such as garages and warehouses. Protected fin-tubed units are also sometimes used in industrial applications.

14.11 TEMPERATURE CONTROL

Steam heating systems can be controlled in various ways. One very common form of central system control is through the use of a "heat timer" which cycles the boiler on and off in response to outside temperature (see Figure 14.16).

The cycle time of the system is then started by an indoor temperature sensor located either on a supply steam pipe or sometimes on the wet return line, or a pressure sensor on the boiler. Heat timers can also be set to maintain slightly lower room temperatures at night. Steam heating systems can also be controlled by a simple thermostat provided it is located in a room that provides a valid indication of building temperature.

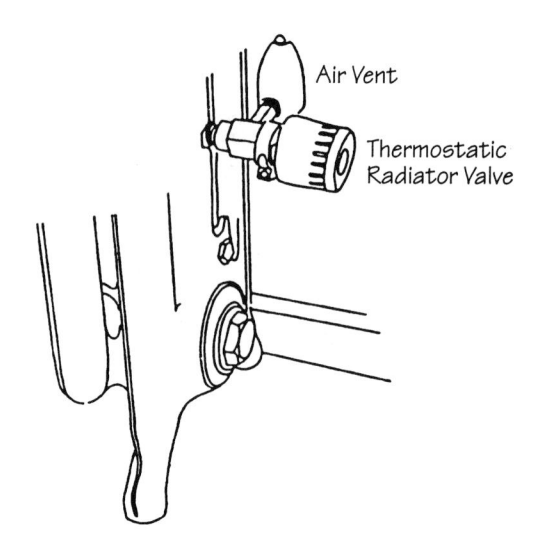

FIGURE 14.17 Thermostatic radiator control.

Temperature control at each radiator can then be provided through the use of orificing, or by thermostatic radiator controls (shown in Figure 14.17).

Thermostatic radiator control valves can be installed on two-pipe steam heating systems and one-pipe systems when in series with the air vent. Such control will produce the most comfortable room conditions without overheating, and improve the overall energy efficiency of the system.

15

HOT WATER HEATING SYSTEMS

Modern hot water heating systems (also known as "hydronic" systems) use a boiler to heat a fluid (usually water) which is then circulated through a closed piping system by a pump (often called a circulator). Recent Census Bureau statistics show that residential hot water heating systems are most popular in the Northeast where they constitute over one-third of new home heating systems. By contrast, only about 2% of new systems in houses in the Midwest and West are hydronic. The national average for hydronic home heating systems is about 7%.

Hot water heating systems are, however, very common in larger buildings throughout the country. In large part this is due to the flexibility provided by small-piped systems which can be easily zoned and controlled.

The basics of hot water heating systems, including system components (i.e., boilers, expansion tanks, and heaters such as baseboard units), piping arrangements, and sizing procedures will be described herein. Avoid oversizing hot water heating system components as this is a common cause of failure. Size all system components for the design conditions which they will meet. For example, pumps are very "happy" when they have to work very hard. Pump motors "burn out" if they have been selected based on excessive frictional assumptions. Perhaps the only hot water system component which can be oversized without detriment is the diaphragm-type expansion tank.

A detailed example of a hot water heating system for "The House" originally presented in Section 5.14 concludes this chapter.

15.1 BASIC SYSTEM DESCRIPTION

Steam heating systems work by releasing latent heat as discussed in Chapter 14. In contrast, hot water heating systems function by having a hot circulating fluid give up some of its heat as it passes through terminal units. Depicted in Figure 15.1 are the

250 HOT WATER HEATING SYSTEMS

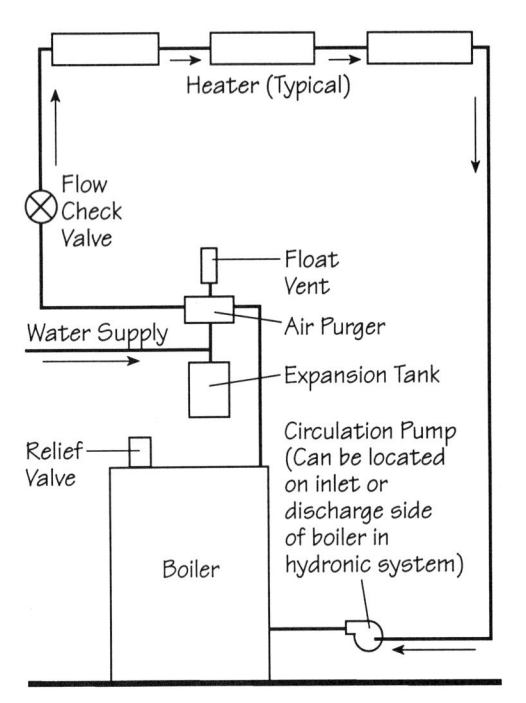

FIGURE 15.1 Hot water heating system components.

basic components of a hot water heating system.

A hot water heating system is a closed system under pressure. Water circulates through the boiler where it is heated (usually to about 180°F in houses and 200 to 220°F in larger systems). It then is pumped through tubing (usually copper) to the units that heat each space (usually "baseboard" radiation in new installations). As the water passes through the terminal heating units, the supply water temperature decreases as it gives up its heat to the space.

Hot water heating systems require an expansion tank since the volume of water will change within the closed system as it is heated and then cools down (i.e., the volume of 50°F water increases almost 4% when it is heated to 200°F). Also note the pressure relief valve (prv) at the top of the boiler in Figure 15.1, which protects the boiler from reaching excessive pressures that may rupture it. The flow check valve on the supply pipe leaving the boiler prevents gravity circulation when the pump is not operating.

Here, the focus will be on hot water heating systems for houses and commercial buildings which use low-temperature water below 250°F (note: boiling will not occur if system pressures are high enough). Applications such as for campuses with many large buildings or industrial processes which employ medium temperature (350°F or less) or high temperature (up to 500°F) water at very high working pressures are not covered.

System Advantages

Hot water heating systems offer the following advantages:

- They are capable of very quiet operation
- Pumps use a small amount of electricity compared to fans in a warm air system
- Required pipes are relatively small compared to those for a steam system
- The small pipes are incomparably smaller than ductwork required in a warm air heating system, resulting in great architectural freedom
- They are easy to zone, including individual control at each terminal unit
- They can be easily combined with air handlers to provide modulated heating coils for ducted air systems
- Terminal units are available in many types and designs.

15.2 HOT WATER FLOW

The flow of water in a hot water heating system is typically referenced in gallons per minute (gpm) with a flow rate of 1 gpm considered to be equal to a flow of 500 lb of water per hour (8.33 lb/gal × 1 gal/min × 60 min/hr).

15.2 HOT WATER FLOW

The quantity of heat (in Btu/h) delivered by a hot water system terminal unit depends upon the water flow rate and temperature drop. Depicted in Figure 15.2 is a heater with a temperature drop of 2°F (from 200 to 198). The total heat output from a system depends upon the temperature drop through all of the heaters. For an entire circuit of heaters, it is common to design for a 20°F temperature drop. Realize, though, that designers of larger two-pipe systems will often use higher temperature drops of up to 40°F which result in more economical piping design and lower pumping requirements.

Formula 15A is used to determine the heat potential of a hot water heating system for each gpm of flow.

$$q_{gpm} = 500 \times T_d \quad \text{Formula 15A}$$

where

q_{gpm} is the heat output in Btu/h per 1 gpm flow

500   is the weight of water in pounds which flows in 1 h for each 1 gpm of flow multiplied by the specific heat of water (1.0 Btu/lb·°F)

T_d is the temperature drop of the flowing water in °F

Example 15A: Determine the heat output from a circuit of heaters (q_{gpm}) which has a hot water flow of 1 gpm and a temperature drop of 20°F.

Solution: Use Formula 15A.

$$q_{gpm} = 500 \times T_d \quad \text{(Formula 15A)}$$
$$= 500 \times 20$$
$$= 10,000 \text{ Btu/h}$$

The value of 10,000 Btu/h is an easy to remember "rule of thumb" for a water flow rate of 1 gpm and a temperature drop (T_d) of 20°F. Provided in Figure 15.3 are values of heat output from hot water in Btu/h per gpm for temperature drops from 15 to 40°F.

The flow rate of hot water needed in gallons per minute (gpm) to satisfy a given heating load is then computed using Formula 15B.

$$\text{gpm}_{hw} = \frac{q_{tot}}{q_{gpm}} \quad \text{Formula 15B}$$

where

gpm_{hw} is the hot water flow rate required in gallons per minute

q_{tot} is the total heating load which needs to be satisfied by a hot water heating circuit in Btu/h

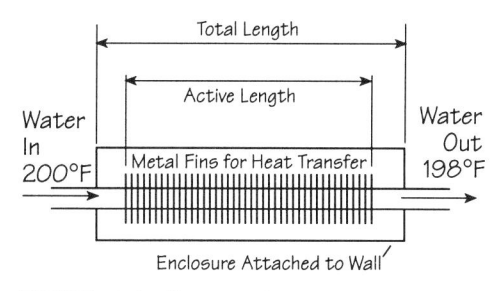

FIGURE 15.2 Hot water heating system—typical temperature drop of 20°F commonly used for design.

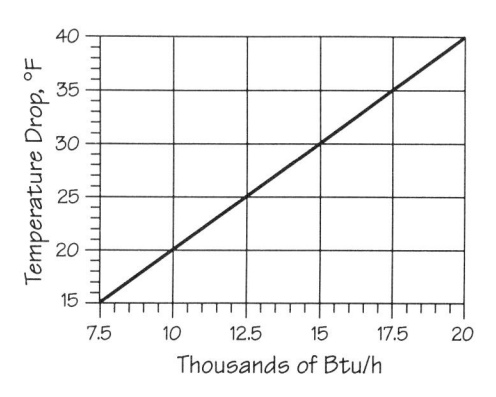

FIGURE 15.3 Heat output per 1 gpm of hot water flow.

q_{gpm} is the heat output per gpm of water flow in Btu/h (by Formula 15A)

Example 15B: A house with a total design heat loss of 70,000 Btu/h (q_{tot}) uses a hot water heating system designed for a 20°F temperature drop. Determine the hot water flow rate required in gallons per minute (gpm$_{hw}$) to serve the various baseboard heaters in the house.

Solution: Use Formula 15B and a value of 10,000 Btu/h (from Figure 15.3) for q_{gpm} since the temperature drop is 20°F.

$$\text{gpm}_{hw} = \frac{q_{tot}}{q_{gpm}} \quad \text{(Formula 15B)}$$

$$= \frac{70,000}{10,000}$$

$$= 7 \text{ gpm, the required system flow rate}$$

Freeze Protection

In Formula 15C, the universal method for computing the required gpm for fluids (including antifreeze mixtures) is provided.

$$\text{gpm}_{fl} = \frac{q_{tot}}{60 \times W \times C \times (t_1 - t_2)}$$

Formula 15C

where

gpm$_{fl}$ is the required fluid flow rate in gallons per minute
q_{tot} is the total heating load which needs to be satisfied by a piping circuit in Btu/h
60 is the conversion of 60 min in 1 h
W is the density of the circulating fluid in lb/gal
C is the specific heat of the circulating fluid in Btu/lb·°F
t_1 is the initial temperature of the circulating fluid in °F
t_2 is the final temperature of the circulating fluid in °F

Example 15C: The house in the previous example is being built as a vacation house in a cold climate. Assume the same design heat loss of 70,000 Btu/h (q_{tot}) and a 20°F fluid temperature drop. Determine the flow of an antifreeze solution (50% by weight water–ethylene glycol mixture) in gallons per minute which will be required to serve the various baseboard terminal devices in the house. [Note: A 50% by weight water–ethylene glycol mixture has a density of 8.76 lb/gal (W), a specific heat of 0.83 Btu/lb·°F (C), and a freezing point of −33°F.]

Solution: Use Formula 15C.

$$\text{gpm}_{fl} = \frac{q_{tot}}{60 \times W \times C \times (t_1 - t_2)}$$

(Formula 15C)

$$= \frac{70,000}{60 \times 8.76 \times 0.83 \times 20}$$

$$= \frac{70,000}{8,725}$$

$$= 8.0 \text{ gpm, system flow rate}$$

Note how this compares to the required flow rate for water in the previous example of 7.0 gpm. For small applications like this, the additional flow will be of minor consequence. Still, large hydronic systems which circulate an antifreeze solution require larger pumps which use more energy.

15.3 PIPING

Most piping used in residential and commercial hot water heating systems is copper tubing. Copper tubing comes in types M,

L, and K. Although Type M has the thinnest walls, it has adequate pressure resistance for most applications and can often be used when available. Interestingly, soft-temper Type M tubing is now being used for new systems in some existing buildings since its flexibility allows for easier routing through existing construction.

Type L copper tubing is readily available and used in the great majority of hot water heating systems since it provides the required pressure resistance and durability. Type K copper tubing has the thickest walls and is used for severe service conditions.

Copper tubing is easy to work in the field, and is not subject to oxidation and scaling as is steel pipe. Copper tubing also offers slightly less frictional resistance than steel, allowing the use of smaller circulator pumps and less expensive operation. See Table 10.1 for Type L copper tubing data.

Maximum Btu Capacity

Flow rate and pipe size have a direct bearing on the maximum amount of heat which a supply line can deliver. Formula 15D is used to determine the flow rate in gpm at a given velocity through pipe.

$$\text{gpm} = 60 \times \text{fps} \times V_{ft} \quad \text{Formula 15D}$$

where

gpm	is the fluid flow rate in gallons per minute
60	is the conversion of 60 sec/min
fps	is the velocity of fluid flow in feet per second
V_{ft}	is the volume capacity inside the pipe in gallons per foot of length

Example 15D: Determine the gpm flow rate through 3/4 in. Type L copper tubing. Assume a water velocity of 3 fps (the velocity used to rate finned tube radiation as discussed below).

Solution: 3/4 in. Type L copper tubing holds 0.025 gal of water per linear foot (as per Table 10.1). Use Formula 15D.

$$\text{gpm} = 60 \times \text{fps} \times V_{ft}$$

(Formula 15D)

$$= 60 \times 3 \times 0.025$$

$$= 4.5 \text{ gpm}$$

A 20°F temperature drop yields heat transfer of approximately 10,000 Btu/h per 1 gpm of hot water flow as indicated using Formula 15A. Therefore, the 4.5-gpm flow rate for 3/4 in. copper tubing equates to a heat transfer of approximately 45,000 Btu/h. Using the same basis of calculation, the capacity of various copper tubing sizes in serving baseboard radiators where the actual heat transfer takes place is depicted in Figure 15.4.

Perhaps of most importance is the heat capacity of 3/4 in. copper tubing typically used in residential applications. Many houses will have design heat losses which

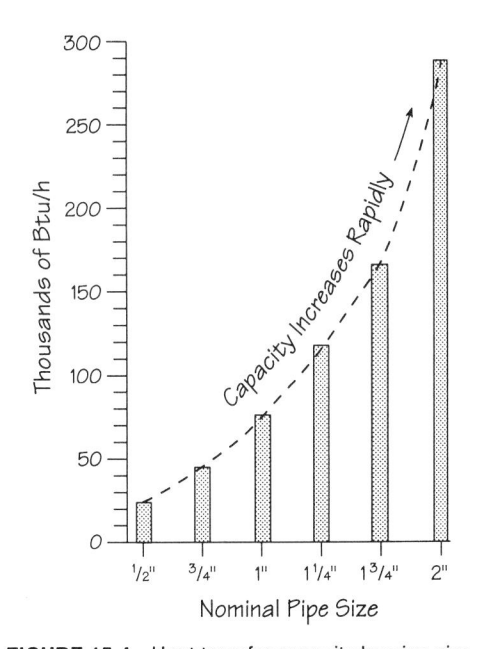

FIGURE 15.4 Heat transfer capacity by pipe size.

far exceed the 45,000 Btu/h capacity. This means that if 3/4 in. tubing is used with a 20°F temperature drop, more than one supply loop will be required (such an arrangement is illustrated in Figure 15.11 below).

Pipe Insulation

Pipes used in hydronic heating systems should be insulated to reduce heat loss and to provide protection from burns. Local codes normally dictate pipe insulation thickness based on pipe size and fluid temperature. The most typical insulation for low temperature hot water systems (below 250°F water temperature) is wrapping with one-half inch fiberglass. The installation of additional insulation thickness should be based on a realistic economic evaluation which considers the operational profile and actual fluid temperatures.

15.4 BOILERS

A boiler is used to produce hot water or steam for distribution to terminal units. The term "boiler" is defined broadly. It is truly a "boiler" when used as part of a steam heating system, since it does in fact boil water and convert it to steam. In contrast, a boiler in a hot water heating system merely heats the water to a high temperature (typically 160–220°F), but does not actually cause it to boil because the water is under pressure. Low-pressure hot water boilers can handle working pressures of up to 160 psi and temperatures of up to 250°F, and are built in accordance with The American Society of Mechanical Engineers (ASME) Boiler and Pressure Vessel Code, Section IV.

Most small to medium sized boilers (up to about 2,000,000 Btu/h) are made of cast iron because they can be cast into shape using precise geometries for effective performance, are resistant to corrosion, and have a long service life. The small cast iron sections also allow for field assembly of boilers in spaces which are inaccessible to package boilers. Larger boilers are often fabricated of steel which is welded or rolled into a single unit. For most residential and small commercial installations, the boiler is sold as a packaged unit which includes the burner, circulating pump, internal wiring, and boiler operating and safety controls.

Fuels and Combustion

Boilers can be fired by many fuels including electricity. Units fired by gas (natural or propane), oil, or a solid fuel (wood or coal) are common. Dual fuel boilers which enable one to choose between two fuels (such as gas or oil) based on availability and cost are also available.

Most small gas boilers employ "atmospheric" burners (exposed to and getting combustion air from the atmosphere of the boiler room) similar to a burner in an oven. Larger gas boilers often use gun-type burners, which are similar to oil burners. In oil-fired boilers, the burner nozzle sprays small droplets of oil into the boiler's combustion chamber (furnace) where it mixes with air and is ignited. Most oil-fired boilers now use "flame-retention" burners which make the flame more compact and improve combustion efficiency (see Figure 15.5).

The heat exchanger can be of various types of design (tubing, castings, etc.) but it must have a large surface area. The heating surface required depends on whether the surface "sees" the fire (getting heat through radiation) or not.

The combustion efficiency (fuel input minus on- and off-cycle losses—see Chapter 17) for most conventional noncondensing boilers ranges from 75 to 86%. Overall system efficiency will be less, depending upon system size and methods of temperature control.

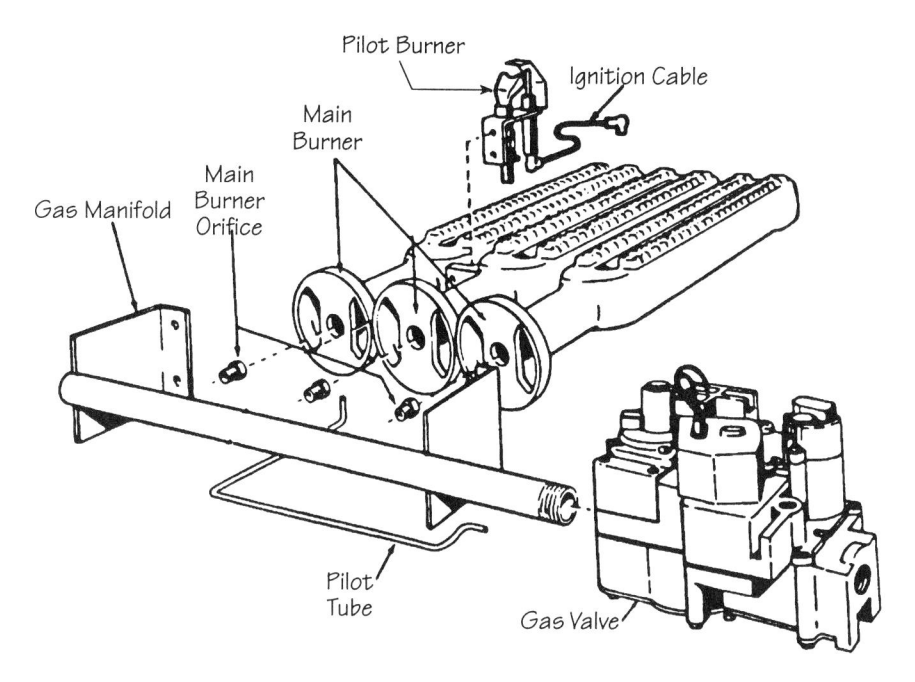

a) Electronic Ignition Gas Burner

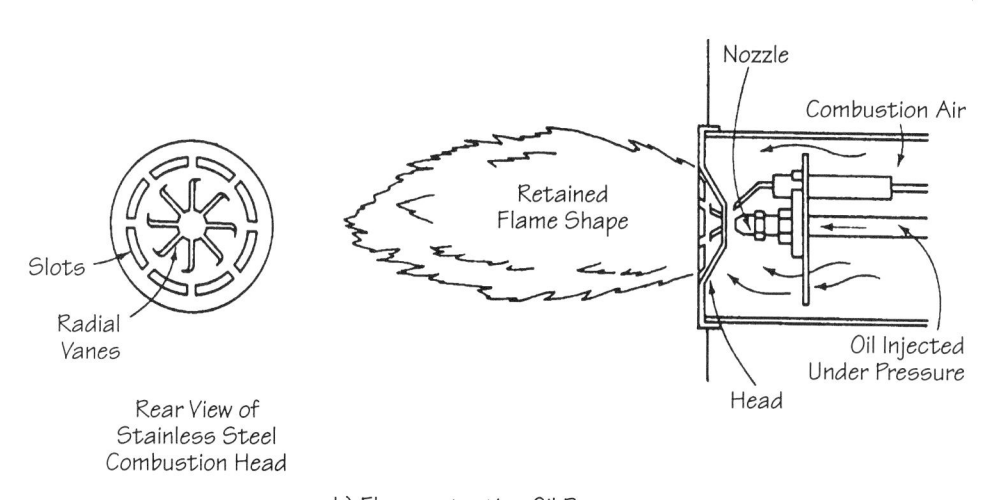

b) Flame-retention Oil Burner

FIGURE 15.5 Typical gas-fired and oil-fired burners for hot water boilers. a) Based on product literature of HB Smith. b) Based on product literature of R. W. Beckett, Inc.

Temperature Control

Large hot water heating systems typically "reset" the temperature of circulating water in response to outside temperatures. Depicted in Figure 15.6 is an application needing a system temperature of 210°F at the full design heating load when it is 0°F outside (Point A). Then when the outside air temperature warms up to 40°F the heating load can be satisfied if the system temperature is only 130°F (Point B).

Such a system is often referred to as a 2:1 reset system since the supply water temperature is reduced by an amount equal to twice the rise of outdoor air temperature, in this case 40°F. Lowered supply water temperature then reduces the Btu/h output of terminal heaters as discussed below. Reset control minimizes equipment cycling and improves system efficiency.

Water Expansion

Very simply, as the boiler heats the system water it expands in volume. To accommodate this expansion, two basic design approaches can be used. One approach is to provide a compression tank with an air cushion located above the boiler along with an air separator. It works by having the pressure on the air cushion increase when the water rises in temperature and increases in volume. A concern about this approach to expansion is the direct contact of water and air. This is because air in a closed hot water system can produce noisy flow and pipe corrosion.

To avoid the direct contact of water and air, most modern systems now use expansion tanks with flexible diaphragms which expand as the system heats up and water volume increases. As shown in Figure 15.7, expansion tanks for small systems are often installed directly on the piping system in combination with an air purger and automatic vent which isolates and expels any air bubbles trapped in the circulating hot water. Expansion tanks for small systems in one- and two-story houses can be selected from manufacturers' literature based upon the size of the heating system in Btu/h and the type of system (e.g., series loop).

Expansion tank selection for larger and taller buildings is based upon the volume of water in the system, operational temperatures, and the acceptable pressure rise. It is important to acknowledge that the capacity of a given expansion tank is decreased as the height of closed system water piping above it increases. Therefore, when acces-

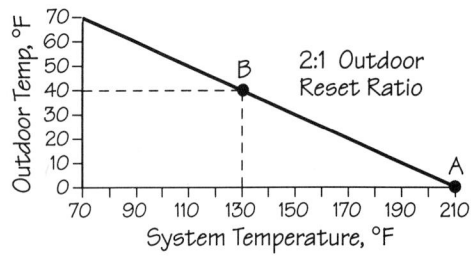

FIGURE 15.6 A 1:1 outdoor reset ratio.

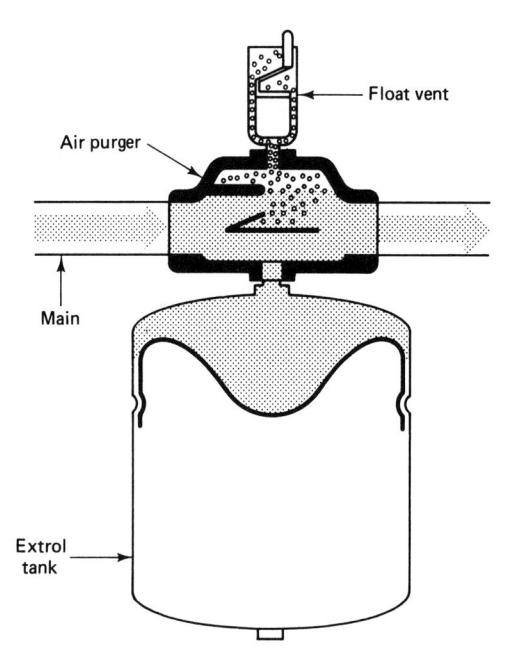

FIGURE 15.7 Expansion tanks to accommodate increased volume of heated water. Drawing courtesy of AMTROL Inc., W. Warwick, R. I. Extrol is a registered trademark of AMTROL Inc.

sible space can be found for the expansion tank near the top of a tall building, this is to be preferred.

15.5 SMALL CAPACITY BOILERS

Modular Boilers

Heating loads can also be met by "gangs" of relatively small capacity modular boilers, controlled and sequenced to provide a hotter circulating fluid as outdoor temperatures fall. As shown conceptually in Figure 15.8, one of the four boilers will be required to produce 140°F hot water in response to an outdoor temperature of 45°F. Then on a day when the outdoor temperature is 30°F, two boilers will be needed to produce the required flow of 160°F hot water. Three of the modular boilers will be needed when the outdoor temperature is 15°F to produce 180°F hot water. Finally, when the outdoor temperature is 0°F, all four boilers will be working to produce 200°F hot water for circulation.

The efficiency of a system using modular boilers is affected by the piping arrangement used and the mode of control. Some modular boilers are piped in parallel so that all boilers remain hot whether they are being fired or not. This results in standby losses and potentially large stack losses (if stack dampers are not installed or are not functioning properly). Another approach to piping is to use a "primary/secondary" piping arrangement which isolates boilers which are not in service. This approach improves system efficiency since only those boilers needed to meet the heating load will be activated and experience on-cycle and off-cycle losses.

Finally, a note of caution regarding stepped firing of modular boilers is in order. In mild weather, a low boiler water temperature can result in flue gases condensing in the boiler and chimney which can lead to possible damage or corrosion.

Wall-Mounted Boilers

Many manufacturers now produce small capacity wall-mounted gas-fired boilers with all required parts assembled in one unit. Combustion air arrives directly through the outside wall of the building for "sealed combustion" with exhaust gases vented directly outside as shown in Figure 15.9.

Wall-mounted boilers are becoming very popular in residential and commercial condominium projects since each unit has control and responsibility for their own small heating system. "Designer" wall-mounted boilers are even available for mounting in conjunction with kitchen cabinets.

15.6 BOILER SIZING

The amount of heat delivered by a boiler when firing continuously at full input is known as the gross output. Residential boilers are available for heating output capacities ranging from about 30,000 to 250,000 Btu/h. Sizing is based upon the amount of heat which the boiler must provide when operated constantly, to offset building heat loss under winter design conditions. Sizing should be based on a careful heat loss calculation as outlined in Section 5.7.

To assure adequacy, sizing also may involve the use of an oversizing factor (OSF) for some occupancy types. Heavily occupied buildings (e.g., houses and apartment buildings) should be engineered closely with a small oversizing factor of 1.05 (5% oversizing) to 1.1 (10% oversizing). For other buildings which are often unoccupied (i.e., offices and schools) an OSF of 1.1 to 1.25 should be used to achieve a quicker pick-up of indoor air temperature. The widely adopted ASHRAE/IES Standard 90.1/1989 "Energy Efficient Design of New Buildings Except Low-Rise Residential Buildings" states that the rated heating or cooling capacity of equipment shall not be more than 125% of the calculated design

258 HOT WATER HEATING SYSTEMS

Note: Combustion gases may condense in boilers with low temperature supply water.

a) 60°F Outdoor Temperature

b) 40°F Outdoor Temperature

c) 20°F Outdoor Temperature

d) 0°F Outdoor Temperature

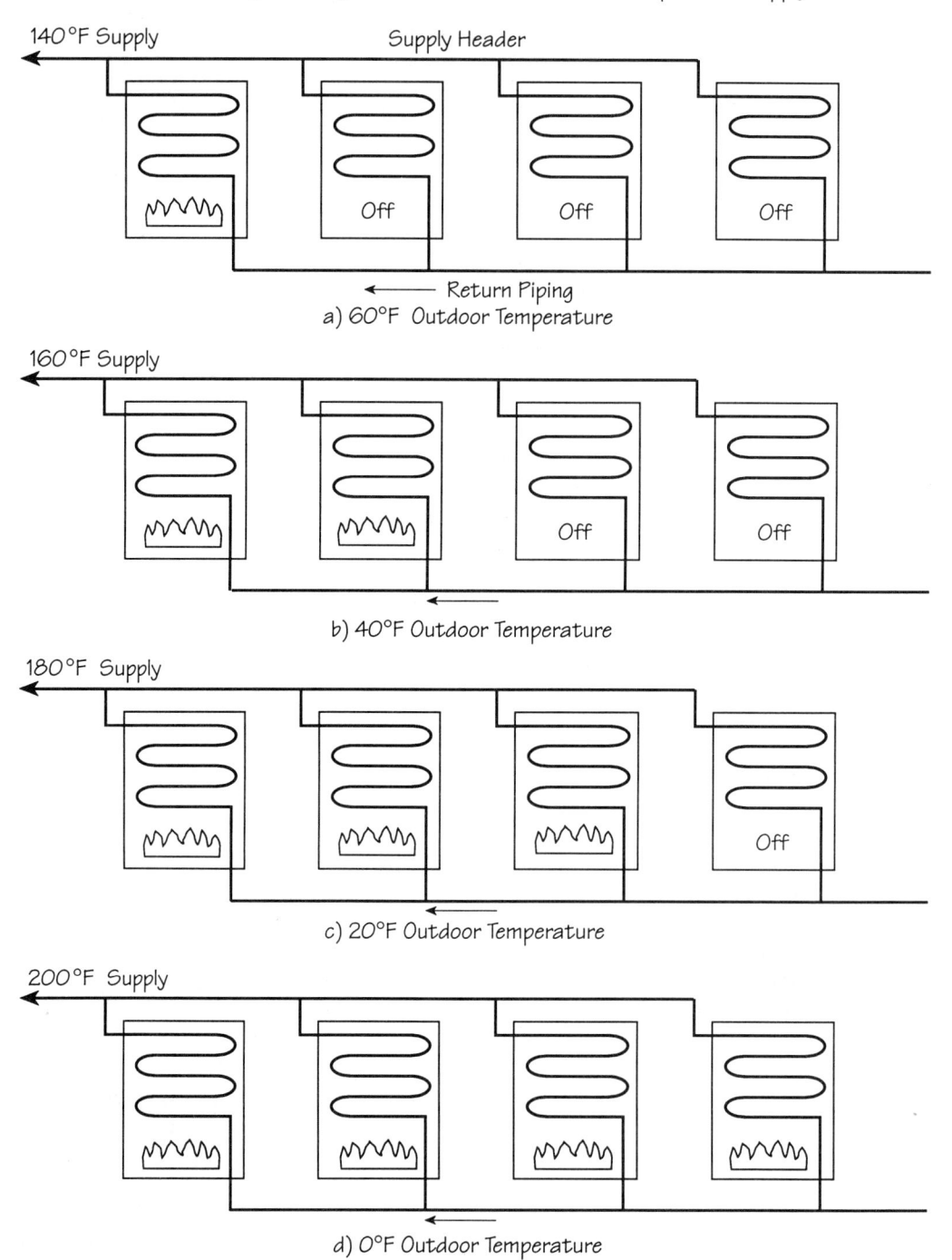

FIGURE 15.8 Sequenced firing of modular boilers.

15.7 FRICTION IN PIPING

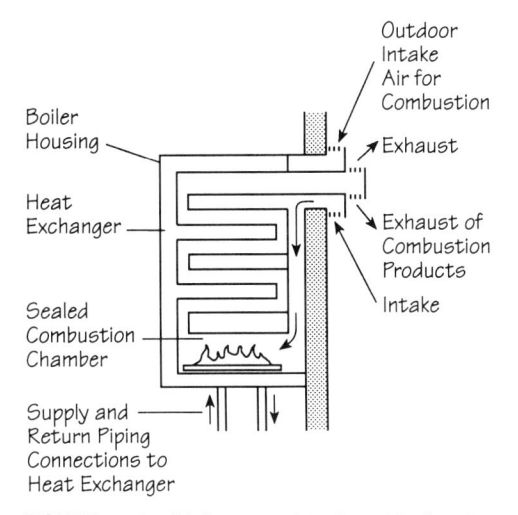

FIGURE 15.9 Wall-mounted boiler with "sealed combustion."

load. This equates to an OSF of 1.25. Also be sure to check for any applicable code requirements regarding system sizing.

Formula 15E is used to determine the required heat output of a boiler.

$$\text{BOIL}_{out} = q_{tot} \times (t_i - t_o) \times \text{OSF}$$

Formula 15E

where

BOIL$_{out}$	is the required boiler heat output in Btu/h
q_{tot}	is the total winter design heat loss in Btu/°F · h
t_i	is the inside winter design temperature in °F
t_o	is the outside winter design temperature in °F
OSF	is the oversizing factor (1.0 to 1.25)

Example 15E: The design heat loss of a house to be built in Denver, Colorado has been calculated to be 800 Btu/°F·h (q_{tot}). Determine the boiler output (BOIL$_{out}$) size required. Assume an indoor design temperature (t_i) of 68°F. Assume an oversizing factor (OSF) of 1.05 (an increase in capacity of 5%).

Solution: The Winter Design dry-bulb temperature (t_o) for Denver is 1°F (97.5% value), as shown in Table 2.1. First calculate the heat output required to meet the design condition. The design delta T is the difference between the indoor design temperature and the Winter Design dry-bulb temperature. Now use Formula 15E.

$$\text{BOIL}_{out} = q_{tot} \times (t_i - t_o) \times \text{OSF}$$

(Formula 15E)

$$= 800 \times (68 - 1) \times 1.05$$

$$= 56{,}280 \text{ Btu/h}$$

In this example let's assume that the choice is between two gas-fired units, one which will deliver an output of 55,000 Btu/h and the other an output of 62,000 Btu/h. Note that the size requirement of 56,280 Btu/h is just barely above 55,000 Btu/h. The smaller unit can be used if a slightly smaller oversizing factor is acceptable for this application. One benefit of the smaller (and less oversized) unit will be improved energy efficiency because it will not "cycle" as often as a larger boiler would in the same system.

Finally, note that this simple sizing example is for a heating system only. If this boiler were also being used to supply domestic hot water, that load must also be considered in the sizing calculation as shown in Section 22.3. Boiler combustion efficiency is discussed in Chapter 17.

15.7 FRICTION IN PIPING

Hot water heating systems are closed piping systems. Therefore, the weight of the column of water going up in the supply pipes is balanced by the weight of the water coming down the return pipes. As a result, the pump in such a system only has to overcome the friction caused by flow through the pipes and fittings.

Hydronic piping systems typically experience a frictional loss of between 2 and 15 ft of water head loss through the system.

Pumps in Hot Water Heating Systems

The liquid in hydronic systems (usually water) is moved by one or more pumps (often known as "circulators") which overcome system friction. Pumps for hydronic applications are usually "in-line" pumps, which means that they are directly supported by the piping and need no base. See Chapter 10 for detailed information on friction and pumps.

15.8 PIPING ARRANGEMENTS AND TEMPERATURE CONTROL

Various piping arrangements are possible for hydronic systems with selections based upon initial cost, building type, and need for precise control and zoning of heat.

Series Loops

Least costly is a simple one-pipe "series loop." As water travels from the boiler and passes through heaters, the temperature of the supply water within the piping system decreases. Therefore, to provide heaters with a similar output, "baseboard C" must be longer than "baseboard B," which must be longer than "baseboard A," as shown in Figure 15.10. This type of system is common for residences and small buildings.

Baseboard heaters with 3/4 in. piping is very common for small residential systems. As was shown in Figure 15.4, 3/4 in. piping has a limitation of delivering only about 45,000 Btu/h of heat (with a 20°F temperature drop), an amount of heat loss exceeded by many houses and small buildings. One very common design approach to

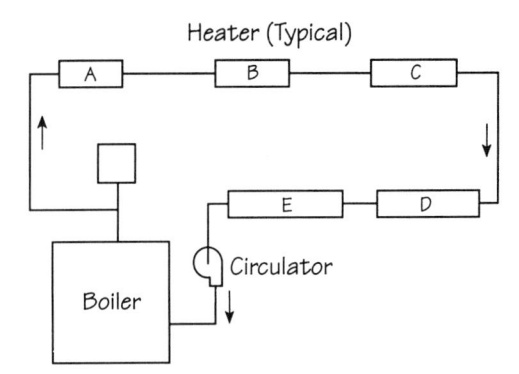

FIGURE 15.10 One-pipe series loop (for small heat loss applications).

overcome this limitation is a series loop which splits as shown in Figure 15.11.

One disadvantage of the simple series loops illustrated in Figures 15.10 and 15.11 is that individual baseboard heaters cannot be isolated for temperature control. This limitation can be easily overcome by using a piping arrangement as illustrated in Figure 15.12 which consists of a loop of hot water off which individual baseboard units are tapped. Proper supply water flow to the heaters is facilitated by using special diversion ("monoflow") fittings. After passing through the heaters, the reduced-temperature water returns to the loop and mixes with the hotter supply water.

A disadvantage of series loops and one-pipe systems is the gradual lowering of supply water temperature as it travels through the loop. This factor can be significant on large loops and requires larger and larger heating units downstream in the loop. It also may lead to the decision to employ multiple loops. Many systems in large houses have one or two "daytime loops" and a "bedroom loop" (see isometric in Figure 15.13). Each can have its own temperature control with circulation provided by either multiple small pumps (as shown) or zone valves.

Two-Pipe Systems

One-pipe systems are inherently limited in their ability to deliver heat since the supply

15.8 PIPING ARRANGEMENTS AND TEMPERATURE CONTROL

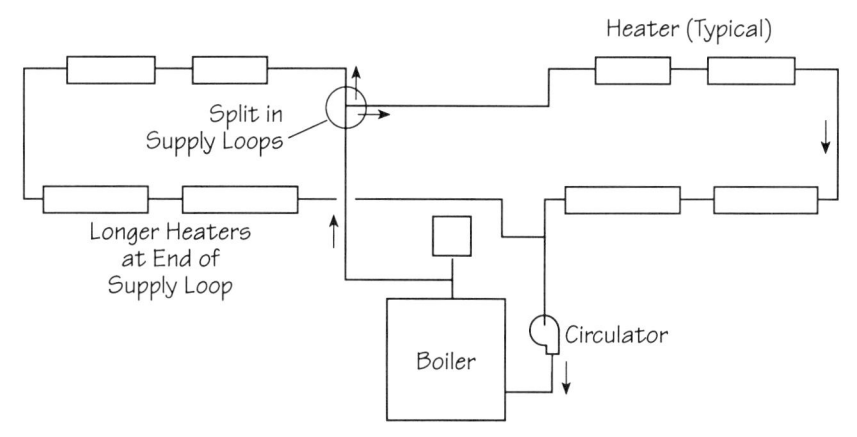

FIGURE 15.11 Split series loop (for applications with a larger design heat loss).

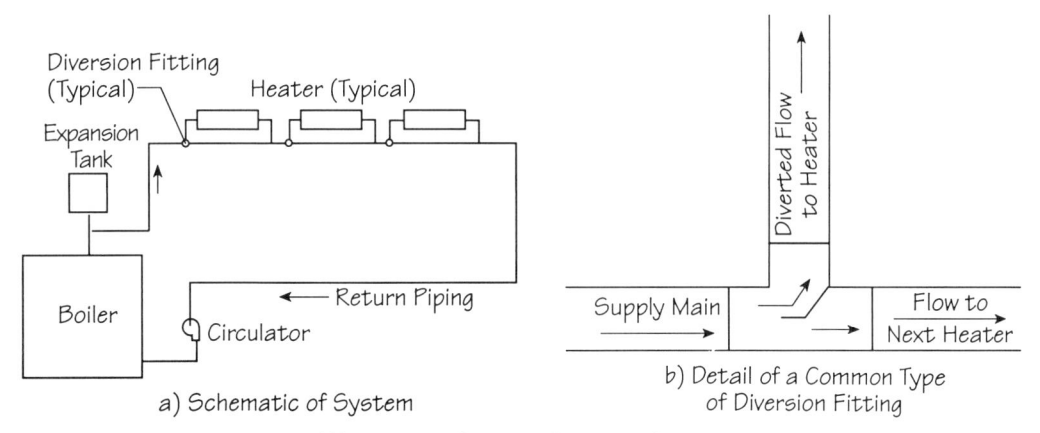

FIGURE 15.12 One-pipe "monoflow" systems.

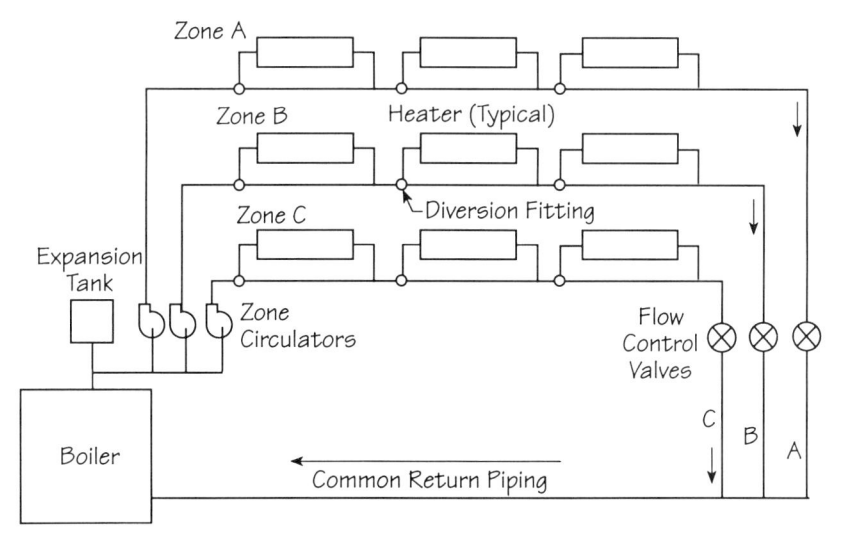

a) Schematic of System

b) Detail of a Common Type of Diversion Fitting

FIGURE 15.13 Zoning through use of multiple one-pipe circuits.

water temperature is always being lowered as more heaters are encountered. This limitation is overcome by introducing a second pipe to carry "return" water which exits the heaters. Two alternative approaches to return water piping are shown in Figure 15.14.

The two-pipe direct-return approach in Figure 15.14a minimizes the amount of piping but can be difficult to balance on long pipe runs. This is because the lengths of piping to and from the first heater (e.g., baseboard unit) are less than the length to and from heaters farther down the line. As a result, heaters furthest from the boiler can be starved of heat if the heaters have a low pressure drop compared to the pressure drops in the pipes.

The two-pipe reverse-return piping arrangement shown in Figure 15.14b equalizes the length of piping to and from all units. This equalizes friction and is one commonly used way to produce balanced and predictable system flow. The disadvantage of the two-pipe reverse-return method is, of course, the need and expense associated with installing additional piping.

"Orificing" is another way of producing predictable flow through all heaters without going to the expense of reverse-return pip-

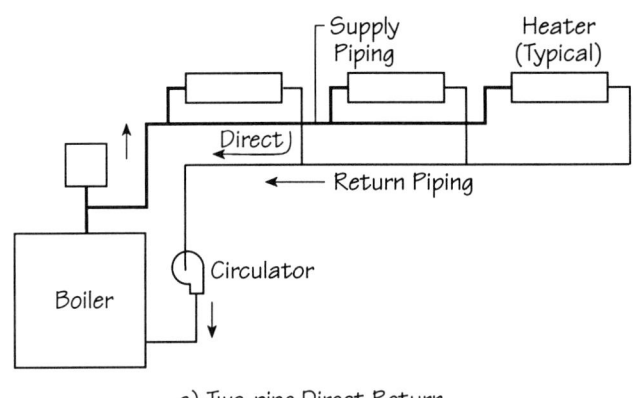

a) Two-pipe Direct Return

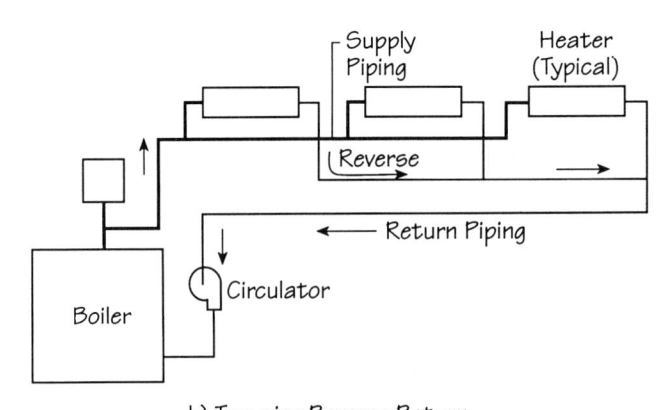

b) Two-pipe Reverse Return

FIGURE 15.14 Two-pipe arrangement alternatives.

ing. Orifice plates installed at the heater inlets provide small openings and high resistance to flow. By installing orifice plates at the inlets to all heaters one is able to produce high pressure drops through the heaters. This compares to a relatively low pressure drop in the hot water supply main. As a result, the great (but relatively even) resistance through all of the heaters produces much more balanced flow.

In Figure 15.15a we see a piping circuit where the resistance through heater A (0.5 ft of water column) is much less than the resistance through heater D (4 ft of water column). This results in a ratio of frictional resistance of 8 (4/0.5) and the source of very unbalanced flow.

In Figure 15.15b orifice plates are added to the inlet of each heater which provide an additional resistance equivalent to 3 ft of water column. Now the resistance through heater A (3.5 feet of water column) is only one-half of the resistance through heater D (7 ft of water column).

Combination Piping Systems

The simple descriptions of direct-return or reverse-return generally become inappropriate when applied to larger buildings

Resistance Through Heaters; feet of water column

0.5 1.7 3.0 4.0
 A   B   C   D

Boiler

Note: On long pipe runs the resistance through the heaters will vary.

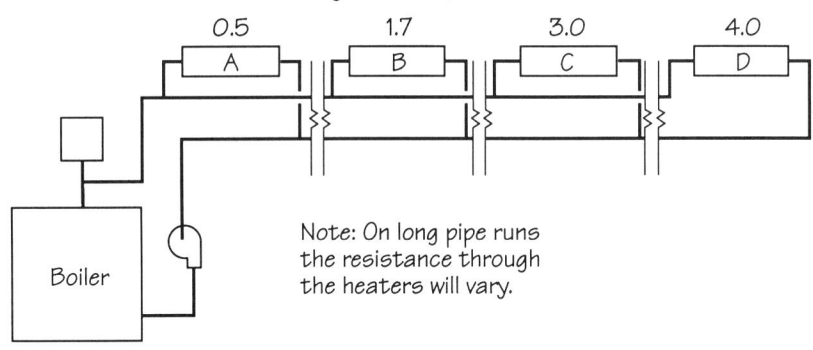

a) Resistance Through Heaters Without Orificing

Resistance Through Heaters Including Orifice Plates; feet of water column

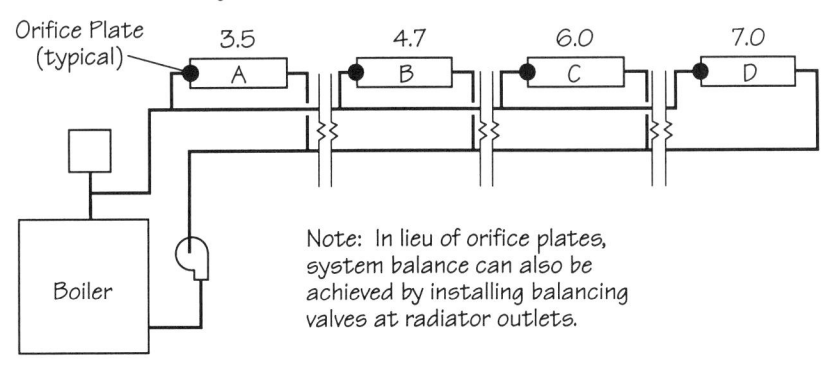

Orifice Plate (typical)
3.5 4.7 6.0 7.0
 A   B   C   D

Boiler

Note: In lieu of orifice plates, system balance can also be achieved by installing balancing valves at radiator outlets.

b) Resistance Through Heaters With Orifice Plates Added

FIGURE 15.15 Using orificing to balance hot water flow through heaters.

which use a combination of piping designs to lessen initial cost. Do not get lost in a desire to "label" or categorize a system. Instead design a system so that the friction through all system elements allows for determination of a predictable flow of water of known temperature to all heaters.

Air Venting

Regardless of the piping arrangement, air must not be allowed to accumulate within a closed hot water piping system. Modern expansion tanks (usually installed above the boiler) generally include an air purger. In addition to this, air vents should be provided at high points of piping and at the heaters to relieve possible air pockets which can render portions of the system "air bound" and inoperative.

15.9 BASEBOARD AND FINNED-TUBE RADIATION

"Baseboard" heaters are the most common type of heater used to deliver the heat from a hot water loop. These units are called "baseboard" simply because they are normally located on the floor and against the base of the wall.

Heat is obtained from the baseboard by a combination of radiation and convection. Baseboard units come in various designs but generically they consist of a copper tube to which square aluminum "fins" are attached to increase surface area and promote heat transfer to the space. Most manufactured baseboard units typically have from 25 to 60 fins per foot of "active" length with the total length of units being 3–6 in. longer than the active length.

The heat transfer fins and tube are usually mounted in architectural metal enclosures designed to maximize convection and to prevent contact with hot surfaces. Some units come with a damper cover which can be closed to provide some local control by minimizing convection. Illustrated in Figure 15.16 is baseboard radiation of the type typically used in residential and small commercial applications.

Baseboard radiation is rated in Btu/h per linear foot depending upon average water temperature and water flow rate (1 gpm is typically assumed). Table 15.1 provides ratings for "typical" residential-type baseboard radiation with 3/4 in. copper tubing (and aluminum fins).

To achieve the rated performance, it is very important that baseboard heaters be mounted in their enclosures at the height specified by the manufacturer. If the unit is mounted too low it may choke off entry of inlet air; if the element is too high inside the enclosure it reduces the stack effect. In either case heater output can be reduced drastically (up to 50%).

Commercial Applications

In basic construction, baseboard radiation and "finned-tube radiation" are similar. The term finned tube is generally used for commercial applications. When the length of available wall space is limited, units that are two or even sometimes three tiers high can be used to concentrate the heat output in one location (see Figure 15.17).

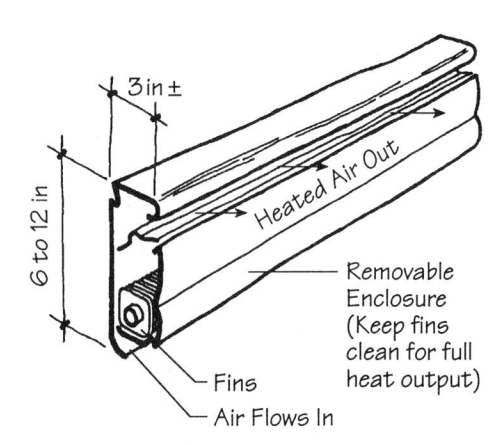

FIGURE 15.16 Typical residential baseboard radiation.

15.9 BASEBOARD AND FINNED-TUBE RADIATION

TABLE 15.1 TYPICAL RESIDENTIAL BASEBOARD RADIATION RATED OUTPUT AT 1 GPM WATER FLOW RATE

Average Water Temperature (°F)	Rated Output (Btu/h)
220	840
215	810
210	770
200	710
190	640
180	580
170	510
160	450

Source: Ratings shown are those of "Fine/Line 30" as manufactured by the Slant/Fin Corporation and are reprinted by permission.

Performance data for a variety of finned tube radiation per foot of active length is provided in Table 15.2. Units are listed for both copper (C) tubing and steel (S) piping. Fin dimensions are given for height (H) and width (W).

A few items to note relative to Table 15.2:

- 3/4 in. copper (c) tubing is the standard normally used in residential and small commercial applications
- Finned-tube unit capacities shown for $\frac{3}{4}$ in. tubing, 1 in. tubing and $1\frac{1}{4}$ in. tubing vary just slightly when other design features are identical
- With all other factors remaining the same, unit ratings increase when a taller enclosure is used
- With all other factors remaining the same, $1\frac{1}{4}$ in. units with 2 tiers and steel tubing have a capacity which is approximately 50% greater than units with 1 tier of tubing and fins.

Published unit ratings are based upon steam or 215°F hot water and 65°F inlet air. Factors in Table 15.3 are then used to convert unit output ratings for other hot water temperatures.

Standard unit ratings are based upon a water velocity of 3 fps. Ratings for lower velocities need to be adjusted by factors shown in Table 15.4.

In examining Table 15.4, you will notice that the heat output from a unit changes only slightly at different flow rates. This is because the controlling factor for heat transfer is the ability of the air to absorb additional heat, a factor which is not significantly affected by the water side (flow rate) coefficient. It is interesting to note that units will provide about 90% of their rated output even when flow rates are as low as 0.25 fps.

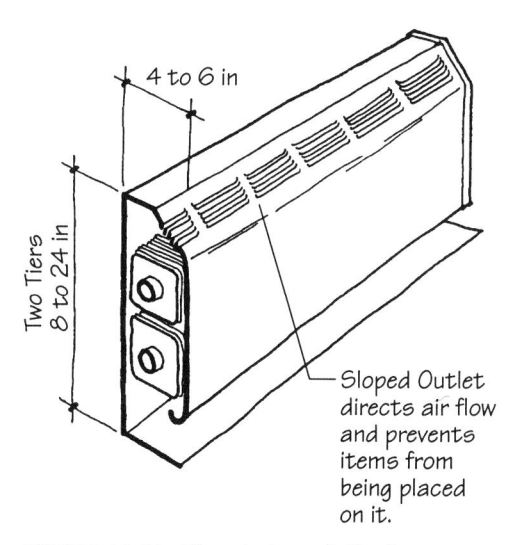

FIGURE 15.17 Finned-tube radiation for commercial applications.

266 HOT WATER HEATING SYSTEMS

TABLE 15.2 TYPICAL "FRONT OUTLET" FINNED-TUBE RADIATION RATINGS

Tube Size (in.)	Mat.	Fins H&W (in.)[a]	Per Foot	Tiers	Actual Enclosure Height (in.)	Rating Btu/h per ft
$\tfrac{3}{4}$	C	$3\tfrac{1}{4}$	35	1	10	1080
$\tfrac{3}{4}$	C	$3\tfrac{1}{4}$	35	1	14	1140
$\tfrac{3}{4}$	C	$3\tfrac{1}{4}$	35	1	18	1170
$\tfrac{3}{4}$	C	$3\tfrac{1}{4}$	50	1	10	1210
$\tfrac{3}{4}$	C	$3\tfrac{1}{4}$	50	1	14	1350
$\tfrac{3}{4}$	C	$3\tfrac{1}{4}$	50	1	18	1450
$\tfrac{3}{4}$	C	$3\tfrac{1}{4}$	58	1	10	1280
$\tfrac{3}{4}$	C	$3\tfrac{1}{4}$	58	1	14	1460
$\tfrac{3}{4}$	C	$3\tfrac{1}{4}$	58	1	18	1560
1	C	$3\tfrac{1}{4}$	35	1	10	1090
1	C	$3\tfrac{1}{4}$	35	1	14	1150
1	C	$3\tfrac{1}{4}$	35	1	18	1190
1	C	$3\tfrac{1}{4}$	50	1	10	1200
1	C	$3\tfrac{1}{4}$	50	1	14	1350
1	C	$2\tfrac{1}{4}$	50	1	18	1460
1	C	$3\tfrac{1}{4}$	58	1	10	1260
1	C	$3\tfrac{1}{4}$	58	1	14	1450
1	C	$3\tfrac{1}{4}$	58	1	18	1600
$1\tfrac{1}{4}$	C	$3\tfrac{1}{4}$	35	1	10	1100
$1\tfrac{1}{4}$	C	$3\tfrac{1}{4}$	50	1	10	1200
$1\tfrac{1}{4}$	C	$3\tfrac{1}{4}$	58	1	10	1250
$1\tfrac{1}{4}$	S	$2\tfrac{1}{2} \times 5\tfrac{1}{8}$	40	1	16	1300
$1\tfrac{1}{4}$	S	$2\tfrac{1}{2} \times 5\tfrac{1}{8}$	40	2	16	1980
$1\tfrac{1}{4}$	S	$2\tfrac{1}{2} \times 5\tfrac{3}{16}$	52	1	16	1510
$1\tfrac{1}{4}$	S	$2\tfrac{1}{2} \times 5\tfrac{3}{16}$	52	2	16	2060

Source: Excerpted with permission from "I=B=R Ratings for Boilers, Baseboard Radiation and Finned Tube (Commercial) Radiation," January 1, 1993 Edition, published by The Hydronics Institute. Ratings shown are for the products of the Trane Company.

[a] Unless otherwise indicated, fin dimensions are $3\tfrac{1}{4}$ in. $\times$ $3\tfrac{1}{4}$ in.

By the way, be thankful for this relative insensitivity to flow rate since many systems still provide adequate heat output even though actual flow rates vary greatly from original design values. The length of heater required is determined using Formula 15F.

$$\text{LF}_{\text{heat}} = \frac{q_{\text{tot}}}{(q_{\text{lf}} \times C_{\text{t}} \times C_{\text{f}})} \quad \text{Formula 15F}$$

where

LF_{heat} is the required length of heater(s) in feet

q_{tot} is the total heating load which needs to be satisfied by the heater(s) in Btu/h

q_{lf} is the rated heat output of baseboard or finned-tube heater per linear foot in Btu/h·ft

C_{t} is the hot water temperature correction factor from Table 15.3 (if applicable)

C_{f} is the water flow rate correction factor from Table 15.4 (if applicable)

Example 15F: The total design heat load

TABLE 15.3 FACTORS USED TO CONVERT I=B=R STEAM RATINGS TO HOT WATER RATINGS AT TEMPERATURE INDICATED

Average Radiator Temperature (°F)	Correction Factor
100	0.15
120	0.26
150	0.45
170	0.61
175	0.65
180	0.69
185	0.73
190	0.78
195	0.82
200	0.86
205	0.91
210	0.95
215	1.00
220	1.05
230	1.14
240	1.25

Source: Excerpted with permission from "I=B=R Ratings for Boilers, Baseboard Radiation and Finned Tube (Commercial) Radiation," January 1, 1993 Edition, published by The Hydronics Institute.

of a 10 ft × 15 ft office with three windows is found to be 7500 Btu/h (q_{tot}). Use Table 15.2 to determine the length in feet of 3/4 in. copper baseboard needed (LF$_{heat}$) if the average water temperature is 180°F, and the water flow rate will be 2 fps. For architectural reasons, the enclosure height should be as small as possible.

Solution: To satisfy the stated architectural criteria, 3/4 in. copper units which have an enclosure height of 10 in. and the largest number of fins per foot (58) will be used. These units have a rated heat output (q_{lf}) of 1280 Btu/hr per foot of length (when the average water temperature is 215°F).

Use Formula 15F. The hot water correction factor (C_t) from Table 15.3 for 180°F water is 0.69. The flow rate correction

TABLE 15.4 FACTORS FOR DETERMINING OUTPUTS AT WATER FLOW RATES OF LESS THAN 3 FEET PER SECOND

Flow Rate (fps)	Correction Factor
3.0	1.000
2.5	0.992
2.0	0.984
1.5	0.973
1.0	0.957
0.5	0.931
0.25	0.905

Source: Excerpted with permission from "I=B=R Ratings for Boilers, Baseboard Radiation and Finned Tube (Commercial) Radiation," January 1, 1993 Edition, published by The Hydronics Institute.

268 HOT WATER HEATING SYSTEMS

factor (C_f) from Table 15.4 for 2 fps is 0.984.

$$\text{LF}_{\text{heat}} = \frac{q_{\text{tot}}}{(q_{\text{lf}} \times C_t \times C_f)}$$

(Formula 15F)

$$= \frac{7500}{(1280 \times 0.69 \times 0.984)}$$

$$= \frac{7500}{870}$$

$$= 8.6 \text{ ft or } 9 \text{ ft}$$

Therefore, a 3-ft finned-tube unit is required below each of the three windows for a total length within the room of 9 ft.

Convectors

Large capacity compact units with finned-tube elements are often referred to as "convectors." Various designs allow air to enter and then rise through one or more "tiers" of finned hot water tubes. They can be floor or wall mounted. And they can either be free-standing, semirecessed, or fully recessed. In general, fully recessed convectors should be avoided on exterior walls unless the wall construction is of such depth that adequate insulation can be installed behind it. Depicted in Figure 15.18 are free-standing and recessed units. Semirecessed units are also available when depth is limited.

15.10 OTHER HYDRONIC TERMINAL DEVICES

Radiators

Cast-iron radiators were the first terminal devices employed in residential hot water heating systems as an outgrowth of the evolution from steam to hot water. Radiators are still used in some new installations, and are retained in many conversions of systems from steam to hot water. Radiators have no minimum flow requirements because they do not require turbulent flow for heat transfer which is critical in finned-tube radiation.

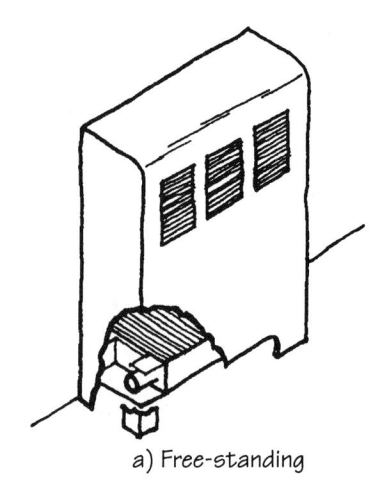

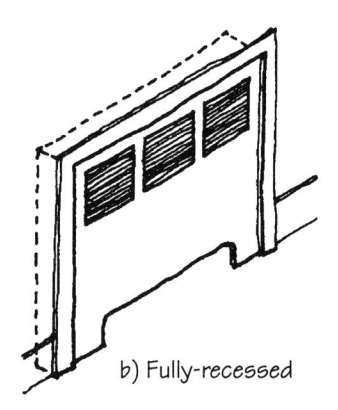

a) Free-standing

b) Fully-recessed

FIGURE 15.18 Convector design alternatives.

Unit Heaters

The natural convective air flow of convectors and resultant heat transfer can be bolstered by the installation of a fan, resulting in what are known as "unit heaters," as shown in Figure 15.19. With unit heaters, the relatively low air flow of natural convection is no longer a limiting factor since a fan is present. Therefore, various design types of unit heaters are available for

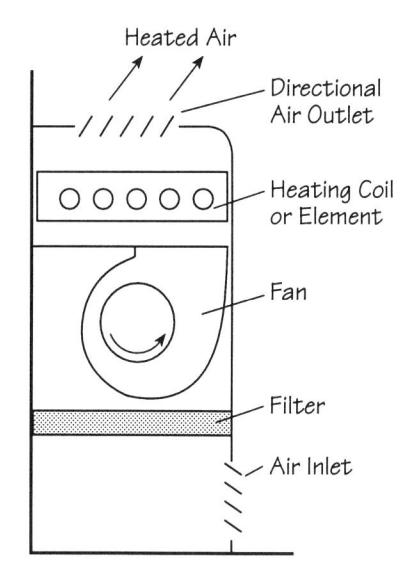

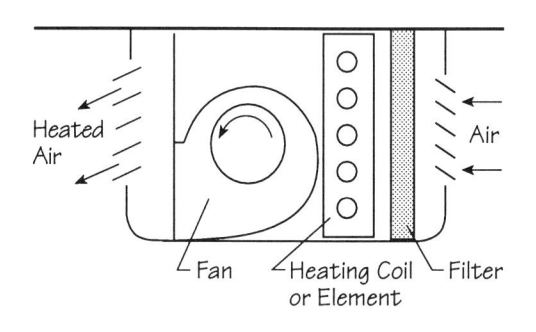

FIGURE 15.19 Forced air flow with unit heaters.

FIGURE 15.20 Space heaters used in large, partially occupied spaces.

mounting high on walls or even in ceilings (e.g., the ceiling of a vestibule). Unit heaters and other units which use a fan to distribute heat rather than depend upon natural convection and radiant heat transfer are wasteful of energy since they require a higher air temperature to achieve a perceived level of comfort.

One type of unit heater is known as a "space heater" since its primary function is to keep a semioccupied space (e.g., a below-building parking garage or a warehouse) somewhat warm (see Figure 15.20).

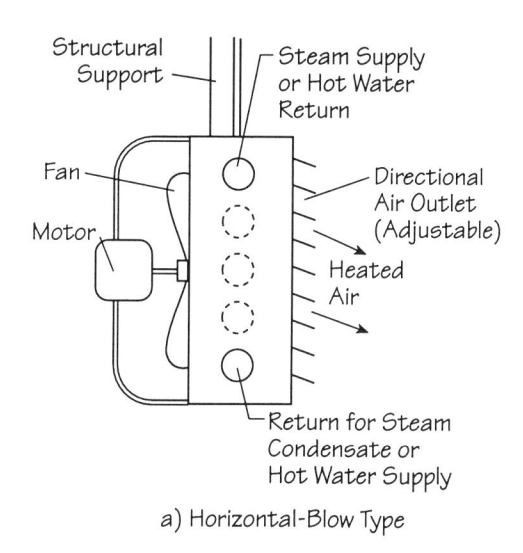

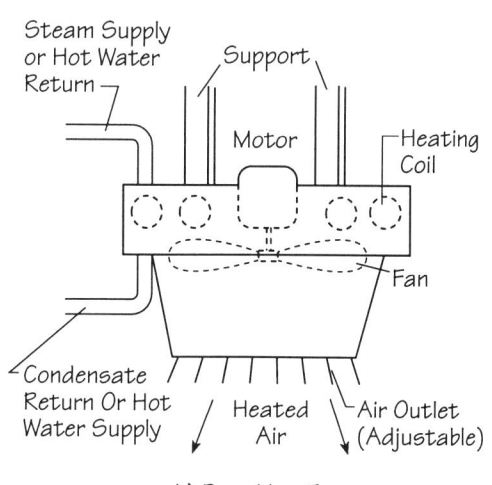

These units can be supplied by hot water (or steam) and can be mounted high on walls or suspended from ceilings and use their relatively powerful fans to distribute heated air throughout the space.

Unit Ventilators

Unit ventilators as shown in Figure 15.21 are large capacity terminal devices which

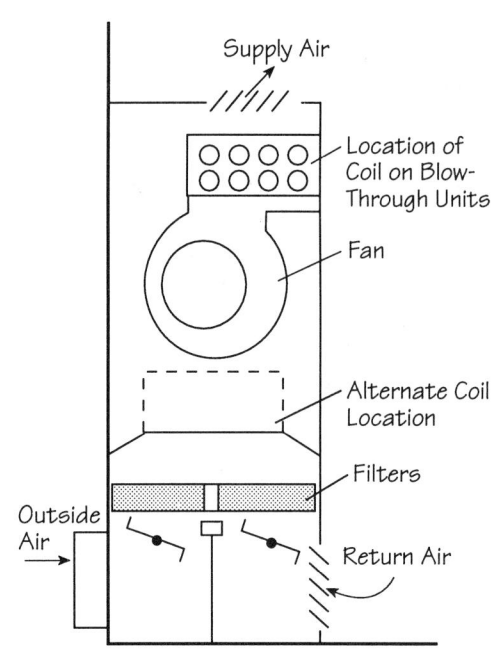

FIGURE 15.21 Unit ventilators can provide heat along with a large quantity of outside air.

can provide heating just like unit heaters, or heating and cooling. In addition, they can also provide large quantities of outside ventilation air. Unit ventilator use is appropriate in spaces with large ventilation requirements such as classroom buildings. Also, since a large quantity of outside air is available, units can be controlled to allow the entry of cool outside air for cooling without the use of mechanical refrigeration.

While very similar to fan-coil units, unit ventilators are physically much larger and the fans are more powerful to handle the larger air quantities. Units should be designed with outside air openings which are resistant to damage from wind and rain. And if they are subject to freezing conditions, an antifreeze mixture should be used rather than plain water.

Other hydronic terminal devices provide not only heating, but cooling and sometimes ventilation air as well. These include fan coils, packaged terminal air conditioners, and heat pumps.

15.11 HOT WATER HEATING OF "THE HOUSE"

Example 15G: Design a series loop hot water heating system for "The House." Start with a design temperature drop of the hot water of 20°F (assumed from 200 to 180°F) which circulates at a flow rate of 3 fpm. Use baseboard radiation with 3/4 in. copper tubing which has rated output per linear foot for average water temperatures as follows:

SUMMARY: RATED BASEBOARD OUTPUT PER LINEAR FOOT

Average Water Temperature (°F)	Rated Output (q_{lf}) (Btu/h)
205	640
200	600
195	570
190	550
185	510
180	480

Baseboard units are available with active lengths of from 2 to 8 ft. Custom lengths can also be easily cut.

Solution:

Step 1: "The House" has a total heat loss for winter design conditions of 33,150 Btu/h (as per Section 5.14). In Figure 15.4, it is noted that a single supply loop with 3/4 in. copper tubing has a heat transfer capacity of 45,000 Btu/h for a temperature drop of 20°F. Therefore, one series loop can be used.

Step 2: Summarized below are the system requirements on a room-by-room basis beginning with Bedroom #2. Average water temperatures (inlet temperature + outlet temperature/2) are listed for the baseboard in each space as are the rated outputs per linear foot.

The length of baseboard radiation required to meet the load in each room is determined by using a simplification of Formula 15F. Since the rated baseboard output at various temperatures is given, factor C_t should be considered as equal to 1.0. Factor C_f is also taken to be 1.0 since the water flow rate is 3 fpm. As an example, the calculation of required baseboard length for Bedroom #2 follows based upon: a heat loss of 4680 Btu/h (q_{tot}) and the baseboard heat output of 585 Btu/h (q_{lf}) per foot for an average water temperature of 197°F.

$$LF_{heat} = \frac{q_{tot}}{(q_{lf} \times C_t \times C_f)} \quad \text{(Formula 15F)}$$

$$= \frac{4680}{585 \times 1 \times 1}$$

$$= 8.0 \text{ ft}$$

ues counter-clockwise around the perimeter of the house.

Step 4: Size the boiler. Since the design heat loss (q_{tot}) is already known in units of Btu/°F·h, a simplification of Formula 15E is used to determine the required boiler output in Btus per hour (Btu/h). Assume an oversizing factor of 1.05.

$$BOIL_{out} = q_{tot} \times OSF \quad \text{(Formula 15E)}$$

$$= 33{,}150 \times 1.05$$

$$= 34{,}810 \text{ Btu/h, the required boiler output which is also greater than the output of the baseboard radiation required}$$

SUMMARY: BASEBOARD SYSTEM REQUIREMENTS

Room	Room Design Heat Loss (q_{tot}) (Btu/h)	Average Basebd. Water Temp. (°F)	Basebd. Output per foot (q_{lf}) (Btu/h)	Rated Req. Length of Baseboard (Ft)	Active Heat Output (Btu/h)
Bedroom #2	4680	197	585	8.0	4680
Ext. bath.	1340	195	570	2.5	1425
Bedroom #1	6440	192	560	11.5	6440
Living room	8100	190	550	15.0	8250
Dining room	5480	187	530	10.5	5565
Kitchen	3030	186	515	5.0*	2575
Family room	2620	185	510	5.5	2805
Vest./hall	1460	183	495	3.0	1485
	33,150				33,225

*Due to space limitations in Kitchen only 5.0 feet of baseboard is provided which produces slightly less heat than the room design heat loss.

It should be noted that some judgment is required in selecting baseboard lengths. For example, baseboard lengths in bedrooms can be slightly undersized since those rooms generally accommodate lower temperatures. Also lengths in kitchens (as in the above summary) can usually be reduced because of internal heat gain.

Step 3: Illustrated in Figure 15.22 is a one-pipe series loop for "The House" which begins at Bedroom #2 and contin-

Step 5: Use Formula 15B to determine the required water flow rate based upon the heat output of the baseboard radiation of 33,225 Btu/h (q_{tot}) and a heat output for each gpm (q_{gpm}) of 10,000 Btu/h.

$$gpm_{hw} = \frac{q_{tot}}{q_{gpm}} \quad \text{(Formula 15B)}$$

$$= \frac{33{,}225}{10{,}000}$$

$$= 3.3 \text{ gpm}$$

272 HOT WATER HEATING SYSTEMS

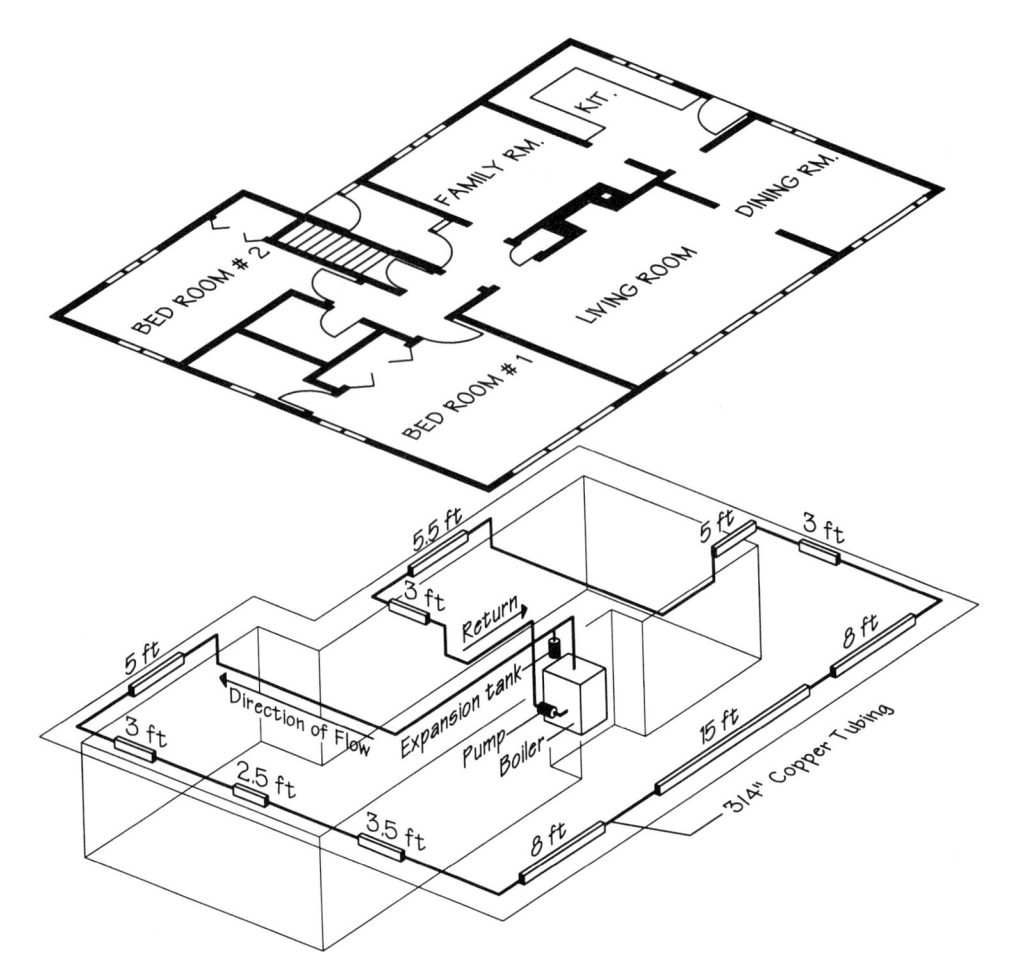

FIGURE 15.22 Piping and baseboard layout for "The House."

The length of the piping circuit to and from the boiler and around the perimeter of "The House" is approximately 200 ft. Therefore, the friction loss through the 3/4-in. series loop will be approximately 7 ft of water (as per Figure 10.3) for a 3.5-gpm flow rate. Once allowing for resistance through the boiler, elbows, and other fittings the total friction loss will be approximately 12–14 ft of water. Based upon this friction loss, and the required flow rate of 3.5 gpm, a small circulator pump (powered by only a 60- to 100-W motor) is then selected.

This series loop system is the simplest and least costly hot water system which could be installed. Other system alternatives include:

- The use of a one-pipe monoflow loop with each baseboard unit equipped with thermostatic valves to provide room-by-room temperature control
- Multiple loops to allow for zoning within the house
- Use of a split loop (as shown in Figure 15.11) so that the two loops will each meet about one-half of the heat load and have fewer heaters. With such an approach the length of baseboard heaters at the ends of the loops can usually be a foot or two shorter

16

ELECTRIC HEATING SYSTEMS

Census bureau statistics show that some form of electric heating system is used for space heating in over one-third of new single family homes and one-half of new multifamily homes. This is particularly true in the South and the Pacific Northwest, where heating needs and/or electricity costs are relatively low and systems such as heat pumps can serve both heating and cooling needs.

Natural gas is currently the dominant energy source for new heating systems in most colder climates while oil remains widely used in the Northeast. Still there are applications where some form of electric heat can be used advantageously. Where permissible by code, these include:

1. In areas where base electric rates are relatively low or where attractive off-peak electric rates are available.
2. In climates where the need for heat is very slight such as in Florida, Southern California, and Hawaii.
3. In buildings rarely used during cold periods such as a summer vacation house.
4. In buildings where the heated area is small, such as a small office area in an unheated warehouse building.
5. In heat pump systems which use electricity more efficiently for both heating and cooling.
6. For special applications such as grandstands and store marquees where electric radiant heating can be employed.

Electric resistance heat is often referred to as being 100% efficient. This is true—but only when considered from the point of use. The efficiency of electricity is actually only about 30% once losses are taken into account at the generating plant and because of transmission over high-voltage power lines. It is because of this "actual efficiency" that electricity is usually such a relatively expensive energy source.

16.1 RESISTANCE HEAT

The operation of electric resistance heat is similar to a household toaster which produces heat from the resistance to the flow of electrical current through an element. The various types of electric resistance heaters are shown in Figure 16.1.

Most common applications are small baseboard units with one tier of finned elements to improve heat exchange (similar to hot water baseboard heaters). Electric baseboard units are typically about 6–8 in. high and about 3 in. deep (see Figure 16.2).

Sizing Electric Baseboard Heaters

Sizing (i.e., length in feet of electric resistance heating) depends upon needed output in Btu/h and is based upon the electrical conversion of 3413 Btu/h for each kilowatt or "kW" (1000 W). Ratings of electric baseboard typically range from about 100 to 350 W per linear foot. The required length of a baseboard heater is determined using Formula 16A.

$$LF_{elec} = \frac{q_{tot}}{q_{lf}} \quad \text{Formula 16A}$$

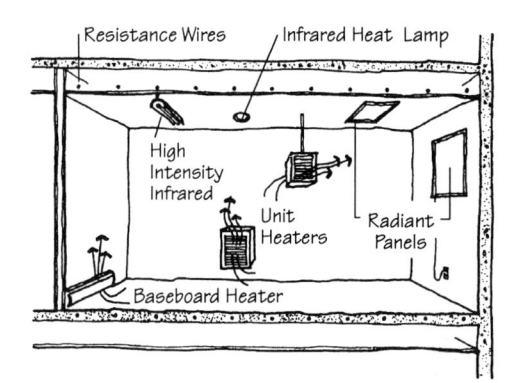

FIGURE 16.1 Types of electric resistance heaters.

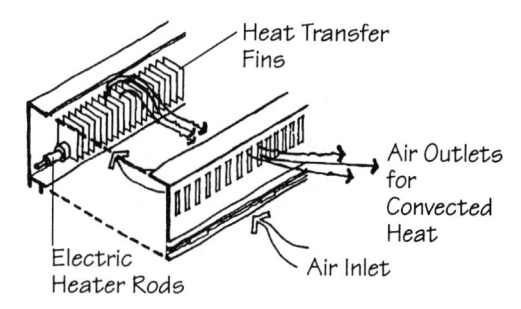

FIGURE 16.2 Typical electric baseboard heater.

where

LF_{elec} is the required length of electric baseboard heater in feet
q_{tot} is the total design heat loss of a space in watts
q_{lf} is the electric heat output per foot of length in watts per foot

Example 16A: The design heat loss for a room (q_{tot}) has been calculated to be 2700 Btu/h. Determine the required length of electric baseboard with an output (q_{lf}) of 200 W per linear foot.

Solution: First convert the heat loss to watts using the conversion that 1000 W is equal to 3413 Btu/h.

$$2700 \text{ Btu/h} \times \frac{1000 \text{ W}}{3413 \text{ Btu/h}} = 791 \text{ W}$$

Then obtain the required length in feet using Formula 16A:

$$LF_{elec} = \frac{q_{tot}}{q_{lf}} \quad \text{(Formula 16A)}$$

$$= \frac{791}{200}$$

= 3.9 or 4 ft of electric resistance baseboard heat is required

Radiant Panels

As discussed in Chapter 1, human exposure to warm surfaces (radiant heat) can provide comfort. Electric radiant panels (with backing insulation) can be used on wall or ceiling surfaces to satisfy several special heating applications. These include spaces requiring absolute quiet, or spaces where the aesthetics of the panels may be preferred over an electric baseboard heater. Radiant ceiling panels can also be finished to look like acoustical ceiling tiles and fit in a grid (see Figure 16.3).

Radiant panels are easy to zone and control and typically provide a heat output of 50–100 W/ft^2 of surface area. Wall panels operate at a surface temperature of about 100°F, with ceiling panels operating at a slightly higher surface temperature of about 125°F.

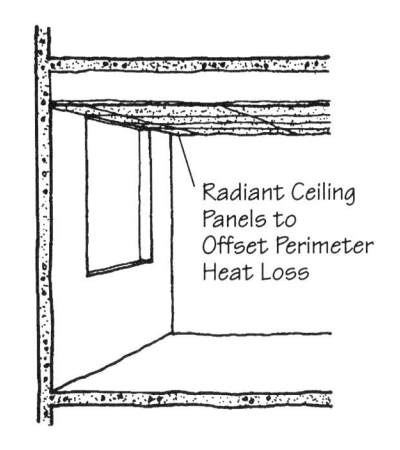

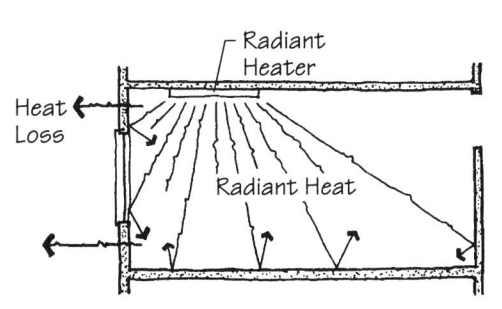

FIGURE 16.3 Radiant panel heaters.

Spot Heaters

Infrared heaters are special devices used to provide "spot" heating for occupants and contents of large commercial/industrial spaces such as garages, bus stations, and repair shops. When infrared spot heaters are used, there is no attempt to heat the air in the entire space, only to improve comfort at select locations. Another form of radiant electric spot heaters are warming lights found in many fast-food restaurants.

Electrical Consumption and Operational Cost

Formula 16B is used to compute yearly consumption of electric resistance heat.

$$\text{kWh}_{yr} = \frac{Q_{tot}}{3413} \quad \text{Formula 16B}$$

where

kWh$_{yr}$ is the quantity of heating electricity required per year in kilowatt hours
Q_{tot} is the total amount of heat required in Btus per year
3413   is the quantity of Btus per kilowatt hour

Formula 16C then is used to compute the yearly operational cost.

$$\$_{yr} = \text{kWh}_{yr} \times \$/\text{kWh} \quad \text{Formula 16C}$$

where

$\$_{yr}$ is the yearly operational cost in dollars and
$/kWh   is the average dollar cost for electricity per kilowatt hour.

Example 16B: "The House," if located in St. Louis, requires 46,546,200 Btus of heat per year (see Section 5.14). Determine the cost to heat such a house with electric resistance heat costing $0.08/kWh.

Solution: First use Formula 16B to determine the required consumption in kilowatt hours of electricity.

$$\text{kWh}_{yr} = \frac{Q_{tot}}{3413}$$

$$= \frac{46{,}546{,}200}{3413} \quad \text{(Formula 16B)}$$

$$= 13{,}640 \text{ kilowatt hours per year}$$

Formula 16C is then used to estimate the yearly operational cost.

$$\$_{yr} = \text{kWh}_{yr} \times \$/\text{kWh}$$

$$\text{(Formula 16C)}$$

$$= 13{,}640 \times 0.08$$

$$= \$1090 \text{ per year using electric baseboard heat}$$

This compares very poorly to an estimated heating season operational cost for a gas-fired furnace of $580 per year (based on an assumed AFUE of 80% and a cost of $1.00 per therm). Clearly, the use of electric resistance heat may not be wise under these circumstances, and in fact many energy codes actually prohibit or greatly limit the use of electric resistance heat. The improved operational economics of satisfying this heating load with an air-to-air heat pump system is considered below.

16.2 HEAT PUMPS, GENERAL

From an economic standpoint, a heat pump offers the best use of electricity. In the 1960s and 1970s there was some concern about the reliability of heat pumps. Since then though the industry has responded with reliable products, and over 800,000 were sold annually during the years 1985 through 1990.

Heat pumps can best be thought of as "reverse air conditioners" which use the vapor compression refrigeration cycle operated in reverse to "pump" heat from some outside source to produce a heating effect in the living space. The heat is pumped from the outside air, the ground, or some water body, which is known as the "heat source." The indoor space to which the heat is pumped is known as the "heat sink." In the reverse (cooling) cycle, the living space is the "heat source" and the outside medium the "heat sink."

Air-to-Air Heat Pumps

By far the most common type of heat pump is an "air-to-air" unit. That is, the "air" outside the house/building is the heat source for the evaporator, and the "air" inside the house/building is the heat sink for the condenser (see Figure 16.4).

Heat pumps are often chosen because they become air conditioners during the warmer months. This feature also makes heat pumps a common selection for hotel/motel guest rooms, although the noise attributed to frequent cycling of the compressor must be considered.

Other Heat Pump Types

Although air-to-air heat pumps are the most common, they are not the only type. Figure 16.5 indicates alternative heat pump sources and sinks. An important advantage of ground-source and well-water heat pumps is improved efficiency because these sources usually have higher and more constant temperatures than outside air during the heating season.

Ground-source heat pumps use water or water–glycol loops buried several feet below the earth. From a cost and environmental point of view, one interesting approach is to bury the loop piping beneath parking lots and driveways during construction while the earth is already disturbed. Heat pumps using well, ground, or site water as the source or sink may require environmen-

16.2 HEAT PUMPS, GENERAL

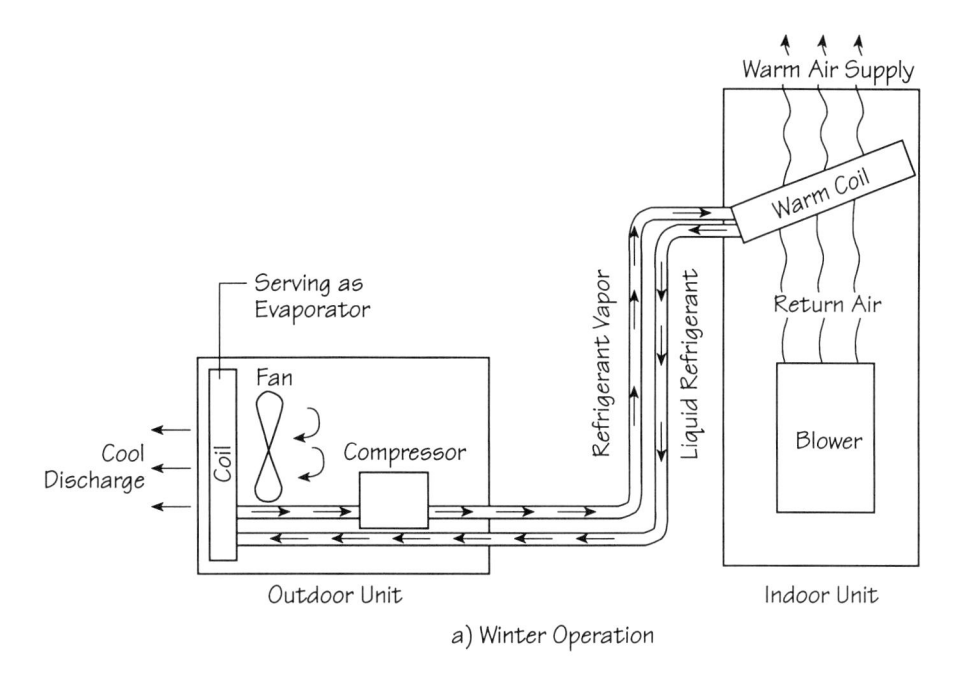

a) Winter Operation

b) Summer Operation

FIGURE 16.4 Air-to-air heat pumps—reversal of component operation depending upon season.

278 ELECTRIC HEATING SYSTEMS

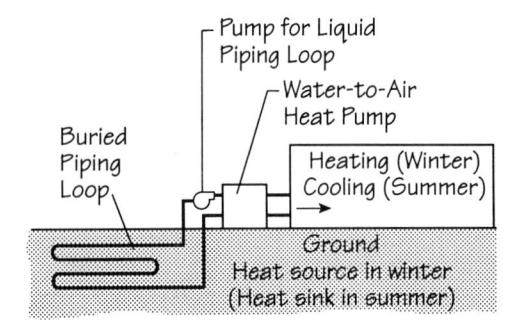

a) Ground-source Heat Pump

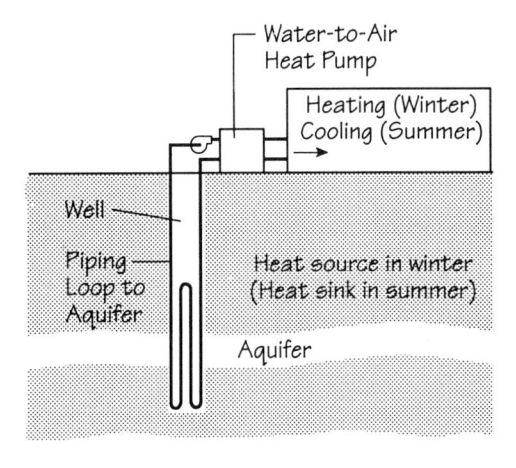

b) Well-water Heat Pump

FIGURE 16.5 Heat pump sources other than air.

tal analysis and/or permits. Also see Section 21.1, where closed-loop heat pumps are discussed for use in commercial buildings.

16.3 HEAT PUMPS, OPERATIONAL PERFORMANCE

Coefficient of Performance

The instantaneous efficiency at a given source temperature is known as the heating coefficient of performance (COP_h), a term which compares heat output with electrical energy input under a given set of operating conditions.

A COP_h of 3.00 means that for every unit of electrical input (in watt-hours), the heat pump will produce 3 units of heat output (also in watt-hours). Since the units cancel, COPs are unitless numbers.

Provided in Table 16.1 are some illustrative heating COPs for an air-to-air heat pump over a range of outdoor (source) temperatures. The operational COP of heat pumps is determined by tests in accordance with Standard 240 of the Air-Conditioning & Refrigeration Institute (ARI). Performance is rated for an indoor (sink) temperature of 70°F and the high and low outdoor conditions of 47°F db/43°F wb and 17°F db/15°F wb.

Note that as outdoor (source) temperatures increase, the heating COP increases. This is why heat pumps are more appropriate in moderate climates where winter temperatures are relatively high (40°F or more) most of the time.

Heat pump COP ratings do not include the energy which may be needed in some climates to prevent the outside coil from freezing. When air temperatures fall below about 40°F, the surface temperature of the outside coil (which in winter serves as the evaporator) will fall below the freezing point and will need to be defrosted. This is accomplished by periodic operation of the unit for several minutes in reverse mode to heat the outside coil (temporarily making it the condenser). During this period, supplemental electric resistance heaters add heat to the air stream to nullify the chilling effect produced.

Heating Seasonal Performance Factor (HSPF)

Regarding heating season performance, much more meaningful than the COP at any specific temperature is the "average" seasonal heat pump performance including supplemental heat, and an allowance for on/off system cycling (known as a "duty factor"). Manufacturers must test and rate their equipment for the Heating Seasonal

TABLE 16.1 ILLUSTRATIVE HEAT PUMP COPs–HEATING MODE INDOOR TEMPERATURE, 70°F

Outdoor Temp. (°F db)	COP_h
−13	1.05
−8	1.30
−3	1.55
2	1.80
7	2.00
12	2.20
17-Low	2.40
22	2.60
27	2.75
32	2.90
37	3.00
42	3.10
47-High	3.15
52	3.20
57	3.25
62	3.30

Performance Factor (HSPF) in accordance with ARI (Air-Conditioning & Refrigeration Institute) Standard 240, and assuming a climate which averages about 5800 HDD_{65} (Department of Energy Region IV). The HSPFs are published in units of Btus per watt-hour.

EER and SEER

Heat pumps, of course, can also function as a cooling system. The Energy Efficiency Ratio (EER) is used for consumer products such as window air-conditioning units and refrigerators. An EER is the ratio of cooling capacity in Btus per hour (Btu/h) divided by the electrical power input in watts at any given set of rating conditions. Remembering that 1 W equals 3.413 Btu/h, a heat pump operating at a COP of 2.5 can also be said to be operating at an EER of 8.53 (2.5 × 3.413).

More meaningful is the Seasonal Energy Efficiency Ratio (SEER) which reflects the cooling performance over the entire cooling season. The SEER is the total cooling provided by an air conditioner or heat pump in Btus over the course of its normal annual usage period divided by the total electric energy input in watt-hours over the same period.

Typical Residential Equipment

Contained in Table 16.2 is a sampling of typical residential heat pump performance data. Note that ratings are published for cooling capacity and seasonal performance (SEER), and high temperature (47°F) heating capacity and the heating seasonal performance factor (HSPF). Models listed in Table 16.2 actually consist of outdoor and indoor units which must be properly matched to satisfy the design loads.

Note that Unit A in Table 16.2 has a HSPF of 7.20. When divided by 3.413 this yields a value of 2.11 which can be thought of as the "seasonal COP" of the system.

Operational Cost during the Heating Season

The HSPF of a heat pump along with other factors which relate to climate and house heat loss are used to estimate the yearly

TABLE 16.2 PERFORMANCE DATA FOR CERTIFIED UNITARY AIR-SOURCE HEAT PUMPS

	Cooling		Heating	
Model	Capacity (MBtu/h)	SEER	High Temp. Capacity (MBtu/h)	HSPF
A	35.0	10.55	35.2	7.20
B	40.0	10.20	41.0	7.25
C	49.5	10.55	48.5	7.55

Source: Extracted with permission from the Directory of Certified Unitary Air-Source Heat Pumps effective August 1, 1992 to January 31, 1993 sponsored and administered by the Air-Conditioning & Refrigeration Institute (ARI). Ratings shown are excerpted from the units of Lennox Industries, Inc.

heating energy electrical consumption in Formula 16D.

$$kWh_{h\text{-}yr} = \frac{0.77 \times HLH \times q_{tot}}{1000 \times HSPF}$$

Formula 16D

where

$kWh_{h\text{-}yr}$ is the quantity of electricity required by the heat pump per heating season in kilowatt hours

0.77 is a correlation factor (required to make the design heating load hours correlate with the effective heating load hours experienced under actual operating conditions)

HLH is the quantity of heating load hours from Figure 16.6 (map from ARI standard 240) or as calculated by Formula 16E

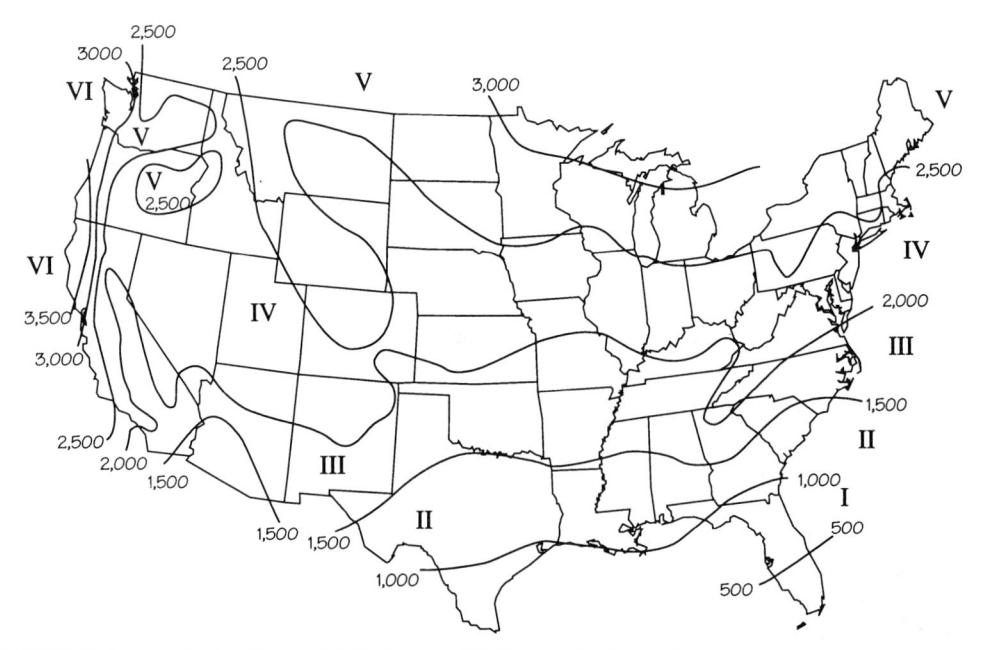

FIGURE 16.6 Heating load hours (HLH). *Source:* ARI Standard 240 and ARI Unitary Directory. Reproduced by courtesy of the Air-Conditioning and Refrigeration Institute, ARI.

q_{tot}	is the total design heat loss in Btu/h	24	is the conversion of 24 h in 1 day
1000	is the conversion to obtain a result in kilowatt hours	T_{wdt}	is the outside winter design temperature in °F
HSPF	is the rated heating seasonal performance factor of the heat pump		

Provided in Figure 16.6 are heating load hours (HLH) for the continental United States. A total of 3,500 HLH is assumed for Alaska and 0 HLH for Hawaii and US territories.

Heating load hours (HLH) can also be computed by using Formula 16E.

$$\text{HLH} = \frac{\text{HDD}_{65} \times 24}{(65 - T_{wdt})} \quad \text{Formula 16E}$$

where

HLH   is the quantity of heating load hours

HDD_{65} is the total of heating degree-days, base 65

Operational Cost During the Cooling Season

$$\text{kWh}_{c\text{-}yr} = \frac{\text{CLH} \times q_c}{1000 \times \text{SEER}}$$

Formula 16F

where

$\text{kWh}_{c\text{-}yr}$ is the quantity of electricity required by the heat pump per cooling season in kilowatt hours

CLH   is the quantity of cooling load hours from map in ARI standard 240 (Figure 16.7) or as calculated more accurately by using Formula 16G

q_c is the design cooling load in Btu/h

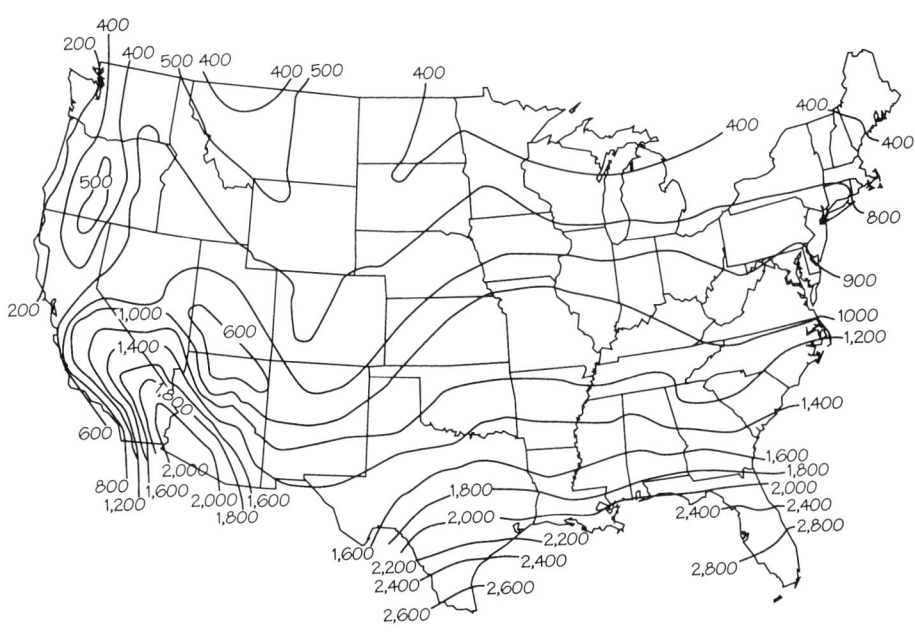

FIGURE 16.7 Cooling load hours (CLH). *Source:* ARI Standard 240 and ARI Unitary Directory. Reproduced by courtesy of the Air-Conditioning and Refrigeration Institute, ARI.

282 ELECTRIC HEATING SYSTEMS

SEER is the seasonal energy efficiency ratio rating of the heat pump during the cooling season

Provided in Figure 16.7 are cooling load hours (CLH) for the continental United States. The CLH values for other locations are: 0 for Alaska, 2300 for Hawaii, 6000 for Puerto Rico and the Virgin Islands, 6600 for Guam and Samoa.

Cooling load hours (CLH) can also be computed by using Formula 16G.

$$\text{CLH} = \frac{\text{CDD}_{65} \times 24}{(T_{sdt} - 65)} \quad \text{Formula 16G}$$

where

CDD_{65} is the total of cooling degree-days, base 65 and
T_{sdt} is the outside summer design temperature in °F.

Example 16C: Determine the electrical consumption and operational cost for "The House" if located in St. Louis as described in Example 16B (and Section 5.14). Assume that it is conditioned by an air-to-air heat pump Model A in Table 16.2. Also assume an electric rate of eight cents per kilowatt hour ($0.08/kWh) during the heating season and ten cents per kilowatt hour ($0.10/kWh) during the cooling season.

Heating Season Factors

A total design heat loss (q_{tot}) of 33,150 Btu/h

St. Louis has an outside winter design temperature (97.5% value) of 8°F (t_{wdt}), and experiences 4750 HDD$_{65}$ (see Table 2.1)

HSPF of 7.20 (Model A, Table 16.2)

Cooling Season Factors

A total design cooling load (q_c) of 24,560 Btu/h

St. Louis has an outside summer design temperature (2½% value) of 94°F (t_{sdt}), and experiences 1500 CDD$_{65}$ (see Table 2.7)

SEER of 10.55 (Model A, Table 16.2)

Solution

Heating Season: First use the climate data and Formula 16E to determine the heating load hours (HLH) for St. Louis as follows:

$$\text{HLH} = \frac{\text{HDD}_{65} \times 24}{65 - t_{wdt}}$$

(Formula 16E)

$$= \frac{4750 \times 24}{65 - 8}$$

$$= \frac{114{,}000}{57}$$

= 2000 heating load hours

Then use Formula 16D to estimate annual electrical consumption for heating by the heat pump.

$$\text{kWh}_{h\text{-yr}} = \frac{0.77 \times \text{HLH} \times q_{tot}}{1000 \times \text{HSPF}}$$

(Formula 16D)

$$= \frac{0.77 \times 2000 \times 33{,}150}{1000 \times 7.20}$$

= 7090 kilowatt hours per year

Formula 16C is then used to estimate the yearly heating cost.

$$\$_{yr} = \text{kWh}_{yr} \times \$/\text{kWh}$$

(Formula 16C)

= 7090 × 0.08

= $570 per year for heating using the heat pump

Therefore, the heat pump costs about the same as a gas-fired furnace which has an estimated operational cost of $580 per year (based on an assumed AFUE of 80% and a cost of $1.00 per therm).

Cooling Season: First use the climate data and Formula 16G to determine the cooling load hours (CLH) for St. Louis as follows:

$$\text{CLH} = \frac{\text{CDD}_{65} \times 24}{(t_{\text{sdt}} - 65)}$$

(Formula 16G)

$$= \frac{1500 \times 24}{(94 - 65)}$$

= 1240 cooling load hours

Then use Formula 16F to estimate annual electrical consumption for cooling by the heat pump.

$$\text{kWh}_{\text{c-yr}} = \frac{\text{CLH} \times q_c}{1000 \times \text{SEER}}$$

(Formula 16F)

$$= \frac{1240 \times 24{,}560}{1000 \times 10.55}$$

= 2890 kilowatt hours per year

Formula 16C is then used to compute the yearly cooling cost.

$$\$_{\text{yr}} = \text{kWh}_{\text{yr}} \times \$/\text{kWh}$$

(Formula 16C)

= 2890 × 0.10

= $290 per year for cooling using the heat pump

Therefore, the yearly operational cost for heating and cooling for the heat pump is about $860 ($570 plus $290). It should also be noted that the heat pump model assumed in this example was not ideal since the cooling capacity of the unit far exceeds the cooling load. This indicates a problem that exists in selecting heat pumps which must be chosen to meet both heating and cooling loads.

As illustrated in this example, air-to-air heat pumps can offer an attractive heating and cooling system alternative in moderate climates (i.e., less than about 6000 heating degree-days). Yet, in colder climates, the low heating COPs often will result in high heating costs.

"Heating-only" heat pumps are also available which are engineered to improve efficiency in the heating mode. Such units can be used to particular advantage in climates such as the Pacific Northwest, where low-priced electricity is available and where less than 500 cooling load hours occur.

16.4 HEAT PUMP SIZING

The operational cost estimated in Example 16C assumed a heat pump sized to meet the entire heating load. In practice, many heat pumps are not sized to meet the entire heating load, particularly for very cold climates.

Air-to-air heat pump performance (COP_h) and capacity to meet a heating load (btu/h) decrease as temperatures decrease, as illustrated in Figure 16.8. This means that these heat pumps can meet less and less of a heating load just when capacity is most needed—namely, at lower outdoor air temperatures.

Also shown in Figure 16.8 is the "crossover" point which is the temperature at which the heat pump alone cannot meet the heating load. Beyond this point (at lower temperatures) some form of supplemental heat (usually electric resistance heat within the heat pump equipment) will be required.

16.5 ELECTRIC BOILERS

Hot water boilers which use electricity as the heat source have somewhat limited application. They do, nevertheless, offer the

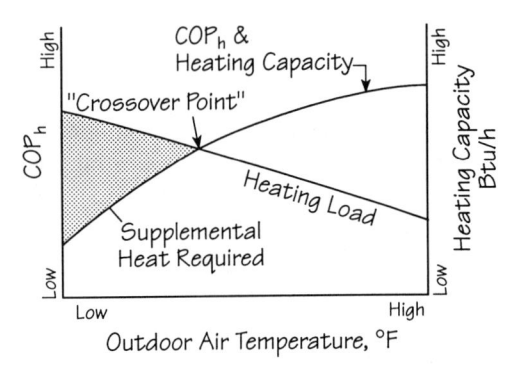

FIGURE 16.8 Heat pump capacity as a function of outdoor (source) air temperature.

advantages of not needing a chimney (flue), and the potential to produce hot water using lower cost "off-peak" electricity.

Appropriate applications include buildings and campuses with a large demand for electricity during the day and greatly reduced needs for electricity during the evening and night-time hours. During the night, "off-peak" electricity can be used to provide and store the heat needed for use the next day. With such a design, the electric boiler does not add to the electric peak demand (in kilowatts), which can be very costly under many utility billing structures.

Electric boilers and indeed all forms of electric heating should be considered very carefully. For example, are the "off-peak" rates which make electric boilers plausible guaranteed, or might future utility rate structures ruin the long-term economic advantage?

17

HEATING FUELS AND COMBUSTION

In previous chapters, heating systems using different fuels have been described. Heating fuel type often plays a fundamental role in system selection. For example, the high cost of electricity in most locations generally makes electric resistance heat a poor choice. System selection should be based on many factors, including an evaluation of operational costs. This may justify additional initial investment in more efficient equipment.

Various terms used to describe combustion efficiency, Annual Fuel Utilization Efficiency (AFUE) values actually achieved by equipment, and venting requirements are discussed herein.

17.1 HEAT CONTENT OF FUELS

Presented in Table 17.1 is the total heat potential for various heating system fuels. Useful energy obtained from a fuel is then determined by applying an efficiency factor. For example, a gallon of #2 oil burned at 80% efficiency will produce 112,000 useful Btus (140,000 × .80).

It is also common to refer to natural gas in terms of 100,000 Btu being equal to a therm of gas.

17.2 STEADY-STATE EFFICIENCY

There are several measures of efficiency for fossil fuel burning heating appliances. The first is known as "steady-state" combustion efficiency. It represents the percentage of the fuel content input into the appliance (at full, continuous output) after subtracting losses for exhaust products.

Manufacturers test their equipment in accordance with ASHRAE/ANSI Standard 103-88 and publish combustion efficiencies. The vent loss and resultant steady-state efficiency of installed equipment can be field-determined using test equipment which is inserted into a small (about 1/4 in. diameter) test hole drilled or punched into the vent. The tests measure the amount

286 HEATING FUELS AND COMBUSTION

TABLE 17.1 APPROXIMATE HEATING VALUE OF FUELS

Fuel Type	Heat Content
Natural gas (methane)	1030 Btu/ft^3
Propane	2500 Btu/ft^3
#2 Oil	140,000 Btu/gal
#4 Oil	145,000 Btu/gal
#6 Oil	153,000 Btu/gal
Electricity	3413 Btu/h per kW
Anthracite (hard) coal	14,000 Btu/lb
Bituminous (soft) coal	12,000 Btu/lb
Hardwoods	24,000,000 Btu/cord
Softwoods	15,000,000 Btu/cord

of carbon dioxide and/or oxygen in the air being vented, along with temperatures in the stack. A smoke spot test is also used to determine if the combustion is clean, which it must be if the results are to be considered valid. Based upon these tests, the steady-state efficiency and amount of excess air can be determined.

"Excess air" is air which has circulated through the combustion chamber above that actually needed for complete combustion. Some excess air is needed as a safety factor to assure that complete combustion takes place. For example, incomplete combustion of methane (natural gas) will result in the creation of potentially deadly carbon monoxide, rather than carbon dioxide, the product of complete combustion, as shown in Figure 17.1. When methane is burned in air, it reacts only with the oxygen (about 21% of the volume) and not the nitrogen, which comprises about 78% of air by volume. The nitrogen is simply carried along with the excess air and carbon dioxide and is vented from the system.

17.3 ANNUAL FUEL UTILIZATION EFFICIENCY (AFUE)

Appliances rarely operate at steady state for sustained periods of time since heating appliances and systems are typically designed to meet 97.5% winter design conditions which are exceeded only 54 h/yr on average (see Section 2.2). Therefore, part-load conditions generally exist, which results in frequent cycling of the equipment (on for a period, off for awhile, on again, etc.). So, a meaningful measure of equipment efficiency should take into account not only the vent (flue) losses, but also the "on-cycle" and "off-cycle" losses, which can be substantial.

On-cycle losses may arise from burners using too much excess air which results in a great deal of heat going up the vent. Also as equipment ages, the heat exchanger

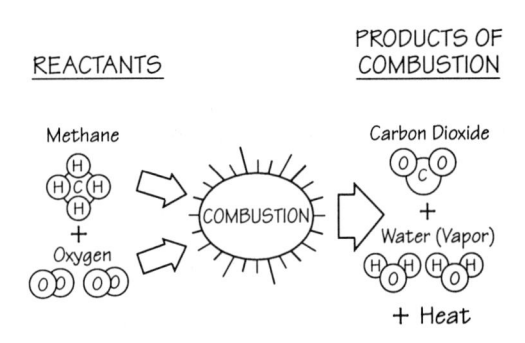

FIGURE 17.1 Combustion of natural gas (methane).

within the unit may become sooty and degrade in performance.

Off-cycle losses occur between the heating cycles. Potentially most wasteful is the loss of heat that continues to be sucked up the chimney because of draft. To limit off-cycle draft, dampers can be used to either close off the chimney or prevent air from getting to the combustion chamber when the appliance is not producing heat.

AFUE stands for Annual Fuel Utilization Efficiency. AFUE differs from Steady State Efficiency since it employs an empirical equation to deduct all anticipated operational losses. These include vent losses, cyclic effects, part-load operation, and continuously burning pilot lights (now rare).

Since the beginning of 1992, Federal law requires that all new furnaces produced must have a minimum AFUE of 78% when tested by the isolated combustion system (ICS) method. The ICS method takes into consideration jacket losses and assumes an indoor location for the heating appliance with all combustion air admitted directly to it through grilles or ducts. Provided in Table 17.2 are typical AFUE values (both Indoor % and ICS) for various types of furnaces located in conditioned spaces. Obviously, many types of furnaces which are still in common use do not meet the new Federal law. The term "atmospheric" simply means that the combustion process uses air which comes from the space (or room) where the equipment is placed.

AFUEs for boilers are about 78–82% for most conventional (noncondensing) oil- or gas-fired boilers in the marketplace. Condensing boilers have AFUEs of 90% or more.

A gap exists in Table 17.2 between equipment with AFUEs in the low 80's, and condensing furnaces (appliances) with AFUEs of over 90%. This is because combustion heating equipment having AFUEs in the mid to high 80's are considered "near condensing." This means that they will produce relatively low flue gas temperatures which in cold weather could lead to condensation within the flue (chimney) and resultant damage to masonry and/or the heating appliance.

TABLE 17.2 TYPICAL FURNACE AFUE VALUES (FURNACES LOCATED IN CONDITIONED SPACE)

	Indoor (%)	ICS
Type of Gas Furnace		
Atmospheric with standing pilot	64.5	63.9
Atmospheric with intermittent ignition	69.0	68.5
Atmospheric with intermittent ignition and auto. vent damper or power vent	78.0	68.5
Atmospheric with intermittent ignition and auto. vent damper or power vent and improved heat transfer	80 to 82	78 to 80
Direct vent with standing pilot, preheat	66.0	64.5
Direct vent, power vent, and intermittent ignition	80.0	78.0
Power burner (forced-draft)	80.0	78.0
Condensing	93.0	91.0
Type of Oil Furnace		
Standard (range depends on heat exchanger)	71 to 76	69 to 74
Standard, good heat exchanger, and automatic vent damper	83.0	74.0
Condensing	91.0	89.0

Source: 1992 ASHRAE Handbook—HVAC Systems and Equipment, Chapter 29, Table 1.

17.4 MEASURING IMPROVED EFFICIENCY

Comparative annual energy savings derived from using an appliance with an improved AFUE can be estimated using Formula 17A.

$$\%ES_{yr} = \frac{AFUE_1 - AFUE_2}{AFUE_1} \times 100$$

Formula 17A

where

- $\%ES_{yr}$ is the percentage energy savings per year
- $AFUE_1$ is the efficiency of the heating appliance with the higher Annual Fuel Utilization Efficiency
- $AFUE_2$ is the efficiency of the heating appliance with the lower AFUE

Example 17A: Determine the percentage energy savings if a heating appliance with an AFUE of 70% ($AFUE_1$) replaces a unit with an AFUE of 60% ($AFUE_2$).

Solution: Use Formula 17A.

$$\%ES_{yr} = \frac{AFUE_1 - AFUE_2}{AFUE_1} \times 100$$

(Formula 17A)

$$= \frac{70 - 60}{70} \times 100$$

$$= 14.3\%$$

In a similar manner it can be shown that an AFUE improvement from 70 to 80% represents a percentage energy savings of 12.5%; and an AFUE improvement from 80 to 90% represents a percentage energy savings of only 11.1%. Note how the percentage savings diminish as efficiencies improve.

The quantity of Btus required to produce 1,000,000 Btus of useful heat for any AFUE value can be determined by using Formula 17B.

$$Btu_{mil} = \frac{MMBtu_{useful}}{AFUE}$$ Formula 17B

where

- Btu_{mil} is the quantity of Btus needed from a fuel source to provide 1,000,000 Btus of useful heat
- $MMBtu_{useful}$ is one million (1,000,000) Btus of useful heat
- AFUE is the Annual Fuel Utilization Efficiency of a heating appliance, as a decimal

Example 17B: Determine the quantity of Btus required to provide 1,000,000 Btus of useful heat with a heating appliance which has an AFUE of .80 (80%).

Solution: Use Formula 17B.

$$Btu_{mil} = \frac{MMBtu_{useful}}{AFUE}$$ (Formula 17B)

$$= \frac{1,000,000}{0.80}$$

$$= 1,250,000 \text{ Btus}$$

The quantity of source Btus required to produce 1,000,000 Btus of useful heat for any AFUE value (including those shown in Table 17.3) can be computed in a similar manner.

Table 17.3 lists the quantity of source heating fuel Btus needed for AFUE values in 5% increments to deliver 1,000,000 Btus of useful heat.

Note in Table 17.3 that a 5% improve-

17.4 MEASURING IMPROVED EFFICIENCY

TABLE 17.3 Btus SAVED BY INCREASED EFFICIENCY

AFUE (%)	Heating Btus Needed from Fuel Source to Meet Heating Load of 1,000,000 Btus	Btus Saved (per Million) Compared to 5% Lower AFUE
95	1,052,630	58,480
90	1,111,110	65,360
85	1,176,470	75,530
80	1,250,000	83,330
75	1,333,330	95,240
70	1,428,570	109,890
65	1,538,460	128,210
60	1,666,670	151,520

ment in AFUE saves 58,480 Btus per million needed if the AFUE is improved from 90 to 95%. Similarly, a much higher quantity of 128,210 Btus are saved per million needed if the AFUE is improved 5% from 60 to 65%. Through examination of Table 17.3 one can clearly see how Btu savings from AFUE improvements diminish. Not shown in Table 17.3 is an AFUE value for 100% since this is not possible even if the most efficient combustion heating equipment is used.

Improvements in AFUE are used to evaluate decisions regarding choices in heat-producing equipment.

Example 17C: A large old house requires about 2000 therms of gas (current cost = $0.75 per therm) each heating season. The existing heating plant which has an estimated AFUE of 65% is in poor condition and needs replacement.

Consideration is being given to the replacement of the existing unit by either a gas-fired condensing furnace with an AFUE rating of 90% at a cost of $2700, or a conventional furnace with an AFUE rating of 80% at a cost of $2000. Evaluate the economic advisability of these two alternatives.

Solution: Potential savings can be computed for each option using Formula 17A.

$$\%ES_{yr} = \frac{AFUE_1 - AFUE_2}{AFUE_1} \times 100$$

(Formula 17A)

$$= \frac{90 - 65}{90} \times 100$$

$$= 27.8\%$$

$$\%ES_{yr} = \frac{AFUE_1 - AFUE_2}{AFUE_1} \times 100$$

(Formula 17A)

$$= \frac{80 - 65}{80} \times 100$$

$$= 18.8\%$$

The difference in potential savings is 9% (27.8% − 18.8%). Additional gas savings derived from investment in the higher efficiency equipment would then be 0.09 × 2000 therms, or 180 therms per year. Based upon a gas cost of $0.75 per therm, a yearly savings of $135 is estimated.

The additional cost for the high efficiency condensing furnace was noted to be $700. Therefore, the "simple payback" in years can be computed to 5.2 years ($700/$135). (Note: For more information on simple payback calculations see Formula 24C.)

Such a simple payback should then be weighed against the homeowner's investment criteria, and satisfaction that the higher efficiency condensing equipment is of proven quality and under warranty. Also keep in mind that this house requires a lot of heating fuel each year. The simple payback period will be longer for high-efficiency heating equipment in small or energy-conserving buildings.

17.5 VENTING OF COMBUSTION HEATING APPLIANCES

After combustion takes place in a heating appliance, the waste products (water vapor, carbon dioxide, and excess air) must be vented or exhausted to the outdoors by a system designed and constructed to safely develop a positive flow. Depending upon the type and efficiency of the appliance, the exhaust will typically be at a temperature of 300–600°F, whereas the surfaces of the vent may be very cold (close to the outside air temperature). Therefore, it is safe to assume that some water vapor within the waste gas may be reduced in temperature to the point that it condenses to liquid water within the vent. This is acceptable as long as vent design and sizing allows the vent surfaces to heat up to the point that any liquid water will again become vaporized. "Wet-time" is the term used for this transient period (usually a few minutes) when condensation is present. However, if the vent does not heat up adequately, condensation may continue to form, which can cause corrosion or damage to masonry.

"Type B" refers to vents constructed of noncombustible, corrosion resistant material of adequate thickness to assure a long life (there are no "Type A" vents). Most type B vents are made of metal with double wall construction. They are typically used in new construction since they are less expensive to install than a masonry chimney with a liner.

Sizing of vents for gas-fired appliances (furnaces, boilers, and hot water heaters) has been successfully governed by the National Fuel Gas Code (NFPA 54 and ANSI 223.1) since the 1950s. The original vent tables were developed for appliances which are naturally vented through a draft hood and vent connection such as the heating appliance shown in Figure 17.2. Vent height (H) and length of the lateral connector (L), both in feet, are of importance to vent sizing.

The draft hood protects the flame from excessive draft or wind effects. It also allows additional air (known as dilution air) to enter and mix with the vent gases. In the process, the relative humidity is lowered, which in turn reduces potentially damaging condensation within the vent.

Use of the original or "old" vent tables can lead to condensation problems in newer fan-assisted heated appliances (without draft hoods) designed to perform at higher efficiencies. To address this issue, new vent tables developed by the American Gas Association (AGA) have been incorporated in

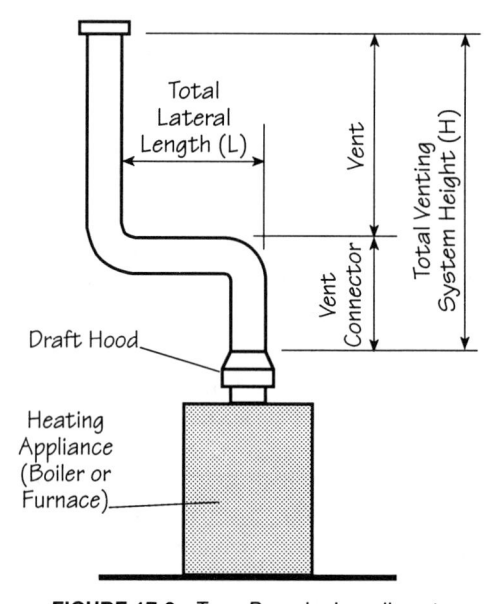

FIGURE 17.2 Type B or single-wall vent.

TABLE 17.4 NATURAL DRAFT APPLIANCES[a]

		Maximum Appliance Input Ratings (Thousands of Btu/h)		
		Vent and Connector Diameter (in.)		
Height (H) (ft)	Lateral (L) (ft)	3	4	5
10	0	53	100	166
	2	42	81	129
	5	40	77	124
	10	36	70	115
15	0	58	112	187
	2	48	93	150
	5	45	87	142
	10	41	82	135
20	0	61	119	202
	2	51	100	166
	5	48	96	160
	10	44	89	150

[a] Capacity of Type B double-wall vents with Type B double-wall connectors serving a single appliance.

Source: Tables 17.4 and 17.5 are excerpted and reprinted with permission from NFPA 54, *National Fuel Gas Code*, Copyright © 1992, National Fire Protection Association, Quincy, MA 02269. This reprinted material is not the complete and official position of the National Fire Protection Association, on the referenced subject which is represented only by the standard in its entirety.

TABLE 17.5 FAN-ASSISTED APPLIANCES[a]

		Appliance Input Ratings (Thousands of Btu/h)					
		Vent and Connector Diameter (in.)					
		3		4		5	
Height (H) (ft)	Lateral (L) (ft)	Min	Max	Min	Max	Min	Max
10	0	0	88	0	175	0	295
	2	12	61	17	118	23	194
	5	23	57	32	113	41	187
	10	30	51	41	104	54	176
15	0	0	94	0	191	0	327
	2	11	69	15	136	20	226
	5	22	65	30	130	39	219
	10	29	59	40	121	51	206
20	0	0	97	0	202	0	349
	2	10	75	14	149	18	250
	5	21	71	29	143	38	242
	10	28	64	38	133	50	229

[a] Capacity of Type B double-wall vents with Type B double-wall connectors serving a single appliance.

the revised 1992 National Fuel Gas Code. Excerpted and listed in Table 17.4 are maximum appliance input ratings in thousands of Btu/h for naturally drafted residential-scale appliances with draft hoods.

For fan-assisted appliances, the revised tables provide the minimum and maximum appliance input ratings in thousands of Btu/h for a given vent height and lateral condition. Minimum ratings are to prevent condensation. Maximum ratings are to avoid positive static pressure. For most residences (with Type B venting), Table 17.5 can be used to determine the size of vents for fan-assisted gas heating appliances (referred to as Category I).

Example 17D: Determine the proper vent size for a fan-assisted mid-efficiency furnace (AFUE = 83.7%) based upon the following design factors:

A Type B double-wall vent and connector serves a single appliance

Appliance input rating of 75,000 Btu/h
Vent height of 15 ft
Vent lateral of 5 ft

Solution: Table 17.5 is the correct one to use for this appliance. Locate the line which corresponds with a 15-ft vent height and 5-ft lateral. It indicates that a 4-in. vent will be adequate since the minimum appliance input must be between a minimum of 30,000 Btu/h and a maximum of 130,000 Btu/h.

Similar tables are published for common vent systems which vent two or more appliances (i.e., both the furnace or boiler and a domestic hot water heater). Vent requirements for high-efficiency condensing appliances (known as Category IV) shall conform with manufacturer recommendations which have been tested and approved.

18

COOLING OF HOUSES

While comfort conditions for humans can be defined rather precisely (see Chapter 1), the duration of "discomfort" which one will choose to endure will depend upon attitude, economic resources, local practice, and of course local climate. Where possible, the need for mechanical cooling equipment should be minimized or eliminated by utilizing natural cooling systems such as air motion and evaporative cooling processes (in hot arid climates). Some cooling of houses can also be achieved through use of ceiling and whole-house fans which require only a small amount of electrical power.

Air conditioning has in the last one hundred years gone from nearly nonexistent to nearly mandatory for most areas of North America. Not because of a changing climate (or at least not yet vis a vis possible "global warming"), but rather as a matter of societal choice for improved comfort. Recent statistics show that over 75% of new homes in America now come with some form of central air conditioning; more specifically, 50% of new homes in the Northeast, about 75% in the Midwest, 95% in the South, and about 60% in the West.

Various natural and mechanical systems available for cooling of houses are presented below. Also included is a simplified cooling load calculation for "The House" originally presented in Section 5.14.

18.1 CLIMATIC NEED FOR COOLING

A useful measuring stick of the need for mechanical cooling in residences is the cooling degree-day (CDD), which measures temperature occurrences above 65°F as fully defined in Section 2.9 and listed in Table 2.8 for 40 United States cities. Summarized in Figure 18.1 is cooling degree-day data (base 65°F) for ready reference.

Listed in Table 2.8 are wet-bulb degree-hours above 65°F which are indicative of the humidity which occurs during the cooling season. A summary of wet-bulb degree-hours above 65°F is provided in Figure 18.2.

294 COOLING OF HOUSES

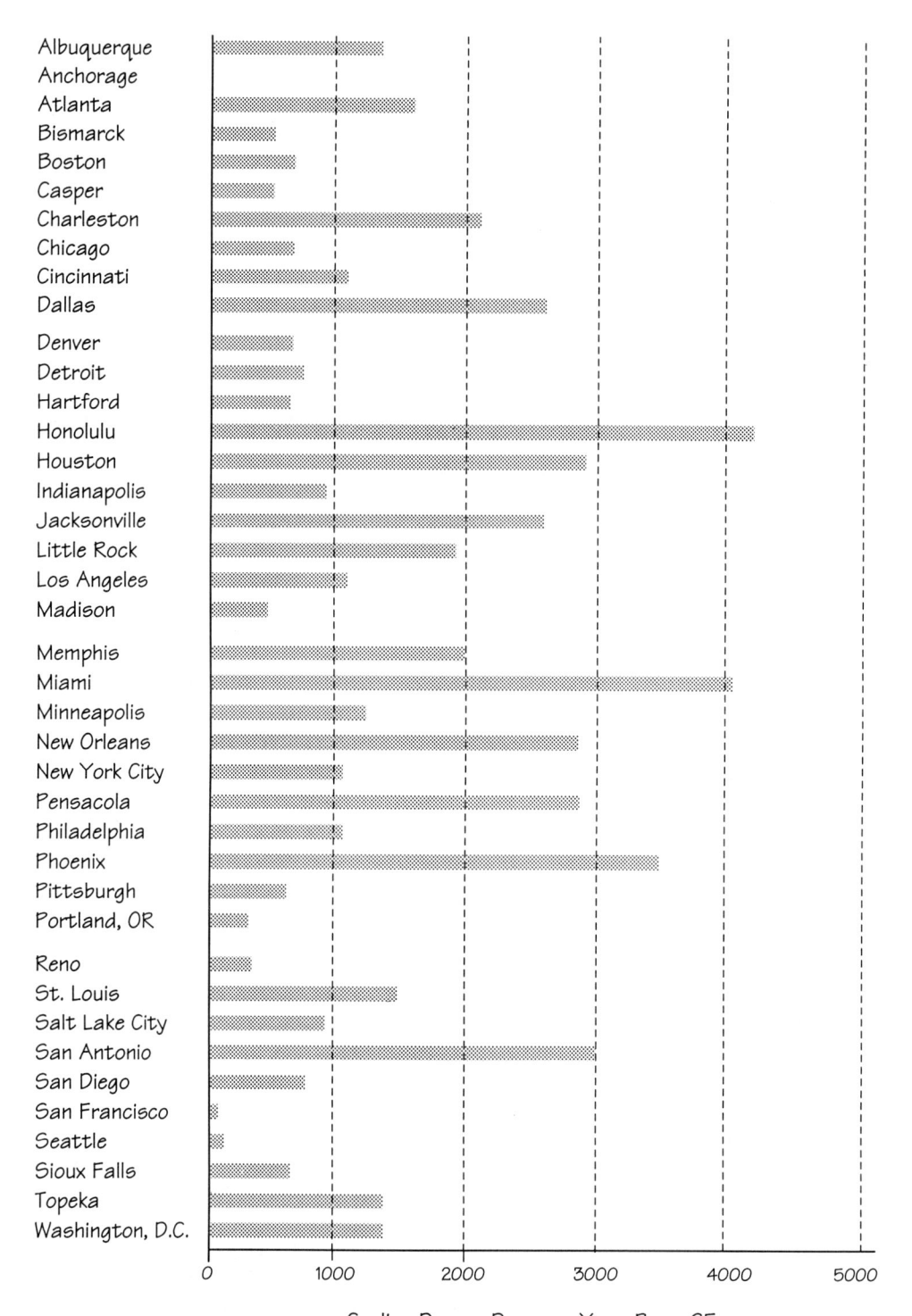

FIGURE 18.1 Cooling degree-days—base 65°F.

18.1 CLIMATIC NEED FOR COOLING

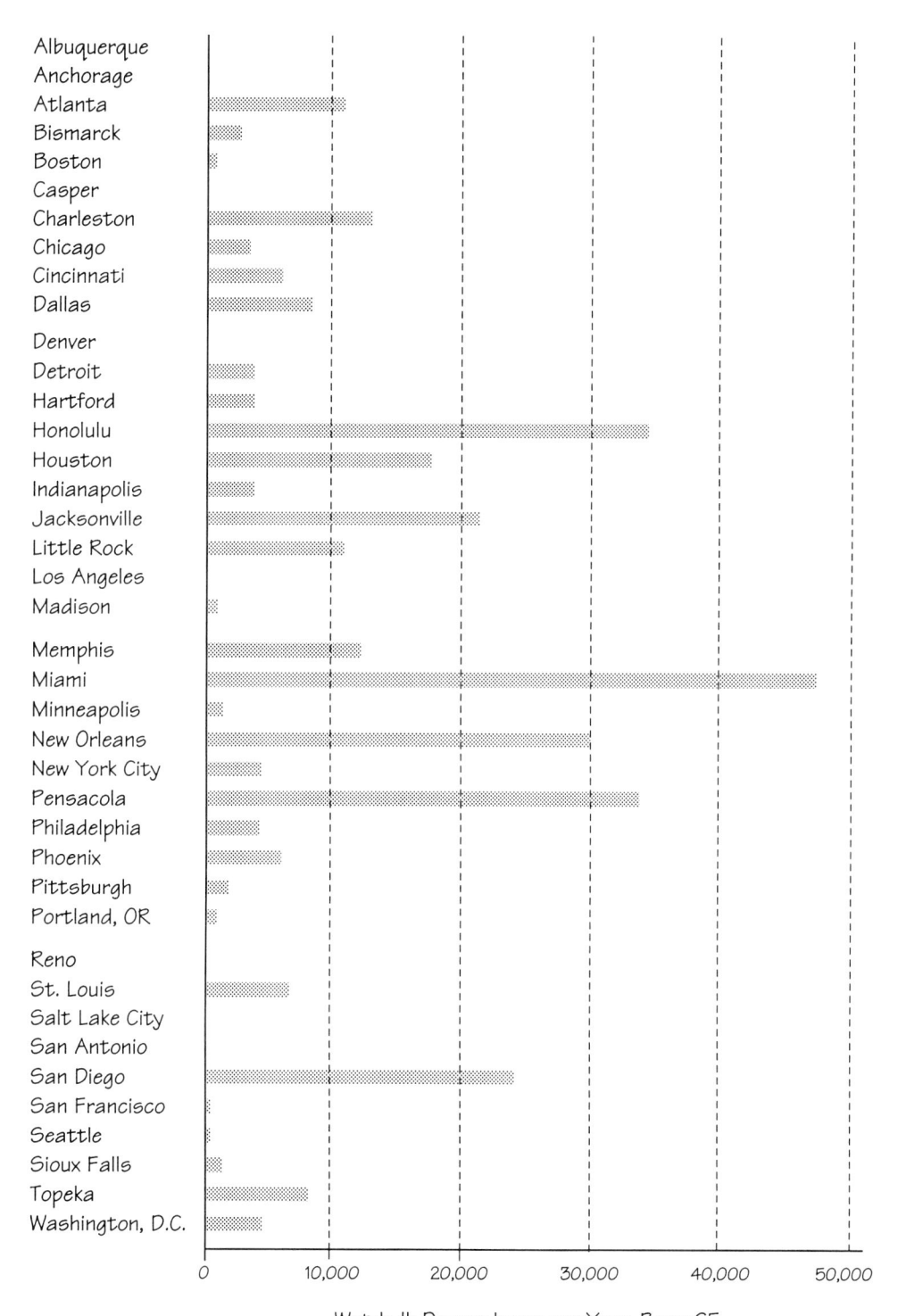

FIGURE 18.2 Wet-bulb degree-hours above 65°F.

296 COOLING OF HOUSES

Some guidance can be drawn from Figures 18.1 and 18.2. Residences in locations with less than 1000 or even 1500 cooling degree-days can probably forego mechanical cooling unless the humidity is also high as indicated by at least 5000 wet-bulb degree-hours. For example, residences in Salt Lake City with 950 cooling degree-days but with 0 wet-bulb degree-hours can usually be built without mechanical cooling, whereas residences in Indianapolis with approximately the same number of cooling degree-days are much more likely to require mechanical cooling because of the higher humidity (4000 wet-bulb degree-hours).

18.2 COOLING THROUGH AIR MOTION

In Chapter 5, infiltration heat loss and its calculation procedures were covered. We saw that the driving forces of infiltration are temperature difference and wind speed.

During the heating season, infiltration of large quantities of outside air is unwanted since that air must be heated to maintain comfort conditions. During warmer weather, however, the entry of larger quantities of outside air through operable windows can be very beneficial in providing natural cooling. Cooling through design which optimizes natural ventilation is appropriate in many climates for residences, and small commercial buildings where noise, air quality (dust etc.), and security are not of concern.

To design for natural cooling, one must understand the basic effects of wind pressure on a building. These effects were dramatically illustrated in wind tunnel tests of a simple flat-roofed building form conducted over 30 years ago at Princeton University. The results (including 26 very informative photographs) were published in *Design with Climate, Bioclimatic Approach to Architectural Regionalism* by Victor Olgyay, a book which provided the cornerstone for much of the current understanding of environmentally responsive design.

Illustrated in Figure 18.3 is the pattern of air flow (left to right) around a building. The positive pressure (+) on the front of the building is created by the moving air which piles up and then pushes out from the front of the house. Note that negative pressure exists on the back side of the building, forming a "wind shadow" (area of calm). The wind shadow on the back, by the way, would be even more pronounced if the building had a pitched roof.

Cooling of a building with natural ventilation, therefore, involves controlled manipulation of positive wind pressures to allow for controlled entry, distribution, and exiting of the moving air stream. Openings (usually windows) through which air enters are known as "inlets," while openings through which air exits are referred to as "outlets."

Guidelines for Inlets and Outlets

The series of wind tunnel tests yielded the following general guidelines regarding the

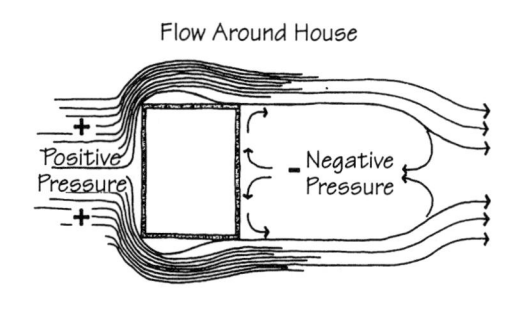

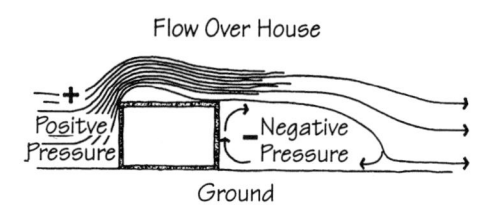

FIGURE 18.3 Winds and pressure around a building.

need for, size, and placement of inlet and outlet openings:

- For flow to occur, a building must have both inlets where high pressure occurs, and outlets in areas of low pressure. If only inlets exist, air that enters will be bounced back to oppose the entry of additional air.
- Large inlets and small outlets produce slow air velocities within a building and minimal cooling effect. In contrast, the highest wind speeds are produced just inside a small inlet combined with a large outlet opening.
- Maximum air flows (but limited distribution) occur when large openings are placed opposite each other.
- Overhangs should either have a vertical separation above window inlets, or a slot separation from the house in order to equalize external pressures (see Figure 18.4a).
- Interior partitions which are perpendicular to the air flow must be minimized to maintain a cooling flow (see Figure 18.4b).
- Directional elements within the inlet (pivot windows, venetian blinds) should be tilted downward to achieve a well-directed flow (see Figure 18.4c).
- Flows are altered (made asymmetrical) by items that unbalance the pressure such as casement windows and shrubs.
- Air flows within rooms depend a great deal on the height of the inlet and very little on the height of the outlet above the floor. Low inlets result in a pleasant downward flow pattern of outside air.

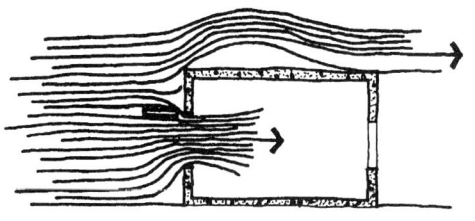

a) Overhang with slot equalizes the external pressures. Results in desirable air flow into room.

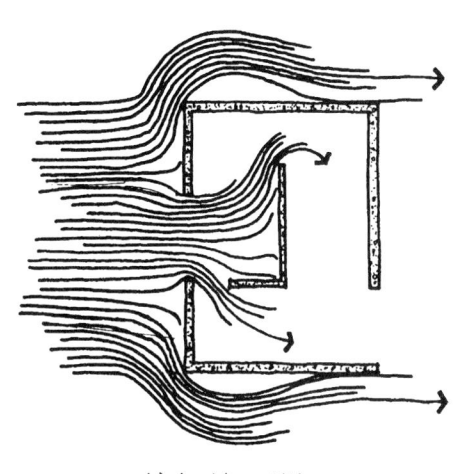

b) Avoid partitions perpendicular to air flow.

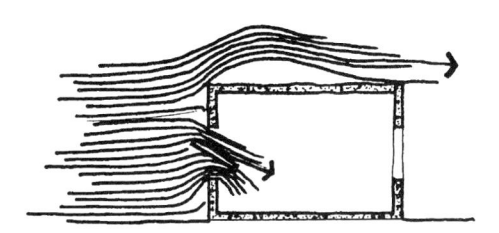

c) Direct air flow downward at inlet for desirable air flow into room.

FIGURE 18.4 Impacts of inlets and outlets on air flow.

Approximate Rate of Air Flow

The *1993 ASHRAE Handbook of Fundamentals* contains a calculation procedure (in Chapter 23) for air flow caused by wind as indicated by Formula 18A. The term "free" area means the area of openings that are totally unobstructed (except for screens that will reduce wind velocities only slightly).

298 COOLING OF HOUSES

$$\text{cfm}_{af} = 88 \times V_{mph} \times A_i \times C_v$$

Formula 18A

where

cfm_{af} is the air flow in cfm (ft^3/m)
88   is the conversion factor of 88 feet per minute being equal to 1 mile per hour
V_{mph} is the velocity of wind in miles per hour
A_i is the free area of inlets in ft^2
C_v is the effectiveness of openings and is assumed to be 0.50 to 0.60 for perpendicular winds and 0.25 to 0.35 for diagonal winds

Example 18A: Determine the approximate air flow through the main living area of a house which is free of obstructing partitions when the perpendicular wind velocity (V) is 5 mph. The house has free inlet and outlet areas of windows equal to 10 ft^2 each.

Solution: Approximate the air flow rate (cfm$_{af}$) using Formula 18A. Utilize a C_v factor of 0.60 to estimate the flow for perpendicular winds.

$$\text{cfm}_{af} = 88 \times V_{mph} \times A_i \times C_v$$

(Formula 18A)

$$= 88 \times 5 \times 10 \times 0.60$$

$$= 2640 \text{ cfm}$$

The Example 18A result agrees with the result which would be obtained by using the mathematical expression put forth in *Design with Climate*. This expression in slightly modified form is shown as Formula 18B. It also allows for different ratios of outlet and inlet areas.

$$\text{cfm}_{af} = O/I_v \times V_{mph} \times A_i \quad \text{Formula 18B}$$

where

cfm_{af} is the air flow in cfm (ft^3/m)
O/I_v is the Outlet/Inlet value from Table 18.1
V_{mph} is the wind velocity in miles per hour
A_i is the free area of inlets in ft^2

Note that O/I_v values are largest when the outlet area is a high ratio of the inlet area. Such a design creates what is known as a "venturi effect" which produces maximum air speeds within a structure which are conducive to producing comfortable cooling conditions.

TABLE 18.1 OUTLET/INLET VALUES

Ratio of Outlet Area To Inlet Area (O:I)	O/I_v
1:1	52.5
2:1	66.7
3:1	70.8
4:1	72.5
5:1	73.3
3:4	45.0
1:2	33.3
1:4	18.3

Example 18B: Determine the approximate air flow through the main living area of a house, free of obstructing partitions, when the perpendicular wind velocity is 5 mph (440 fpm). In this case, the house has a free inlet area of 20 ft² and a free outlet area of 40 ft².

Solution: Obtain the outlet/inlet value of 66.7 for an outlet/inlet ratio of 2:1 (40 ft²/20 ft²). Then use Formula 18B.

$$\text{cfm}_{af} = O/I_v \times V_{mph} \times A_i$$

(Formula 18B)

$$= 66.7 \times 20 \times 5$$

$$= 6670 \text{ cfm}$$

Heat Removal Potential

Once an air flow rate in cfm has been determined, one may determine the amount of heat which can be removed through natural ventilation (q_{nv}) when a lower outside temperature exists. Use Formula 18C which is similar to Formula 6C in the ventilation chapter.

$$q_{nv} = 1.10 \times \text{cfm}_{af} \times (t_i - t_o)$$

Formula 18C

where

q_{nv}	is sensible heat removed by natural ventilation in Btu/h
1.10	is a constant (see Formula 6C for details)
cfm$_{af}$	is the flow rate of outside air entering the building in cfm (ft³/m)
t_i	is the inside air temperature in °F
t_o	is the outside air temperature in °F

Example 18C: Determine the quantity of heat which can be removed from the house in the previous example which has a flow rate of 6670 cfm if the indoor temperature (T_i) is 80°F and the outdoor temperature (T_o) is 75°F.

Solution: Use Formula 18C to compute the heat removed by natural ventilation of the house in Btu/h.

$$q_{nv} = 1.10 \times \text{cfm}_{af} \times (t_i - t_o)$$

(Formula 18C)

$$= 1.10 \times 6670 \times (80 - 75)$$

$$= 36{,}685 \text{ Btu/h}$$

This is the removal of a significant amount of heat, about equal to a 3-ton air conditioner.

Cooling Effect of Moving Air

Air motion can also produce a cooling effect as discussed in Section 1.4. Use Formula 18D to compute air speed at the inlet. Realize, though, that the velocity of air within a room will be much slower depending upon room geometry.

$$V_{fpm} = \frac{\text{cfm}_{af}}{A_i} \quad \text{Formula 18D}$$

where

V_{fpm}	is the indoor air velocity at the inlet in feet per minute
cfm$_{af}$	is the flow rate of outside air entering the building in cfm (ft³/m)
A_i	is the free area of inlets in ft²

Example 18D: Determine the indoor air velocity (V_{fpm}) at the inlet for the previous example which has a flow rate (cfm$_{af}$) of 6670 cfm and an inlet area (A_i) of 20 ft².

Solution: Use Formula 18D.

$$V_{fpm} = \frac{cfm_{af}}{A_i} \quad \text{(Formula 18D)}$$

$$= \frac{6670}{20}$$

$$= 333 \text{ fpm}$$

Although indoor air motion above about 200–300 fpm is potentially drafty and annoying, the result obtained here should be fine. First, the velocity obtained is at the inlet. Once the air travels into a room, the velocity decreases markedly. In addition, when natural cooling is desired, slightly higher speeds will normally be tolerated and even enjoyed. And in this example another factor bears mention. The outlet area is twice as large, or 40 ft². Therefore, air motion at the outlet will be one-half as fast, or only about 167 fpm.

18.3 AIR-CIRCULATION FANS

As discussed previously, air motion can produce conditions for human comfort even when dry-bulb temperatures are elevated. In many situations, natural ventilation is not a realistic or dependable design option. In these cases, air-circulation fans (ceiling or portable) which create an airspeed of approximately 150–200 fpm can compensate for a temperature increase of about 4°F with no loss of comfort.

Ceiling fans should be mounted at least 10 in. below a normal 8-ft ceiling height, or at a height of about 7 ft 6 in. to 8 ft in rooms with high ceilings. Guidelines for sizing ceiling fans are provided in Table 18.2.

18.4 WHOLE-HOUSE FANS

Another alternative to large inlets for natural ventilation is the use of a large whole-house fan (WHF) located centrally in the attic space. The fan draws in air through partially open windows and exhausts it through the ceiling and attic.

Whole-house fans should be sized to provide about 20 changes of house air per hour using Formula 18E.

$$cfm = 0.33 \times H_{vol} \quad \text{Formula 18E}$$

where

cfm is the required fan flow rating in cubic feet per minute at 0.1 in. wg static pressure

0.33 is the product of 20 air changes per hour and the conversion of 1 hour equals 60 minutes

H_{vol} is the conditioned house volume in ft³

TABLE 18.2 CEILING FAN SIZING CHART

Largest Dimension of Room	Minimum Fan Diameter
12 feet or less	36 in.
12 to 16 feet	48 in.
16 to 17.5 feet	52 in.
17.5 to 18.5 feet	56 in.
18.5 feet or more	Two fans

Source: Cooling with Ventilation, published by the Solar Energy Research Institute.

Example 18E: Select a whole-house fan (WHF) for "The House" which has a conditioned volume (H_{vol}) of 11,536 ft^3.

Solution: Use Formula 18E.

cfm = 0.33 × H_{vol} (Formula 18E)

= 0.33 × 11,536

= 3807 cfm

A review of one manufacturer's literature indicates that a 24 in. WHF has an air flow rate of 3600 cfm (at 0.1 in wg), and that a 30-in. WHF delivers 5100 cfm. Either could be selected. The smaller fan would change the air about 19 times (which is acceptable), whereas the larger fan operates more slowly (lower rpm) and thus would be slightly quieter.

In selecting fans from manufacturers' data be sure to check the operational conditions. Some manufacturers rate their fans for free-air operation without any frictional resistance. In such cases, cfm fan flow ratings should be reduced by 20% to determine the rating for the 0.1 in. wg static pressure resistance assumed for whole-house fan applications. For example, a fan which has a free-air rating of 5000 cfm would have a rating of 4000 cfm (a 20% reduction) when used as a whole-house fan.

Inlet and Outlet Areas

In order for a whole-house fan to work there must be a "free area" of space within air inlets (windows) and air outlets (attic vents). The required free areas (required for air inlets and also required for attic vents) are determined by using Table 18.3.

18.5 MECHANICAL COOLING

Individual Units

The simplest form of mechanical cooling in a house is an individual unit which either fits into part of a window opening or is placed in a preplanned sleeve (see Figure 18.5). Individual units are of small capacity and capable of removing between 3,000 and 18,000 Btus of heat per hour ($\frac{1}{4}$ to $1\frac{1}{2}$ tons). Window units are either controlled manually (they are turned either on or off), or by a thermostat.

The units run on electricity, utilize the vapor compression refrigeration cycle (discussed in detail in Section 20.2), and are relatively noisy.

Central Air Conditioning

As noted earlier, more than 75% of new homes in America now come with some form of central air conditioning. The most common form in northern American climates is to install a cooling coil in the plenum space of a warm air furnace. With such a system, the ductwork which delivers hot air in the winter does double duty by also delivering cool air to spaces in the summer.

TABLE 18.3 WHOLE-HOUSE FAN REQUIRED OPENINGS

Whole-house Fan Diameter	Required "Free Areas" for Air Inlets and Attic Vents
20 in.	628 in.^{2} (4.36 ft^2)
24 in.	904 in.^{2} (6.28 ft^2)
30 in.	1413 in.^{2} (9.81 ft^2)
36 in.	2034 in.^{2} (14.1 ft^2)

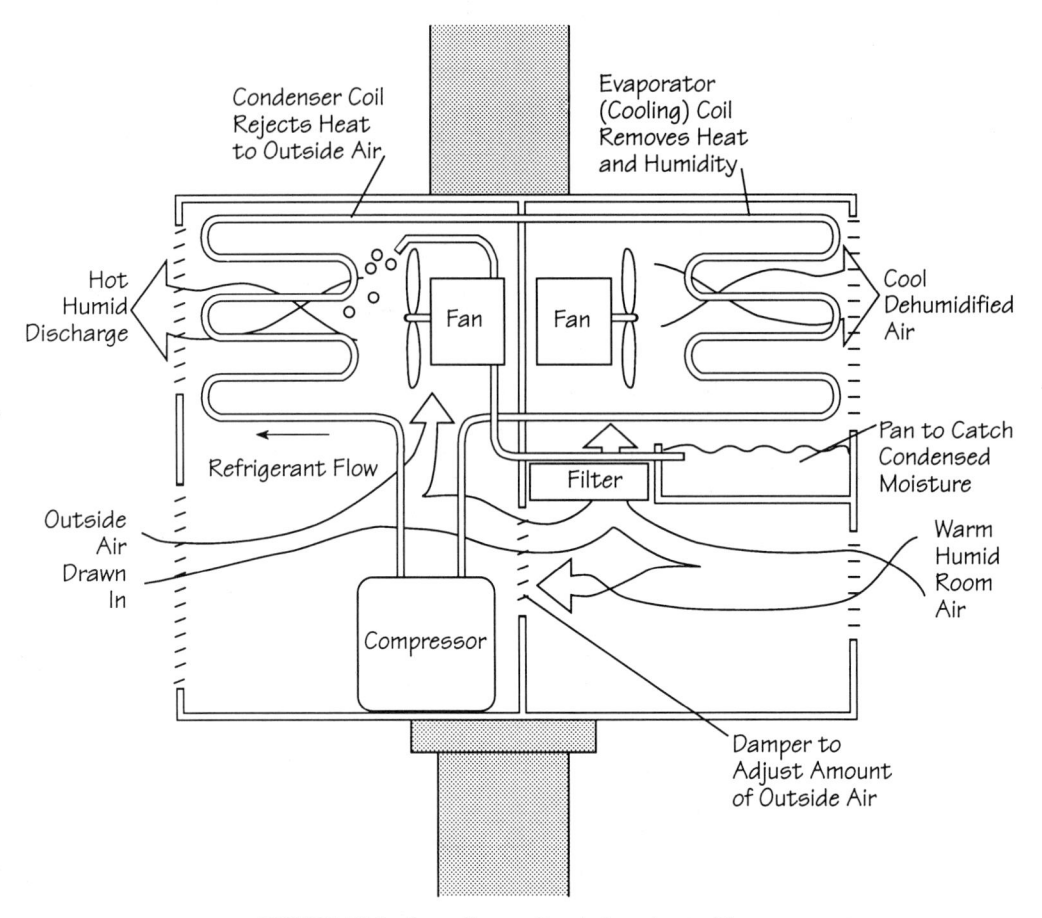

FIGURE 18.5 A small capacity window air conditioner.

The cooling coil is, of course, actually the "evaporator" in a vapor compression refrigeration cycle. The compressor and (air-cooled) condenser are located nearby outside in what is known as a "condensing unit" (see Figure 18.6).

Condensing units should be located near the evaporator (usually within 50 ft) to minimize pipe friction, and positioned to allow for air motion around them. In some cases, concern for compressor noise may also be a factor in siting units. Information on condensing unit performance and selection is contained in Chapter 20.

The cooling coil removes both sensible and latent heat (moisture) from the air stream. The surface temperature of the coil is low (50°F or lower) to receive warm return air at approximately 75–80°F and sensibly cool it about 25°F (to 50–55°F) through contact with its large surface area. The low coil temperature will also remove latent heat (moisture) from air which passes over it which has a high relative humidity.

As shown on the psychrometric chart in Figure 18.7, when relative humidities are over about 40%, the cool (50–55°F) surface of the cooling coil will be below the dew point temperature of the return air. As a result, moisture will condense out

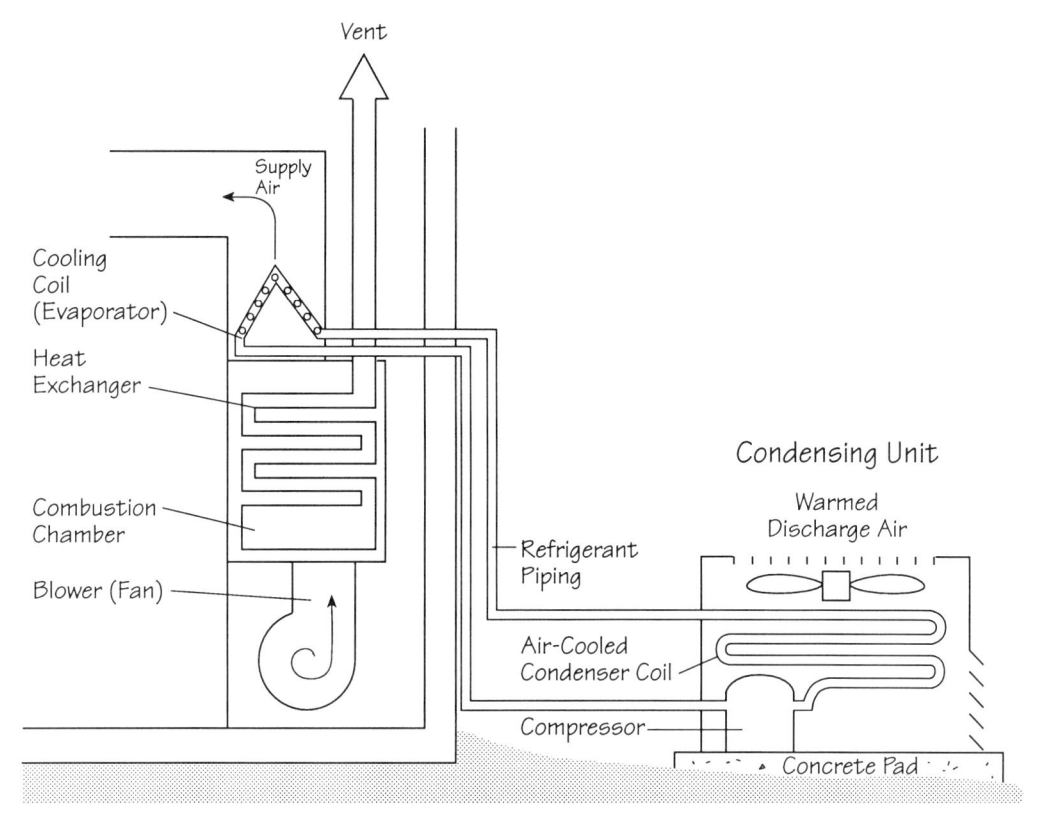

FIGURE 18.6 Cooling coil located in supply air stream and condensing unit to reject heat outside.

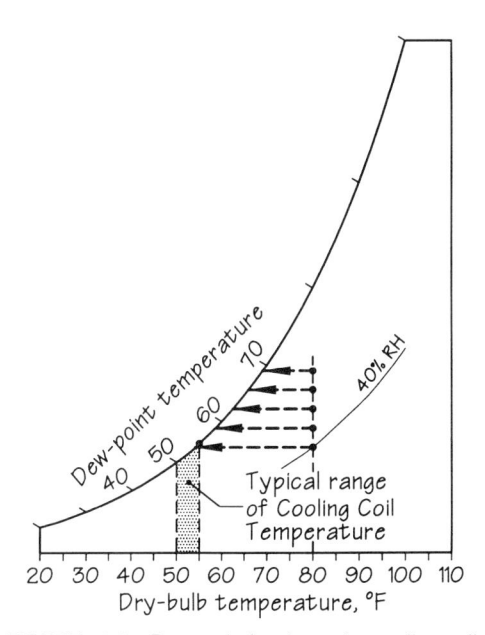

FIGURE 18.7 Removal of moisture by cooling coil.

of the air into a drain pan which should be served by a properly pitched pipe to a drain.

Heat Pumps, Cooling

Heat pumps also use the vapor compression refrigeration cycle to provide mechanical cooling in the summer. Very importantly, heat pumps contain a valve which reverses the cycle during the heating season. As a result, the inside coil becomes the condenser and dispels heat to the ducted airstream within the house. At the same time the outside coil becomes the evaporator and "pumps" heat from the outdoor air.

Since heat pumps use electricity to provide for both heating and cooling, there is no need for a furnace or boiler. Their use should in general be limited to mild or at worst moderate heating climates (about

304 COOLING OF HOUSES

6,000 HDD) where electricity charges are low. See Section 16.3 for detailed information and examples on year-round heat pump performance.

Required Air Quantity

Use Formula 18F to determine the required air flow rate to meet a given cooling load.

$$\text{cfm} = \frac{q_{\text{tot}}}{1.10 \times (t_i - t_s)} \quad \text{Formula 18F}$$

where

- cfm   is the air flow requirement in cubic feet per minute
- q_{tot} is the total quantity of sensible heat needing to be removed, in Btu/h
- 1.10   is a constant based on the specific heat and density of air (see Formula 14D for details)
- t_i is the inside design temperature in °F
- t_s is the supply air temperature in °F

Example 18F: The living room in a house has a sensible cooling load (q_{tot}) of 6000 Btu/h. Determine the quantity of supply air if the inside design temperature (t_i) is 80°F and the supply air temperature (t_o) is 55°F.

Solution: Use Formula 18F.

$$\text{cfm} = \frac{q_{\text{tot}}}{1.10 \times (t_i - t_o)}$$

(Formula 18F)

$$= \frac{6000}{1.10 \times (80 - 55)}$$

$$= \frac{6000}{27.5}$$

$$= 218 \text{ cfm}$$

18.6 EVAPORATIVE COOLING

In dry climates such as in the Southwest, it is possible to use an evaporative cooling process which adds humidity to an airstream and lowers the dry-bulb air temperature. This is an "adiabatic" process since the total heat of the air (enthalpy) is unchanged. Figure 18.8 indicates locations in the United States with low humidity conditions which favor the use of evaporative coolers.

The adiabatic process is shown on the psychrometric chart in Figure 18.9 and can simply be thought of as a "trade" of sensible for latent heat. At condition "A," air is at an uncomfortable dry-bulb air temperature of 95°F and 15% relative humidity. Moisture is then added to the air bringing it to comfortable condition "B" which has a dry-bulb temperature of 75°F and 50% relative humidity.

Residential evaporative cooling units (colloquially known as "swamp coolers") are typically located centrally to minimize ductwork. They include a pump which supplies water to a saturated pad over which the supply air is blown. As a result, they can lead to bacterial contamination if they are not properly maintained.

Residential evaporative coolers can be seen on the roofs of many houses in dry, hot climates in the Southwest. A limitation for their use in existing houses is the fact that they are ineffective on the occasional days when the outdoor air does have a high relative humidity. To overcome this problem, in many new houses even in dry areas, the more energy-intensive compression refrigeration systems are installed. Others employ some form of assisted evaporative cooling such as using a rock bed under the house to store the coolness of the night air

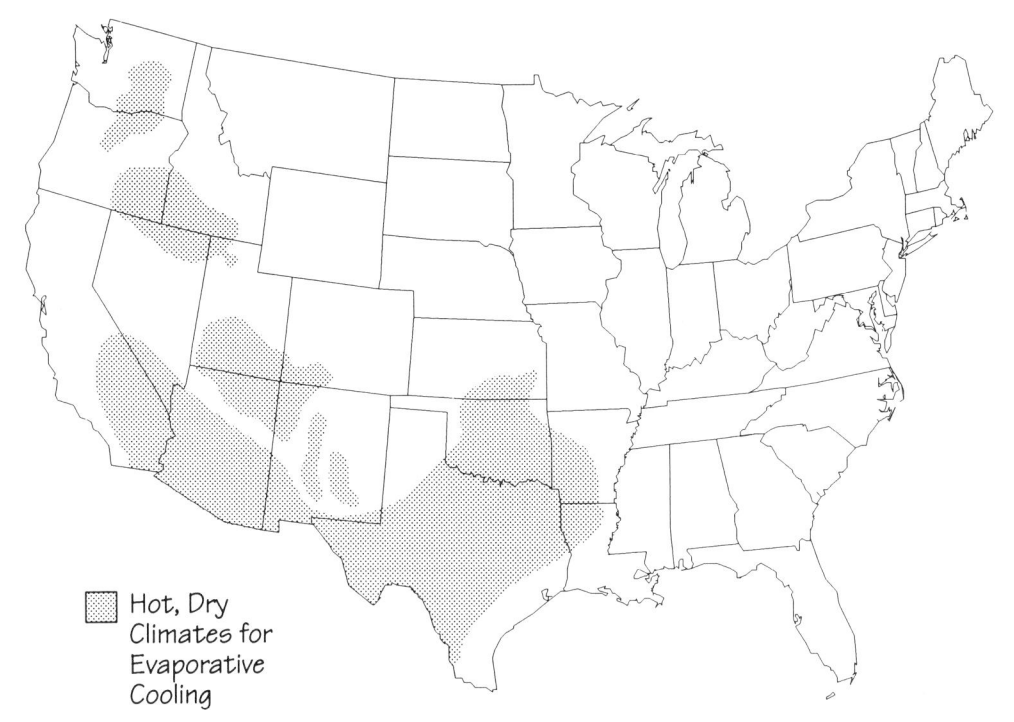

FIGURE 18.8 Arid regions that can often use evaporative cooling.

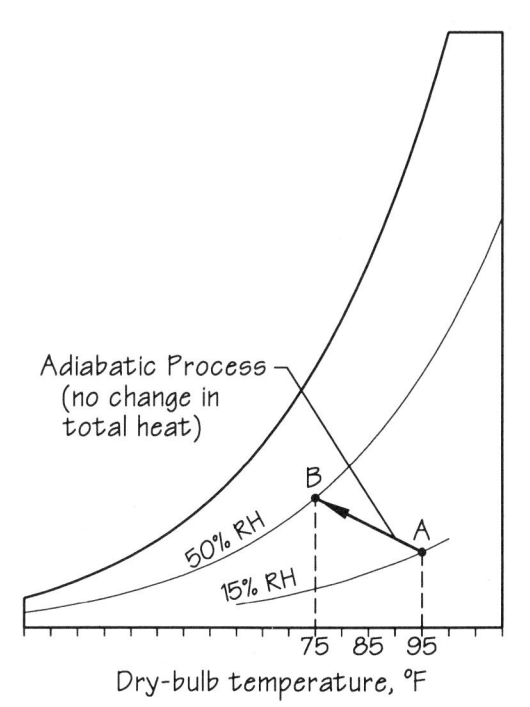

FIGURE 18.9 Evaporative cooling process—an exchange of sensible and latent heat.

which then can help to prechill supply air the following day.

18.7 RESIDENTIAL COOLING LOAD CALCULATIONS

When a room air conditioner is desired, it may be acceptable to size it using some local rule of thumb or simply heeding the recommendation of a salesperson. However, the design of a central air-conditioning system for a residence requires a much more detailed calculation procedure. Outlined on these pages is the simplified ASHRAE cooling load calculation method for Single Family Detached Residences.

The items that contribute to residential cooling loads (glass area, wall area, ceiling area, infiltration, etc.) are discussed below. The cooling loads resulting from these items are illustrated in a series of examples for the "Living Room" in "The House" if lo-

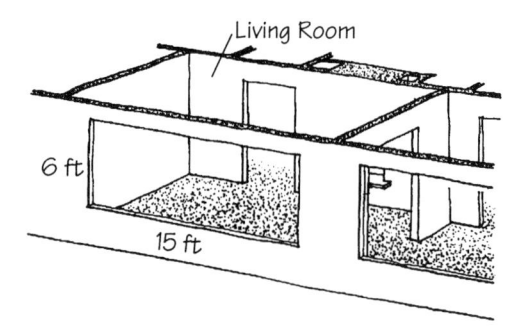

FIGURE 18.10 "The Living Room."

cated in St. Louis, Missouri, as depicted in Figure 18.10.

Glass and Window Area

Heat gain through window areas (both transmission heat and solar) is typically the largest load in residential cooling load applications, as computed using Formula 18G.

$$q_g = \text{GLF} \times A \quad \text{Formula 18G}$$

where

- q_g is the sensible cooling load due to glass in Btu/h
- GLF is the Glass Load Factor in units of Btu/h · ft^2 from Table 18.4 or 18.5
- A is the area of glass in ft^2

Provided in Tables 18.4 and 18.5 are GLF values for clear regular double glass and clear triple glass respectively, both with no inside shading. Table 18.5 can also be used with acceptable accuracy for clear low-E glazing. The *1993 ASHRAE Handbook of Fundamentals* also contains values for single glazing, heat absorbing double-glass, and various shade-producing elements including draperies, venetian blinds, and roller shades.

ASHRAE notes that values shown in Tables 18.4 and 18.5 for SE, SW and South apply for 40°N latitude. Values in the Tables should be increased by 30% for 48°N latitude, and decreased by 30% for 32°N latitude (linear interpolation should be used for intermediate latitudes).

Although ASHRAE doesn't say it, a review of solar heat gain factors for 24°N suggests that the values for 32°N latitude can also be used for lower latitudes in the United States. Only a very minor error will result unless there are large areas of southeast or southwest facing unshaded glass (which would be very poor design for such locations).

Example 18G: Determine the solar cooling load due to fully sun-lit glazing for the living room of "The House" based upon the following conditions:

TABLE 18.4 WINDOW GLASS LOAD FACTORS (GLF) (Btu/h·ft^2) SINGLE-FAMILY DETACHED RESIDENCES AT 40°N LATITUDE—REGULAR DOUBLE GLASS, NO INSIDE SHADING

Window Orientation	Outside Design Temperature (°F)					
	85	90	95	100	105	110
North	30	30	34	37	38	41
NE & NW	55	56	59	62	63	66
East and West	77	78	81	84	85	88
SE & SW	69	70	73	76	77	80
South	46	47	50	53	54	57
Horiz. Skylight	137	138	140	143	144	147

Source: Extracted with permission from the *1993 ASHRAE Handbook of Fundamentals*, Chapter 25, Table 3.

TABLE 18.5 WINDOW GLASS LOAD FACTORS (GLF) (Btu/h·ft²) SINGLE-FAMILY DETACHED RESIDENCES AT 40°N LATITUDE—CLEAR TRIPLE (OR LOW-E) GLAZING NO INSIDE SHADING

Window Orientation	Outside Design Temperature (°F)		
	85	90	95
North	27	27	30
NE & NW	50	50	53
East & West	70	70	73
SE & SW	62	63	65
South	42	42	45
Horiz. Skylight	124	125	127

Source: Extracted with permission from the *1993 ASHRAE Handbook of Fundamentals*, Chapter 25, Table 3.

Glass area of 90 ft² of low-E double glazing

Summer dry-bulb outside design temperature for St. Louis of 94°F (2.5% value from Table 2.7)

Room faces south

38°N latitude

Solution: Using Table 18.5, obtain the Glass Load Factor (GLF) for 40°N latitude, south orientation, and the closest outside design temperature which is 95°F (Note: also to be used in other examples for "The House"). The GLF value obtained is 45 Btu/h·ft².

Now modify the GLF to a 38°N latitude value (for St. Louis) by first reducing the original amount by 30% to obtain a value of 31.5 (45 × 0.7) for 32°N latitude.

Then by linear interpolation determine the result for 38°N latitude as follows:

$$31.5 + \left[\frac{38-32}{40-32} \times (45 - 31.5)\right]$$

$$31.5 + \left[\frac{6}{8} \times 13.5\right]$$

$$31.5 + 10.1 = 42 \text{ Btu/h} \cdot \text{ft}^2$$

It is interesting to note that the interpolated value of 42 for south-facing glass at 38°N latitude is less than the value of 45 for 40°N latitude. This is because the summer sun shines more directly on east and west facades at latitudes closer to the equator. Now use Formula 18G.

$$q_g = \text{GLF} \times A \quad \text{(Formula 18G)}$$

$$= 42 \times 90$$

$$= 3780 \text{ Btu/h}$$

This will be the solar cooling load for the south glass of the Living Room if it is unshaded. If, however, the glass area is partially shaded (i.e., by an overhang), then Shade Line Factors need to be employed, as described below.

Shading from Overhangs

Shading of window glazing by overhangs will reduce cooling loads in houses. The procedure that follows first determines the area of glazing which will be shaded by an overhang. It is then easy to determine the unshaded area that remains, and the solar heat gains that pertain to each area of glazing.

Table 18.6 provides Shade Line Factors (SLF) for overhangs which are used to determine shaded glass area using Formula 18H. Note that Formula 18H includes al-

308 COOLING OF HOUSES

TABLE 18.6 SHADE LINE FACTORS (SLF)[a]

Window Orientation	Latitude, Degrees North						
	24	32	36	40	44	48	52
East and West	0.8	0.8	0.8	0.8	0.8	0.8	0.8
SE and SW	1.8	1.6	1.4	1.3	1.1	1.0	0.9
South	9.2	5.0	3.4	2.6	2.1	1.8	1.5

[a]Values are averages for the 5 hours of greatest solar intensity on August 1.

Source: Extracted with permission from the *1993 ASHRAE Handbook of Fundamentals*, Chapter 25, Table 6.

lowance for a "separation" of opaque wall area above the south glazing and beneath the overhang. This is to allow for the entry of winter solar heat gain through the high portions of south-facing windows when solar altitude angles are low.

$$A_{sh} = [(SLF \times OH_w) - SEP] \times W_g$$

Formula 18H

where

A_{sh} is the area of shaded glass in ft^2
SLF   is the Shade Line Factor from Table 18.6
OH_w is the overhang width in feet
SEP   is the height of the overhang above the top of the glazing known as the "separation" in feet
W_g is the width of glazing being shaded in feet

The area of glass which is found to be shaded is then treated as if it receives the same solar heat gain as north glass. Formula 18I is then used to compute the area of unshaded glass which continues to receive the full amount of solar heat gain for its orientation.

$$A_{unsh} = A_{tot} - A_{sh} \quad \text{Formula 18I}$$

where

A_{unsh} is the area of unshaded glass in ft^2
A_{tot} is the total glass area under consideration in ft^2
A_{sh} is the area of shaded glass in ft^2

Example 18H: The south-facing glass area of the "Living Room" is actually shaded by a 1.83-ft wide overhang (OH_w) which extends out above a 1.00-ft high separation (S) as shown in Figure 18.11. The

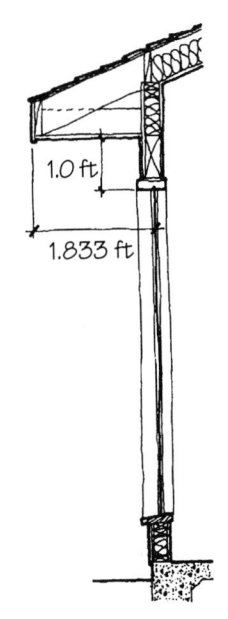

FIGURE 18.11 Shading from overhang, Example 18H.

total width of glazing (W_g) being shaded by the overhang is 15 ft, and the height of glazing is 6 ft. Compute the glazing heat gain of the room after consideration is given to the overhang.

Solution: Shade Line Factors (SLF) are obtained from Table 18.6. For south glass, SLFs of 3.4 for 36°N latitude and 2.6 for 40°N latitude are listed in Table 18.6. Interpolate linearly to obtain an SLF of 3.0 for south-facing windows at 38°N latitude. Then use Formula 18H to determine the area of shaded south-facing glass area in the "Living Room."

$$A_{sh} = [(SLF \times OH_w) - SEP] \times W_g$$

(Formula 18H)

$$= [(3.0 \times 1.833) - 1] \times 15$$

$= 67.5$ ft^2 is the shaded glass area

And the unshaded glass area is found by using Formula 18I. The total window area (A_{tot}) is 90 ft^2 (15 × 6).

$$A_{unsh} = A_{tot} - A_{sh} \quad \text{(Formula 18I)}$$

$$= 90 - 67.5$$

$= 22.5$ ft^2 is the unshaded glass area

Formula 18G can now be used twice to compute the solar heat gains for the shaded and unshaded glass areas. For the area of shaded glass (A_{sh}), the GLF is 28 Btu/h · ft^2 (the value obtained from Table 18.5 for an outside design temperature of 95°F and for north (shaded) glass of 30 Btu/h · ft^2 modified for 38°N latitude).

$$q_g = GLF \times A \quad \text{(Formula 18G)}$$

$$= 28 \times 67.5$$

$$= 1890 \text{ Btu/h}$$

And for the area of unshaded glass (A_{unsh}), the GLF is still 42 Btu/h · ft^2 (see previous example).

$$q_g = GLF \times A \quad \text{(Formula 18G)}$$

$$= 42 \times 22.5$$

$$= 945 \text{ Btu/h}$$

These results for the shaded and unshaded glass areas are then added to get the room total of 2835 Btu/h. This compares to the result of 3780 Btu/h for the unshaded window.

Walls, Roofs, and Other Exposed Opaque Surfaces

Cooling loads due to exposed opaque building elements such as exterior walls, doors, ceilings and roofs, and exposed floors are computed using Formula 18J (which is identical to Formula 7D):

$$q_{cond} = U \times A \times CLTD \quad \text{Formula 18J}$$

where

q_{cond}	is the cooling load due to heat gain through an opaque surface in Btu/h
U	is the component U-factor in Btu/ft^2 · °F · h
A	is the component area in ft^2
CLTD	is the appropriate cooling load temperature difference in °F from Table 18.7

In order to use Table 18.7, first obtain the summer design dry-bulb temperature (for the 2.5% condition) for the location of interest and the mean daily dry-bulb temperature range from Table 2.7. Designations for daily range are: L (low: less than 16°F); M (medium: 16–25°F); or H (high: greater than 25°F).

TABLE 18.7 CLTD VALUES FOR SINGLE-FAMILY DETACHED RESIDENCES[a]

Daily Temperature Range	85 L	85 M	90 L	90 M	90 H	95 L	95 M	95 H	100 M	100 H	105 M	110 H
All walls and doors:												
North	8	3	13	8	3	18	13	8	18	13	18	23
NE and NW	14	9	19	14	9	24	19	14	24	19	24	29
East and West	18	13	23	18	13	28	23	18	28	23	28	33
SE and SW	16	11	21	16	11	26	21	16	26	21	26	31
South	11	6	16	11	6	21	16	11	21	16	21	26
Roofs and ceilings: Attic or flat built up	42	37	47	42	37	51	47	42	51	47	51	56
Floors and ceilings: Under conditioned space, over unconditioned room, over crawl space	9	4	12	9	4	14	12	9	14	12	14	19
Partitions: Inside or shaded	9	4	12	9	4	14	12	9	14	12	14	19

Design Temperature (°F) across top.

[a] Cooling Load Temperature Differences for single-family detached houses, duplexes, or multifamily with both east and west exposed walls or only north and south exposed walls in degrees F.

Source: Extracted with permission from the *1993 ASHRAE Handbook of Fundamentals*, Chapter 25, Table 1.

Example 18I: Determine the cooling loads for the opaque elements of the living room in "The House" based upon the following:

Opaque wall area of 70 ft^2 facing south; $U = 0.05$

Ceiling area of 300 ft^2; $U = 0.03$

Summer design dry-bulb temperature of 95°F

Daily temperature range of 18°F (Medium)

Solution: Use Formula 18J to compute the load for each element. Obtain applicable CLTD values from Table 18.7.

Opaque Walls: The CLTD value from Table 18.7 for a dry-bulb design temperature of 95°F, and a medium (M) daily temperature range for south walls is 16°F.

$$q_{cond} = U \times A \times CLTD$$

(Formula 18J)

$$= 0.05 \times 70 \times 16$$

$$= 56 \text{ Btu/h}$$

Roof: The CLTD value from Table 18.7 for a dry-bulb design temperature of 95°F, and a medium (M) daily temperature range for roofs is 47°F.

$$q_{cond} = U \times A \times CLTD$$

(Formula 18J)

$$= 0.03 \times 300 \times 47$$

$$= 423 \text{ Btu/h}$$

It is important to realize that the cooling load from these opaque elements (exterior wall and roof) add up to only 479 Btu/h which is less than 20% of the cooling load due to the shaded window (2835 Btu/h) as determined in Example 18G.

Infiltration

Infiltration of warm outside air also needs to be considered in a cooling load calculation. Use Formula 18K.

$$q_{inf} = 0.018 \times ACH \times V \times (t_o - t_i)$$

Formula 18K

where

q_{inf} is sensible infiltration heat gain, in Btu/h
0.018   is the heat capacity of air (at warm outside air temperatures) in Btu/°F·ft³
ACH   is the air change per hour rate as a function of outside design temperature from Table 18.8
V is the room volume in ft³
t_o is the outdoor design temperature in °F
t_i is the indoor design temperature in °F

Table 18.8 provides summer infiltration air change rates for various construction classes. Summer ACH rates are low when compared to winter values due to lower wind speeds along with a smaller difference between indoor and outdoor air temperatures.

Example 18J: Determine the infiltration cooling load for the living room in "The House" based upon the following:

Room volume (V) of 2400 cubic feet (ft³)
Outdoor design temperature (t_o) of 95°F
Indoor design temperature (t_i) of 75°F
The construction "airtightness" class shall be considered to be "Medium"

Solution: Obtain from Table 18.8 an air change per hour rate (ACH) of 0.50 for an outdoor design temperature of 95°F and Medium construction. Then use Formula 18K.

TABLE 18.8 SUMMER INFILTRATION AIR CHANGES PER HOUR (ACH) RATES AS A FUNCTION OF OUTDOOR DESIGN TEMPERATURE[a]

Airtightness Class	Outdoor Design Temperature (°F)					
	85	90	95	100	105	110
Tight	0.33	0.34	0.35	0.36	0.37	0.38
Medium	0.46	0.48	0.50	0.52	0.54	0.56
Loose	0.68	0.70	0.72	0.74	0.76	0.78

[a]Values for 7.5 mph wind and indoor temperature of 75 degrees F.
Source: Extracted with permission from the *1993 ASHRAE Handbook of Fundamentals*, Chapter 25, Table 8.

$$q_{inf} = 0.018 \times ACH \times V \times (t_o - t_i)$$

(Formula 18K)

$$= 0.018 \times 0.50 \times 2400 \times (95 - 75)$$

$$= 432 \text{ Btu/h}$$

Internal Heat Gains

Internal heat gains in houses come from people, lights, and appliances. It normally is difficult to accurately quantify these gains. A "rule of thumb" which can be used for a typical house is to assume heat gain of approximately 2200 Btu/h for the entire house (based on 1000 Btu/h from people and 1200 Btu/h from lights and appliances). This total can then be apportioned to the various rooms based on either a percentage of floor area or judgment.

Example 18K: Determine the cooling load for the living room in "The House" due to internal heat gains.

Solution: The "rule of thumb" for internal heat gains of 2200 Btu/h for typical houses will be utilized. The living room comprises about 20% of the floor area of "The House." Nonetheless, since more people and more appliances (e.g., television) are likely to be there, 30% of the total internal heat gain will be assumed for the living room, or 660 Btu/h.

Latent Cooling Loads

House occupancy also results in moisture (latent heat) from infiltration, cooking, bathing, laundry, and the occupants themselves. Since precise calculation of these factors is impossible, it is common to assume that the latent cooling load (q_{lcl}) for houses will be 30% of the computed sensible cooling load as computed by Formula 18L.

$$q_{lcl} = SCL \times 0.30 \quad \text{Formula 18L}$$

where

q_{lcl} is the latent cooling load in Btu/h
SCL is the sensible cooling load in Btu/h
0.30 is the assumption that the latent cooling load of residences will be 30% of the sensible cooling load

By the way, the 30% latent load assumption results in a sensible-heat ratio of 0.77 (1/1.3) which means that latent heat is assumed to be 23% (100 − 77) of the cooling load. A slightly higher latent heat assumption of 35% or more may be used for cooling system design and equipment selection for houses in very humid climates.

Example 18L: Determine the latent cooling load for the "Living Room."

Solution: First sum up the sensible cooling loads for the living room as computed in the previous examples in this section and as shown in the following summary:

Sensible Cooling Load Component	Total Load (Btu/h)
Windows (shaded)	2835
Opaque wall	56
Roof	423
Infiltration	432
Internal heat gains	660
Total (SCL):	4406 or 4400

The latent cooling load can now be computed by using Formula 18L.

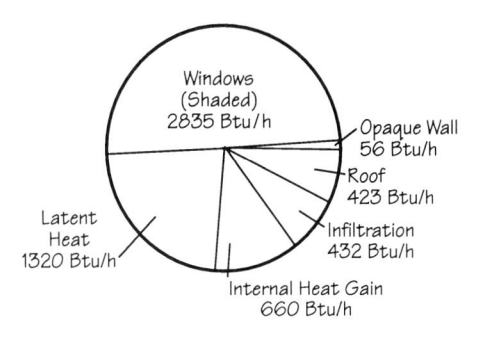

FIGURE 18.12 Cooling load summary for "The Living Room."

Windows (Shaded) 2835 Btu/h
Opaque Wall 56 Btu/h
Roof 423 Btu/h
Infiltration 432 Btu/h
Internal Heat Gain 660 Btu/h
Latent Heat 1320 Btu/h

q_{lcl} = SCL × 0.30   (Formula 18L)

= 4400 × 0.30

= 1320 Btu/h of latent heat gain

Living Room Load Summary

The cooling load calculation for the "Living Room" is now complete and a summary of the loads is provided in Figure 18.12. Note the significant cooling load impact of the south-facing windows even though they are partially shaded.

The total room cooling load of 5720 Btu/h represents a need for approximately 1/2 ton of refrigeration. Now that the cooling load is known, the required supply air flow to the "Living Room" to offset the sensible heat gains can be determined.

Example 18M: Determine the required supply air flow of cool air from a central air-conditioning system to meet the sensible cooling load of 4400 Btu/h in the living room. (Remember: the latent cooling load is removed at the cooling coil when the return air passes over it.) Assume an indoor setpoint temperature of 78°F (t_i) and a supply air temperature of 58°F (t_s).

Solution: Use Formula 18F.

$$\text{cfm} = \frac{q_{tot}}{1.10 \times (t_i - t_s)}$$

(Formula 18F)

$$= \frac{4400}{1.10 \times (78 - 58)}$$

= 200 cfm

18.8 MECHANICAL COOLING OF "THE HOUSE"

Example 18N: Determine the supply air requirements for the various rooms of "The House."

Solution: Using similar calculation procedures to that used for the Living Room above, the various cooling loads for "The House" were computed as follows:

Room	Sensible Cooling Load (Btu/h)	Latent Cooling Load (Btu/h)	Total Cooling Load (Btu/h)
Dining room	3190	960	4150
Living room	4400	1320	5720
Bedroom #1	3890	1170	5060
Exterior bathroom	1310	390	1700
Bedroom #2	2290	690	2980
Vest./hall	670	200	870
Family room	1340	400	1740
Kitchen	1800	540	2340
Totals:	18,890	5670	24,560

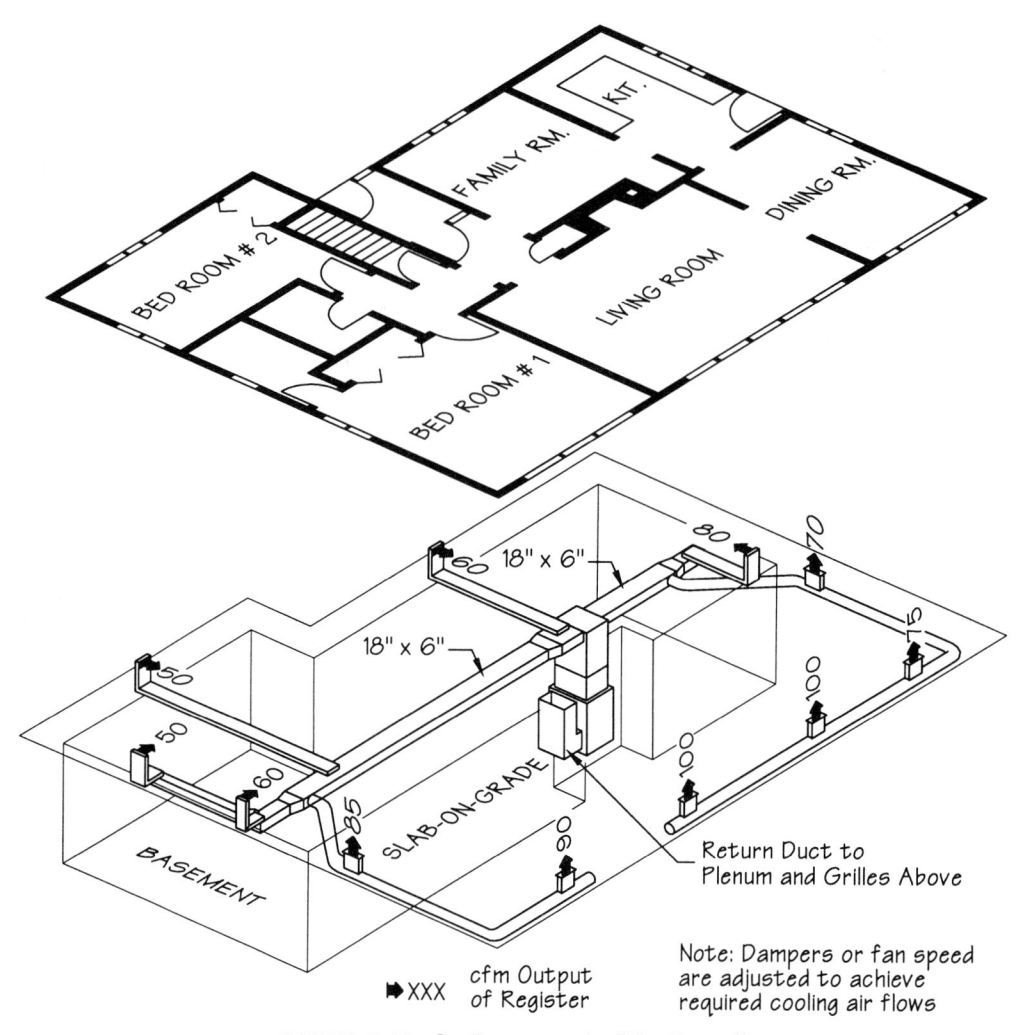

FIGURE 18.13 Cooling system for "The House."

The total cooling load indicates a need for cooling equipment with a capacity of about 2 tons (24,000 Btu/h) and a sensible heat ratio (SHR) of 0.77 since the latent load was assumed to be 30% of the sensible total. In choosing air conditioning equipment do not oversize to be on the safe side. Such an approach can be counterproductive since oversized equipment may cycle too frequently to properly address the latent cooling load.

Now that sensible cooling loads for all spaces are known, supply air requirements for cooling can be computed in a manner similar to that detailed for the Living Room. Rounded results are noted in the following summary which also lists warm air supply air requirements (for winter) which were computed in Section 13.11.

As indicated below, the required cooling and heating cfm are fairly close (only the bathroom has a need for a greater supply of

SUMMARY: ROOM-BY-ROOM SUPPLY AIR REQUIREMENTS

Room	Required Cooling (cfm)	Required Heating (cfm)
Dining room	145	170
Living room	200	250
Bedroom #1	175	200
Exterior bathroom	60	40
Bedroom #2	100	150
Vest./hall	30	50
Family room	60	80
Kitchen	80	100
Total:	850	1040

cool air) for "The House" when located in St. Louis. Therefore, the air delivery system for heating (shown in Figure 13.12) can also be used for cooling with only minor adjustments as shown in Figure 18.13.

Often, house construction and climate will result in widely different summer and winter air flow requirements for some or all spaces. As a result, systems must be designed to operate flexibly. This can be accomplished by having furnaces with multi-speed blowers and dampers within ductwork sections and at air outlets.

19

APPLIED PSYCHROMETRICS

Surely you have noticed fogging of a mirror after a shower, condensation on the inside of a single-glazed window on a cold day, or moisture on the outside of a cold drink glass on a hot humid day. This chapter will show how the same very simple basic process is used by cooling coils to dehumidify air. Also discussed are alternative methods to control humidity such as the use of desiccants.

Did you know that air at dry-bulb temperatures ranging from 60 to 108°F can actually contain the same amount of total heat? It can when the amount of latent heat (humidity) is also considered. This is known as "enthalpy"—a term which is fundamental to many HVAC processes such as evaporative cooling.

Throughout this book (and in particular Section 9.2) we have seen how the properties of air-moisture mixtures can be plotted on a psychrometric chart. The following pages will illustrate how the "psych" chart is actually used by a designer to plot HVAC processes for design conditions in order to supply air which can provide for human comfort.

19.1 AIR CONDITIONING PROCESSES

The basic air-moisture processes of air conditioning are illustrated on the psychrometric chart in Figure 19.1. These processes will be evaluated to illustrate how HVAC systems and their apparatus (e.g., cooling coils) are designed to supply air which can satisfy desired indoor conditions of temperature and humidity.

For simplicity, the processes shown on Figure 19.1 such as "cooling and dehumidification" are represented as straight lines. Realize, however, that the actual HVAC processes involved to achieve such a change in the conditions of air are often more complex and may involve several steps as discussed below.

A basic psychrometric chart was shown in Figure 9.5 and explained. Figure 19.2

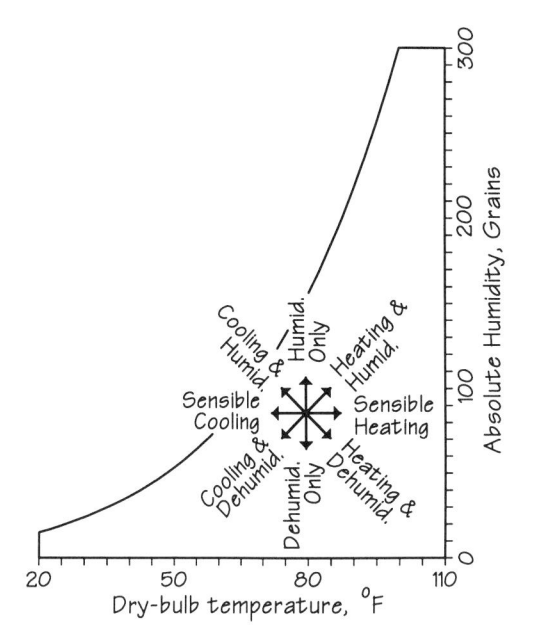

FIGURE 19.1 Air conditioning processes.

provides a far more detailed psychrometric chart for sea level. A few items to note on the detailed psychrometric chart.

Indicated above the top curve of the psychrometric chart are values of "enthalpy" at saturation (meaning for 100% RH) in Btu per pound. Lines of Constant Enthalpy are then shown on the Psychrometric Chart as diagonal lines to the lower right.

On the vertical axis, grains of moisture per pound of dry air range from 0 grains at the bottom to 300 at the top. Also indicated along the vertical axis are "humidity ratio" values ranging from 0 at the bottom to more than 0.040 lb of moisture per pound of dry air. Note that for air with 70 grains, the humidity ratio is 0.010 since there are 7000 grains in a pound (70/7000 = 0.010).

The volume of air per pound depends upon temperature and humidity. These values are shown on the chart as steeply sloped (almost vertical) straight lines. For example, the line which begins along the base line near 75°F dry-bulb, and extends up toward 65°F wet-bulb, represents a volume of air of 13.5 ft^3/lb.

The sensible heat factor lines indicated to the left on the figure are used to plot HVAC processes as shown in Example 19C.

19.2 TOTAL HEAT (ENTHALPY)

Enthalpy? A strange and often difficult concept to grasp. The term enthalpy refers to the "total heat" (both sensible and latent) present in an air–moisture mixture. Sensible heat is heat in the air as measured by a conventional thermometer. Latent heat refers to the heat which was required to produce water vapor (moisture) from liquid water.

Enthalpy is expressed relative to a value of 0 Btu/lb for dry air at a dry-bulb temperature of 0°F. Then, as the dry-bulb temperature increases, the enthalpy (due to sensible heat) simply increases at a rate based on the specific heat of air (which remains relatively constant at 0.24 Btu/lb·°F). Therefore, the enthalpy of 50°F "dry" air (0% relative humidity) would simply be 12 Btu/lb (i.e., 50 × 0.24), which can be verified on the Psychrometric Chart (see Figure 19.2).

As indicated in Figure 19.3, 70°F db air when saturated at 100% RH (Point A) has about the same enthalpy as 84°F db air at 50% RH (Point B), and 97°F db air at 25% RH (Point C). Each air–moisture mixture has an enthalpy of about 34.1 Btu/lb, although the contributions and ratios of sensible heat and latent heat are of course different.

The enthalpy of a particular air–vapor mixture can be calculated by using Formula 19A.

$$h = (0.24 \times t) + [W \times (1061 + 0.45t)]$$

Formula 19A

where

h is the enthalpy of the air–vapor mixture in Btu/lb of dry air

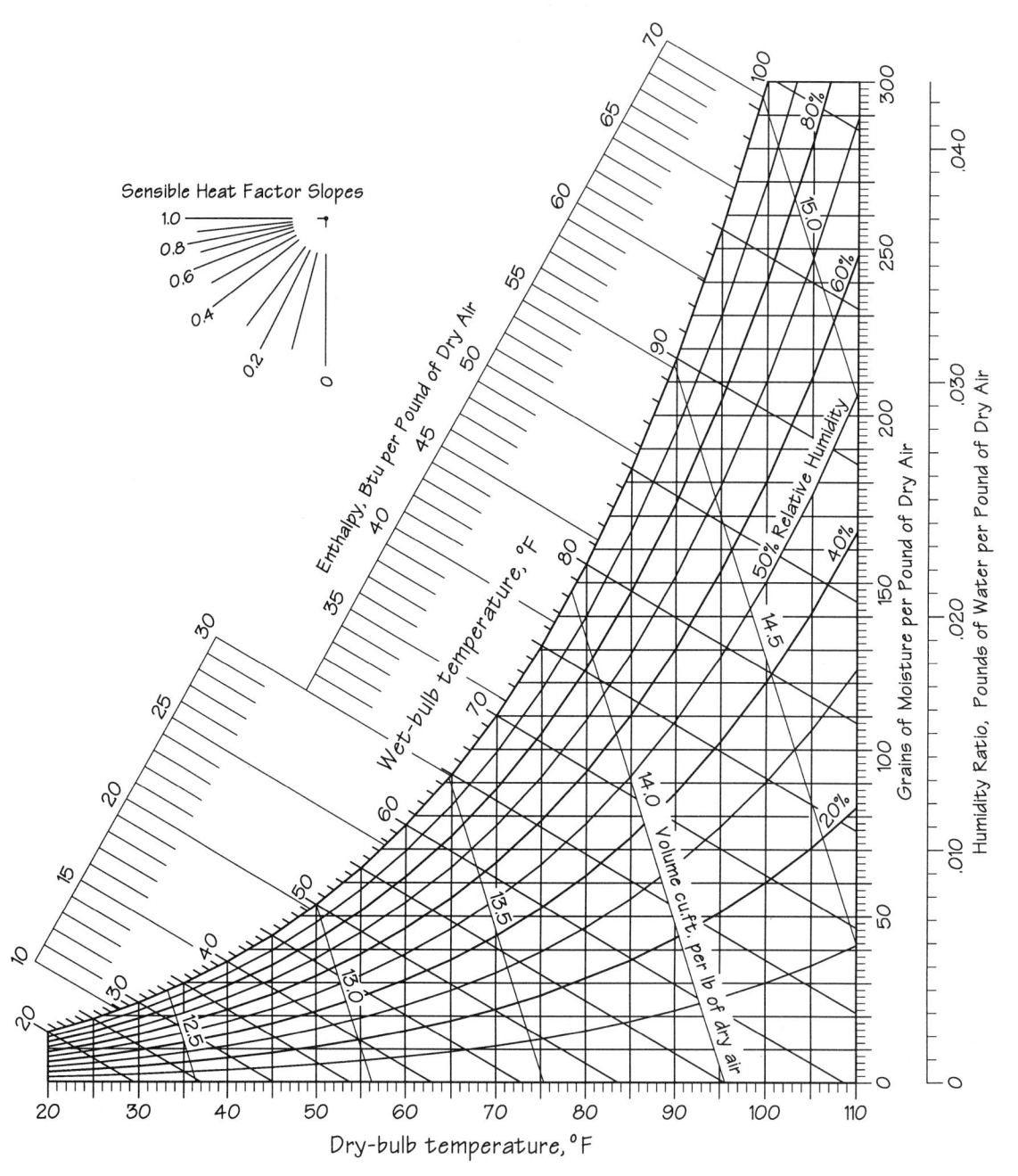

FIGURE 19.2 Psychrometric chart for sea level.

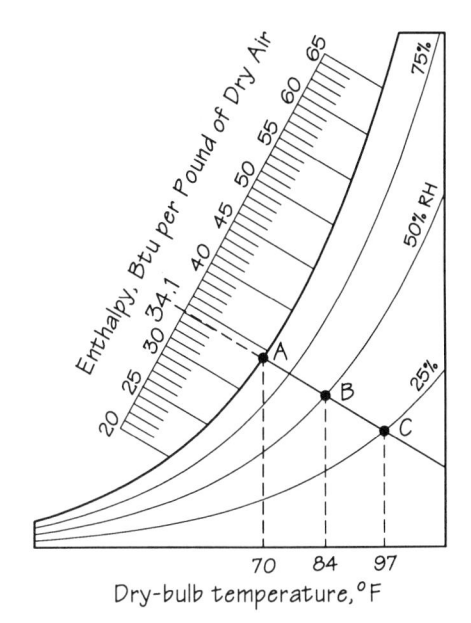

FIGURE 19.3 Air–moisture mixtures of equal enthalpy.

0.24 is the specific heat of air in Btu/lb·°F (at normal HVAC system temperatures)
t is the dry-bulb temperature in °F
W is the humidity ratio in lb of water per lb of dry air as defined below
1061 is the latent heat value of water in Btu/lb
0.45 is the specific heat of water vapor in Btu/lb·°F

Example 19A: Determine the enthalpy of 72°F db air with 50% RH.

Solution: Use Formula 19A. Note on the Psychrometric Chart that the humidity ratio for this temperature and relative humidity is 0.008.

$$h = (0.24 \times t) + [W \times (1061 + 0.45t)]$$
(Formula 19A)

$$= (0.24 \times 72) + (0.008 \times 1093.4)$$

$$= 17.28 + 8.75$$

$$= 26.03 \text{ Btu/lb of dry air}$$

Values of enthalpy shown in Table 19.1 were computed using Formula 19A. Compare these results to enthalpy values on the psychrometric chart (Figure 19.2). Note how air at these five conditions which vary greatly in temperature and relative humidity have enthalpies which are nearly the same.

19.3 HUMIDITY CONTROL

For human comfort it is desirable to limit relative humidity during the cooling season to 50–60% in most types of buildings. Regardless of whether humidity is being controlled by a central system or an individual air conditioner, the basic process is the same. Namely, when humid air passes over cool surfaces which are below the dew point temperature of the circulating air, condensation will take place and the relative humidity of the air will decrease.

In many air-conditioning systems, the

TABLE 19.1 ENTHALPY OF AIR

Air		Sensible Heat (Btu/lb)	Latent Heat (Btu/lb)	Enthalpy (Total Heat) (Btu/lb)
°F	RH			
60	100%	14.40	12.01	26.41
64	80%	15.36	10.90	26.26
72	50%	17.28	8.75	26.03
80	30%	19.20	7.70	26.90
108	0%	25.92	0.00	25.92

cool surface which dehumidifies the air is the evaporator coil itself. In perimeter cooling devices such as fan coil units, the cool surface is the chilled water coil. In either case, the liquid condensate must be collected in a pan and drained off. The only exception is in very small window units where the condensed moisture is simply blown outside of the unit by the small condenser fan.

The dew point temperatures for various dry-bulb air temperature and relative humidity conditions are presented in Table 19.2. For example, the table shows that 78°F air with a relative humidity of 50% has a dew point temperature of 58°F. Therefore, a coil which has a surface temperature below that temperature will cause moisture to condense out of any air which comes into contact with it. Cooling coils typically have temperatures ranging from 47 to 55°F.

Desiccant Dehumidification

As discussed above, humidity is commonly controlled by passing supply air over cooling coils which have surface temperatures lower than the dew point temperature of the air. Such an approach works well for most applications. Nevertheless, in special circumstances where very low relative humidity is required (e.g., electronics manufacturing or pharmaceutical preparation) a different approach is needed. Otherwise the cooling coil would need to have very low temperatures (approaching 32°F) which could result in condensed moisture freezing on the coil. In addition, once the dehumidification has occurred a large amount of energy would be needed to reheat the supply air to an acceptable temperature.

The alternative approach which can be used for dehumidification uses materials known as "desiccants." A desiccant is a liquid or solid which has an affinity to absorb water or water vapor and thus act as a drying agent. The most common type of desiccant dehumidifier uses a rotating wheel which contains an absorbent material (usually lithium bromide). As the wheel rotates, it absorbs moisture from the supply air stream in one chamber. The wheel then rotates so that the moisture-laden absorbent material enters a second chamber where a hot air stream is blown over it to remove moisture from the wheel (see Figure 19.4). In the process the desiccant is "regenerated," which allows for its continued use.

19.4 COOLING COILS

To be effective in removing heat and moisture from air, cooling coils have to operate at a suitable temperature while having a significant amount of surface area in contact with the circulating air being conditioned.

Cooling coils are most effective if they are several rows deep to increase the contact area (known as a low "bypass fac-

TABLE 19.2 DEW POINT TEMPERATURES (°F)

Room or Return Air Temp. (°F)	\multicolumn{5}{c}{Relative Humidity (RH)}
	50%	60%	70%	80%	90%
74	54	59	64	68	72
76	56	61	65	69	73
78	58	63	67	72	75
80	59	65	70	73	77
82	61	67	72	75	78

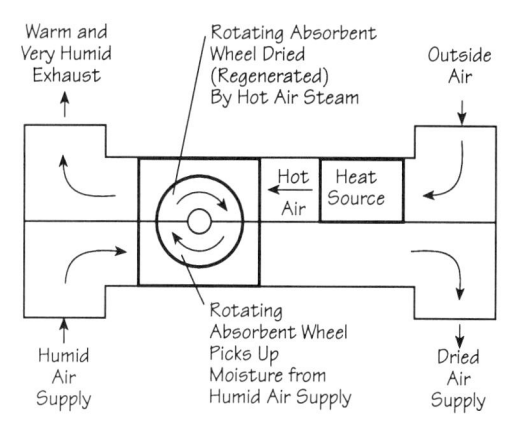

FIGURE 19.4 Desiccant dehumidification.

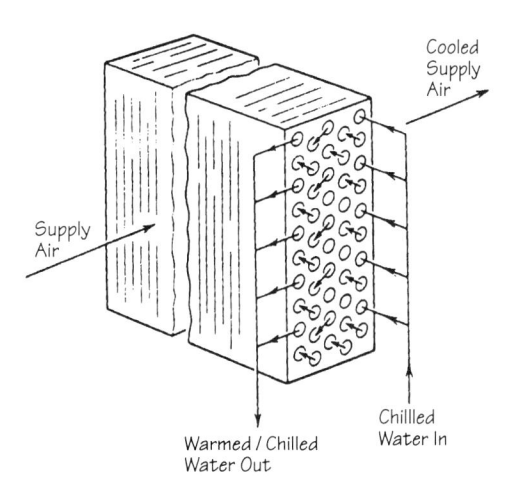

FIGURE 19.5 Cooling coil with counterflow air flow.

tor''—see below). Cooling coils are typically available with 2, 4, 6, or 8 rows. A cooling coil is shown in Figure 19.5 with a counterflow heat transfer arrangement where the cold water enters the coil at the point where the coldest air is leaving the coil.

Cooling coils are typically selected by computerized analysis based on design cooling loads and system operating parameters (water supply and return temperatures, allowable pressure drop due to friction, etc.). Yet an understanding of the psychrometric fundamentals of coil performance and terminology is helpful to understanding the data and computer selections. Hence their inclusion in this section.

Bypass Factor (BF)

The fraction of the air which passes through a coil without being conditioned (i.e., not cooled or dehumidified), is known as the bypass factor. Bypass factors depend upon apparatus design and the velocity of the airstream. Typical bypass factors for various applications are noted in Table 19.3.

Apparatus Dewpoint (adp)

Heat exchange conditions in air-conditioning processes vary as the air passes through the equipment (i.e., the first row of a cooling coil changes the condition of air which comes into contact with the second row and so on). For simplicity sake the term ''effective surface temperature'' (EST) is considered to be the constant temperature which would produce the same conditioning result as the actual processes in the apparatus. With regard to equipment which cools and dehumidifies air, the EST is commonly referred to as the apparatus dewpoint (adp) of the coil, and is used in calculation procedures to determine required supply air quantities.

19.5 HEAT RATIO TERMINOLOGY

The goal of an air-conditioning process is actually rather simple, that is, to produce and deliver air to a space at a temperature and humidity which will create comfortable conditions for the occupants.

However, to design a system to do this one needs to know the properties of the air which will enter the cooling apparatus (usually a mixture of recirculated room air and outside ventilation air) and the quantity and ratios of sensible and latent heat in the air

TABLE 19.3 TYPICAL BYPASS FACTORS

Type of Cooling Load Application	Coil Bypass Factor
Residential scale loads	0.30 to 0.50
Small commercial loads (small retail shop)	0.20 to 0.30
Typical commercial buildings	0.10 to 0.20
Buildings with high internal sensible heat loads	0.05 to 0.10
100% outdoor air applications	0 to 0.10

Source: Courtesy of Carrier Corporation.

being conditioned. The following factors need to be determined and applied.

Room Sensible Heat Factor

The "room sensible heat factor" (RSHF) (also known as the room sensible-heat ratio) is found by using Formula 19B. A discussion of applicable sensible and latent heat loads can be found as follows: internal loads from lights, people, and equipment (Chapter 4); heat gains from outside ventilation and/or infiltration air (Chapter 6); and for envelope-related solar cooling loads (Chapter 7).

$$\text{RSHF} = \frac{\text{RSH}}{\text{RSH} + \text{RLH}} \quad \text{Formula 19B}$$

where

RSHF   is the room sensible heat factor
RSH   is the room sensible heat in Btu/h
RLH   is the room latent heat in Btu/h

Example 19B: Determine the room sensible heat factor (RSHF) of a classroom in New York City which has 40,000 Btu/h of room sensible heat (RSH) and 10,000 Btu/h of room latent heat (RLH).

Solution: Use Formula 19B.

$$\text{RSHF} = \frac{\text{RSH}}{\text{RSH} + \text{RLH}}$$

(Formula 19B)

$$= \frac{40,000}{40,000 + 10,000}$$

$$= 0.80$$

This means that 0.80 (or 80%) of the room cooling load is sensible heat.

19.6 GRAPHING HVAC PROCESSES

The room sensible heat factor is used to determine supply air requirements to satisfy a given set of conditions. This will be illustrated in Example 19C, which graphs a typical cooling process on a psychrometric chart to determine the required air flow in cfm, and the total heat which must be removed in Btu/h.

Example 19C: Determine the required flow of cool air (in cfm) to meet the classroom cooling load described in Example 19B and based on the following:

Indoor Design Conditions (t_A): 78°F db, 50% RH

Outdoor Design Conditions (t_B): 89°F db, 73°F wb (the 2½% design condition for New York City from Table 2.7)

Ventilation: 25% outside air

Room sensible heat (RSH) of 40,000 Btu/h and room latent heat (RLH) of 10,000 Btu/h. This results in a RSHF of 0.80 (see Example 19B) and a room total cooling load (q_{tot}) of 50,000 Btu/h

Neglect the possible influences of fan and duct heat gain and duct leakage

Solution: The psychrometric chart is used to plot the conditioning process required. The goal of the graphical procedure is to determine the amount of total heat (enthalpy) which the cooling coil must remove in conditioning the supply air.

Step 1: On a psychrometric chart, locate Point A (which represents the indoor design conditions) and Point B (which represents the outdoor design conditions). Connect these points by a line as shown in Figure 19.6a.

Step 2: Outside ventilation air is stated to be 25%. Now locate Point C which represents the entering air condition of the air going into the cooling apparatus. Point C is 25% or one-quarter of the distance along line A-B from Point A to Point B as shown in Figure 19.6b.

The dry-bulb temperature at Point C (db_C) can also be determined by Formula 19C.

$$db_C = (db_A \times .A) + (db_B \times .B)$$

Formula 19C

where

db_C is the dry-bulb temperature of the mixed air in °F

db_A is the inside design dry-bulb temperature in °F

.A   is the fraction of inside air being recirculated

db_B is the outside design dry-bulb temperature in °F

.B   is the fraction of outside ventilation air being mixed with the recirculated air

For this example the dry-bulb temperature of the mixed air (db_C) can be determined by Formula 19C as follows:

$$db_C = (db_A \times .A) + (db_B \times .B)$$

(Formula 19C)

$$= (78 \times .75) + (89 \times .25)$$

$$= 80.8°F$$

Step 3: Then draw line "X" from Point A to the left through the rest of the chart at a slope representing a sensible heat factor (SHF) of 0.80 (e.g., parallel to the 0.80 SHF marking found on the chart) (Figure 19.6c). The supply air conditions must fall somewhere on this line "X" which represents the proportion of sensible and latent room cooling loads.

Step 4: Now draw freehand line "Y" from Point C which represents the conditioning process which the air undergoes while passing over the cooling coil (Figure 19.6d). This line first goes to the left horizontally as only sensible cooling takes place. Then as the cooling air stream approaches saturation (above about 95% RH) a portion of the air begins to reach the dewpoint temperature and contact with the cool coil surface causes moisture (latent heat) to condense out. The line then curves downward in the region of very high relative humidity (about 95% RH) until it intersects with the sloping sensible heat factor line "X" drawn in Step 3. The intersection is Point D, a temperature of 55°F which is the supply air temperature.

Step 5: Values of constant enthalpy lines

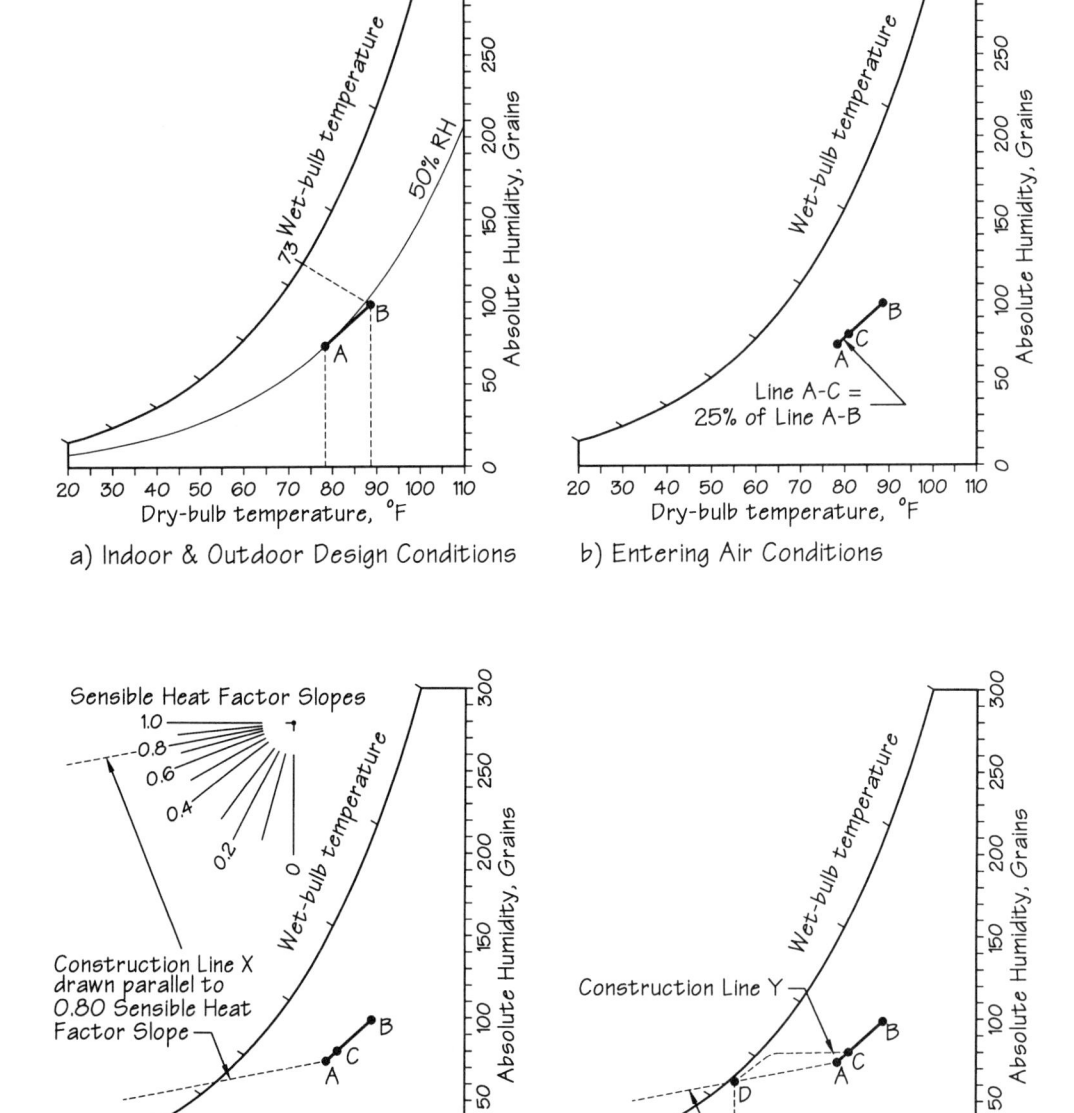

324 APPLIED PSYCHROMETRICS

a) Indoor & Outdoor Design Conditions

b) Entering Air Conditions

c) Sensible Heat Factor, Line X

d) The Cooling Process, Line Y

FIGURE 19.6 Psychrometrics for Example 19C.

perature (t_A) is 78°F and the supply air temperature (t_B) is 55°F.

$$\text{cfm} = \frac{q_{tot}}{1.10 \times (t_A - t_B)}$$

(Formula 6C)

$$= \frac{50,000}{1.10 \times (78 - 55)}$$

$$= 1980 \text{ cfm}$$

Step 7: And finally, the total heat removed by the coil can be determined by using Formula 6E and the enthalpies of the air which enters the cooling coil at Point C (H_C) and leaves the coil at Point D (H_D) at the supply air temperature.

$$q_{oa\text{-}tot} = 4.5 \times \text{cfm} \times (H_C - H_D)$$

(Formula 6E)

$$= 4.5 \times 1980 \times (32.1 - 23.0)$$

$$= 81,080 \text{ Btu/h}$$

Congratulations! If you were able to follow Example 19C you should now be able to use the psychrometric chart and applicable formulas to solve a wide range of HVAC design applications. You also should be able to refine the analysis by considering additional adjustments, for example, a degree or two of heat picked up in the return air plenum, another few degrees of heat from supply and return air fans, and perhaps another degree of heat pickup by supply ducts passing through the plenum space.

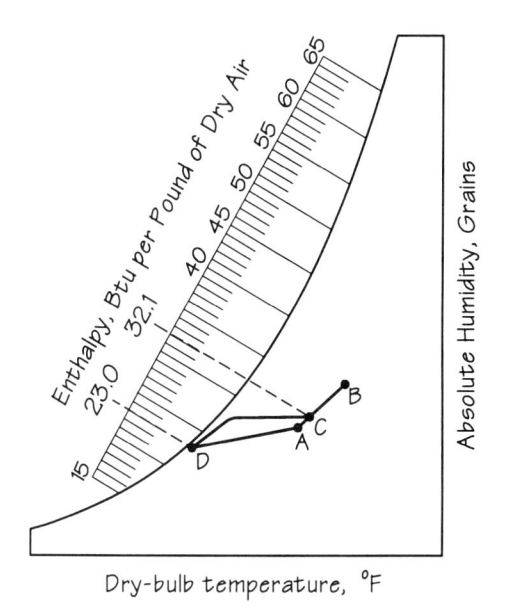

FIGURE 19.7 Total heat removed by the sensible cooling and dehumification process.

which slope downward to the right are indicated above the top curved line (see Figure 19.7). The enthalpy of the air which enters the cooling coil (Point C) can now be seen to be 32.1 Btu per pound of dry air. The enthalpy of the air which exits the cooling coil is found to be 23.0 Btu per pound of dry air. The difference is the amount of total heat which must be removed by the cooling coil.

Step 6: Formula 6C can now be reworked to determine the required air flow in cfm. The room total cooling load (q_{tot}) is 50,000 Btu/h. The indoor design tem-

20

COOLING SYSTEM EQUIPMENT

In the middle of the nineteenth century the use of ice for food preservation purposes was already common. During the summer of 1881, naval engineers attempted to provide relief for mortally wounded President Garfield. They devised a system to cool one room in the White House, which used about 8000 pounds of ice per day.

While cooling from melting ice is simple to understand, maintenance of a constant supply of natural ice can be difficult and expensive. The first ice-making machine was patented in 1851 while the first compressor using ammonia as the refrigerant was patented in 1872. These inventions and others led to development of the repeatable cooling process or cycle, known as the vapor compression cycle, which is widely used today by most mechanical cooling equipment.

This chapter provides information on major pieces of cooling system equipment including unitary air conditioners, condensers, cooling towers, chillers, and the various types of compressors. Particular emphasis will be placed on the types of "package units" used and the basis for their selection. Also covered are absorption refrigeration machines which offer an alternative to the use of refrigerants which are potentially damaging to the environment.

Once you understand the choices available in cooling system equipment you will be able to make design decisions based on building type, climate, relative energy costs, client expectations, maintenance implications, and many other factors.

20.1 BASIC EQUIPMENT OPTIONS

The most familiar type of cooling equipment is the individual room air conditioner unit as described in Chapter 18. It uses the vapor compression refrigeration cycle to absorb heat inside a space, and then reject it to the outside air. While window units are small in size and cooling capacity, the basic principles of heat transfer used also apply to equipment which cools very large buildings. In a window unit, all of the mechanical components (including the compressor, evaporator, condenser, fan, and controls) are contained in one enclosure.

Central cooling of houses and small buildings is often accomplished by using what is known as a "split system" (see Figure 20.1).

In this case the evaporator coil remains inside the house within the ducted air stream. Within the evaporator the circulating refrigerant then absorbs heat and turns into a vapor. It is then piped outside to a condensing unit (which contains the condenser, compressor, and controls). Within the condenser, the refrigerant circulates through a coil, is cooled by outside air which is blown over it by a fan, and condenses to a liquid. Hence, the name "air-cooled" condenser.

In very large commercial systems heat is rejected to a water-cooled condenser (usually mounted as a unit with the compressor and evaporator and known as a "chiller"). The warmed condenser water is then pumped to a cooling tower which is often mounted on a rooftop (see Figure 20.2).

At the cooling tower the condenser water is atomized into little droplets which then reject the heat to the outside air. The water then recirculates back to the condenser to pick up more heat before being pumped back to the cooling tower.

Equipment Packages

For all but the largest cooling applications, HVAC designers usually select factory-assembled equipment packages which include controls. As such, the manufacturer as-

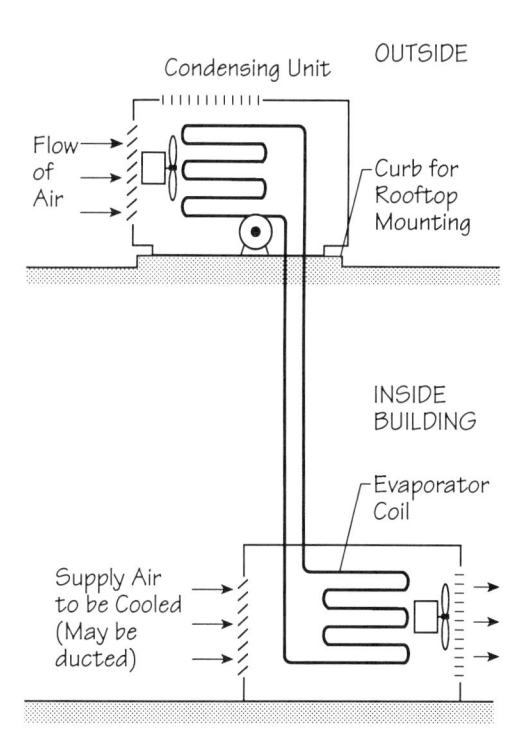

FIGURE 20.1 Central cooling with a "split system" where the evaporator coil is inside and the condensing unit is outside the building.

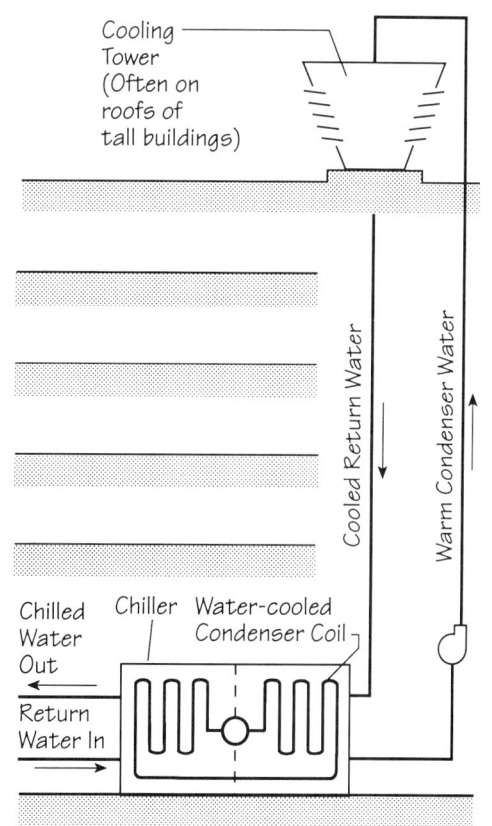

FIGURE 20.2 Large cooling system that rejects heat outside by a cooling tower.

sumes responsibility for compatibility of components and field labor is minimized. Typical packages include:

Unitary Air Conditioners This term fits many pieces of equipment as discussed below. The main point is that in addition to a compressor, evaporator, and sometimes a condenser, unitary air conditioners also include a fan to deliver air directly to a space or distribute it through ductwork. Units of capacities of more than 60,000 Btu/h (5 tons) normally come with multiple compressors so that light loads can be met efficiently with one compressor.

Condensing Units Contain the compressor and air-cooled condenser in one unit. Used in residential and small-scale commercial systems.

Packaged Chillers Produce chilled water for circulation to cooling coils. Contain the compressor, evaporator (chiller), and controls in one unit. Packaged chillers are available with and without integral condensers.

20.2 VAPOR COMPRESSION CYCLE

Most common types of cooling equipment use the vapor compression refrigeration cycle and a refrigerant. The refrigerant is a material which absorbs heat and in the process changes from a liquid to a gas in one location and then dispels the heat elsewhere as it changes back from a gas to a liquid. To be effective, refrigerants need to have high latent heats of vaporization (i.e., require large amounts of heat to change state from a liquid to a vapor).

Various substances have been and are being used as refrigerants. These range from toxic and explosive ammonia and potentially corrosive sulfur dioxide in early systems, to "freon" used with a vapor compression cycle, to water in the absorption refrigeration cycle.

"Freon" actually refers to a chemical family of halocarbon refrigerants including chlorofluorocarbons (CFCs) and hydrochlorofluorocarbons (HCFCs). These include refrigerants CFC-11 (CCl_3F), CFC-12 (CCl_2F_2), and HCFC-22 ($CHClF_2$) which until the late 1980s were the working fluids of unquestioned choice in compression refrigeration systems for air conditioning applications. These refrigerants possess the key properties of being nontoxic, nonflammable, and noncorrosive while being able to absorb a large amount of heat when changing state from a liquid to a gas.

Currently, serious environmental concerns of global warming and ozone depletion have implicated some of the CFC and HCFC refrigerants, and federal legislation (Clean Air Act Amendment of 1990) establishes a schedule for manufacturers to phase out production of these refrigerants.

The Clean Air Act Amendment of 1990 has led to an HVAC industry-wide trend towards development of nonozone depleting replacement refrigerants for new and existing air conditioning units, such as HFC-134a which is already being used in new automobile air conditioners.

Description of Cycle

Most current mechanical cooling systems employ a vapor compression refrigeration cycle driven by electricity to remove unwanted heat. Alternatives include steam and gas-fired turbine driven machines for large capacity units of 800–5000 tons. Gas-fired reciprocating engine driven vapor compression units in the range of 50–500 tons have recently begun to enter the commercial equipment market.

The various components of the vapor compression refrigeration cycle are shown in Figure 20.3. Temperatures and pressures

20.2 VAPOR COMPRESSION CYCLE

shown are representative only, and will depend upon the refrigerant and equipment selected.

The goal of the cooling cycle is the removal of heat from where it is unwanted, and the transport of that heat to a location where it can be dispelled. The heat is absorbed in the evaporator, which is so named because it is here that the refrigerant absorbs heat and "evaporates" from a low pressure liquid/vapor mixture to a low pressure vapor (gas). To effectively accomplish this, evaporators are heat exchangers which absorb heat either from air or water and in so doing create a cooling effect. Systems where the evaporator comes into direct contact with the air being conditioned are known as direct expansion or "DX" systems. Other systems use the evaporator to produce chilled water which is then piped to cooling coils.

Upon leaving the evaporator, the refrigerant should be a low-pressure vapor. To assure this, evaporators are usually designed to add some "superheat" (heat in addition to the amount needed to fully vaporize the refrigerant) since refrigerant in a liquid state can damage the compressor. Then the compressor presses the gas into a smaller volume and in so doing raises both the pressure and temperature of the still gaseous refrigerant. This part of the cycle is called the "high side" since the refrigerant is now under high pressure (see Figure 20.3).

Next, heat is rejected to another heat exchanger known as the condenser. First any superheat is removed (known as desuperheating). Then the refrigerant releases the rest of the heat it absorbed by condensing back to a high-pressure liquid.

After leaving the condenser, the refrigerant passes through a flow control device which lowers the pressure. This is the "low side" of the cycle. In this low-pressure liquid condition, the refrigerant is again ready to enter the evaporator for another cycle.

FIGURE 20.3 Vapor compression refrigeration cycle components.

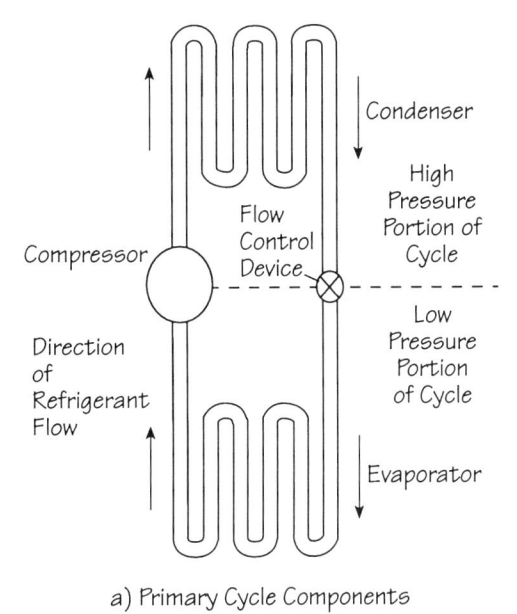

Coefficient of Performance

Various performance indicators for equipment which uses the vapor compression cycle were first introduced in Section 16.3 including the coefficient of performance (COP). COP can also be expressed in thermodynamic terms as indicated in Formula 20A. The "refrigerating effect" is the increase in enthalpy by the refrigerant in the evaporator. The heat of compression is the increase in enthalpy of the refrigerant in the compressor.

$$COP = \frac{R.E.}{H.C.} \quad \text{Formula 20A}$$

where

COP   is the coefficient of performance
R.E. is the refrigerating effect in Btu/lb
H.C. is the heat of compression in Btu/lb

Example 20A: Determine the COP for a vapor compression refrigeration process where the refrigeration effect (R.E.) is 48 Btu per pound of refrigerant and the heat of compression is 16 Btu per pound of refrigerant.
Solution: Use Formula 20A

$$COP = \frac{R.E.}{H.C.} \quad \text{(Formula 20A)}$$
$$= \frac{48 \text{ Btu/lb}}{16 \text{ Btu/lb}}$$
$$= 3.00$$

20.3 UNITARY AIR CONDITIONERS

The simplest type of unitary air conditioner is the individual (window) unit. Other small self-contained unitary air conditioners are known as packaged terminal air conditioners (PTAC) and packaged terminal heat pumps (PTHP) as described fully in Section 21.1. Larger types of unitary air conditioners include the following:

Floor-Mounted Package Cooling Units

Several types of floor-mounted package cooling units are available as shown in Figure 20.4. Some contain all of the vapor compression refrigeration system components including an air-cooled condenser section which needs a source of outside air. Other units contain the compressor(s), evaporator, controls, and either a water-cooled condenser or a remote (but nearby) air-cooled condenser. In this case, the unit itself is called "condenserless."

Recirculated supply air can be delivered through ductwork or by the free air discharge of the fan(s). Units usually do not satisfy any ventilation air requirements. Units typically come in capacities ranging from 3 to 50 tons. Due to compressor noise, these units are usually limited to use in noise-tolerant commercial and industrial applications. They are often installed as water-cooled units in groupings connected to a cooling tower circuit.

Rooftop Units

A familiar sight is air-conditioning units mounted on the roofs of stores and rela-

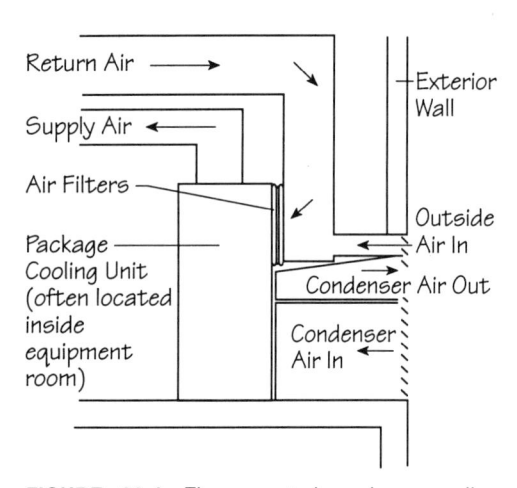

FIGURE 20.4 Floor-mounted package cooling units.

tively small (less than about 50,000 ft^2) buildings of three stories or less. Units are available in capacities of 5–30 tons of refrigeration. Rooftop mounting is relatively inexpensive. For one-story buildings, rooftop mounting requires a minimum of ductwork to supply and return air through the ceiling below as shown in Figure 20.5.

Rooftop units can supply air to one zone, or up to 12 zones with a "multizone" rooftop unit. Disadvantages of rooftop mounting include more difficult (and perhaps neglected) servicing and maintenance (such as filter replacement), possible damage to the roofing system during installation or servicing, and possible noise transmission to the building. One way to dampen noise and vibration transmission from rooftop units on lightweight (metal decked) buildings is to place the equipment on a concrete mounting pad and provide sound traps or acoustic lining in the supply and return ductwork.

Selection Procedure for Unitary Air Conditioners

Units are rated by the Air-Conditioning and Refrigeration Institute (ARI) based on a standard of 80°F db, 67°F wb air entering the evaporator and a 95°F db condenser air temperature. Presented in Table 20.1 is performance data for a typical air-cooled unit to meet cooling loads with the heat of the compressor(s) factored in. Performance values are based upon the following:

- A return air temperature of 80°F db entering the evaporator
- The wet-bulb temperature of the return air entering the evaporator (i.e., 72, 67, or 62°F wb representative of relative humidities of 70%, 50%, and 35% respectively)
- The dry-bulb temperature of the air entering the condenser
- A variety of air flow rates in cfm. Note how the bypass factor increases with an increase in air flow for each unit
- Total cooling capacity (TC) in thousands of Btu/h
- Sensible heat capacity (SHC) in thousands of Btu/h

Note in Table 20.1 that for a given entering condenser temperature the total cool-

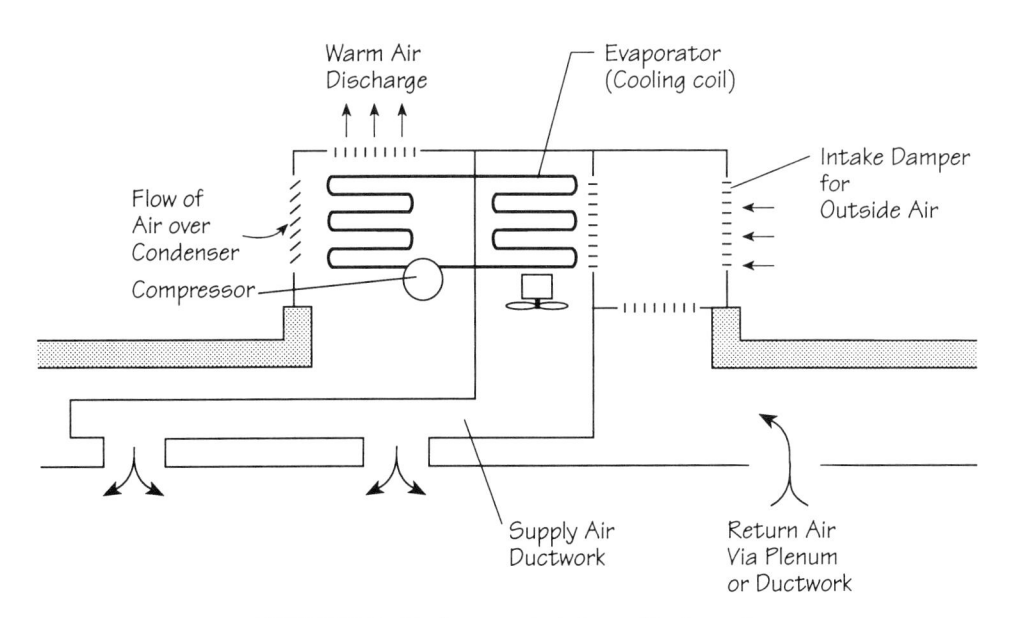

FIGURE 20.5 Rooftop-mounted air conditioning unit.

332 COOLING SYSTEM EQUIPMENT

TABLE 20.1 TYPICAL AIR-COOLED PACKAGE UNIT PERFORMANCE DATA[a]

Condenser, Entering Air Temp. (°F)		4500 cfm			6000 cfm			7500 cfm		
		\multicolumn{9}{c}{Evaporator Air Entering Wet-bulb Temp. (°F)}
		72	67	62	72	67	62	72	67	62
85	TC	198	182	166	207	190	176	212	196	184
	SHC	99	122	145	109	139	167	119	155	184
95	TC	188	173	158	196	181	167	201	186	176
	SHC	95	119	141	106	135	163	115	151	176
100	TC	183	168	154	191	176	163	196	181	172
	SHC	93	117	139	104	133	160	113	148	172
105	TC	178	164	150	186	171	159	190	176	168
	SHC	92	115	137	102	131	157	111	147	168
Bypass Factor		\multicolumn{3}{c}{0.15}	\multicolumn{3}{c}{0.19}	\multicolumn{3}{c}{0.22}

Source: Data extracted with permission courtesy of the Carrier Corporation.

[a] Table based upon an 80-degree Fahrenheit dry-bulb air temperature entering the evaporator coil. A calculation procedure is used to correct the SHC for higher or lower entering air temperatures.

ing capacity (TC) and the sensible heat capacity (SHC) increase with air flow rate. Also note that the latent heat capacity (LHC), which is the difference between the TC and the SHC, decreases as the flow rate in cfm increases. For example, for a 95°F db condenser air temperature and a 67°F wb evaporator entering air temperature the latent capacity is: 54,000 Btu/h (LHC = 173 − 119) at 4500 cfm; 46,000 Btu/h (LHC = 181 − 135) at 6000 cfm; and only 35,000 Btu/h (LHC = 186 − 151) at 7500 cfm.

Example 20B: Select a package cooling unit from Table 20.1 to satisfy the following:

Installation in Dallas, Texas
Required supply (evaporator) air quantity: 6000 cfm
Sensible cooling load: 125,000 Btu/h
Latent cooling load: 30,000 Btu/h
An entering return (evaporator) air temperature of 67°F wet-bulb

Solution: Obtain the outdoor summer design temperature ($2\frac{1}{2}\%$ condition) for Dallas of 100°F db from Table 2.7. This is the entering air temperature required for selection of the air-cooled condenser.

The total cooling load (sensible + latent) is computed to be 155,000 Btu/h (125,000 + 30,000). This is the total capacity (TC) required of the unit while 125,000 Btu/h is the required sensible heat capacity (SHC).

Table 20.1 shows that for the design conditions (6000 cfm, 67°F entering wet-bulb, and outdoor condenser entering air temp. of 100°F dry-bulb), that the unit has a total heat capacity (TC) of 176,000 Btu/h (176). This compares to the total cooling load of 155,000 Btu/h. It also has an adequate sensible heat capacity (SHC) of 133,000 Btu/h (133) to handle the sensible load of 125,000 Btu/h.

20.4 AIR-COOLED CONDENSERS

Every house and apartment has at least one air-cooled condenser in it—the coil (usually on the back of the refrigerator) which is rejecting the heat removed from inside the box to the air in the kitchen. Many houses also have an air-cooled condenser as part of a central air conditioning system to similarly reject heat from inside the house during warm weather.

Air-cooled condensers (see Figure 20.6) are also very popular as part of the air conditioning system for commercial buildings, and are available in capacities of up to about 500 tons. Their normal application, though, is in smaller buildings where cooling loads are less than 150 tons or so, since larger loads are normally met more cost-effectively by water-cooled condensers (and associated cooling towers) as discussed below.

Cooling of the refrigerant requires air motion over the condenser coil of about 1000 cfm per ton of refrigeration. Therefore, the condenser is most appropriately located in an exposed outdoor location although an indoor location with a ducted air flow is possible if demanded by site constraints. Whether located indoors or out, most air-cooled condensers used in HVAC systems employ fans with power requirements of 0.1–0.2 horsepower per ton (75–150 W per ton) to assure air flow.

Perhaps the greatest advantage to using air-cooled condensers is their relatively maintenance-free operation. The chief disadvantage in using them in large systems is their intensive use of electricity which is often expensive and may increase peak demand charges (see Section 24.8).

Air-Cooled Condensing Units

Often the air-cooled condenser is packaged with the compressor in what is known as a "condensing unit." Therefore, condensing units must be located relatively close to the evaporator coil (normally within 50 ft) to minimize the pressure drop in the refrigerant discharge line between the compressor and evaporator.

The ability of an air-cooled condenser to reject heat to the outside depends upon the outside air temperature. As shown in Figure 20.7, as the ambient air temperature de-

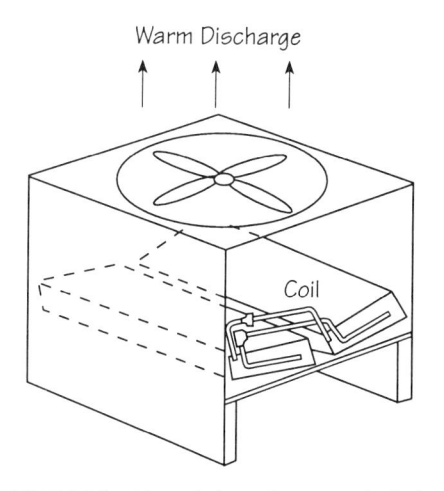

FIGURE 20.6 Air-cooled condenser—rejects heat to outside air blown over the coil.

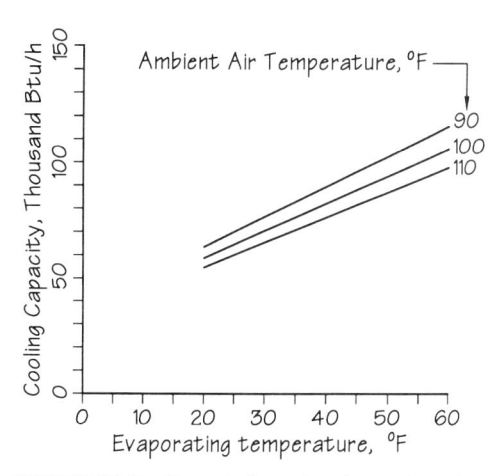

FIGURE 20.7 Air-cooled condensing unit performance.

334 COOLING SYSTEM EQUIPMENT

creases the capacity of the air-cooled condenser in Btu/h increases. Also note that air-cooled condenser capacity increases as the evaporating temperature of the refrigerant coming out of evaporator (and under suction by the compressor) increases.

Table 20.2 presents the operational characteristics of a typical 2-ton residential-scale condensing unit and matched evaporator coil. "Adp" refers to apparatus dewpoint of the cooling coil (see Section 19.4).

Example 20C: The peak cooling load calculation for a Philadelphia, PA house results in a sensible cooling load of 17,000 Btu/h and an assumed latent cooling load of 5100 Btu/h (based upon 30% of sensible). Determine whether the condensing unit (and matched evaporator coil) performance listed in Table 20.2 will be adequate to maintain an indoor air temperature of 80°F db and a relative humidity of approximately 50% (entering air of 67°F wb).

Solution: The summer dry-bulb design temperature for Philadelphia is 90°F (2½% value from Table 2.7).

Refer to Table 20.2, where an outside air temperature of 90°F and an entering air temperature of 67°F wb, corresponds to a total capacity of 23,700 Btu/h. The sensible capacity for an entering db temperature of 80°F is 17,600 Btu/h which would leave a latent capacity of 6,100 Btu/h (23,700 − 17,600). Therefore, this unit is well sized

TABLE 20.2 TYPICAL PERFORMANCE DATA FOR A "2 TON" AIR-COOLED CONDENSING UNIT AND INDOOR COIL[a]

Outside Air (°F db)	Entering Air (°F wb)	Total Cap.	Sensible Cap. at Entering db Temp. of 78°F	Sensible Cap. at Entering db Temp. of 80°F	Comp. KW	adp
85	63	22.5	19.6	21.2	1.75	51.8
	67	24.3	16.2	17.8	1.81	55.9
	71	26.2	12.7	14.3	1.87	60.1
90	63	21.9	19.4	21.0	1.82	52.1
	67	23.7	16.0	17.6	1.88	56.2
	71	25.5	12.5	14.1	1.94	60.4
95	63	21.4	19.2	20.8	1.89	52.3
	67	23.1	15.8	17.4	1.95	56.5
	71	24.9	12.3	13.9	2.01	60.7
100	63	20.7	19.0	20.6	1.96	52.7
	67	22.4	15.5	17.1	2.02	56.9
	71	24.2	12.0	13.6	2.09	61.1
105	63	20.1	18.7	20.2*	2.04	53.0
	67	21.7	15.3	16.9	2.10	57.2
	71	23.4	11.7	13.4	2.17	61.5
115	63	18.8	18.2	19.1*	2.19	53.7
	67	20.3	14.7	16.3	2.25	57.9
	71	21.8	11.2	12.8	2.32	62.2

[a] Capacities in thousands of Btu/h; airflow: 900 cfm.
Dry coil condition (Total capacity = Sensible capacity).
Source: The Trane Company.

for this application. Also note that the compressor power requirement will be 1.88 kW for this nominal 2-ton unit.

20.5 OTHER CONDENSER TYPES

Water-Cooled Condensers

Water can also be used to condense the refrigerant. Water-cooled condensers are normally used for larger applications in commercial buildings (50 tons or more), and are normally supplied as a piece of equipment which includes the compressor. Design types include shell-and-tube, shell-and-coil, and tube-in-tube water-cooled condensers to achieve the required heat exchange (see Figure 20.8). The selection of condenser type is normally left to equipment manufacturers and is not of primary concern to the HVAC building designer.

The use of a water-cooled condenser and a cooling tower yields a recirculating method of heat rejection. Without the cooling tower, the water which cooled the condenser would have to be discarded and new make-up added to the system. In most cases such a once-through system is not economical and furthermore is prohibited by law.

Double-Bundle (Heat Recovery) Condenser

Heat absorbed by water in a water-cooled condenser does not have to be discharged to the atmosphere through the cooling tower circuit. Large buildings that need heat for any purpose during a significant portion of the time when the mechanical refrigeration equipment is also needed for space cooling can use the otherwise wasted heat with a "double-bundle condenser" as shown in Figure 20.9.

In a double-bundle condenser, one tube bundle is connected to the cooling tower in the conventional manner. The other bundle is connected to the system needing heat (i.e., a perimeter heating loop or the domestic hot water system). Heat can be rejected to either condenser or to both condensers simultaneously.

Evaporative Condensers

An evaporative condenser is a hybrid of an air-cooled condenser and a cooling tower. These condensers have nozzles which spray water over the condenser coil while a centrifugal fan circulates air through the unit. Heat is thus lost from the refrigerant by evaporating the sprayed water.

Similar to cooling towers, evaporative condensers must be designed with "water eliminating baffles" to minimize the amount of liquid water (known as "drift") being blown out of the unit. Unvaporized liquid should flow by gravity over the coil,

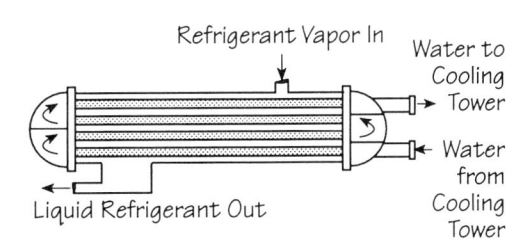

FIGURE 20.8 Water-cooled condenser.

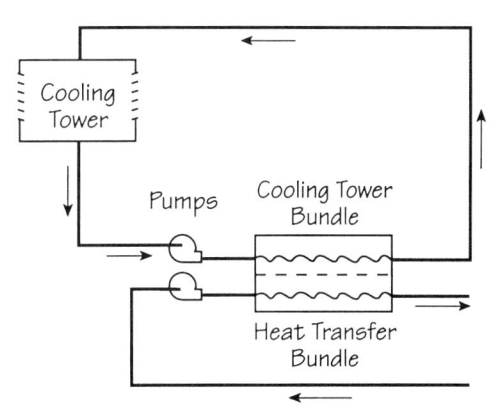

FIGURE 20.9 Heat recovery through use of a double bundle condenser.

be collected in a drain pan, and be recirculated by a small pump. One advantage of using evaporative condensers is their slightly smaller size and greater energy efficiency when compared to air-cooled condensers.

20.6 COOLING TOWERS

In large cooling systems, the water which cools a condenser must then be cooled itself if it is to be reusable. Cooling towers work by causing a portion of the warm water which cooled the condenser to evaporate in a system open to the atmosphere. This is accomplished by pumping the water to the top of a tower, and then spraying the water over tower packing known as "fill."

Cooling tower fill can either be a series of small splash bars, or a series of closely spaced vertical plastic sheets. The purpose of the tower fill is to create a very large surface area for the water droplets, and in so doing cause some of the water to evaporate. Of course, for evaporation to occur, heat is needed. In a cooling tower this heat is obtained from the rest of the circulating water which does not evaporate, and is thus cooled in the process.

Cooling towers used for HVAC applications are the mechanical draft type which use fans to either induce or force outside air to flow through the tower fill. Illustrated in Figure 20.10 are two common types of cooling towers. The tower in Figure 20.10a uses a propeller fan at the top to induce air flow from one (single-entry) or two (double-entry) sides of the tower (as shown). The tower in Figure 20.10b uses a centrifugal fan to force (blow) air by counterflow motion through the falling tower water and fill.

Since cooling towers are open to the environment, they require a significant amount of monitoring and maintenance to prevent the buildup of scale-forming salts (which can clog condenser tubes and spray heads), corrosion, and control of biological growth.

To control the buildup of salts, a small portion (usually less than 1%) of the circulating cooling tower water is continuously drained and replaced by pure make-up water in a process known as "blowdown." Blowdown also helps to limit corrosion since fewer dissolved solids limit conductivity and corrosion-producing electrolytic action. Corrosion control is also achieved through the intermittent addition of corrosion inhibitors. To control the formation of slime and algae small quantities of chlorine-containing compounds are also sometimes added.

Cooling towers coupled with a vapor compression refrigeration system typically circulate approximately 3 gal of water per minute (gpm) per ton of refrigeration. "Drift" eliminators near the air outlet are designed to minimize the amount of water vapor carried off by the heated air.

Unlike air-cooled condensers, cooling towers can be located at almost any distance from the condenser and other components of the cooling system. Installation on rooftops is common in urban locations, although tower weight (approximately 40–60 lb per ton of capacity) must be considered. For rural buildings, location of the tower on the site away from the main building is sometimes preferred. When this is done, try to locate the tower so that it is the highest point of the condenser water system, to simplify the pumping and piping system. Towers must also be installed where the air flow in and out is unrestricted.

Tower location should always be selected recognizing that there can be a misty discharge (known as a plume) when the cooling towers are operated in cool weather. Plumes can be particularly annoying, and are considered unacceptable if they produce local at-grade fog conditions.

Plumes from cooling towers can usually be minimized by employing tall stacks. In

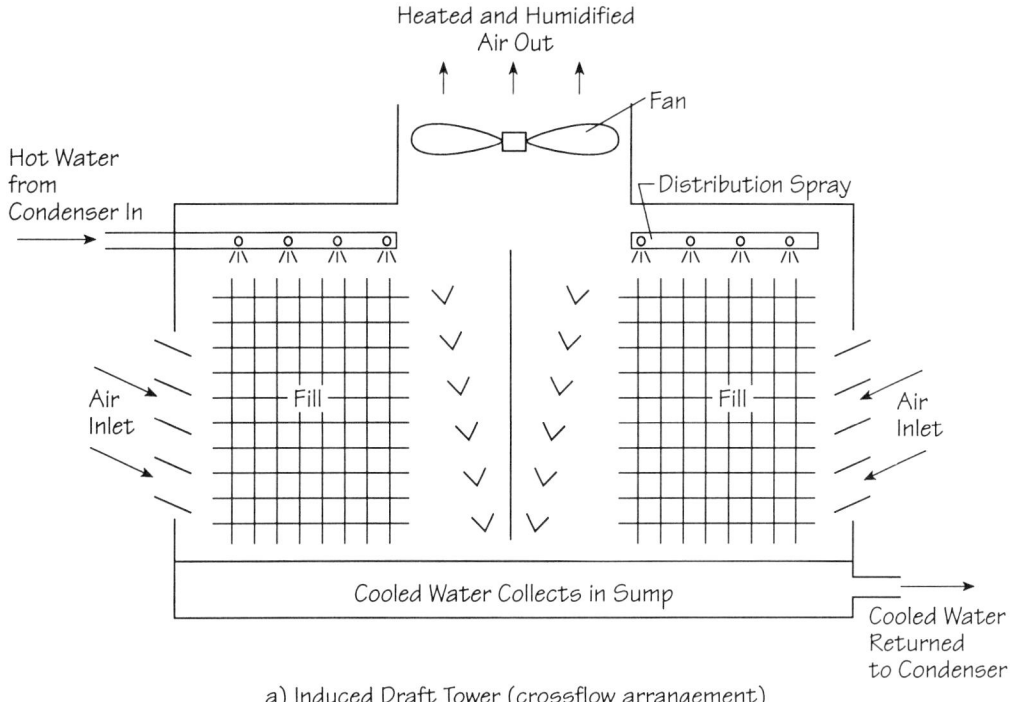

a) Induced Draft Tower (crossflow arrangement)

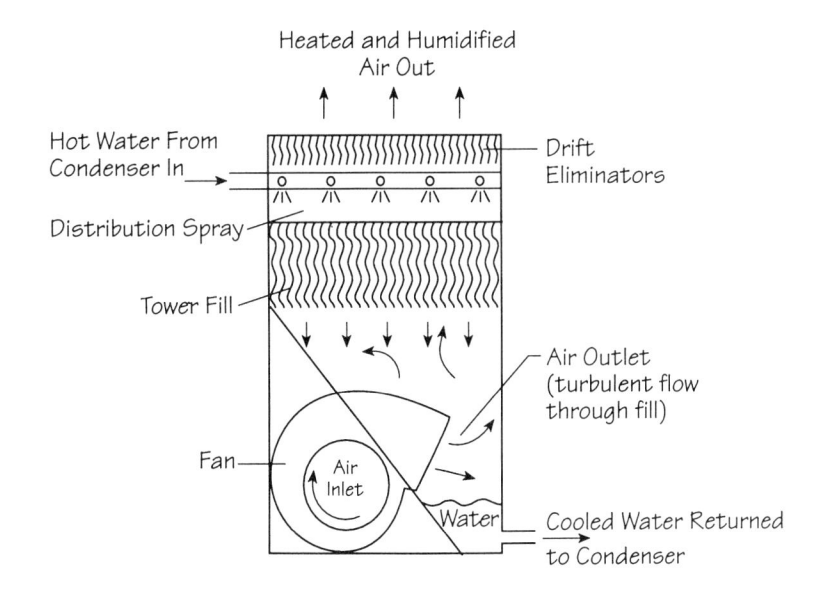

b) Forced Draft Tower with Centrifugal Fan

FIGURE 20.10 Mechanical cooling towers—heat rejection by condenser water in an "open" piping circuit.

338 COOLING SYSTEM EQUIPMENT

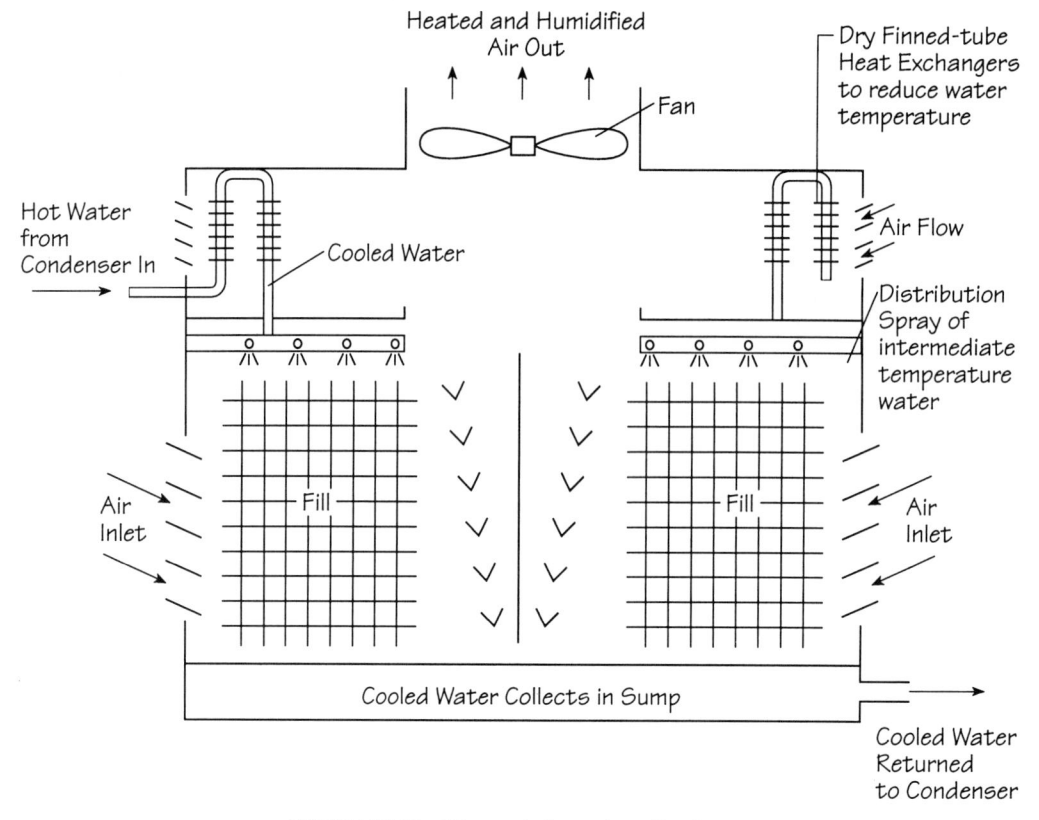

FIGURE 20.11 Plume abatement cooling tower.

addition, plumes can be all but eliminated through employment of a plume abatement tower design which first circulates water from the water-cooled condenser through dry heat exchangers exposed to the cool outdoor conditions before then being released in the wet (evaporative) fill section in the normal fashion (see Figure 20.11).

Cooling Tower Water Flow Rate

It was shown that a water flow rate of 1 gpm is equal to approximately 500 lb of water per hour (Formula 10A). Based upon this factor, and the range (R) of a cooling tower, Formula 20B can be used to determine the required water flow rate in gallons per minute necessary to reject the total amount of heat required (the cooling load plus the heat added in the chiller).

$$\text{gpm}_{ct} = \frac{(q_{bcl} + q_{ch})}{R \times 500} \quad \text{Formula 20B}$$

where

gpm_{ct} is the capacity of a cooling tower in gallons per minute

q_{bcl} is the building cooling load in Btu/h

q_{ch} is the heat added in the chiller (see "rules of thumb" below)

R is the range of the cooling tower (the difference in temperature between entering and exiting water)

500 is the number of pounds of water per hour at a flow of 1 gpm

Rule-of-thumb values typically used for the heat added in the chiller (q_{ch}) are: 3,000

Btu/h per ton for the heat of compression in electric chillers; 18,000 Btu/h per ton for low-pressure and 10,000 Btu/h per ton for high-pressure absorption units; and 12,000 Btu/h per ton for steam turbines. Use of Formula 20B will be illustrated in Example 20D.

Cooling Tower Performance

"Range" is the term used for the temperature reduction for water which passes through a cooling tower. Thus, a tower that cools condenser water from 95 to 85°F is said to have a range of 10°F.

The lowest temperature which water being cooled in a cooling tower can attain is the outside wet-bulb temperature. The "approach" of a cooling tower is the difference between the temperature of the leaving water and the outside wet-bulb temperature. Thus, a tower that discharges water at 85°F when the wet-bulb temperature is 78°F has an approach of 7°F.

The capacity of a cooling tower to reject heat is sometimes called the tower "thermal loading." Illustrated in Table 20.3 is the performance of several typical cooling tower designs under a range of design conditions. The capacities shown are in gpm of condenser water.

The outdoor design wet-bulb temperature is fundamental to cooling tower capacity. A low wet-bulb design temperature such as 65°F would apply to dry climates such as Albuquerque and Salt Lake City. Whereas a 75°F wet-bulb temperature would be used for cities such as New York, Minneapolis, and Madison, Wisconsin. And a 78°F wet-bulb design temperature would be used for design in more humid areas such as Dallas and Richmond.

Scan the tower ratings in Table 20.3 for a particular outdoor wet-bulb temperature and note how much the gpm ratings vary. For example, for an outdoor wet-bulb temperature of 75°F, the rating for Model A varies from 115 to 195 gpm depending upon the range and entering temperatures. Also note how markedly tower capacity increases as the "approach" increases.

TABLE 20.3 TYPICAL COOLING TOWER RATINGS

Outdoor Wet-bulb Temp (°F)	Entering Dry-bulb Water Temp. (°F)	Exiting Dry-bulb Water Temp. (°F)	Range (°F)	Approach (°F)	GPM of Water Models A	B	C	D
65	90	80	10	15	165	320	470	700
70	90	80	10	10	140	260	405	610
70	95	85	10	15	200	380	560	840
72	90	80	10	8	130	250	370	550
72	95	85	10	13	185	255	530	790
75	95	85	10	10	165	320	470	700
75	97	87	10	12	195	370	555	830
75	100	85	15	10	130	250	370	560
75	102	85	17	10	125	240	360	540
75	105	85	20	10	115	220	330	490
78	95	85	10	7	135	260	390	580
78	100	85	15	7	115	220	330	490
78	102	85	17	7	110	210	310	460
78	105	85	20	7	100	190	280	420

Example 20D: Select a cooling tower model from Table 20.3 which will reject a total of 2,200,000 Btu/h ($q_{bcl} + q_{ch}$) for an office building in New York City. Design for a reduction in water temperature of 10°F (the "range") from 95°F to 85°F.

Solution: The wet-bulb design temperature for New York City is 75°F ($2\frac{1}{2}$% value from Table 2.7). Formula 20B is then used to solve for the required condenser water flow rate to the cooling tower (gpm_{ct}).

$$gpm_{ct} = \frac{(q_{bcl} + q_{ch})}{R \times 500} \quad \text{(Formula 20B)}$$

$$= \frac{2,200,000}{10 \times 500}$$

$$= 440 \text{ gpm}$$

Table 20.3 shows that Model C has a capacity of 470 gpm for a 75°F wet-bulb temperature, an entering water temperature of 95°F, and a range of 10°F. The approach will be 10°F. This is an acceptable choice. Other towers may, of course, also be used if operated with a different entering water temperature and range. Tower selection will be based on many factors including desired capacity, fan power, tower size, pump requirements, noise, aesthetics, and the impact of entering water temperature on other system components and operational efficiency.

"Water-side" Economizer Cycle

When building cooling is needed and the outside wet-bulb temperature is below about 45°F, cool water from the cooling tower can be used in lieu of operating the chiller. Two basic approaches are available for use. Cooling tower water can be piped directly through the chilled water piping circuit after straining (or filtering) it to eliminate impurities. Such a "strainer cycle" (shown in Figure 20.12a) was much more popular from about 1975 to 1985. Today many designers feel that it is prudent to keep the chilled water piping circuit closed to the environment so as to minimize potential corrosion problems. To accomplish this a heat exchanger transfers the cooling effect derived from the cooling tower water circuit to the chilled water circuit within a building as shown in Figure 20.12b.

20.7 COOLING PONDS AND SPRAYS

It is also possible to reject the heat from a mechanical cooling system through the use of natural or man-made bodies of water. Cooling of the water in the pond occurs when surface water evaporates. The cooler pond water can then be used to remove heat in the condenser. Indicated in Table 20.4 is the evaporation of water per square foot of free water surface assuming still air, an air temperature of 75°F db, and 50% relative humidity.

As you can see, a relatively small amount of heat is transferred through evaporation in still air at typical outdoor temperatures. Moreover, an outdoor water body is likely to gain sensible heat due to the outdoor air temperature and solar radiation. Evaporation will, of course, increase substantially when winds blow but such unpredictability is an unwelcome feature in an HVAC system. Evaporation can also be greatly enhanced if water is sprayed to increase the exposed surface area. However, spraying results in much greater water loss and mists which may become annoying depending upon wind speed and direction.

Ponds, sprays, and fountains can provide a wonderfully refreshing psychological benefit on a hot summer day. In spite of that, only in rare circumstances can they reliably play the role of "heat rejector" in an HVAC cooling system.

20.7 COOLING PONDS AND SPRAY PONDS **341**

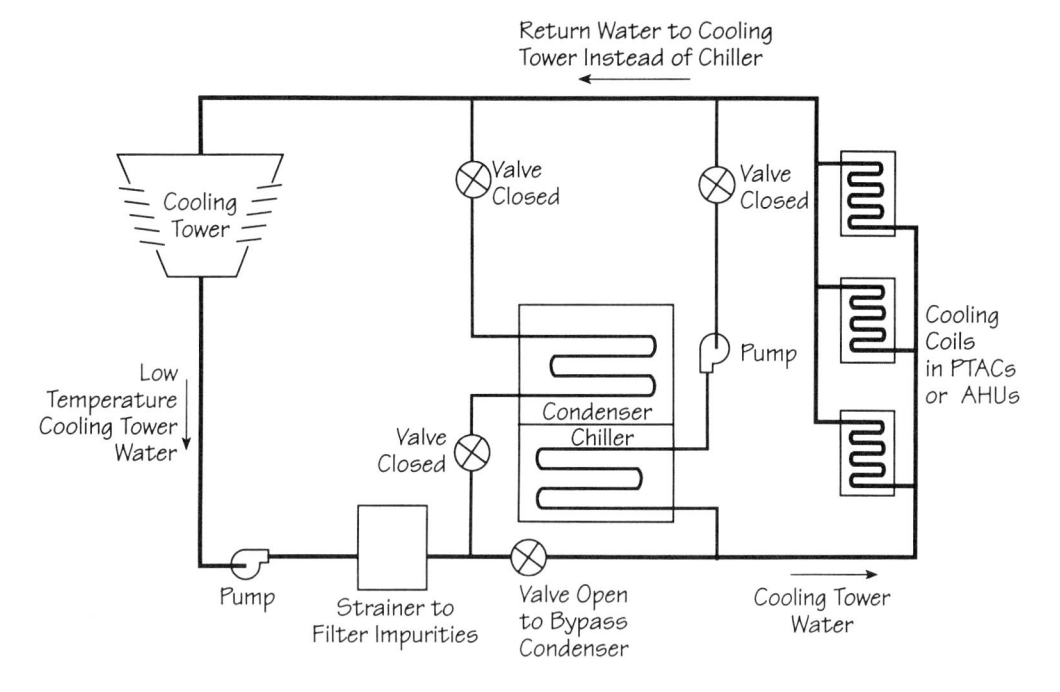

a) Strainer Cycle

b) Using Heat Exchanger

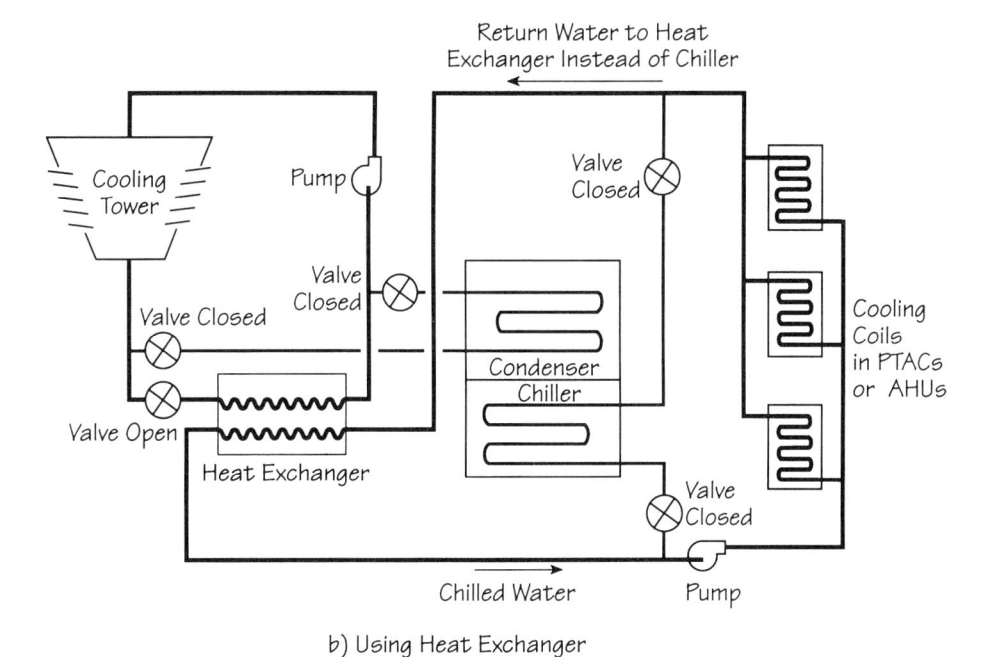

FIGURE 20.12 A water-side economizer cycle allows for energy conserving operation of chilled water circuits during cool outside weather.

TABLE 20.4 HEAT TRANSFER THROUGH EVAPORATION PER SQUARE FOOT OF FREE WATER SURFACE STILL AIR AT 75°F db and 50% RH

Water Temperature (°F)	Heat Transfer Through Evaporation (Btu/h per ft^2)
75	42
100	140
125	330
150	680
175	1260
200	2190

Source: Courtesy of the Carrier Corporation.

20.8 EVAPORATORS

The evaporator (also known as a "liquid cooler") is the heat exchanger where the heat which represents the cooling load is transferred to the refrigerant. The most common type is known as a "dry," "direct-expansion" or "DX" evaporator which is simply a continuous coil or tube which provides a large surface area in a small amount of space. Refrigerant flow within the coil should be turbulent and of sufficient velocity so that it scrubs the coil. In most systems the evaporator "coil" is really a series of parallel circuits of equal length and refrigerant loading.

Another type of evaporator sometimes coupled with helical (screw) and centrifugal compressors is the "flooded coil" evaporator which is in contact with liquid refrigerant under all load conditions. Flooded coil evaporators can operate with a smaller temperature difference between the refrigerant and cooled fluid, and thus require less compressor power for increased energy efficiency.

20.9 ELECTRIC CHILLERS

Chillers include in one piece of package equipment the evaporator, compressor (usually reciprocating), controls, and perhaps a water-cooled condenser. Packaged water chiller selection is based upon the following:

- Total required cooling capacity in tons
- The water temperature which enters the condenser in °F db
- The °F db temperature rise of the condenser water, typically assumed in manufacturer ratings to be 10°F
- The °F db temperature of the leaving chilled water
- The chilled water temperature drop (published ratings are usually for a 10°F drop and are considered applicable for a range of 6 to 14°F)
- The cleanliness of the water in the chiller and condenser known as the "fouling factor" (typically assumed to be .00025 in manufacturers' tables)

Manufacturers publish chiller performance data in various tabular formats including those where the exiting (leaving) chilled water temperature is the dominant design factor based upon supply air conditions and coil requirements. Provided in Table 20.5 is typical data of cooling performance, power input, and condenser

TABLE 20.5 TYPICAL PACKAGED WATER CHILLER RATINGS CAPACITIES IN TONS, KW POWER INPUT, AND GPM

Model	Leaving Chilled Water Temp. (°F db)	Entering Condenser Water Temperature (°F db)							
		80			85			90		
		Tons	KW	gpm	Tons	KW	gpm	Tons	KW	gpm
25T	42	23.0	20	67	22.9	21	65	22.3	21	63
	44	24.4	21	69	23.7[a]	21	68	23.0	22	66
	45	24.8	21	71	24.1	21	69	23.4	22	67
	46	25.2	21	72	24.5	22	70	23.8	22	69
	48	26.0	21	74	25.3	22	73	24.7	22	71
	50	26.8	21	77	26.1	22	75	25.5	23	73
30T	42	27.5	24	80	26.6	25	79	25.9	25	77
	44	28.4	24	83	27.6[a]	25	83	26.7	26	80
	45	29.0	24	84	28.1	25	84	27.3	26	82
	46	29.7	25	85	28.5	25	86	27.8	26	83
	48	30.5	25	88	29.6	26	88	28.7	27	85
	50	31.3	25	90	30.6	26	89	29.7	27	86
40T	42	40.0	36	114	38.9	37	111	37.5	38	109
	44	41.7	37	117	40.4[a]	37	115	39.0	39	113
	45	42.4	37	119	41.1	38	117	39.7	39	115
	46	43.1	37	121	41.8	38	119	40.4	39	117
	48	44.6	38	125	43.3	39	123	41.9	40	120
	50	46.0	38	129	44.8	40	126	43.4	41	124

[a] Indicates ARI base rating conditions.

water gpm for 25-, 30- and 40-ton electric driven packaged chillers.

In viewing Table 20.5, note that the capacity for a given unit is greatest when there is the smallest difference between entering condenser water temperature and leaving chilled water temperature. For example, Model 25T has a capacity of 26.8 tons for an entering condenser water temperature of 80°F and a leaving chilled water temperature of 50°F while the unit capacity is only 22.3 tons when the entering condenser water is 90°F and the leaving chilled water is 42°F.

For packaged chillers, values for condenser water gpm are based upon the total cooling load they can address in tons plus the heat produced by the compressor.

Example 20E: Select a packaged water chiller from Table 20.5, and a cooling tower from Table 20.3, to address a cooling load of 450,000 Btu/h for a building in Topeka, Kansas. Assume the ARI base ratings of an exiting chilled water temperature of 44°F and an entering condenser water temperature of 85°F.

Solution: The building cooling load is 37.5 tons (450,000/12,000). This load can be met by packaged water chiller model 40T (in Table 20.5) which has a capacity of 40.4 tons for the ARI base-rating conditions. The condenser water enters at 85°F db, and leaves at 95°F, at a flow rate of 115 gpm. The chiller power requirement is 37 kW which results in a value of 0.92 kW per ton. The outdoor wet-bulb ($2\frac{1}{2}\%$) design temperature for Topeka is 78°F. For an entering db water temperature of 95°F, and a range of 10°F, cooling tower model A (in Table 20.3) has a capacity of 135 gpm which is adequate to meet the load of 115 gpm.

Chillers for large tonnage applications typically use flooded coil evaporators coupled with centrifugal compressors. Machines are described as "hermetic" if the motor is contained within an integral housing.

20.10 TYPES OF COMPRESSORS

Most compressors utilize some form of "positive displacement" to reduce the volume of vaporized refrigerant (gas) in a confined chamber and thus raise its pressure and temperature. In contrast, some compressors for large cooling applications employ centrifugal force and velocity to increase pressure.

Rotary Compressors

The simplest positive displacement type is the rotary compressor, which has a turning rotor eccentric to the cylinder housing, and blades which slide to form a continuous seal for the refrigerant (see Figure 20.13). At the beginning of the stroke the shaft rotates, and a volume of refrigerant enters the chamber. Then as the stroke continues, the eccentric nature of the rotor within the housing causes the volume available to the refrigerant to be reduced, and thus raises its pressure and temperature.

The simplicity of construction makes rotary compressors relatively quiet and well suited to household cooling applications such as window air conditioners and refrigerator/freezers.

Reciprocating Compressors

Reciprocating compressors (as shown in Figure 20.14) are similar to typical automobile engines and employ a crankshaft to drive reciprocating (move up and down) pistons which then compress the volume of vaporous refrigerant in the cylinders and thus increase pressure and temperature.

Reciprocating compressors are referred to as the "open-type" if the driveshaft extends through a seal in the casing for an external motor-driven drive. In contrast, "hermetic" compressors have the motor sealed within the casing as a protection

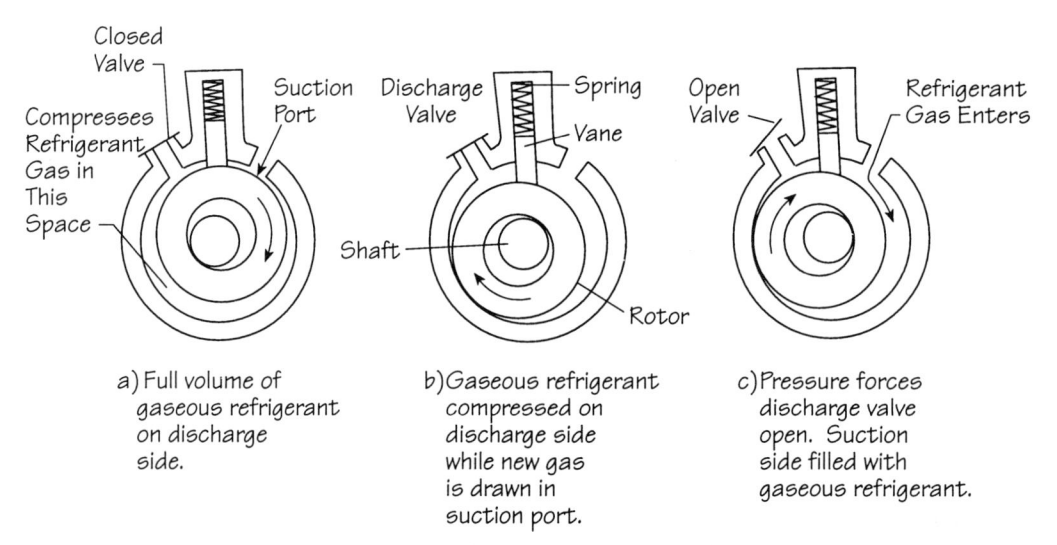

FIGURE 20.13 Rotary compressor operation.

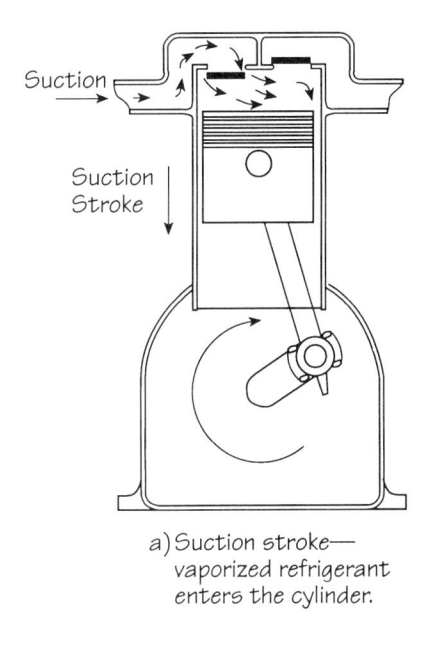

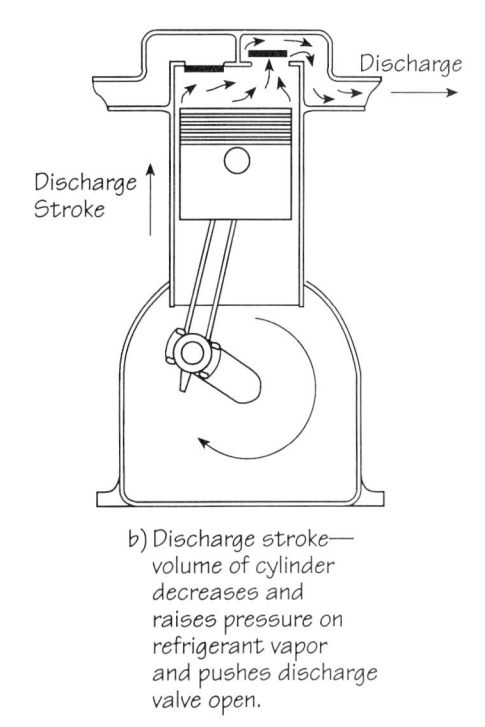

FIGURE 20.14 Reciprocating compressor, open-type.

a) Suction stroke—vaporized refrigerant enters the cylinder.

b) Discharge stroke—volume of cylinder decreases and raises pressure on refrigerant vapor and pushes discharge valve open.

against leakage of refrigerant around the driveshaft. A disadvantage of hermetic compressors is the loss in efficiency (5–10%) since the motor heat is contained.

Reciprocating compressors are very common and are well suited for applications ranging from very small to several hundred tons of refrigeration. Vibration caused by the reciprocating motion can preclude their use for some applications.

Scroll (Orbital) Compressors

Scroll compressors are a relatively new product in the HVAC field, employed since the mid-1980s. Today they appear to be gaining in acceptance as several leading manufacturers now use scroll compressors in many units in lieu of reciprocating compressors.

Scroll compressors use two interlocked spiral-shaped members which enclose the refrigerant gas in spaces (pockets) between the members. One spiral is fixed, the other rotates or "orbits." This causes the refrigerant to be compressed in smaller and smaller pockets until it reaches the center where it is discharged (see Figure 20.15).

Scroll compressor units are currently available for relatively small capacities of up to 15 tons. Advantages include low maintenance, low noise and vibration, and a higher efficiency than reciprocating compressors.

Helical (Screw) Compressors

Again the goal of any positive displacement compressor is to reduce the volume of an initial amount of gaseous refrigerant. The screw compressor does this by having a pair of helical-shaped screws which mesh while rotating and compress the volume of refrigerant gas as it travels from the inlet before being discharged (see Figure 20.16). Screw compressors are available for applications

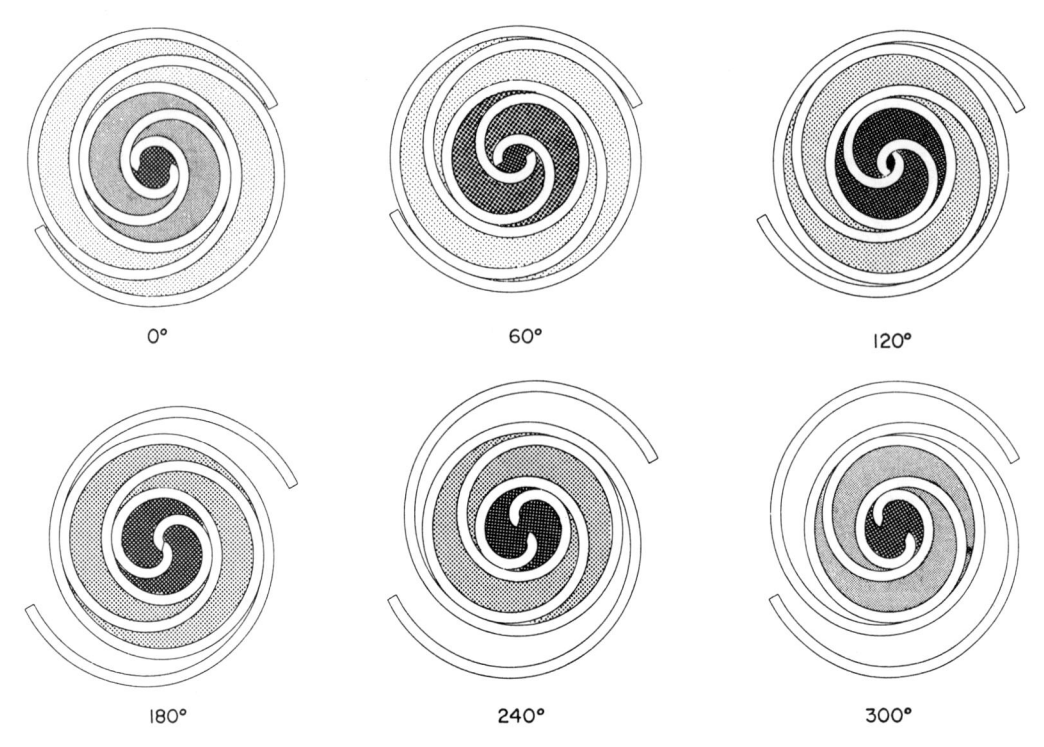

FIGURE 20.15 Scroll compressors are being increasingly used by manufacturers of small capacity package units. Source: Reprinted with permission from the *ASHRAE Handbook*, Equipment Volume, 1988.

ranging from about 20 to 1000 thousand tons.

Centrifugal Compressors

Anyone who has experienced an amusement park ride which rotates quickly and pushes the riders outward with great force can begin to appreciate how a centrifugal compressor works (see Figure 20.17). Instead of employing positive displacement, these compressors increase the kinetic energy (velocity) of the refrigerant gas through centrifugal force before then con-

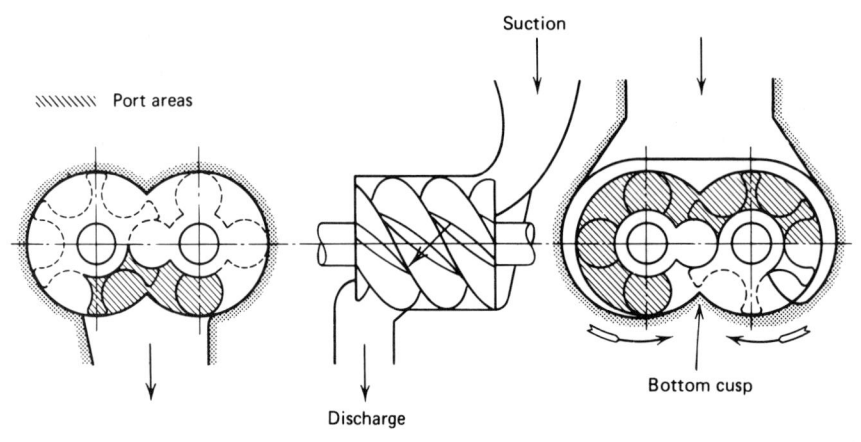

FIGURE 20.16 Helical (screw) compressors are compact in size and operate with a minimum of vibration. Source: Reprinted with permission from the *ASHRAE Handbook*, Equipment Volume, 1975.

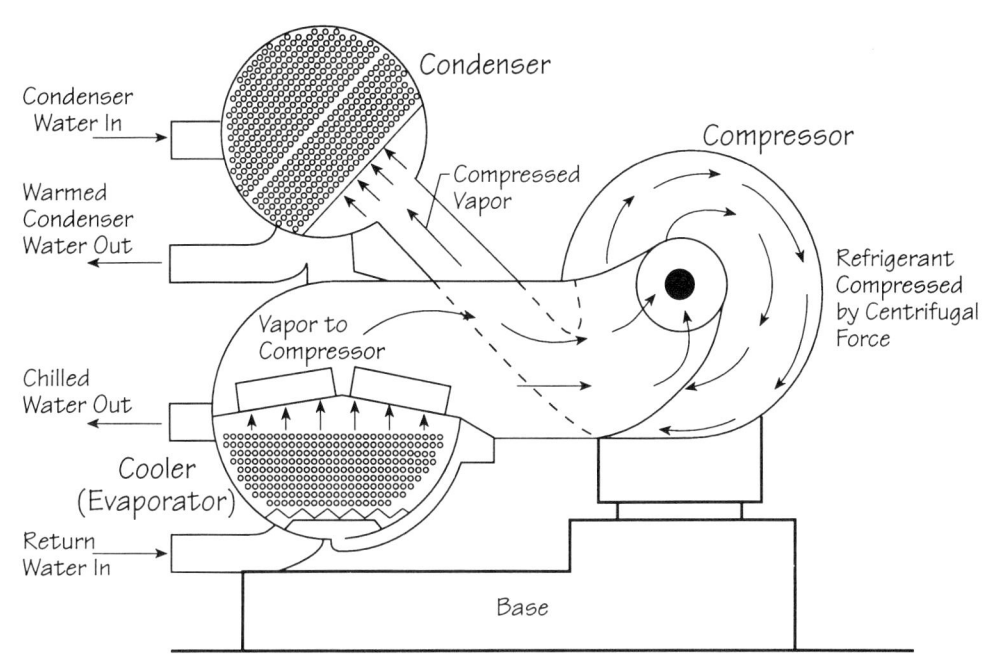

FIGURE 20.17 Centrifugal compressors—used in large applications above 150–200 tons of refrigeration.

verting the energy into increased pressure. This is accomplished by introducing the gaseous refrigerant into the center of rapidly spinning disks (up to 20,000 revolutions per minute) known as impellers to force the vapor outward and increase velocity.

20.11 COMPRESSOR CAPACITY CONTROL

Control approaches which are used to match system cooling capacity to the load are discussed below.

Cycling

Control of cooling capacity by cycling equipment on and off is only well suited to residential and small commercial systems. Frequent cycling will result in a shortened life of the compressor and motor as well as wide temperature swings.

Cylinder Unloading

For systems with reciprocating compressors, an easy and common approach to capacity reduction is known as "cylinder unloading." Although all the pistons continue to reciprocate, the refrigerant is not compressed in the unloaded cylinder(s). This is accomplished by controls which either hold open the cylinder suction valve, or cause the refrigerant to bypass the cylinder(s) entirely.

Hot-Gas Bypass

After the refrigerant is compressed, the normal destination in the vapor compression cycle is the condenser. Not true, though, for some small packaged equipment when a reduced cooling load occurs. In these units some of the "hot-gas" (refrigerant) is controlled to "bypass" the condenser coil and direct it to the evaporator coil inlet as shown in Figure 20.18.

The hot-gas bypass method of capacity

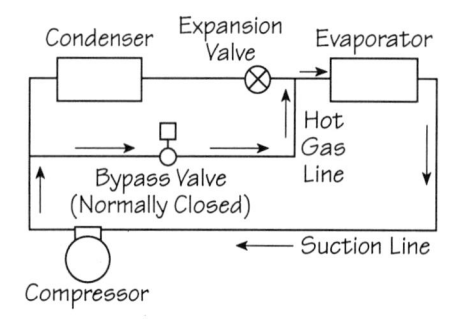

FIGURE 20.18 Bypassing the "hot (refrigerant) gas" past the condenser coil to reduce capacity.

control is wasteful of energy since the compressor continues to operate at some fixed minimum capacity, even when meeting part-load conditions. Hot-gas bypass is also sometimes used by larger equipment to stabilize operation at low loads.

Speed Change

Capacity varies as a function of compressor speed. Systems of moderate to large size often employ solid-state motor controls to reduce compressor speed under part-load conditions.

Multiple Compressors

Capacity control of larger cooling applications can be met by sequencing two or more compressors of equal or varied capacity. Control sequences can be established to have one or more of the compressors operating constantly to meet a base load with the other compressor(s) cycling occasionally. An additional advantage of using multiple compressors is the "stand-by" ability to meet a significant portion of the peak cooling load even if one compressor fails.

20.12 PART-LOAD PERFORMANCE (IPLV)

Although a cooling system must be designed to satisfy the maximum cooling load, it is also true that peak loads occur very infrequently. Far more typical are the many hours of operation when cooling loads are in the range of 25 to 75% of full capacity.

In selecting cooling system equipment, a key criteria should always be the efficiency of part-load performance. ASHRAE/IES Standard 90.1-1989 for "Energy Efficient Design of New buildings except New Low-Rise Residential Buildings" contains provisions for Integrated Part-Load Value (IPLV) performance of certain types of ARI-rated cooling equipment with capacity control capability. IPLVs have the same units as EERs. The IPLV is a weighted average of full-load and three part-load performances when operated in steady-state (to exclude the effects of cycling). The IPLV for chiller performance is determined by Formula 20C which is extracted from ARI Standard 590.

$$\text{IPLV} = (0.17 \times \text{EER}_{100})$$
$$+ (0.39 \times \text{EER}_{75})$$
$$+ (0.33 \times \text{EER}_{50})$$
$$+ (0.11 \times \text{EER}_{25})$$

(Formula 20C)

where

IPLV   is the integrated part-load value, a weighted average of steady-state performance for full-load and three part-load conditions

0.17   is the fraction of chiller operation at full-load (EER_{100}). Note that the four fractions of operational time indicated in the formula (0.17, 0.39, 0.33, and 0.11) add to 100% of operation time

EER_{100} is the EER rating of the equipment when operating in steady state at full load (100%). Similar ratings for steady-state operation at 75%, 50%, and 25% of full load.

Example 20F: Determine the IPLV of a chiller which has tested part-load performance EER values as follows:

Part-Load Rating (%)	EER
100	13.2
75	14.0
50	14.5
25	15.5

Solution: Use Formula 20C to obtain the weighted average of the EER performance for the four load ratings.

$$\text{IPLV} = (0.17 \times \text{EER}_{100})$$
$$+ (0.39 \times \text{EER}_{75})$$
$$+ (0.33 \times \text{EER}_{50})$$
$$+ (0.11 \times \text{EER}_{25})$$

(Formula 20C)

$$= (0.17 \times 13.2)$$
$$+ (0.39 \times 14.0)$$
$$+ (0.33 \times 14.5)$$
$$+ (0.11 \times 15.5)$$
$$= 2.24 + 5.46$$
$$+ 4.79 + 1.71$$
$$= 14.2, \text{ which is equivalent to a COP of } 4.16 \ (14.2/3.413)$$

20.13 ABSORPTION REFRIGERATION

The great, great majority of mechanical air conditioning equipment in use today (as detailed above) employs the vapor compression refrigeration cycle and uses a great deal of (usually expensive) electricity. Luckily there is another cooling "cycle" which can be used. Known as absorption refrigeration, this cycle may seem paradoxical at first since it requires a heat source. Absorption refrigeration can be economical for certain large applications and does not employ ozone-depleting refrigerants.

The absorption refrigeration cycle uses a salt solution to absorb moisture and in the process produce a cooling effect. The cycle begins with the absorption of water vapor by a salt solution (typically lithium bromide). As shown in Figure 20.19, a compartment (the evaporator) contains the refrigerant (water). Pressure in the evaporator is subatmospheric to lower the boiling point of the refrigerant. When the water boils (evaporates) it creates a cooling effect in the remaining water in the tank, just as evaporation of moisture from our skin causes a cooling effect.

The refrigerant (water) which evaporates then moves as a vapor to another compartment (known as the absorber). The salt solution in the absorber picks up the water vapor just as common table salt will absorb moisture when humid conditions exist. A coil in the evaporator then transfers the cooling effect of the evaporated spray to water which is typically chilled about 10°F (from about 55–45°F).

The absorption process described so far would only function for a limited time before the salt solution became saturated. To avoid this a "generator" is added to heat the salt solution so that water will boil off and recharge the solution. The hot dehydrated salt solution is returned to the absorber compartment via a heat exchanger where it transfers some of its heat to the

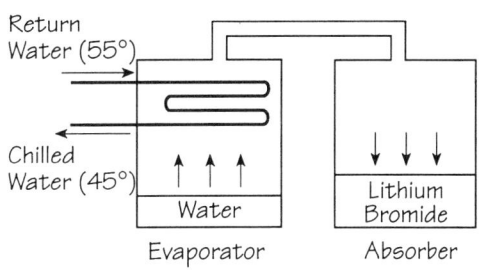

FIGURE 20.19 The initiation of the absorption refrigeration cycle.

water/salt solution being pumped to the generator.

The water vapor driven from the salt solution in the generator flows through a condenser where it changes back to a liquid and is returned to the evaporator to complete the closed cycle. Condensing water also is circulated through a heat exchanger in the absorber to remove any waste heat.

The complete absorption refrigeration cycle is illustrated in Figure 20.20. The basic process is accelerated by using recirculation to spray a mist in each compartment which greatly increases the surface areas of the water and salt solution.

Steam is probably the most commonly used heat source to recharge the salt solution in absorption chillers. Absorption chillers are a good option where utilities sell steam and/or high peak demand rates are in place.

Absorption Chillers

Absorption chillers are package units which employ the absorption refrigeration cycle and require a source of heat to regenerate the salt solution in the absorber.

A problem with absorption cooling has been the low COP (0.5–0.7) of conventional single-stage units which require about 18 lb of steam to produce one ton-hour of refrigeration. Today, most applications which lend themselves to absorption refrigeration are being met by more efficient two-stage absorption chillers. These units have COPs of 1.0 to 1.2 and can be fired by steam, gas, or oil.

Disadvantages of two-stage absorption chillers are increased initial cost (up to double the cost of a comparably sized electric chiller) and the need for a cooling tower. These costs are increasingly being offset by

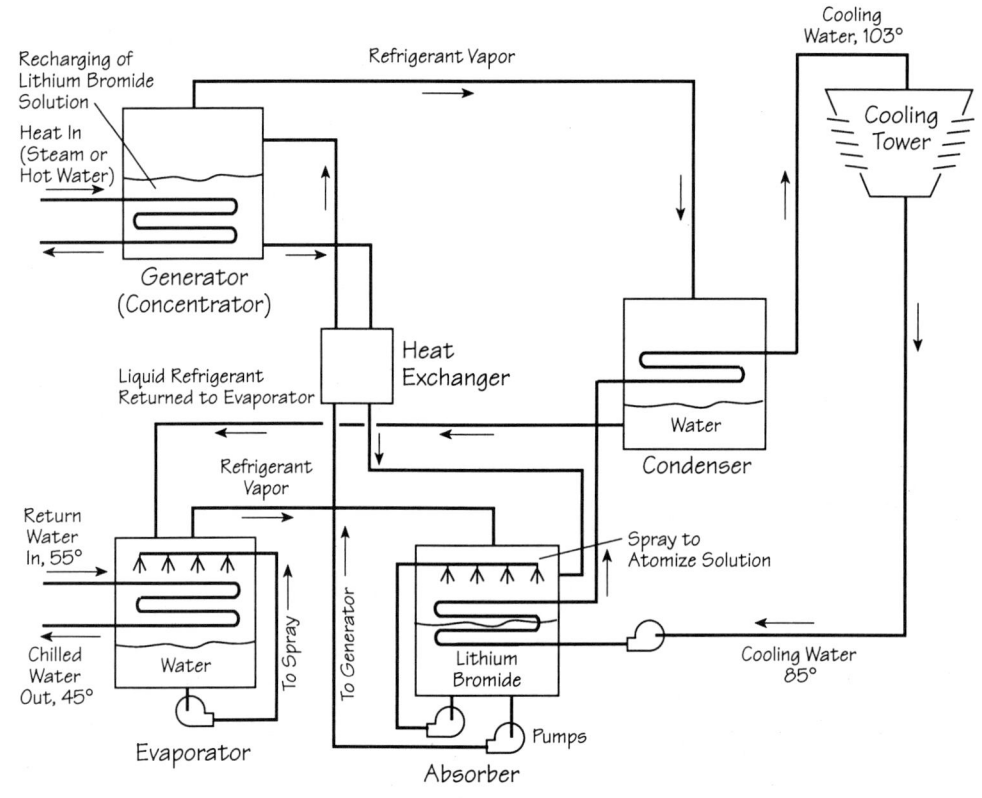

FIGURE 20.20 The complete absorption refrigeration cycle.

rebates offered by many inner-city utilities. This is because they do not add to the summer afternoon peak demand faced by the utility.

20.14 OTHER NONELECTRIC COOLING EQUIPMENT

Gas-Fired Cooling

Absorption refrigeration is not the only alternative to electrically driven cooling equipment. Increasingly, cooling equipment driven by a gas-fired internal combustion engine is being developed and introduced to commercial usage. Some utilities are even offering rebates and other incentives when gas-fired cooling units are installed since they do not add to peak electrical demands.

Gas-fired packaged water chillers which use reciprocating or screw compressors are now available in capacities of 25–500 tons. The basic units have COPs of about 1.6 at full load which improves slightly during part-load performance (through engine speed modulation). A gas-fired internal combustion engine converts about one-third of the fuel consumed to shaft power. The remainder is dissipated as heat, much of which can be recovered and used to produce hot water. Smaller gas-fired cooling units including heat pumps are still in the development and demonstration stage.

Turbine-Drive Chillers

Gas (and steam) turbines can be used to drive large centrifugal compressors (500- to 5000-ton machines). Most turbines used for HVAC cooling applications are of the multistage type with efficiencies ranging from about 55 to 75%. Turbines operate at very high speeds (typically 3,000–10,000 rpm) with speed reduced when part-load conditions exist.

21

COOLING DISTRIBUTION AND DELIVERY SYSTEMS

Heat gains within buildings from internal and external sources can be offset in various ways. Perimeter spaces in some building types (e.g., offices, hotels, motels, and dormitories) often use locally controlled individual units for cooling and heating. These include package terminal air conditioners, package terminal heat pumps, and fan-coil units.

Internal spaces within most commercial buildings usually require mechanical ventilation and have a need for heat removal even in winter. Small buildings often use some type of unitary equipment (often rooftop) for these interior spaces which deliver a constant volume of air to one or more zones. Larger buildings typically have air handling units (AHU) with supply fans which deliver cool air through ductwork installed by "tin knockers" as they are affectionately known by the construction trades.

Discussed on the following pages are the various cooling distribution and delivery systems using chilled water and cool air in common use today.

21.1 LOCAL (INDIVIDUAL) SYSTEMS

The simplest local system to cool a room or small space is the individual unit which is mounted in a window or wall opening (as described in Section 18.5). All components (compressor, evaporator, condenser, etc.) are self-contained in one package with cooling (heat removal) capacities of up to about 30,000 Btu/h ($2\frac{1}{2}$ tons) available. Small commercial buildings which use room air conditioners for cooling usually have some form of central heating system such as baseboard hot water heating.

Packaged Terminal Air Conditioners (PTAC)

This system is a small capacity "through the wall" unit which incorporates all of the features of the individual (window) air conditioning unit while also having the ability to provide heat. When the heat source is electricity, the PTAC needs only an electrical connection. Other PTACs are piped to sources of circulating chilled and hot water. A typical PTAC is shown in Figure 21.1. Some PTACs are actually packaged terminal heat pumps (PTHP) which also satisfy the need for heat in winter once controls reverse the cycle and "reject" heat indoors.

PTACs and PTHPs are common in buildings where each space is a zone under individual control such as in hotels and motels, apartment buildings, and nursing homes. They are also installed in portions of buildings where the major part of the building is conditioned by a central system such as the office areas in churches and theaters.

The terminology of PTAC and PTHP is fairly new. Many still refer to them as "incremental" units since they are available in a range of capacities or increments in cooling capacities from about $1\frac{1}{2}$ to 5 tons.

Fan-Coil Units

Fan-coil units circulate centrally produced chilled and hot water through coils to produce the desired thermal effect. Fan-coil units allow for individual temperature control in spaces such as office buildings where close control of humidity is not required. Although the cooling coils of fan-coil units will cause some moisture to condense, in most cases fan-coil units are not used where significant latent loads occur.

Fan-coil units are available in a variety of configurations including vertical, low-profile vertical, and horizontal ceiling models. Depicted in Figure 21.2 is a typical fan-coil unit including a drain pan connected to a drain line to collect and remove any moisture taken from the air as it flows over the cold coil.

Most fan-coil units have typically been placed below windows to counter heat loss through glass and to offset drafty conditions. Today, though, the use of high-performance "low-E" glazing with its lower U-factor can sometimes allow for units which are located at inner walls or are ceiling mounted.

Simplest and least expensive is the two-pipe fan-coil with one pipe supplying either chilled or hot water and with the second pipe serving as the return. The primary limitation of such a system is that all units served by a single piping system receive

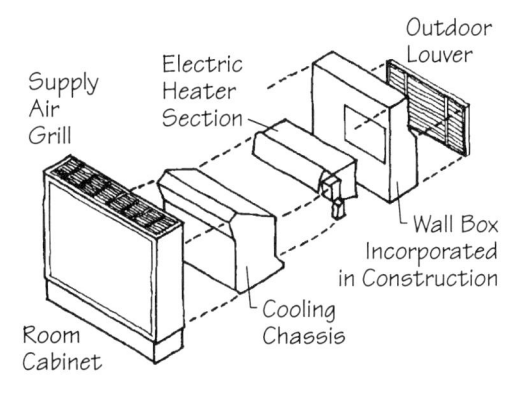

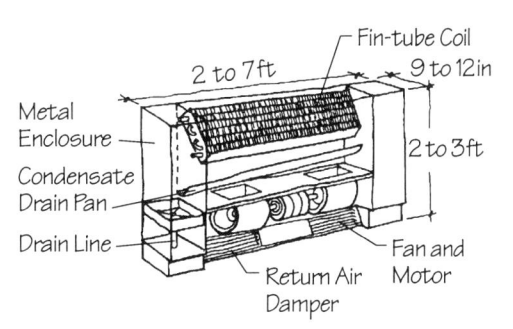

FIGURE 21.1 Packaged terminal air conditioner (PTAC).

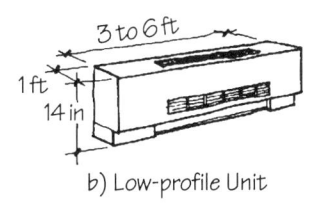

FIGURE 21.2 Fan-coil unit components.

either chilled or hot water—but not both. This presents a problem during the spring and fall in buildings which need heat in some spaces while other spaces need cooling. One solution is to also incorporate an electric resistance heating coil in the fan-coil unit (applicable to mild climates and areas with low electricity rates). The electric heating coil can then provide the limited amount of heat needed by some spaces during the swing seasons when chilled water is being circulated to and from the fan-coil units.

More typically, the problems of changeover are avoided by using four-pipe fan-coil units. These units can provide either heating or cooling to any space year round as required—but of course with an additional piping cost. Also possible, but very uncommon, is a three-pipe system which mixes the hot and cold return water and results in a system which is inherently energy wasteful.

Presented in Table 21.1 is a selection of cooling capacity information for typical vertical fan-coil units. Data shown includes that for the ARI standard rating condition of entering air at 80°F db/67°F wb (50% RH), 45°F entering water temperature, and a 10°F water temperature rise through the coil. TH refers to total heat, and SH refers to sensible heat, both in thousands of Btus per hour (MBtu/h). The latent heat capacity will be equal to TH−SH.

Example 21A: Select two-pipe fan-coil units (FCU) from Table 21.1 to cool the perimeter zones of "The Office" when located on the top floor in Philadelphia and based upon the following:

The perimeter zones are 15-ft wide.
Cooling loads were computed in Example 8A as follows:

Zone	Sensible Heat (Btu/h)	Latent Heat (Btu/h)
South	22,475	4,075
West	23,755	3,105

TABLE 21.1 FAN-COIL UNITS ENTERING AIR: 80°F db/67°F wb (50% RH)

Water Temp. Rise Through Coil (°F)	cfm	40°F TH	40°F SH	40°F gpm	45°F TH	45°F SH	45°F gpm	50°F TH	50°F SH	50°F gpm
8	200	7.9	5.3	2.0	6.3	4.7	1.7	4.8	4.1	1.3
	300	12.2	8.0	3.2	9.7	6.9	2.5	7.4	6.1	2.0
	400	15.3	10.1	3.9	12.3	8.8	3.1	9.4	7.7	2.4
	600	22.1	14.3	5.6	17.5	12.3	4.5	13.5	10.8	3.5
	800	32.9	20.9	8.3	26.1	18.0	6.6	19.5	15.4	5.0
	1000	38.6	24.6	9.8	30.6	21.1	7.8	22.9	18.2	5.8
	1200	48.5	30.9	12.3	38.4	26.6	9.7	28.3	22.6	7.2
10	200	7.3	5.0	1.5	5.3	4.3	1.2	4.2	3.9	0.9
	300	11.3	7.6	2.3	8.1	6.6	1.9	6.8	5.8	1.4
	400	14.2	9.6	2.9	11.4	8.5	2.3	8.6	7.4	1.8
	600	20.5	13.6	4.2	16.4	11.9	3.4	12.5	10.4	2.6
	800	30.9	20.0	6.3	24.2	17.3	4.9	17.9	14.9	3.6
	1000	36.2	23.5	7.3	28.4	20.3	5.8	21.1	17.6	4.3
	1200	45.4	29.6	9.2	35.4	25.4	7.2	25.8	21.7	5.3

Source: The Trane Company.

Solution: Based upon the plan of the space, and to allow for even cooling distribution and local control, it is decided to locate two units in each structural bay (see Figure 21.3). Therefore, a total of four units will be placed below south-facing windows and four units will be placed below west-facing windows.

Each of the six units in the south zone must be able to meet a sensible cooling load of 5620 Btu/h (22,475/4). Similarly, each of the four units in the west zone must be able to meet a sensible cooling load of 5940 Btu/h (23,755/4). Note that the latent cooling loads are small and should be easily met by the units selected for their sensible capacity.

To address the required sensible cooling capacities, the units can be chosen from among those presented in Table 21.1. Ultimate selection will depend upon various factors including chilled water temperature requirements of other equipment and coils in the building.

One possible system choice for the units along the south wall is to use a 45°F entering water temperature and a 10°F water temperature rise through the coil. If 300-cfm units are used each will have a sensible cooling capacity of 6600 Btu/h, and a latent cooling capacity of 1500 Btu/h (TH—SH). Using the same design temperatures for the units along the west wall results in selection of the same capacity 300 cfm units.

In most cases, fan-coil units which are sized to meet sensible cooling loads will have a heating capacity which is more than adequate. This should of course be verified using the sizing procedures presented in Chapter 15 for hot water heating systems. Consideration might also be given to installing units with electric heating coils in mild climates or areas with low electric rates to allow for operational flexibility of the two-pipe system during the intermediate seasons before changeover occurs.

Closed-Loop Heat Pumps

A series of water-to-air pumps can be used for cooling (and heating) of large buildings. Such a system, known as a "closed-loop heat pump" system, is generally integrated with a hot water boiler and cooling tower to assure complete and flexible operation to meet all load conditions as illustrated in Figure 21.4. With a closed-loop heat pump system, excess heat in particular zones of the building (i.e., south and interior) can be utilized in other zones needing heat such as the north zone.

Packaged terminal heat pumps (PTHP) used with closed-loop systems are relatively small, locally zoned units typically located above the ceiling or under windows. At such locations, the impact of compressor noise on occupancy can be objectionable. Detailed information on heat pumps is provided in Chapter 16.

21.2 AIR SUPPLY FUNDAMENTALS

Cooling Through Supply Air

The amount of heat which will be removed by an air supply system depends on supply

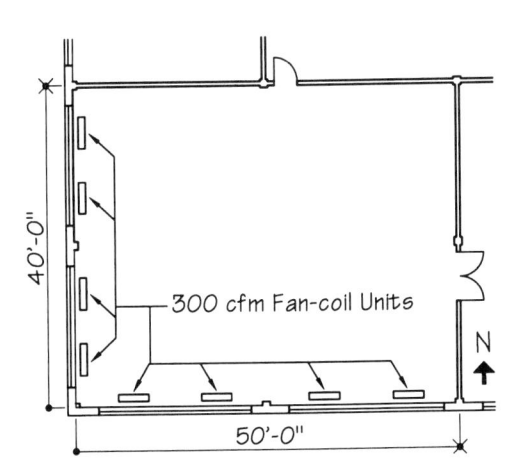

FIGURE 21.3 Fan-coil selections for "The Office."

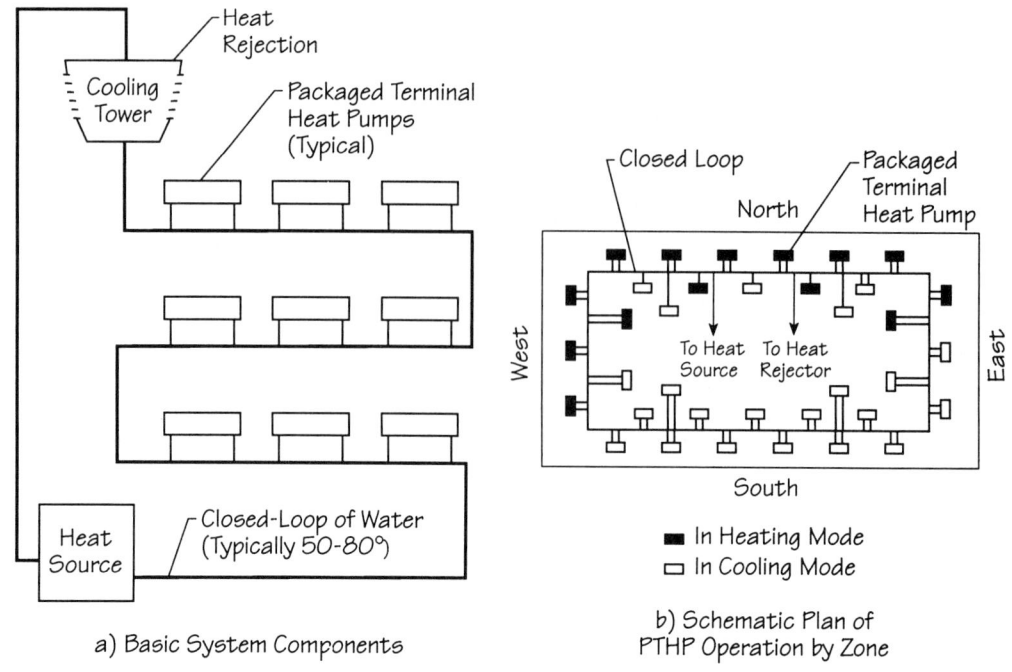

356 COOLING DISTRIBUTION AND DELIVERY SYSTEMS

FIGURE 21.4 Closed-loop heat pumps can provide energy-efficient conditioning of building zones which need simultaneous heating and cooling (heat removal).

air flow rate (cfm) and temperature difference $(t_i - t_s)$. In building cooling applications, a design air temperature difference of approximately 18–20°F is common. In addition, for human comfort, air supplied to a room should not be any colder than about 53°F.

Formula 21A is used (at sea level and up to 2000 ft) to determine the required air flow rate to meet a given sensible cooling load. For higher elevations increase cfm by 4% per 1000 ft because of the lighter air.

$$\text{cfm} = \frac{q_{\text{tot}}}{1.10 \times (t_i - t_s)} \quad \text{Formula 21A}$$

where:

cfm   is the air flow requirement in cubic feet per minute

q_{tot} is the total quantity of sensible heat needing to be removed in Btu/h

1.10   is a constant based on the specific heat and density of air (see Formula 6C for details)

t_i is the inside design temperature in °F

t_s is the supply air temperature in °F

In analyzing Formula 21A, one can see that removal of a given amount of sensible heat can be accomplished in various ways by varying the supply air flow rate and the temperature difference (between the inside design temperature and the supply air temperature).

Nomographs allow for the graphical solution of a mathematical equation. Figure 21.5 is a nomograph for determining the required supply air flow in cfm based upon temperature difference. To obtain an answer enter along the vertical axis (sensible heat removal in Btu/h) and draw a horizontal line to the applicable temperature difference. Then draw a vertical line to inter-

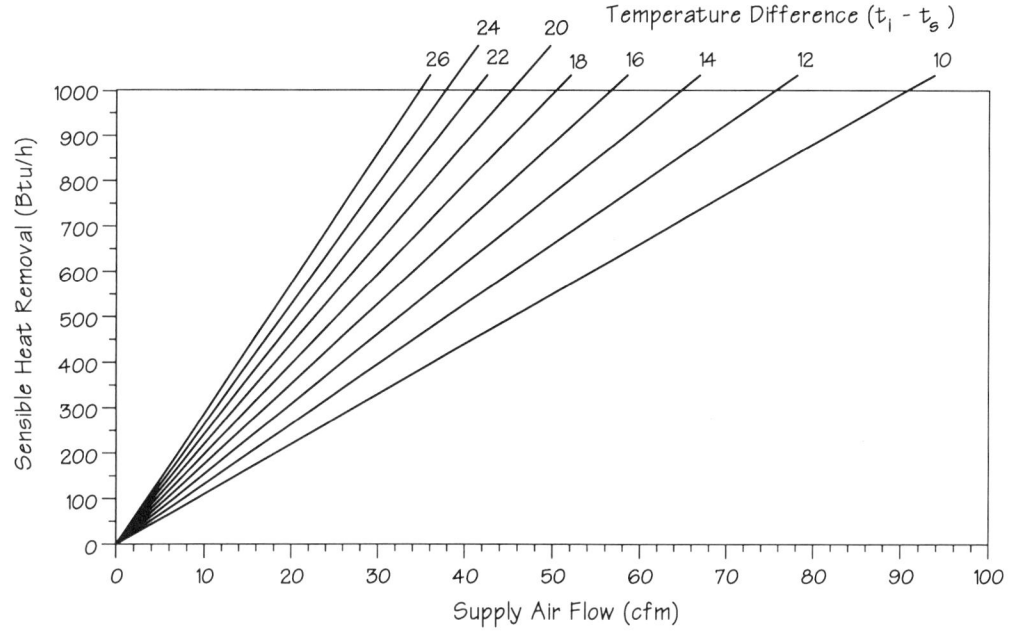

FIGURE 21.5 Nomograph: heat removal by supply air.

cept the horizontal axis and obtain the answer in cfm.

Example 21B: Use the heat removal nomograph to determine three combinations of supply air flow rate and temperature difference between the inside design temperature and the supply air temperature to remove 8000 Btu/h of room heat. Also determine the supply air flow rate for a reduced load of 5000 Btu/h, and a temperature difference of 20°F.

Solution: As shown on Figure 21.6, the nomograph is entered along the vertical axis at 800 Btu/h. Realize that since this value is only 1/10th of the cooling load, the results obtained must then be multiplied by 10.

Move horizontally within the nomograph to bounce off the three values of temperature difference ($t_i - t_s$), in this case 24, 20, and 16°F. Along the horizontal axis, the nomograph indicates the respective results of approximately 303 cfm, 364 cfm, and 455 cfm (remember to multiply answers along the horizontal axis by 10).

The nomograph could also be used to determine the required air quantity for a cooling load of 5,000 Btu/h and a temperature difference of 20°F. For this case the required air quantity would be approximately 225 cfm compared to the 364 cfm air flow required to meet the higher cooling load of 8,000 Btu/h for the same temperature difference of 20°F. This demonstrates how variable air volume (VAV) systems could be designed to meet changing loads as discussed below.

Supply Air Flow Rates: "Rules of Thumb"

Formula 21A clearly shows that the cooling load (q_{tot}) for a given design application determines the required supply air flow rate in cfm. Sometimes it is useful to be able to estimate air flow requirements even before detailed cooling load calculations are performed. For "typical" office buildings a

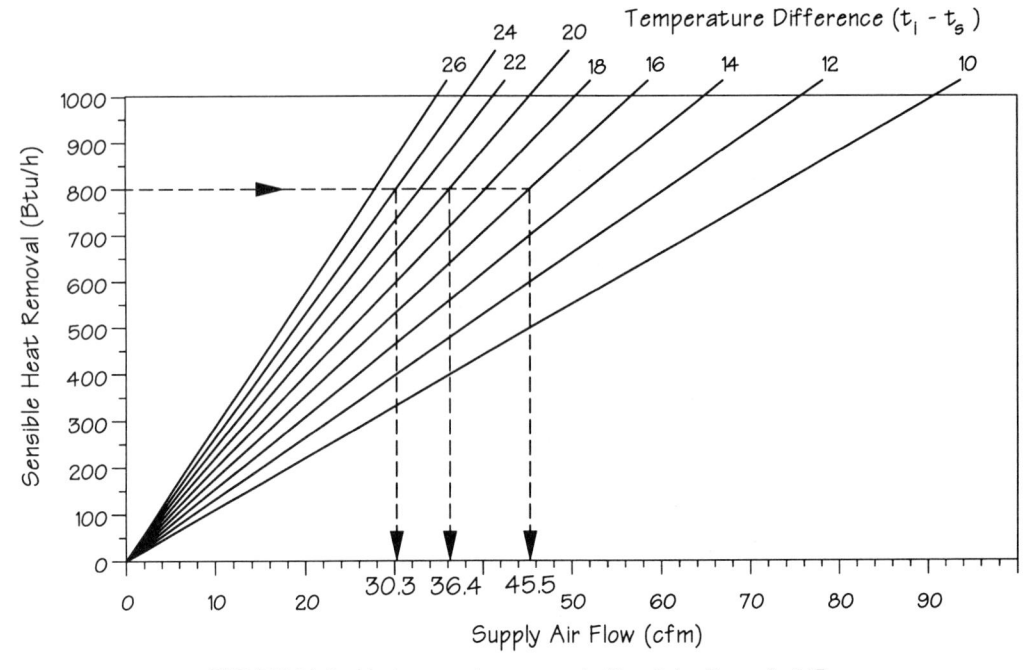

FIGURE 21.6 Heat removal nomograph: Result for Example 21B.

useful rule of thumb is a supply air requirement of approximately 1 cfm per square foot of office area for interior building zones. In contrast, exterior building zones typically have higher cooling loads which typically translate into the need for about 1.5 cfm per square foot of office area (or more). Be sure to only use these rules of thumb for preliminary design of typical office buildings. And with experience you will be able to develop additional rules of your own for other building types.

Air Velocity

Once air flow rates in cfm are known it is possible to design a delivery system of supply air ductwork (see Chapter 9 for detailed information on duct design). Recommended air velocities shown in Table 21.2 vary depending upon application and whether the controlling factor will be noise generation or duct friction (and resulting impact upon the performance of the supply air fan). "Low velocity" systems for commercial buildings refers to air velocities of up to 2500 fpm.

It should be stressed that values in Table 21.2 are maximum air delivery velocities which apply best to ducts supplying large air quantities. For small air quantities lower velocities are often called for to keep developed friction to a manageable amount.

Risers and Ceiling Spaces

Vertical and horizontal distribution of air supply systems is a major design issue requiring coordination between the architect and engineer. During the schematic design phase, basic HVAC system selection must be made since it:

- Will impact upon needed hung ceiling space and resultant floor-to-floor heights
- Often presents architectural limitations at the perimeter

21.2 AIR SUPPLY FUNDAMENTALS

TABLE 21.2 RECOMMENDED MAXIMUM DUCT VELOCITIES FOR LOW VELOCITY SYSTEMS IN FEET PER MINUTE (fpm)

Application	Controlling Factor: Noise Generation Main Ducts	Controlling Factor: Duct Friction Main Ducts Supply	Return	Branch Ducts Supply	Return
Apartments	1000	1500	1300	1200	1000
Private Offices, Libraries	1200	2000	1500	1600	1200
Theaters, Auditoriums	800	1300	1100	1000	800
General Offices, Banks	1500	2000	1500	1600	1200
Average Stores	1800	2000	1500	1600	1200
Industrial	2500	3000	1800	2200	1500

Source: Courtesy of The Carrier Company.

- Will require designation of space for equipment
- Often will require large vertical shafts for risers

Risers are used to deliver air from an air-handling unit to a conditioned space on a different floor. Duct risers that have one dimension which is two to four times the other dimension are usually best in allowing for transition from vertical to horizontal ductwork. To minimize duct size it is usually better to have several well-spaced shafts rather than one large one. Often, however, open office planning limits the locations available for vertical risers. Openings around ducts within shafts should be sealed to prevent any possible spread of fire and smoke and possible stack effect (see Section 6.9).

The required area (A) of a duct riser in square feet (ft^2) can be estimated by using Formula 21B (a simplification of Formula 13C). To use the formula one must know the required air flow rate in cfm (ft^3/m) and air velocity in fpm (ft/m). The value of 1.25 represents a frictional allowance (FA) to account for ducts with a pronounced rectangular shape typical of supply and return shafts.

$$A = \frac{ft^3/m}{ft/m} \times 1.25 \quad \text{Formula 21B}$$

Example 21C: Estimate the size of supply and return risers during the schematic design phase for an office building for a law firm. The riser will be serving 7200 ft^2 of interior space for the library, meeting rooms, and offices.

Solution: Since the riser will serve an interior zone of office space the rule-of-thumb of 1 cfm per square foot can be used. This results in a supply air flow of 7200 cfm (ft^3/m).

Table 21.2 lists maximum recommended air velocities. Since noise control in a law office is very important, a maximum velocity of 1000 fpm (ft/m) will be assumed for the supply and return air risers.

Formula 21B can now be used to estimate the required area for each riser (supply and return) as follows:

$$A = \frac{ft^3/m}{ft/m} \times 1.25 \quad \text{(Formula 21B)}$$

$$= \frac{7200}{1000} \times 1.25$$

$$= 9.0 \; ft^2$$

360 COOLING DISTRIBUTION AND DELIVERY SYSTEMS

Based upon this finding, supply and return risers which are 1.5 ft by 6.0 ft or 9 ft^2 each are incorporated into the design to serve the zone as shown in Figure 21.7. Shafts incorporated into the construction would be slightly larger to allow for insulation on supply risers and construction clearances.

21.3 BASIC AIR SYSTEM ELEMENTS

Many HVAC air supply systems contain the same basic components in varying forms. Therefore, it is appropriate to describe these components and learn the terminology used for "air-handling units" which supply air to satisfy ventilation requirements and space conditioning needs as shown in Figure 21.8.

Outside Air Intake

The outside air intake (OAI) or fresh air intake (FAI) is sized to admit at least the minimum required ventilation air. In many areas, it should also be sized to allow a much greater air flow so that cool outside air can sometimes be used directly for cooling purposes, using what is known as an "economizer cycle." The OAI is normally a louver with a stormproof blade to prevent wind-blown rain from entering while intake air velocities are limited to 400 fpm to avoid sucking in moisture. Screens are mounted on the inside of the louver to prevent entry of insects, birds, and animals.

Motorized Dampers

Motorized dampers are used to open, close, or modulate the outside air supply. It is common for system controls to have dampers which: admit the minimum quantity of

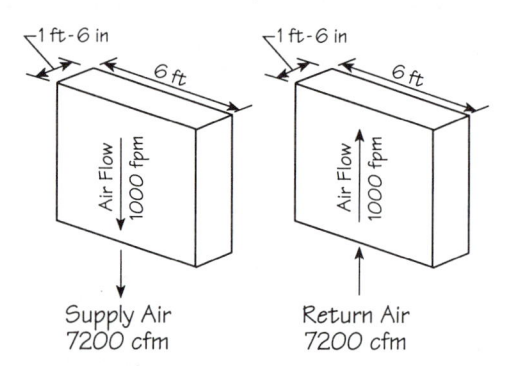

FIGURE 21.7 Duct riser sizes for Example 21C.

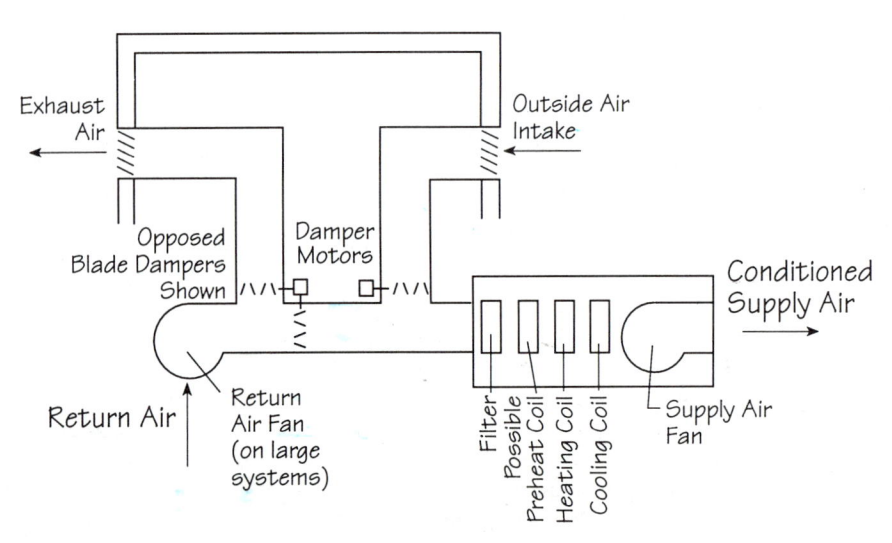

FIGURE 21.8 Air-handling system components.

ventilation air required during very cold weather, open more fully if cool outside air can be used for cooling purposes, and again close to allow a minimum amount of outside air for hot/humid summer conditions. These automatic dampers are controlled to work in concert with dampers which control the recirculating air stream. These dampers also close fully when a freeze condition exists.

Filters

The air stream is filtered before it is conditioned. Filters come in many types and efficiencies depending upon the particular equipment and design application (see Section 6.10). Often several filters are used in combination to achieve a high-quality air stream.

Preheat Coil

In cold climates, preheat coils are installed in systems that serve spaces that require large quantities of outside air. They are designed to heat the cold outside air to about 45°F before it travels further in the system. This prevents freezing of water in the other coils (e.g., the cooling coil).

Conditioning (Heating and Cooling) Coils

The air stream next passes over heating and/or cooling coils which will condition the air as required to satisfy supply air requirements. These coils (and in particular the cooling coil) impose frictional resistance to air flow, which must be overcome by the fan system.

During the cooling season, the cooling coil will cool and dehumidify the supply air. During cold weather, the heating coil is activated to warm the air stream.

Supply Air Fan

In the system being described, the supply air has been drawn through the louver, damper, filter, and coils by the supply air fan. Such a system is known as a "Draw through" fan arrangement, as shown in Figure 21.9. A system where the fan location is before the coils is referred to as a "blow through" fan arrangement.

In either case, the job of the supply fan is to overcome system friction and deliver supply air to each terminal device (diffuser).

Supply Air Ductwork

In a central system, once the air has been conditioned it must be delivered to the air terminal devices through supply air ductwork in as direct a manner as possible. Abrupt changes in direction of air flow should be avoided. To minimize friction, sharp turns in rectangular ductwork should be made with vaned elbows to guide the air around the bend. Offsets, transitions, and other items of supply air ductwork should also be designed to minimize friction (see Section 9.4). The supply air ductwork also

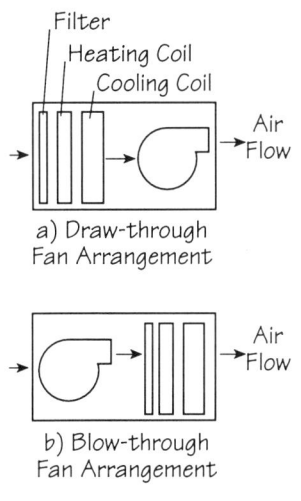

FIGURE 21.9 Fan arrangements in air handlers.

Supply Air Diffusers

Finally the supply air arrives at the supply air diffusers and enters the conditioned space. Most central system terminal devices are in a hung ceiling as shown in Figure 21.10. Depending upon system type, when cooling is required, the supply air may range in temperature from 52 to 70°F. For well-distributed cooling, an air flow pattern needs to be created in the room. Be sure to prevent short-circuiting of supply air by separating the supply of air diffusers and the return air grilles.

Return Air Inlets and Plenum

In modern buildings, a very common path for air to return to the central cooling equipment is in the space between the hung ceiling and the floor structure above—a space known as a "plenum." Return air usually flows into the plenum through large air grilles in the hung ceiling. Spacing of grilles is not as critical as for the supply air diffusers. Plenums should provide a clear and relatively unobstructed path for the return air with a low frictional resistance.

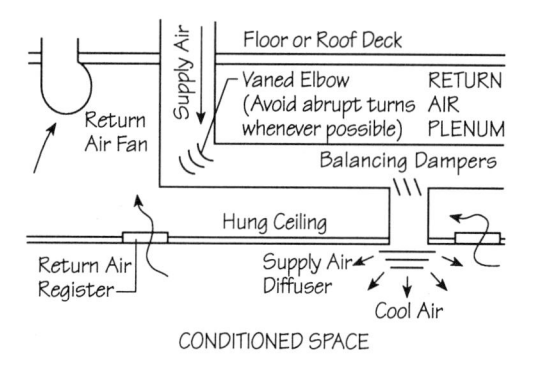

FIGURE 21.10 Supply air flow through ductwork to conditioned space and return through ceiling plenum.

Sometimes a system of return air ductwork is employed in lieu of a plenum. This is particularly true for applications in which air quality control is critical (e.g., operating rooms, laboratories, and clean rooms).

Air must also be returned from spaces (such as conference rooms) with full-height partitions which extend to the structure above. This is accomplished by having short runs of return air ductwork from the plenum which penetrate the partitions and connect to return air grilles.

Exhaust Air

An exhaust or "relief" air outlet dispels the quantity of air which will not be recirculated to allow for entry of new outside air. Motorized dampers are positioned by system controls as discussed in Chapter 22. The exhaust air outlet has a motorized damper and a stormproof louver with screen. Another goal of fan systems in most commercial buildings is to create positive pressure (approximately 0.05 in. wg) within the conditioned space. This will reduce infiltration of outside air which may contain undesirable temperature, humidity, dirt, or pollen. Pressurization is obtained by admitting 5–10% more outside air than is exhausted. A system which admits 5% more outside air than it exhausts is shown in Figure 21.11. Return air fans are used to aid in air return for large systems (about 10,000 cfm or larger) and systems with economizer cycles.

The Complete System

The air-handling system described above includes a fan to move the air, along with cooling and heating coils to condition it as needed. The complete system also includes the equipment which produces the heating effect in the heating coil, the cooling effect in the cooling coil, and a method of heat rejection as shown in Figure 21.12. In this illustration, the heat source is a boiler,

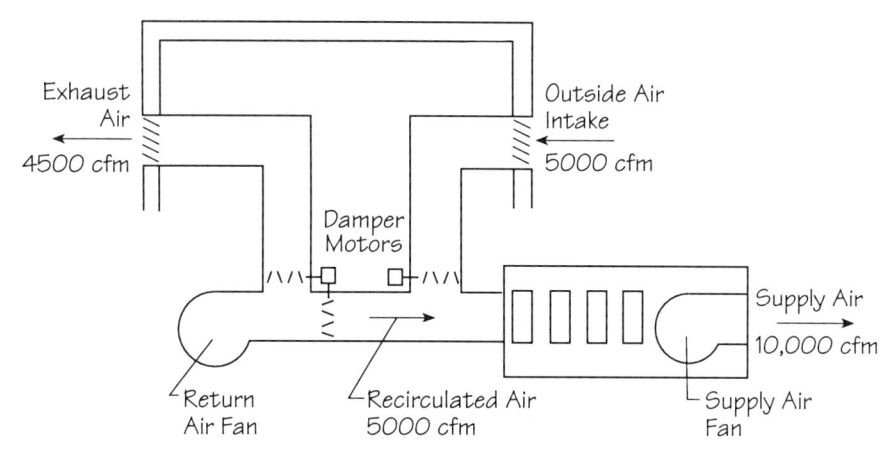

FIGURE 21.11 Air intake and exhaust for building pressurization.

though other sources of heat such as a steam converter or a double-bundle condenser (see Section 20.5) can also be used.

21.4 AIR SYSTEM TYPES

Since ventilation with outside air is required, it is common to provide interior areas of buildings with some type of air supply system. Buildings with large zones having similar thermal loads can often use very simple constant volume systems. Small buildings with several zones (i.e., N, E, S, W, and Interior) may use multizone units, often rooftop mounted. A common system type for spaces with varying cooling loads is some type of variable air volume (VAV) system. Other systems such as multizone, dual duct, and high-velocity induction are very energy intensive and are reserved for special applications.

Constant Volume Systems

The simplest and most common type of air cooling systems deliver a constant volume of supply air which varies in temperature. In DX units this variation in supply air temperature is achieved through compressor capacity control as discussed in Section 20.11. Temperature variation is also often achieved by controls which adjust the flows of hot and cold water through coils. Supply water temperatures of chilled and hot water supplied to the coils are also often varied (known as "reset") throughout the year to conserve energy. Constant volume systems are frequently used for the following applications:

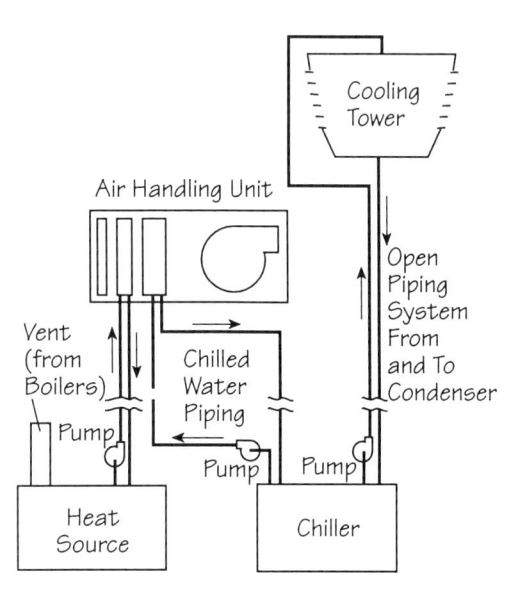

FIGURE 21.12 A complete system—air handling, sources of hot and cold water, and rejection of heat and possible combustion products.

Free-standing units located directly within or adjoining the space and having a limited amount of distributing ductwork (if any) can be used to cool retail shops and other small commercial single-zone spaces.

Larger buildings use central air-handling units to deliver a constant volume of conditioned (filtered, cooled, and humidity controlled) air to spaces or zones with similar cooling loads and space requirements.

Depicted in Figure 21.13 is a constant volume system with terminal reheat coils. The air-handling unit supplies air at 54–58°F, and the reheat coils heat this air as required to maintain proper temperatures in the various spaces. HVAC system design should attempt to minimize the use of "reheat" since cooling air only to reheat it is inherently wasteful of energy. As a result, energy codes sometimes restrict the use of reheat systems for comfort applications to systems using recovered heat only.

Variable Air Volume (VAV) Systems

As shown and discussed above in Section 21.2, an air cooling system can vary either air temperature or the supply air flow rate to meet changing cooling loads. Today variable air volume (VAV) systems (Figure 21.14) are a commonly used type of air distribution system for conditioning and ventilating spaces in large buildings requiring multiple zones of control. In a VAV system, a central air handler delivers supply air to local VAV terminal units (known as "VAV boxes"). These VAV boxes are selected to meet the design cooling load. The boxes then throttle the flow rate of cool air to the space to meet the zone cooling loads as called for by the space thermostat.

VAV systems and boxes generally fall into the types illustrated in Figure 21.15 and described below. Boxes typically have a round supply inlet (for connection to rigid or flexible round ductwork) and round or rectangular air outlet(s).

Throttling VAV Systems. In response to a reduced cooling load, this system con-

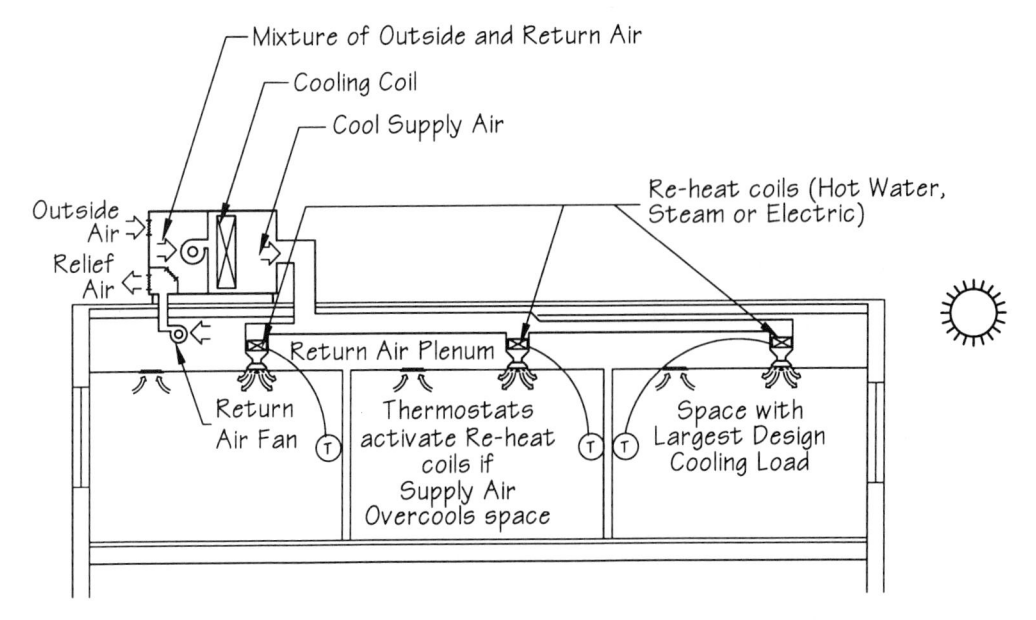

FIGURE 21.13 Constant volume terminal reheat system.

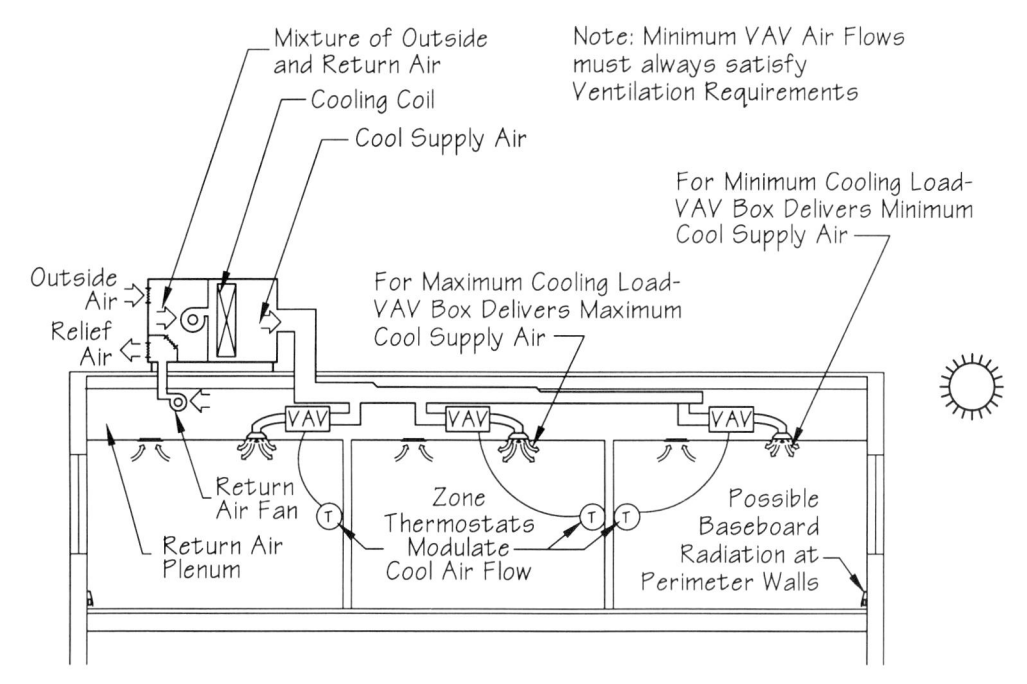

FIGURE 21.14 Variable air volume (VAV) systems.

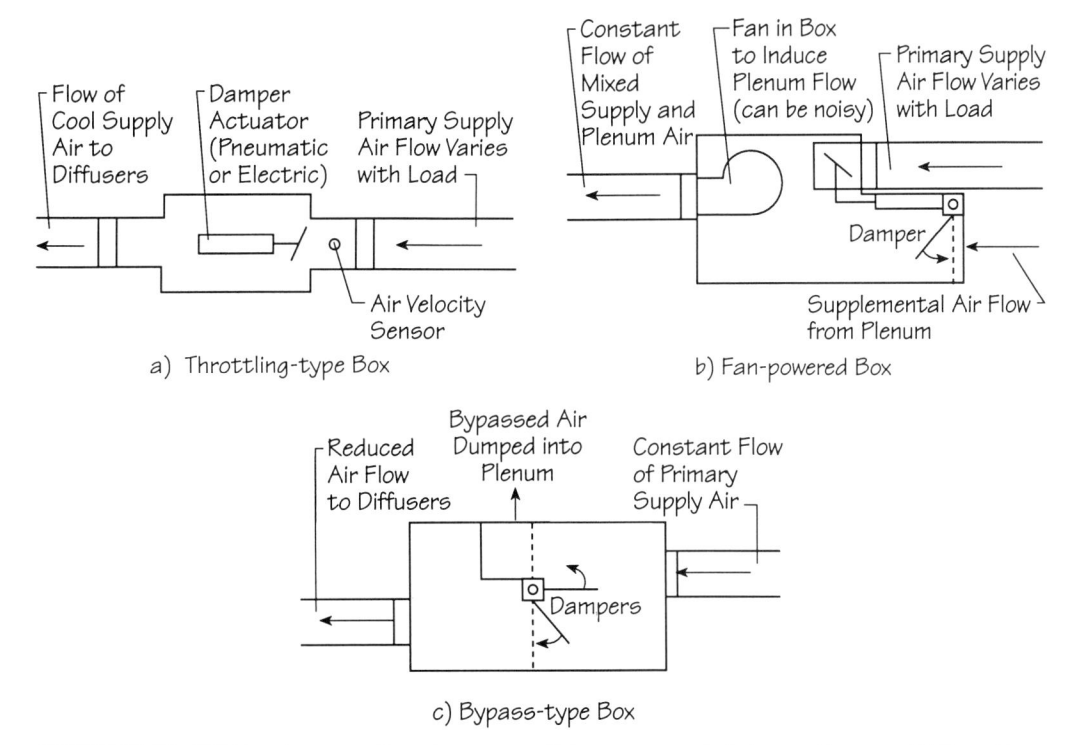

FIGURE 21.15 VAV box types—options for energy efficiency, assured air flow and improved IAQ, and low initial cost.

366 COOLING DISTRIBUTION AND DELIVERY SYSTEMS

tinuously modulates the supply of air at the VAV boxes. Sophisticated systems also modulate the supply fan to maximize energy efficiency.

Fan-Powered VAV Systems. The purpose of these systems is to overcome the perception of poor air quality caused by reductions in supply air to spaces under light cooling load conditions. These systems maintain a constant flow of supply air through a mixture of primary supply air and secondary air which is recirculated from the return air plenum. Fan-powered systems use more energy than throttling VAV systems.

"Bypass" VAV Systems. In response to a reduced demand for space cooling, this type of VAV box dumps excess cool supply air which is not needed into the return air plenum. As a result, temperature control of the space is maintained and the return air is cooled. Unfortunately, the bypass system does not reduce central fan power requirements which can be considerable. Nor does it provide optimal reductions in energy usage to produce the cooling effect.

Supply air temperatures in VAV systems are generally fixed at 55–60°F, although in sophisticated systems the temperature can be reset at the air-handling unit. Reset to a higher supply air temperature during swing seasons will result in increased supply air flow and better room air quality in rooms served by bypass or throttling-type VAV systems.

Presented in Table 21.3 is performance data for VAV boxes with throttling control.

TABLE 21.3 VAV TERMINAL UNITS (BOXES)

Basic Unit
Noise Criteria (NC) Level

Static Pressure Difference from Inlet to Discharge in Inches of Water Column

Unit Inlet Diameter (in.)	cfm	0.50	1.00	2.00	3.00
7	150	—	—	20	23
	300	—	20	27	30
	400	—	22	29	32
	500	21	25	31	34
	600	22	27	32	37
	700	25	29	34	38
8	200	—	—	21	25
	400	—	21	28	31
	600	20	25	31	34
	700	22	27	32	36
	800	23	28	33	37
	900	25	29	34	38
	1000	27	30	36	40
10	300	—	—	24	27
	600	—	24	29	34
	800	—	26	33	37
	1000	22	28	34	40
	1200	24	29	37	41
	1400	25	31	38	42
	1500	25	31	40	43

Source: Data excerpted with permission from product data of the Carnes Company.

Units are sized and selected based on their cfm capacity, pressure drop, and noise criteria (NC) level. The VAV boxes can be noisy. When NC values are below 20 (which is very quiet), they are not shown in the table.

Example 21D: Select VAV boxes to cool the perimeter and interior zones of "The Office" from Table 21.3 when located on the top floor in Philadelphia and based upon the following:

- An inside design temperature (t_i) of 78°F and a supply air temperature (t_s) of 58°F
- The perimeter zones are 15 ft wide
- Cooling loads were computed in Example 8A as follows:

Zone	Sensible Load (Btu/h)	Latent Load (Btu/h)
South	22,475	3,085
West	23,755	2,355
Interior	20,050	4,240

Solution: Begin by determining the maximum required air flow for each zone using Formula 21A. For the south zone the result would be:

$$\text{cfm} = \frac{q_{\text{tot}}}{1.10 \times (t_i - t_s)}$$

(Formula 21A)

$$= \frac{22,475}{1.10 \times (78 - 58)}$$

$$= 1020 \text{ cfm for the south zone}$$

The number of VAV boxes used will depend upon zone areas, initial cost, noise, and desired flexibility. Table 21.3 shows that VAV boxes with a 10-in. inlet diameter can supply 1200 cfm with an NC level of 41 (at 3 in. of wc). Such an NC level is marginally acceptable for open bay office conditions (see Table 21.4), and can be improved with noise attenutation. Therefore, 10-in. VAV boxes can be used to provide the required supply air flows in the south and west zones. For the interior zone, two 7-in. VAV boxes (each supplying about 455 cfm) can be used to lower the sound level and increase flexibility.

Figure 21.16 shows the VAV box arrangement and ceiling diffusers (typically spaced every 140 to 200 square ft^2 depending upon the lighting system and diffuser performance for the ceiling height).

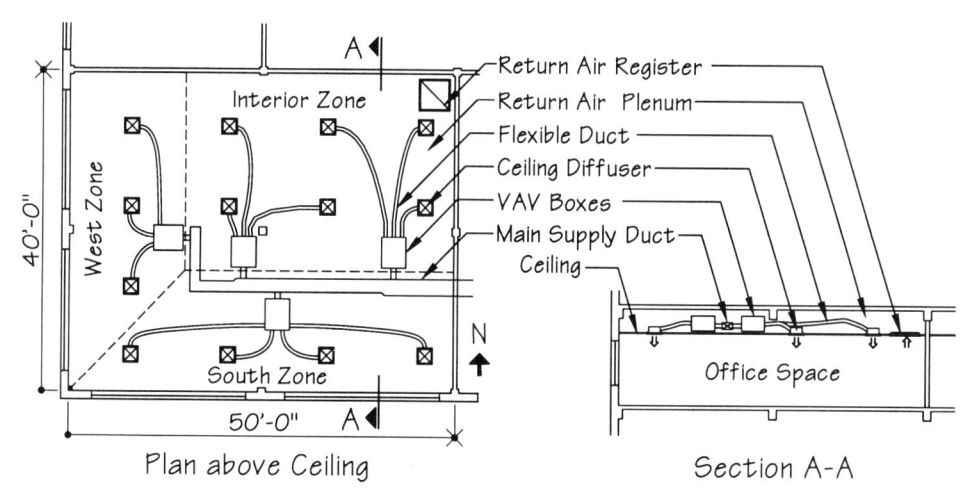

FIGURE 21.16 VAV box selections for "The Office."

It should be realized that when reduced cooling loads exist, the VAV units will throttle down to lower air flows. This potential indoor air quality concern could be avoided if fan-powered VAV units were used to provide a constant volume of supply air. However, the NC rating of fan-powered units must be checked for the office occupancy.

Multizone Systems

Different areas (or zones) of a building often have different types of occupancies and/or cooling loads, and need temperature control by a local thermostat. Multizone air-handling units have a supply fan which delivers a constant volume of air to each zone. Controls determine the required supply air temperature for each zone and adjust the quantities of air which must flow over cooling and heating coils before being mixed and supplied to the zone. The coils are referred to as the "cool deck" and "hot deck" since they are aligned in a linear fashion to serve various zones (usually up to 12 in available packaged equipment). In most applications, zoning will be limited to six or fewer zones based upon orientation and to satisfy the air flow requirements of various spaces. A typical multizone system is shown in Figure 21.17.

Dual Duct Systems

Dual duct systems are those that have two ducted systems of supply air: one with cool air, the other with warm air. Supply air to each space is provided by mixing the two temperature air streams in whatever proportion is required to maintain the indoor design temperature.

The use of dual duct systems in new buildings is rare today because of the inherently wasteful practice of using energy to produce a flow of cool air while also using energy to heat air with which it will be mixed. In addition, the installation cost and clearance problems created by two duct systems are major drawbacks. Applications where dual duct systems are selected tend to use medium- or high-velocity flow to reduce duct size.

On the positive side, dual duct systems offer excellent temperature and humidity control which is well suited to buildings with multiple use spaces demanding careful temperature control and constant flow to maintain proper pressures and air flow relationships such as in hospitals and laboratories.

High-Velocity Induction Systems

Many older high-rise office buildings use induction units to condition perimeter spaces. These units work by supplying units with centrally conditioned primary air at high pressure and velocity (to minimize supply ductwork size). This high-pressure supply air (known as "primary air") flows through induction nozzles which then induce or cause room air to circulate over a secondary coil which contains either chilled or hot water depending upon the season. And when the space is not occupied during the heating season, the primary air supply system can be turned off and the terminal induction units maintain heat as simple convectors.

Induction units were very popular for use in high-rise buildings since they provided required ventilation air at the perimeters of "sealed" office buildings (without operable windows) and minimized shaft space. Today the use of induction units is very limited in new or retrofit applications due to the energy-intensive fan requirements.

21.5 AIR OUTLETS

Supply Air

Supply air in large buildings usually enters spaces through outlets which are either high

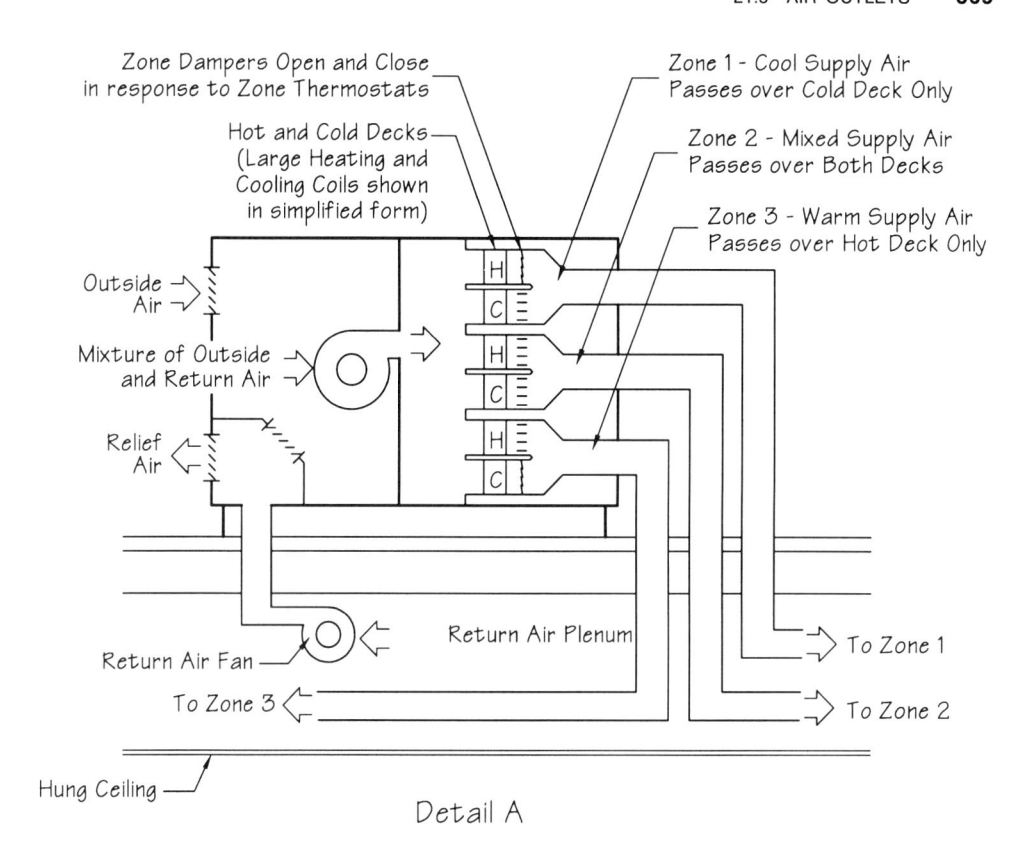

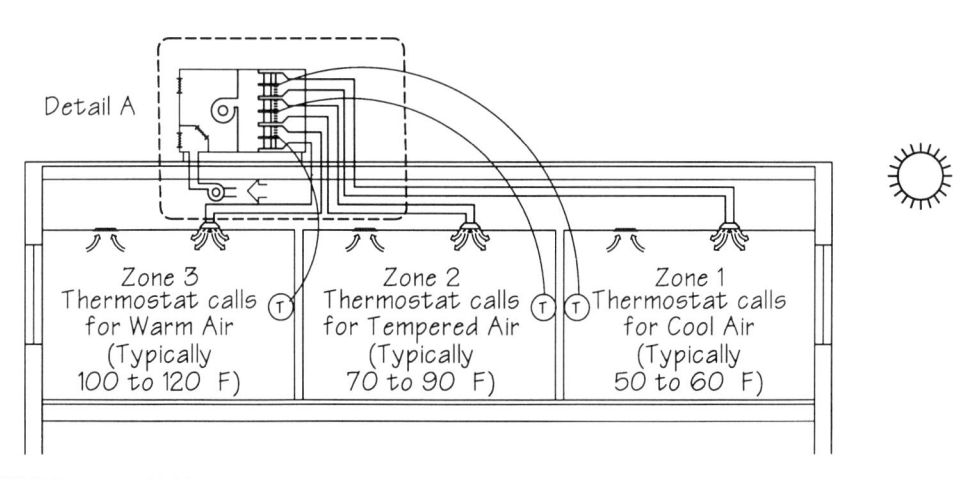

FIGURE 21.17 Multizone system with air flowing over "hot" and "cold" decks to zones with different thermal needs.

on walls (registers) or in the ceiling. Ceiling outlets are known as "diffusers" since their design must facilitate a diffusion or mixing of the cool supply air with the relatively warm room air. Cool supply air may range in temperature from 52 to 70°F. Air at the low end of the supply temperature range can present a problem since it is denser than the room air, and thus has a tendency to fall or dump out (sometimes known as "puddling"). Improper diffuser selection can also lead air at low flows to hug the ceiling (known as the "surface" or "coanda effect"). See Section 13.10 for additional air outlet terminology.

Diffusers in HVAC systems need to be selected for the range of air flows (in cfm) which they will accommodate. Selection of the basic diffuser type (round, square, or linear) is generally an aesthetic architectural decision. Square diffusers which are commonly used are available with a 1-way, 2-way, 3-way, or 4-way throw of cool air depending upon room placement, as shown in Figure 21.18. Round diffusers are available as adjustable units which can provide a horizontal, vertical, or intermediate air flow pattern depending upon room geometry and application. Also available are half-round units for use near walls, as shown in Figure 21.19.

Linear diffusers are often employed by designers to minimize the visual impact of the air distribution system. Typical slot patterns are indicated in Figure 21.20a. Indicated in Figure 21.20b is a laminar flow diffuser which would be used in special applications where turbulent air flow is to be avoided such as in a clean room.

Return Air Outlets

The design of air outlets to return air from the room to central equipment is less critical. Normally they are square well-spaced openings (grilles) to the plenum space. Return air outlets should not be located too

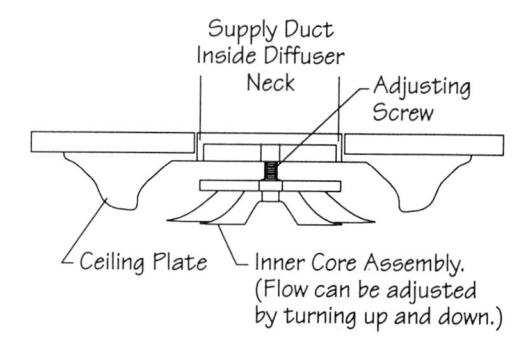

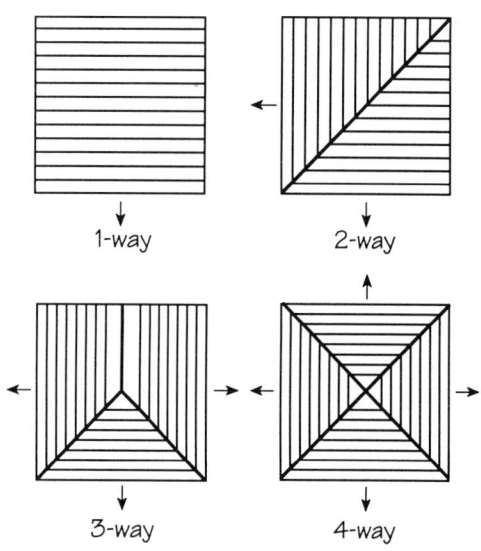

FIGURE 21.18 Throw of supply air from square diffusers.

FIGURE 21.19 Round diffuser options.

a) Section of Adjustable Round Diffuser

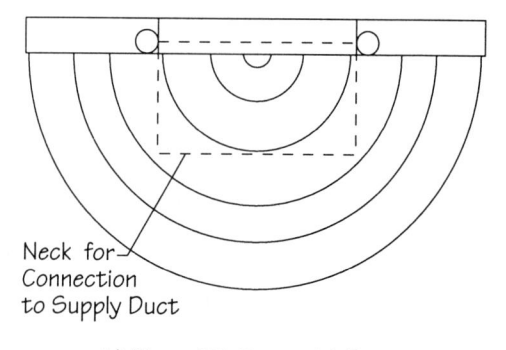

b) Plan of Half-round Diffuser

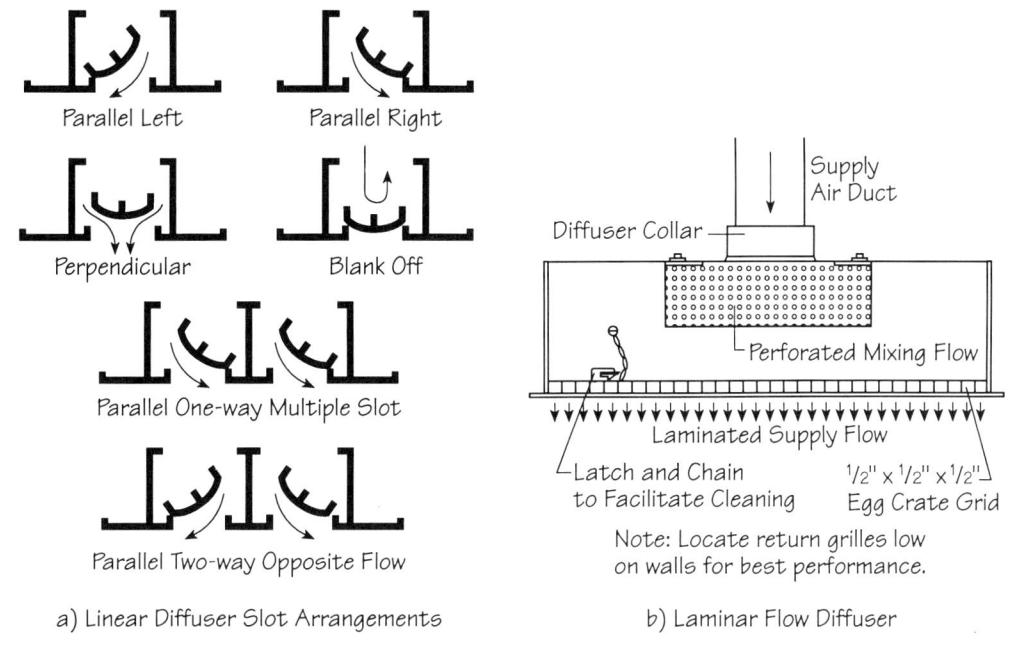

FIGURE 21.20 Diffusers for special applications.

a) Linear Diffuser Slot Arrangements
b) Laminar Flow Diffuser

close to supply diffusers because they might short-circuit the intended air flow. They also should be located to prevent unwanted noise transmission between spaces.

Noise

Sound considerations are often a very important criteria in the selection of air terminal units and diffusers. The terminology used is the "NC" (Noise Criteria) rating method. This method tests outlets under various conditions and a range of sound frequencies, and determines the maximum sound pressure level achieved in decibels (dB).

The decibel scale is logarithmic, like the Richter scale for earthquakes. Therefore, a sound pressure level of 60 dB, which is typical of a hotel lobby or restaurant, is actually 10 times as loud as a noise level of 50 dB typical of a normal home.

The noise criteria (NC) rating for a diffuser is the highest decibel level achieved in tests under different frequencies (within the normal hearing range). For example, a diffuser with the test results shown below for a given air velocity would be assigned a rating of NC-30.

	Frequency (cycles per second)				
	250	500	1000	2000	4000
Decibel level achieved	25	27	28	30	26

Air velocity also has a great bearing on noise produced. A doubling of velocity will increase the sound level by about a factor

TABLE 21.4 RECOMMENDED AIR SYSTEM NOISE LEVELS

Application	NC Level	Maximum Velocity (ft/min)
Residence		
Rural	20 to 25	700
Urban	25 to 30	700
Apartments	30 to 35	700
Offices		
Executive	25 to 30	700
Private	30 to 35	700
Open Bay Areas	35 to 40	800
Concert Halls	20 to 25	700
Classrooms	30 to 35	700
Bank/Post Office	35 to 40	800
Cafeterias	40 to 50	900
Supermarket	40 to 50	900

Source: Extracted from Manual T, as published by ACCA.

of 4. For example, a diffuser which has a noise rating of NC-12 at an air velocity of 400 fpm will have a noise rating of about NC-48 if the air velocity is doubled to 800 fpm.

Supply air outlets are selected based upon the background noise of the building type and the supply air velocity. Provided in Table 21.4 are some recommended values for a range of residential and commercial air system applications.

A review of Table 21.4 indicates the importance of the application in selecting diffusers. For example, sound levels in spaces such as concert halls are, of course, of great concern.

Why not lower the velocity? After all, if the velocity is halved, the noise level will be only about one-fourth as loud. Yes—but ducts would also need to be bigger and construction clearances may not allow for cooling of a large open space such as a concert hall in this manner. This is just another example of how HVAC design often requires a weighing of many factors before final design decisions are made.

22

DOMESTIC HOT WATER

Domestic hot water (DHW), referred to as service hot water in the *ASHRAE Handbook*, is needed for cleaning and bathing as well as laundering and commercial purposes. The most common residential-scale DHW systems directly heat the water within a storage tank by natural gas or electricity. Then as hot water is drawn off and used, more water is heated so that the unit "recovers" its ability to satisfy the load for which it has been sized.

Hot water in larger buildings is often produced by transferring heat from the heating system boiler to a coil. In many older systems hot water from this coil was simply mixed with cold water to temper it before delivery through the hot water supply piping system. Now it is much more common and efficient to transfer heat from the coil within a boiler to a separate well-insulated storage tank.

The rather simple task of heating hot water can also be accomplished through some form of heat recovery or conversion of steam which is available from utilities in many older cities. Also possible is the use of heat pump hot water heaters or active solar collection systems.

22.1 WATER TEMPERATURE NEEDED

Before the energy crisis of 1974, it was common to heat domestic hot water in homes and commercial buildings to about 140°F, an unnecessarily high temperature for most common applications (e.g., hand washing and showers). This was recognized, and in the interest of energy conservation it then became common, and even required by energy codes for some applications, to limit normal domestic hot water temperature to 100 or 120°F.

Then in the 1980s a potentially serious health problem emerged for hot water systems maintained at lower temperatures, known as "Legionnaires' Disease." It was determined in 1984 that Legionella Pneumophila can colonize and breed in hot water

373

systems maintained at 115°F or less. Therefore, as a health precaution, ASHRAE now recommends that the domestic hot water temperature should be maintained at 140°F.

Representative hot water utilization temperatures now recommended by ASHRAE are indicated in Table 22.1.

In many cases, some mixing, or tempering, by cold water is required to achieve utilization temperatures if 140°F water is supplied. This is accomplished by using mixing valves. It is advisable to install mixing valves of high quality since valve failure can result in elevation of supply water temperature and possibly result in scalding. It should also be noted that some designers are skeptical about the 140°F hot water recommendation and are more concerned about minimizing the chances for scalding from hot water. As a result, they continue to design for lower temperatures.

Scale Deposits

There also is another very good reason for domestic hot water systems to operate at relatively low temperatures, and particularly when "hard" water (with calcium and magnesium salts) is used. Scale (in the form of lime deposits) is a much greater problem at higher temperatures. For example, scale deposits in a domestic hot water heater will more than double if the hot water temperature is increased from 120 to 150°F. The problem with scale deposits is that they provide an insulating layer to heating elements (coils) and the heater tank itself. This can reduce efficiency and lead to premature failure of tanks (evidenced by a "crackling" noise) when water gets between the insulating scale deposit and the tank wall. For large, commercial systems always use a water softener if source water is hard.

Local Source Water Temperature

Well water is subject to year-round earth temperatures which are close to the average year-round ambient air temperature in a location, varying from over 70°F in southern Florida to near 35°F in Anchorage, Alaska. Reservoir water has an average yearly temperature similar to well water, although there is some seasonal variation. Approximate ground water (deep earth) temperatures are shown in Figure 22.1 and listed in Table 22.2.

TABLE 22.1 REPRESENTATIVE HOT WATER UTILIZATION TEMPERATURES

Use	°F
Lavatory	
Hand washing	105
Shaving	115
Showers and tubs	110
Therapeutic baths	95
Commercial and institutional laundry	up to 180
Residential dishwashing[a] and laundry	140
Surgical scrubbing	110
Special commercial applications	Varies (up to 195)

[a] Modern dishwashers can use lower temperature DHW and heat it to a higher required temperature internally.

Source: Extracted with permission from the 1991 ASHRAE Handbook, HVAC Applications, Chapter 44, Table 3.

22.1 WATER TEMPREATURE NEEDED   375

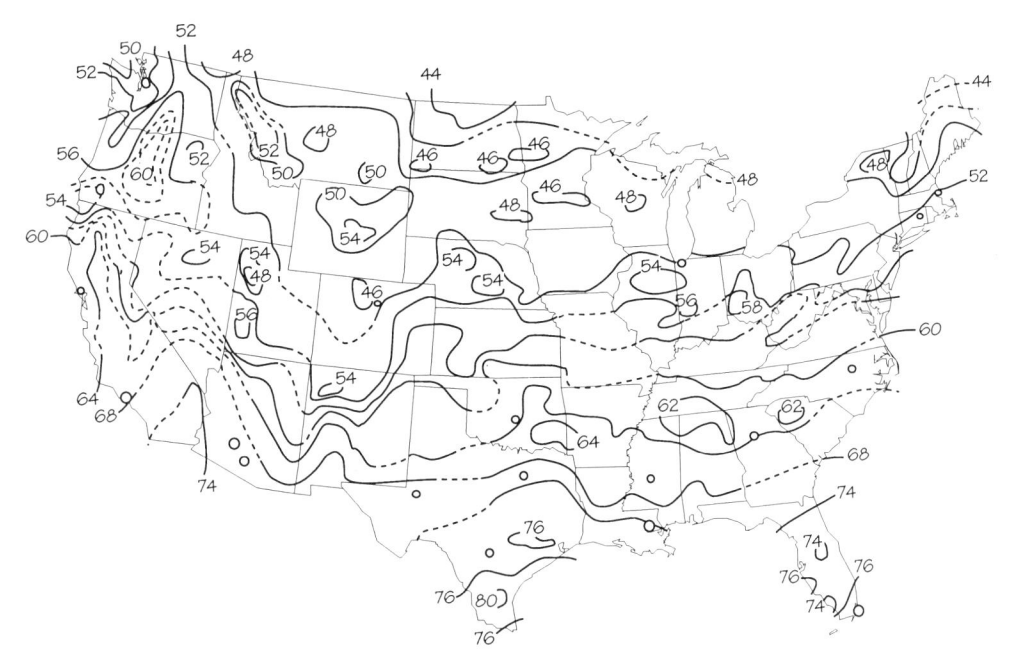

FIGURE 22.1 Well water temperatures for the United States.

TABLE 22.2 APPROXIMATE GROUND WATER TEMPERATURE			
City	°F	City	°F
Albuquerque	56	Memphis	62
Anchorage	35	Miami	76
Atlanta	61	Minn./St. Paul	45
Bismarck	41	New Orleans	68
Boston	51	New York City	54
Casper	45	Pensacola	68
Charleston	65	Philadelphia	54
Chicago	49	Phoenix	70
Cincinnati	53	Pittsburgh	50
Dallas	66	Portland	53
Denver	50	Reno	49
Detroit	49	St. Louis	56
Hartford	50	San Antonio	69
Honolulu	77	San Diego	63
Houston	68	San Francisco	57
Indianapolis	52	Salt Lake City	52
Jacksonville	68	Seattle	53
Little Rock	62	Sioux Falls	46
Los Angeles	65	Topeka	54
Madison	45	Washington, DC	57

Source: 1993 National Oceanic and Atmospheric Administration (NOAA) Annual Summaries.

22.2 QUANTITY OF HOT WATER REQUIRED

Domestic hot water systems are rated in Btu/h to produce the quantity of hot water which the users of the building are likely to demand per hour. The system must also be able to recover this capacity to satisfy the demand for hot water during the next hour as well.

Houses

In residences, the demand for hot water will vary greatly depending upon family composition, lifestyle, and personal habits. A rule of thumb commonly used is to assume a daily hot water usage of 15–20 gal per person per day. And in a "typical" family, a peak requirement of 10–12 gal of hot water per person may be required within one hour (typically for morning showers).

To satisfy residential hot water requirements, minimum hot water heater capacities have been established by the U.S. Department of Housing and Urban Development and the Federal Housing Authority as summarized in Table 22.3.

Note that the requirements for electric water heaters include recovery rates (the ability to make more hot water) which are about one-half of those for gas-fired heaters, and only about one-third of those for oil-fired water heaters.

Commercial Buildings

Buildings of various types require very different quantities of domestic hot water as

TABLE 22.3 HUD–FHA MINIMUM WATER HEATER CAPACITIES FOR ONE- AND TWO-FAMILY LIVING UNITS

Number of Baths	1 to 1.5		2 to 2.5			3 to 3.5		
Number of Bedrooms	2	3	3	4	5	4	5	6
GAS								
Storage, gal	30	30	40	40	50	50	50	50
1000 Btu/h	36	36	36	38	47	38	47	50
1 hour draw, gal	60	60	70	72	90	82	90	92
Recovery, gph	30	30	30	32	40	32	40	42
ELECTRIC								
Storage, gal	30	40	50	50	66	66	66	80
kW input	3.5	4.5	5.5	5.5	5.5	5.5	5.5	5.5
1 hour draw, gal	44	58	72	72	88	88	88	102
Recovery, gph	14	18	22	22	22	22	22	22
OIL								
Storage, gal	30	30	30	30	30	30	30	30
1000 Btu/h	70	70	70	70	70	70	70	70
1 hour draw, gal	89	89	89	89	89	89	89	89
Recovery, gph	59	59	59	59	59	59	59	59
TANK-TYPE INDIRECT								
Manuf.-rated draw, gal in 3 hours, 100°F rise	49	49	75	75	75	75	75	75
Tank capacity	66	66	66	66	82	82	82	82
TANKLESS-TYPE INDIRECT								
Manuf.-rated draw, gal in 5 minutes, 100°F rise	15	15	25	25	35	35	35	35

Source: HUD-FHA Minimum Property Standards for One and Two Family Living Units, No. 4900.1-1982.

indicated in Table 22.4. For example, office buildings require a maximum of 0.4 gallons per hour per person whereas small apartment buildings may require 10 gallons per hour per apartment or more.

For most building types, the hot water demand values found in Table 22.4 should prove to be adequate. With regard to urban apartment houses, however, it is common for more occupants to inhabit apartments than are assumed. This can result in an increased hot water demand of one-third or more.

22.3 SYSTEM SIZING

A properly sized hot water system must have an adequate "minimum recovery rate" in gallons per hour and "usable" storage capacity. The term usable refers to the assumption that only 60–80% of the water within the tank has a high enough temperature. This is because the supply pipe from the tank is high and the water near the bottom of the tank will be lower in temperature due to stratification.

Indicated in Figure 22.2 are the recommended recovery capacities and usable storage capacities for apartments and office buildings (Note: The *ASHRAE Handbook of HVAC Applications* contains similar diagrams for six other building types.)

As indicated in Figure 22.2, as the usable storage capacity is reduced, the recovery capacity must be increased in order to meet the hourly demand for hot water. Presented in Table 22.5 is the minimum rec-

TABLE 22.4 HOT WATER DEMANDS AND USE FOR VARIOUS TYPES OF BUILDINGS

Type of Buildings	Units	Maximum Hour	Maximum Day	Average Day
Men's dormitories	gal/student	3.8	22.0	13.1
Women's dormitories	gal/student	5.0	26.5	12.3
Motels: No. of units[a]				
20 or less	gal/unit	6.0	35.0	20.0
60	gal/unit	5.0	25.0	14.0
100 or more	gal/unit	4.0	15.0	10.0
Nursing homes	gal/bed	4.5	30.0	18.4
Office buildings	gal/person	0.4	2.0	1.0
Food service:				
Type A: Full meal restaurant	gal/maximum meals/hour	1.5	11.0	2.4
Type B: Drive-ins, luncheonettes	gal/maximum meals/hour	0.7	6.0	0.7[b]
Apartment houses[a]				
20 or less	gal/apartment	12.0	80.0	42.0
50	gal/apartment	10.0	73.0	40.0
75	gal/apartment	8.5	66.0	38.0
100	gal/apartment	7.0	60.0	37.0
200 or more	gal/apartment	5.0	50.0	35.0
Elementary schools	gal/student	0.6	1.5	0.6[b]
Junior and senior high schools	gal/student	1.0	3.6	1.8[b]

[a] Interpolate for intermediate values
[b] Per day of operation

Source: Extracted with permission from the *ASHRAE Handbook HVAC Applications*, 1991, Chapter 44, Table 7.

378 DOMESTIC HOT WATER

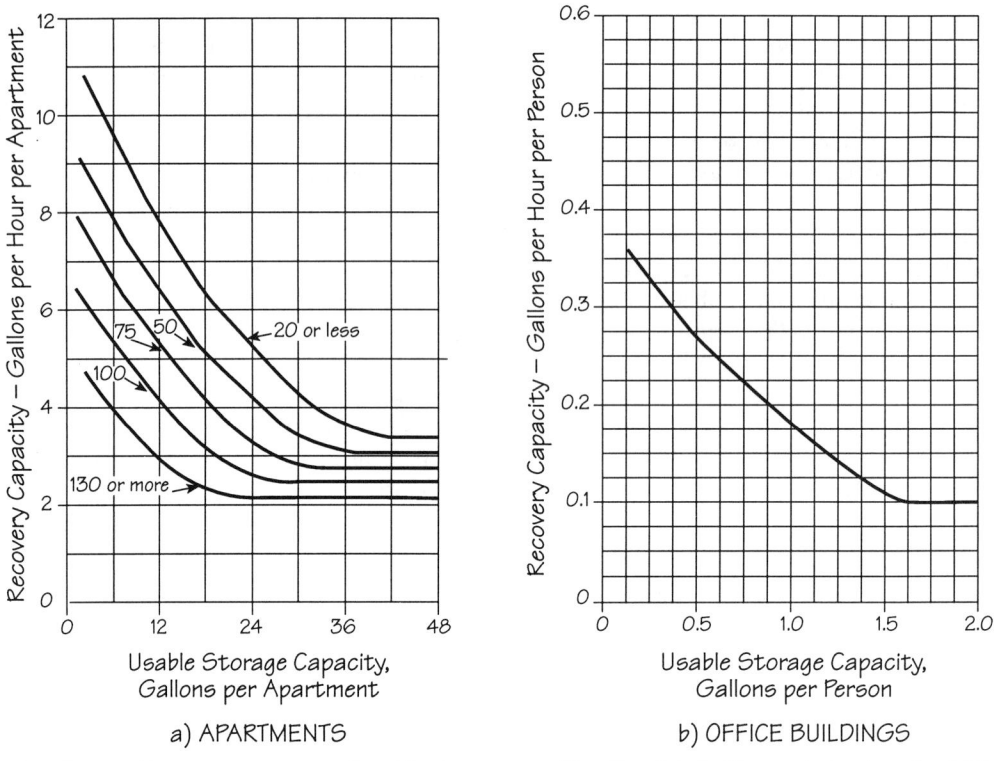

FIGURE 22.2 Hot water recovery and usable storage capacities. Reprinted with permission from the *ASHRAE Handbook HVAC Applications*, 1991, Chapter 44, Figures 16 and 18.

ommended recovery capacity for the various commercial building types along with the corresponding minimum usable storage capacity in gallons per person. Notice how comparatively large the need for hot water is in apartment and other residential-type buildings.

The required recovery rate of a hot water system is determined using Formula 22A.

$$\text{gph} = \text{Rec}_{cap} \times N \quad \text{Formula 22A}$$

where

gph is the required recovery capacity of a water heater in gallons per hour

Rec_{cap} is the recovery capacity required per hour per unit N

N is the number of applicable units (e.g., people)

Formula 22B is used to compute the size of the storage tank required.

$$\text{TANK}_{gal} = \frac{\text{US}_{cap}}{F_{use}} \times N \quad \text{Formula 22B}$$

where

TANK_{gal} is the required size of a hot water storage tank in gallons

US_{cap} is the usable storage capacity required in gallons per person (or other unit)

F_{use} is the fraction of the tank considered usable in supplying hot water

N is the number of applicable units (e.g., people)

TABLE 22.5 MINIMUM RECOVERY RATES AND CORRESPONDING MINIMUM USABLE STORAGE CAPACITY

Type of Building	Minimum Recovery Rates	Corresponding Minimum Usable Storage Capacity
Men's dormitories	0.85 gal/h per student	10 gal per student
Women's dormitories	1.10 gal/h per student	12 gal per student
Motels: No. of units		
20 or less	1.50 gal/h per unit	16 gal per unit
60	1.25 gal/h per unit	14 gal per unit
100 or more	1.00 gal/h per unit	12 gal per unit
Nursing homes	1.25 gal/h per bed	12 gal per bed
Office buildings	0.10 gal/h per person	1.6 gal per person
Food service:		
Type A: Full meal restaurant	0.45 gal per max. # of meals per hour	7.5 gal per max. # of meals per hour
Type B: Drive-ins, luncheonettes	0.25 gal per max. # of meals per hour	2.0 gal per max. # of meals per hour
Apartment houses:		
20 or less	3.40 gal/h per apt.	42 gal per apt.
50	3.00 gal/h per apt.	38 gal per apt.
75	2.75 gal/h per apt.	34 gal per apt.
100	2.40 gal/h per apt.	28 gal per apt.
130 or more	2.10 gal/h per apt.	24 gal per apt.
Elementary schools	0.06 gal/h per student	1.5 gal per student
Junior and senior high schools	0.15 gal/h per student	3.0 gal per student

Source: Based upon information contained in Figures 13 through 20 in the *ASHRAE Handbook HVAC Applications*, 1991, Chapter 44.

Example 22A: Determine the water heater size for a building with 200 apartments based upon the minimum recovery rate. Also compute the minimum size of storage tank assuming that it will contain 70% usable hot water (F_{us}).

Solution: From either Table 22.5 or Figure 22.2a obtain the minimum recovery capacity (Rec_{cap}) rate of 2.1 gallons per hour per person. Use Formula 22A.

$$gph = Rec_{cap} \times N \quad \text{(Formula 22A)}$$
$$= 2.1 \times 200$$
$$= 420 \text{ gal/h}$$

Now compute the required storage tank size using Formula 22B, and a usable storage capacity of 24 gallons per apartment (from Table 22.5 or Figure 22.2a).

$$TANK_{gal} = \frac{US_{cap}}{F_{use}} \times N$$

(Formula 22B)

$$= \frac{24}{0.70} \times 200$$

$$= 6860 \text{ gal (minimum)}$$

Hourly Energy Requirement

The quantity of energy required to heat hot water on an hourly basis can be found using Formula 22C.

$$q_h = \frac{8.33 \times G_h \times (t_o - t_i)}{EFF}$$

Formula 22C

where

q_h	is the heat required to produce the gallons of hot water needed in Btu/h
8.33	is the density of water in lb/gal
G_h	is the recovery rate required for hot water in gph (from Table 22.3)
t_o	is the outlet temperature of hot water in °F
t_i	is the inlet temperature of supply water in °F (see Table 22.2)
EFF	is the efficiency of the hot water producing system as a fraction

Example 22B: Determine the quantity of heat required per hour for a 100-unit apartment house in Denver, CO. The complete system has an efficiency (EFF) of 0.80.

Solution: Table 22.5 indicates that the minimum recovery rate for a 100-unit apartment house is 2.40 gallons per hour per unit, or a total of 240 gal/h. The outlet water temperature (t_o) will be 140°F. Table 22.2 shows that the inlet water temperature (t_i) for Denver will be approximately 50°F. Now use Formula 22C.

$$q_h = \frac{8.33 \times G_h \times (t_o - t_i)}{EFF}$$

(Formula 22C)

$$= \frac{8.33 \times 240 \times (140 - 50)}{0.80}$$

$$= 224{,}910 \text{ Btu/h}$$

Therefore, a hot water heater would have to have a rated output of at least 224,910 Btu/h to satisfy this load. In addition, if this load was being satisfied by a boiler charging a storage tank which also serves to heat the building, then the boiler must have a large enough rated output. For example, if the peak heating load for this building were 1,000,000 Btu/h then the boiler output must be at least 1,224,910 Btu/h.

Yearly Energy Requirement

A slight modification of Formula 22C allows for the quantification of energy required to heat hot water on a yearly basis as shown in Formula 22D.

$$Q_{yr} = \frac{8.33 \times G_{day} \times D_{yr} \times (t_o - t_i)}{EFF}$$

Formula 22D

where

Q_{yr}	is the heat needed to produce the required amount of hot water in Btus per year
G_{day}	is the daily requirement for hot water in gallons per day (from Table 22.4)
D_{yr}	is the days of operation per year

Example 22C: Determine the amount of annual energy needed by a 150-unit apartment house to produce 140°F domestic hot water assuming the following:

A. A building located in Miami, FL
B. A building located in Minneapolis/St. Paul, MN
C. An overall system efficiency (EFF) of 0.70 (70%)

Solution: Use Formula 22D for each location. Table 22.4 indicates that an apartment house of such a size will use approximately 37 gal of hot water for each of the 150 apartments each day for a total of 5550 gallons per day (G_{day}). For an apartment house there are 365 days of operation per year.

A. Miami: From Table 22.2 the average temperature of the water source (t_i) is found to be 76°F.

$$Q_{yr} = \frac{8.33 \times G_{day} \times D_{yr} \times (t_o - t_i)}{EFF}$$

(Formula 22D)

$$= \frac{8.33 \times 5550 \times 365 \times (140 - 76)}{0.70}$$

= 1,542,810,000 Btus per year

B. Minneapolis/St. Paul: From Table 22.2 the average temperature of the water source (t_i) is found to be 45°F.

$$Q_y = \frac{8.33 \times G_{day} \times D_{yr} \times (t_o - t_i)}{EFF}$$

(Formula 22D)

$$= \frac{8.33 \times 5550 \times 365 \times (140 - 45)}{0.70}$$

= 2,290,100,000 Btus per year

It takes about 50% more heat to produce the required quantity of hot water in Minneapolis/St. Paul than in Miami just because of the difference in water source temperature.

22.4 DIRECT-FIRED DHW STORAGE HEATERS

For residences, and small commercial buildings, the most common type of DHW system incorporates the burner or heating element(s), insulated tank storage, and controls in a single stand alone unit. Depending upon house size, automatic storage heaters for residences have tanks which store from 30 to 80 gal of hot water depending upon fuel source (see Table 22.3). Most DHW heaters are either gas-fired or heated by electricity. Gas-fired DHW heaters often are vented to the same chimney used by the heating system boiler. To limit standby heat loss, automatic vent or flue dampers are required by gas-fired heaters which get their combustion air and/or their air for draft hood dilution from a conditioned space (see Figure 22.3).

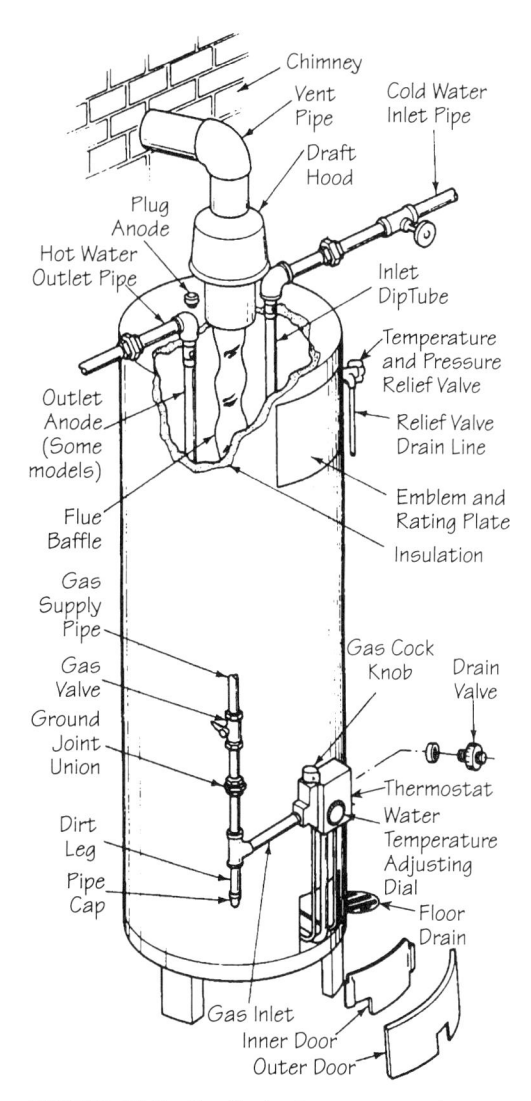

FIGURE 22.3 Gas-fired storage water heater. Based on product literature of A. O. Smith Corporation.

382 DOMESTIC HOT WATER

Electric water heaters have a large capacity primary resistance heating element immersed in the lower portion of the tank. Some electric heaters also have a secondary element immersed in the upper portion of the tank to reheat heated water which may have cooled slightly. Electric water heaters are often used where electric charges are low, for small hot water loads, and in vacation houses.

Direct-fired hot water storage heaters which work hard (such as in an apartment building) often have a short expected life of 10 yr or less. This results from the large quantity of water which is heated and causes minerals to settle out of solution. These minerals then attach themselves to the insides of the tank and cause hot-firing conditions, which shortens the life of the entire heater assembly.

Oil-fired hot water heaters are relatively rare. Yet, some new (or replacement) units are still installed in the Northeast in large commercial buildings which also use oil for the heating system boiler.

22.5 INDIRECT-FIRED DHW SYSTEMS

Tankless Coils

Many older large capacity DHW systems use a "tankless coil" placed high within the heating system boiler (see Figure 22.4). During the heating season, a tankless coil system can deliver hot water without the need for a separate piece of hot water heating equipment. But once the heating season ends, the tankless coil system becomes very inefficient since the large heating system boiler must continue to be heated just to produce domestic hot water.

The use of tankless coils in large buildings can be particularly inefficient. Often, greater efficiency will result if a separate boiler is installed which is sized to meet the domestic hot water heating load only. Then the boiler which heats the building will go off line during the warmer months to allow for servicing, maintenance, and extended life.

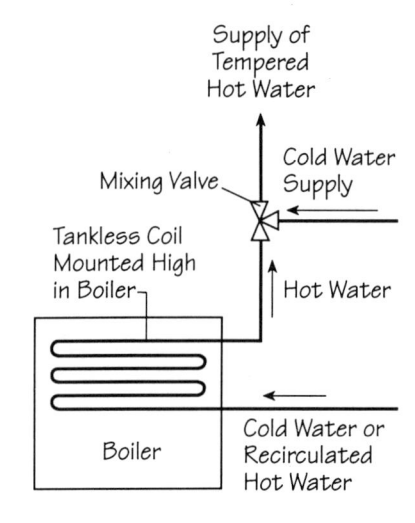

a) System Schematic

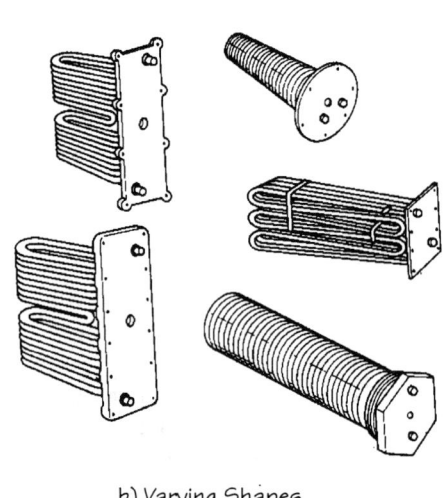

b) Varying Shapes and Configurations of Tankless Coil Assemblies

FIGURE 22.4 Tankless coil heating of hot water.

Charging of Insulated Storage Tanks

The real problem with the tankless coil is not the coil—but the "tanklessness." Boilers can be operated in a manner where they go on for a short period of time to fully heat or "charge" all of the water within one or more well-insulated storage tank(s) as shown in Figure 22.5. By the way, lightweight tanks are available which are made out of stainless steel for very long service life.

The separate storage tank approach can improve system efficiency greatly since the boiler then can go to cold shutdown for an extended period of time before more hot water needs to be produced.

Hot Water Produced Through Heat Exchange

Hot water can be produced with any available source of heat. Some large buildings obtain their domestic hot water from a heat exchanger (known as a "converter") which uses steam or high-temperature hot water (see Figure 22.6). Such an approach is common in urban areas where utilities offer steam for purchase, and on campuses where central plants serve many buildings.

It should be clear that any source of available or "waste" heat can be used to produce domestic hot water. As a result many "heat recovery" possibilities exist on most large projects including:

- Heat otherwise rejected through cooling towers by using a double-bundle condenser (see Section 20.5)
- Heat obtained from desuperheating of refrigerant in systems with air-cooled condensers, including heat pump cycles
- Waste heat from industrial or commercial processes
- Heat from cogeneration systems

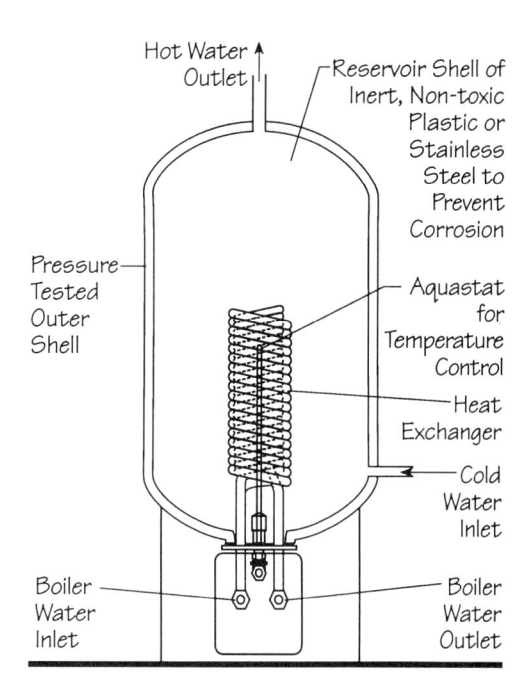

a) Detail of Typical Tank Components

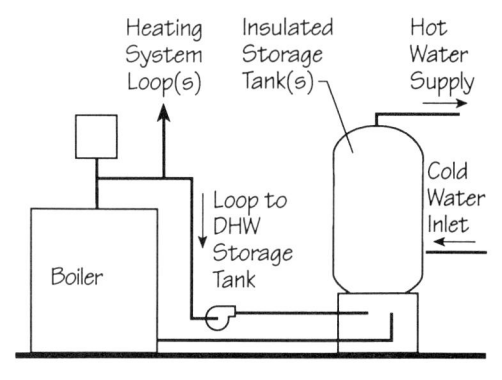

b) Arrangement of Boiler and Tanks

FIGURE 22.5 Separate storage tanks with external water heating.

Point-of-Use Instantaneous Heaters

Small, occasional hot water needs can be met by point-of-use electric heaters. These heaters work as immersion elements which quickly (almost instantly) heat a small

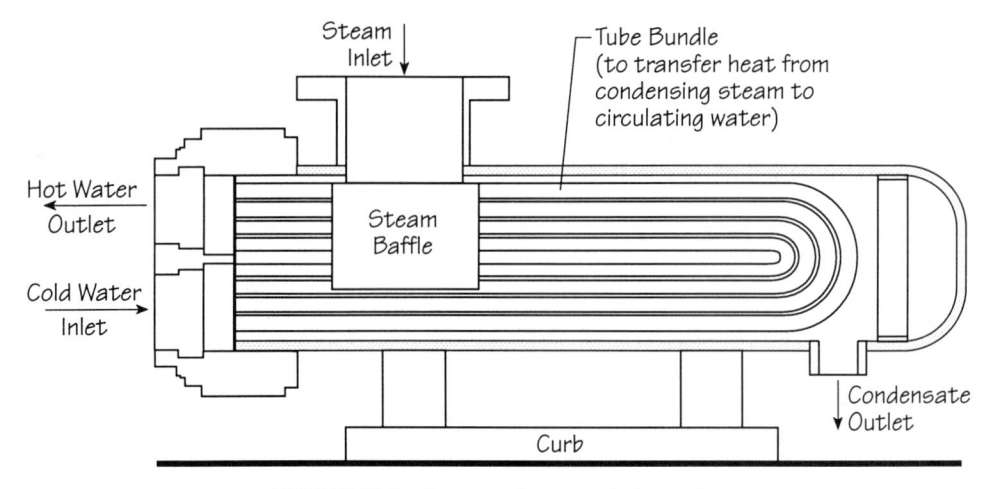

FIGURE 22.6 Converter for steam to hot water.

quantity of water to a carefully controlled temperature.

These types of heaters are normally installed in remote locations of a building in place of long lengths of hot water piping and hot water recirculation. In most locations, operation of such electric heaters will be expensive if used often. Good applications include rarely used and remote kitchenettes in conference rooms, meeting rooms, and so on. Point-of-use gas-fired and electric (up to 3 kW) DHW units are common in Europe.

22.6 HEAT PUMP DHW HEATERS

An air-to-water heat pump can also be used to produce domestic hot water. Such a system extracts heat from the air in a building area (such as a basement) while heating up water. In concept, this is well suited to warm climates. The following example illustrates one of the advantages of such a system.

Example 22D: Determine the quantity of heat which will be extracted from a house by a heat pump hot water heater which can produce 30 gal of hot water per hour (G_h). Assume that water needs to be heated to 140°F (t_o) from a source temperature of 76°F (t_i) in a warm climate such as Miami, Florida. The COP or system efficiency (EFF) is 2.8 after allowing for the heat of the compressor (which is inside the house).

Solution: Use Formula 22C.

$$q_h = \frac{8.33 \times G_{hr} \times (t_o - t_i)}{\text{EFF}}$$

(Formula 22C)

$$= \frac{8.33 \times 30 \times (140 - 76)}{2.8}$$

$$= 5710 \text{ Btu/h}$$

Such a quantity of heat removed from a house is substantial (almost one-half ton of mechanical cooling) and very beneficial in warm climates.

Heat pump hot water heaters have enjoyed only limited popularity to date, perhaps in part because of the undesirability of indoor compressor noise.

22.7 SOLAR DHW

Of the various types of active solar collectors, the most common are "flat plate" collectors with construction similar to that shown in Figure 22.7.

The typical configuration is a flat metal plate painted or coated so that it will absorb and retain the maximum amount of heat when it is installed in a glass-covered box insulated on the back and sides. To facilitate handling and installation, flat plate collectors are typically manufactured in modules, 3–4 ft wide and 6–7 ft high.

Flat plate collector absorber plates are usually made of metal for a good reason—it absorbs heat very well. Metals, nevertheless, are heavy and expensive. As a design alternative some flat plate collectors use lightweight plastic or rubber absorber plates which admittedly absorb much less solar heat on a square foot basis, but offset this shortcoming by being much less expensive to fabricate and lighter, and thus, easier to install.

The heat absorbed by the absorber plate is transferred to attached fluid passages which direct the heated water or antifreeze solution (usually at 140–170°F), to a header where it is pumped to a tank.

Most solar DHW systems in America are "closed loop" systems as shown in Figure 22.8, meaning that the water or antifreeze solution is recirculated, and simply transfers heat to the water in the hot water tank. In addition, a conventional hot water heater and tank is needed to add heat to the water and store it when the solar heating system is unable to heat water to the required temperature. Therefore, the solar tank is used as a "preheat" tank, which is piped in series with the conventional heater.

The closed loop system functions as follows:

A differential thermostat measures the temperature of water in the preheat tank and fluid in the collectors. When the collector fluid temperature is several degrees warmer than the tank temperature, the pump is activated and the heated fluid begins to flow through the coil in the storage tank.

When the fluid temperature in the collectors falls to within a few degrees of the storage tank temperature, the differential thermostat will turn the pump off and stop the system. This will be occasioned by very cloudy weather or the coming of night.

In cold climates an antifreeze solution is used in closed loop systems since fluid remains in the collectors and associated outdoor piping overnight. Other systems known as "draindown" or "drainback" systems leave the collectors high and dry at night when freezing is a possibility by draining the water and either discarding it or storing it in a drainback tank. Finally, domestic hot water systems in hot climates where freezing is not a problem can use a convective loop "thermosyphon design." In thermosyphon systems the storage tank is mounted higher than the collectors. Water

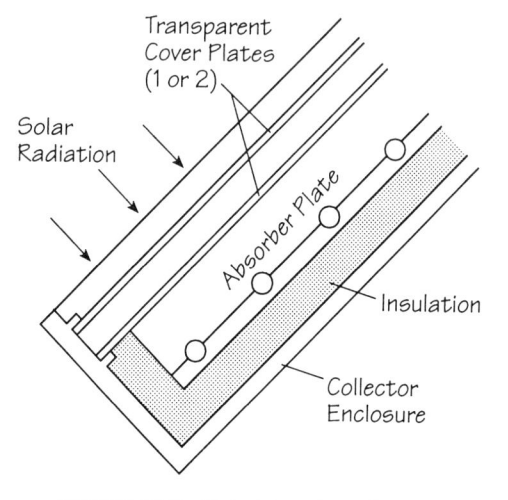

FIGURE 22.7 Flat plate solar collectors.

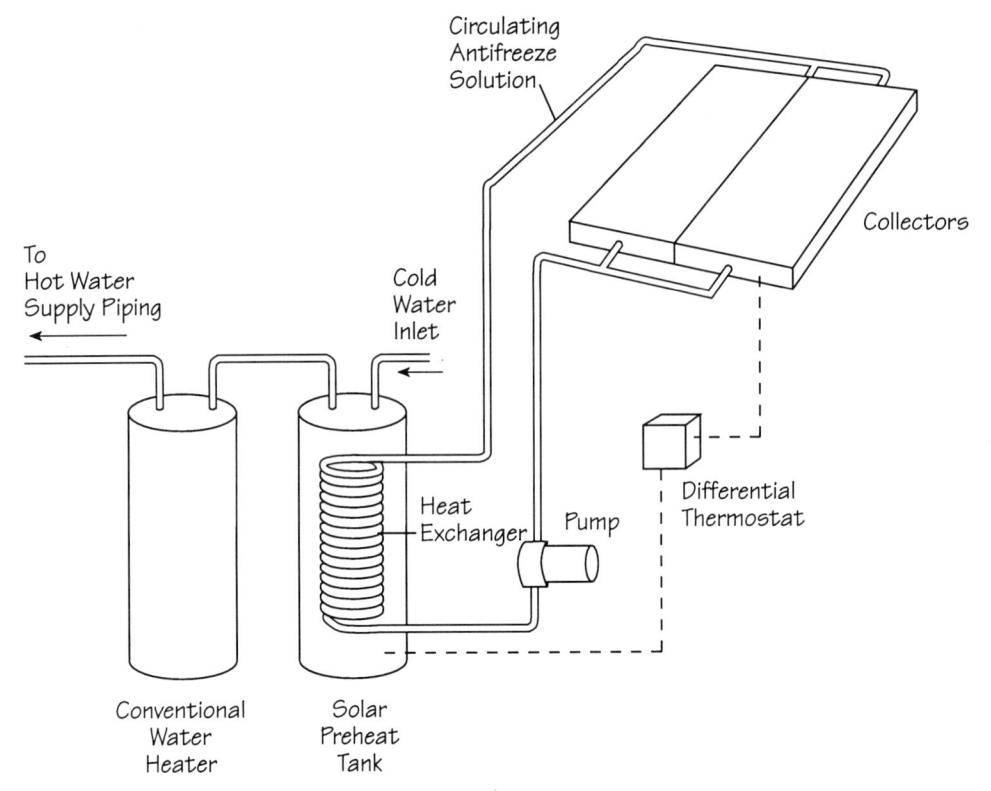

FIGURE 22.8 Closed-loop solar DHW system.

heated in the collectors rises by natural convection to the tank while colder water drops down to the collectors where it is heated and then rises to the tank to complete the loop.

System Sizing and Efficiency

A very rough design rule of thumb is that one ''typical'' collector is needed for each person in the household. Therefore, three or four collectors is the typical residential collector array to provide a meaningful portion (25–80% depending on collector efficiency and climate) of the domestic hot water needed annually.

Solar DHW systems use collectors with two or more cover plates to maintain efficiency during cold weather. Collectors without cover plates, or only a single cover plate, are limited to low-temperature applications such as heating water for swimming pools.

Flat plate collector efficiency depends on many factors, including:

- Number and type of cover plates
- The air spaces between the cover plates
- The infrared emittance of the absorber plate
- The amount of insulation behind the absorber plate and at the sides of the collector

Collector efficiency also depends markedly upon the angle at which they are mounted, as shown in Figure 22.9.

Solar collectors used year round to produce domestic hot water are typically mounted at a tilt approximately equal to the

a) For Year-Round Domestic Hot Water Collection

b) For Winter Space Heating

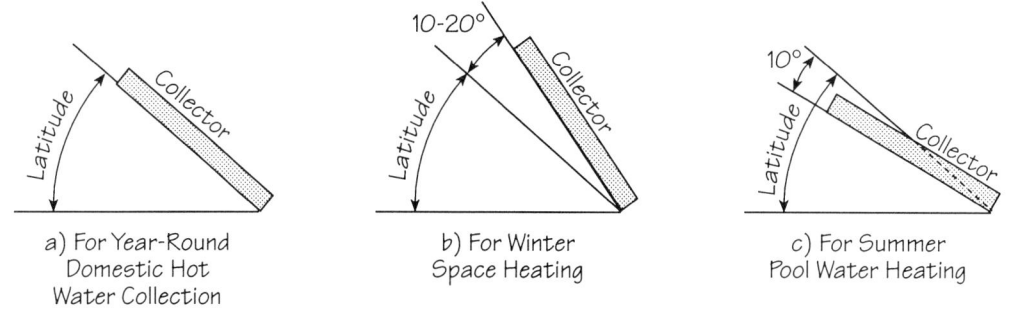

c) For Summer Pool Water Heating

FIGURE 22.9 Mounting angles of solar collectors for various applications.

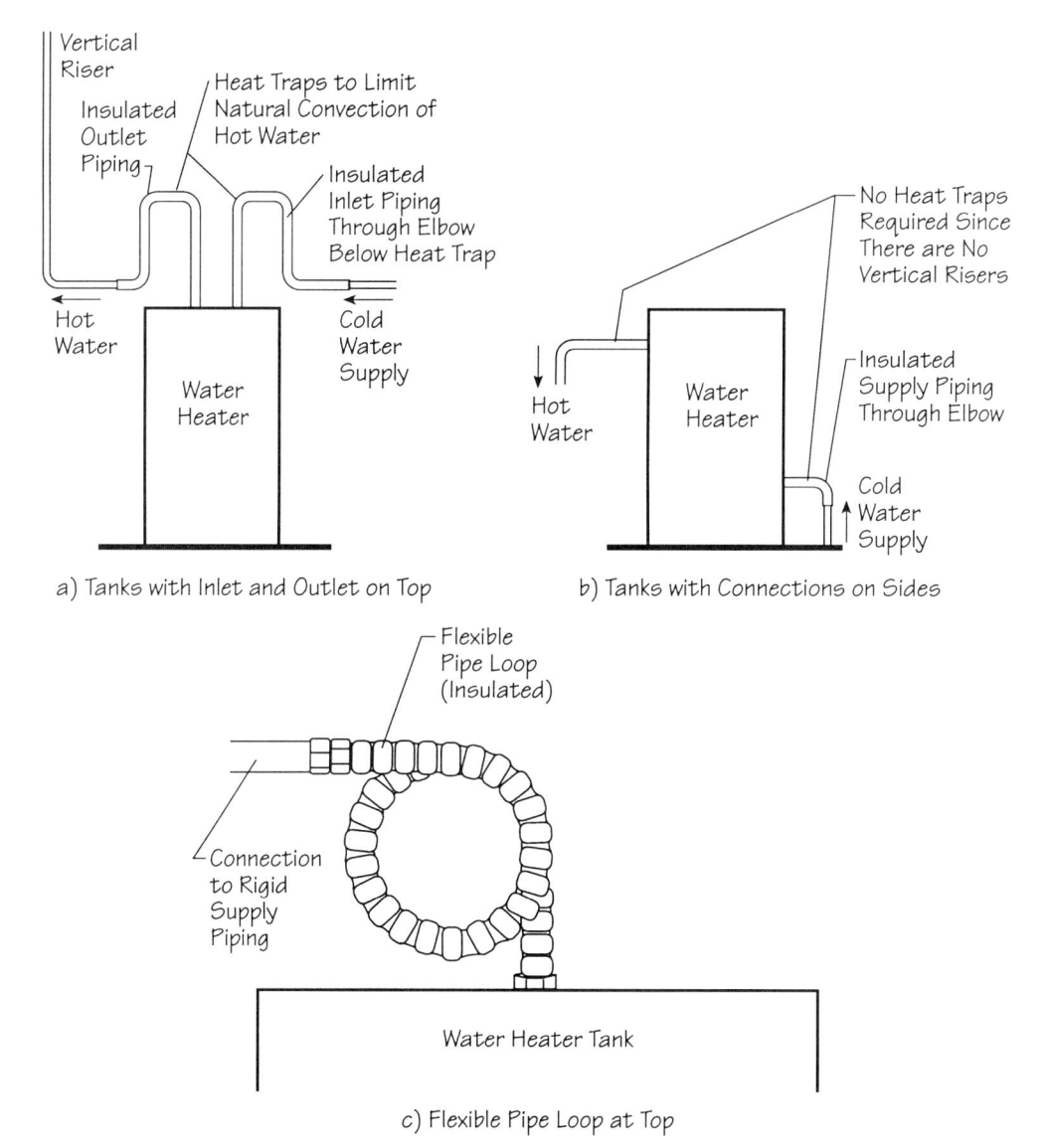

a) Tanks with Inlet and Outlet on Top

b) Tanks with Connections on Sides

c) Flexible Pipe Loop at Top

FIGURE 22.10 Heat traps to limit convection of hot water.

387

latitude angle. In contrast, collectors used for space heating are usually mounted at a tilt of latitude plus 10 to 20 degrees to enhance winter collection when sun angles are low. And collectors for pool heaters are mounted to be exposed to the high summer sun at an angle equal to latitude minus 10 degrees.

22.8 DHW PIPING SYSTEMS

General information for hot water piping systems can be found in Chapter 10. However, several piping-related items pertain uniquely to domestic hot water systems including "heat traps" and possible need for recirculation.

Heat Traps

Piping to and from water heaters and storage tanks without recirculating return piping should have "heat traps" to limit natural convection and contain the hot water within the water heater or storage tank. As a result, standby heat loss is reduced and the energy efficiency of the system is improved. Some storage water heaters are manufactured with internal heat traps (check manufacturer drawings and specifications). If not, then a piping arrangement such as one of those shown in Figure 22.10 is required.

External heat traps should always be insulated and placed close to the tank inlet and outlets.

Recirculation of Domestic Hot Water

Buildings with long runs (over 75 ft or so) of hot water piping may require installation of a recirculating system to assure prompt delivery of domestic hot water. Early rising occupants in buildings without a recirculating system may need to let the "hot" water run for many minutes before truly hot water will come from a shower or faucet. In so doing they will waste the water which cooled off in the pipes overnight.

Recirculating systems are of course more expensive to install since they require insulated return piping from all hot water supply locations. They are controlled by an aquastat which monitors return line water temperature and controls a small in-line circulator pump.

The prime applications are hotels, motels, dormitories, and large multifamily residential buildings. All the same, it may even be wise to install a recirculating hot water system in single-family houses since the circulator pumps typically only use 40–100 W.

23

CONTROL SYSTEMS

HVAC system controls range from simple on/off switches to highly programmed control sequences for major buildings. Controls are also used to provide for safe operation of equipment and to improve operational and energy efficiency and to provide for smoke control and life safety in buildings.

Various types of controls are used depending upon building type, and the size and type of heating/cooling/ventilation system used. These include electric, electronic, pneumatic (compressed air), and computerized DDC (direct digital control).

The field of HVAC controls is complex and continually changing. Perhaps the best way to begin looking at HVAC controls is to assume that technology now allows us to accomplish just about any type of control we can imagine. The real trick is to understand what operational commands are needed to make a system work and then to match these goals with a control system which can reliably and affordably meet them.

23.1 BASIC TERMINOLOGY

The field of HVAC controls is laden with terminology, some of which is essential. Below is just the "tip of the iceberg" necessary for basic understanding.

Analog control Control that is continuously variable (e.g., a valve on a gas stove to adjust the flame). In HVAC systems, pneumatic controls are inherently analog, as their readings and responses are naturally variable. The continuously variable nature of analog control is illustrated in Figure 23.1.

Control point The actual value (e.g., temperature) of a controlled variable at any specific moment.

Controlled variable The condition being controlled, such as air or water temperature.

Cycling A periodic change in the controlled variable from one value to another. Uncontrolled cycling is known as "hunting."

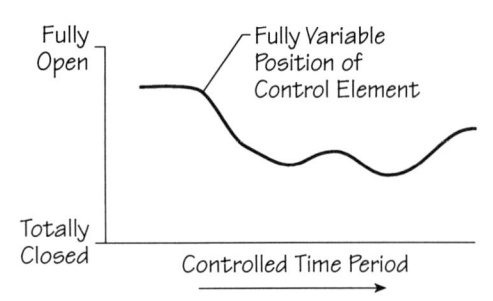

FIGURE 23.1 Continuously variable analog control.

Deadband A range of the controlled variable (e.g., temperature) within which no corrective action (heating or cooling) is made. For example, houses are typically controlled to call for heat when the temperature falls below about 68°F and cooled when the temperature exceeds 80°F. The deadband is between 68 and 80°F, also sometimes known as the "zero energy band."

Deviation The difference between the value of a controlled variable and set point (i.e., the control point at which a controller is set). Also known as the offset.

Digital control Results from a series of on and off pulses (e.g., like the Morse code) arranged to convey information. Computer operation is based on processing of digital language. Digital controllers adjust system functioning over time in a stepped manner (see Figure 23.2) by processing the input of controlled variables, and responding based upon control algorithms.

Droop A sustained deviation between the control point and the set point in a two-position control system due to a change in the heating or cooling load.

Electromagnetic devices See relays and solenoid valves.

Lag The delay of a system in responding to a change called for by a system sensor.

N.C. Normally closed. Describes the closed position for a controlled item when the control system is inoperative such as during a power failure.

N.O. Normally open. Items which should be designed to be normally open when the control system is inoperative such as during a power failure. An example is a three-way mixing valve which should fail with the hot water connection open when serving pipes and equipment subject to possible freezing.

Offset The deviation between the control point and the set point of a proportional control system which is sustained once stable operating conditions are achieved.

Relay An electromechanical switch which opens or closes contacts in response to some controlled action. Relays can have single or multiple contacts, and can be used to isolate systems with different voltages.

Set point The desired control point at which a controller such as a room thermostat is set.

Solenoid valve A valve that contains a plunger which is activated by the principle of electromagnetism when electric current flows.

Transducer A device that converts one energy form to another or amplifies an input or output signal. A transducer can convert a pneumatic signal to an electric signal (P/E transducer) or vice versa

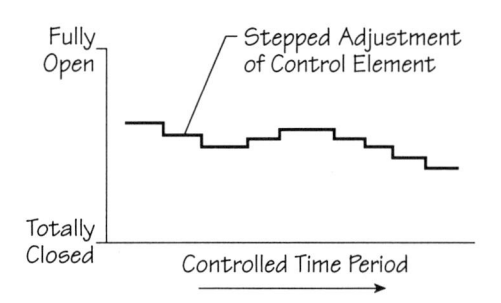

FIGURE 23.2 Stepped operation through digital control.

(E/P transducer). Transducers are also used to amplify a very low voltage or current signal to a higher level.

23.2 MODES OF CONTROL

A wide range of control modes are available. They range from simple controls to turn something either on or off (known as "basic two-position" control) to very sophisticated computer control modes.

Basic Two-Position ("On/Off") Control

Two-position control is a mode where a control circuit is either open or closed. In two-position temperature control, two values are selected to turn an item on and off; for example, a cooling system control which is activated when a thermostat senses a selected high temperature and is turned off when some lower temperature is reached. The difference between the on and off values is known as the "differential."

With such controls, the differential is the minimum temperature swing possible; some amount of temperature "overshoot" and "undershoot" will result from cyclical operation of the equipment. Basic two-position control is illustrated in Figure 23.3a.

Timed Two-Position Control

As shown in Figure 23.3b, undershoot and overshoot fluctuations of basic two-position control can be minimized by using timed two-position control which responds to gradual changes in the average value of the controlled variable (e.g., ambient room temperature).

A type of timed two-position control known as "time proportioning" can be provided by electromechanical, electronic, or computer-based DDC controls which adjust the cycle time of a system in response to changes in loads. Electromechanical controls can also obtain timed two-position control by adding a "heat anticipator" which is energized at the same time as the heating system.

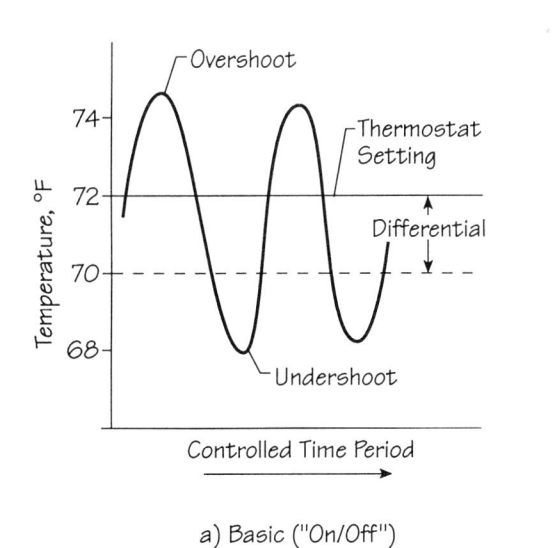

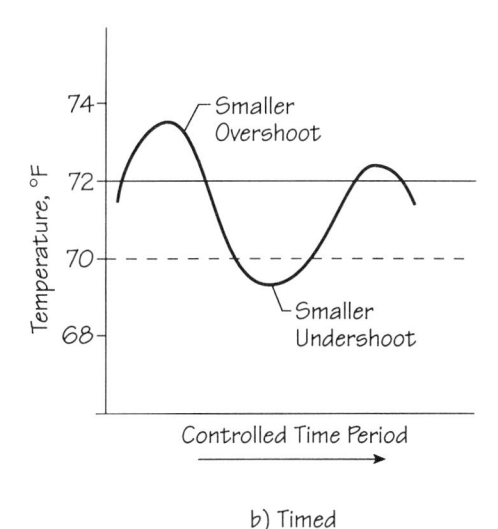

FIGURE 23.3 Two-position "on/off" control.

Stepped Control

Stepped control is used to activate or deactivate stages of equipment (multiple compressors, electric heaters, etc.) so that individual items which can be either on or off (two-position control) can function together to yield a result proportional to the design requirements. Indicated in Figure 23.4 is how a four-stage electric heater can provide stepped control in meeting various loads.

Proportional (P) Control

Proportional controls vary the output of a device (e.g., the amount a valve or damper is open) to satisfy the needs of the controlled variable. As indicated in Figure 23.5, proportional control allows for performance close to a desired set point (e.g., temperature). With proportional controls, the fluctuation which the controlled variable (i.e., temperature) goes through is known as the "proportional," "modulating," or "throttling" range and the sustained deviation from the set point is the "offset."

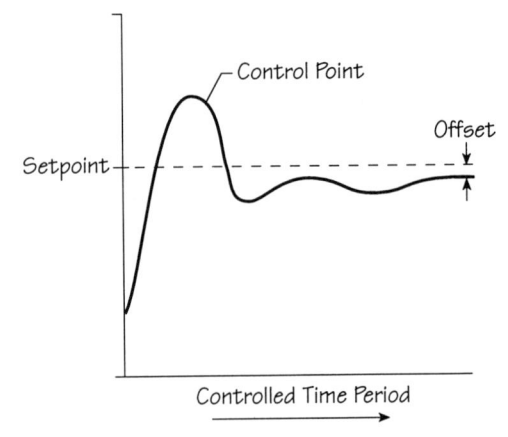

FIGURE 23.5 Proportional (P) control.

Proportional-Integral (PI) Control

PI control is a sophisticated control method which adds a reset response to correct the offset of basic proportional control. PI control reaches the set point after a period of time, with some overshoot, but without offset (see Figure 23.6). PI control is also known as "proportional-plus reset" or "two-mode''' control.

Proportional-Integral-Derivative (PID) Control

PID control enhances PI control by being able to respond quickly to the rate of change

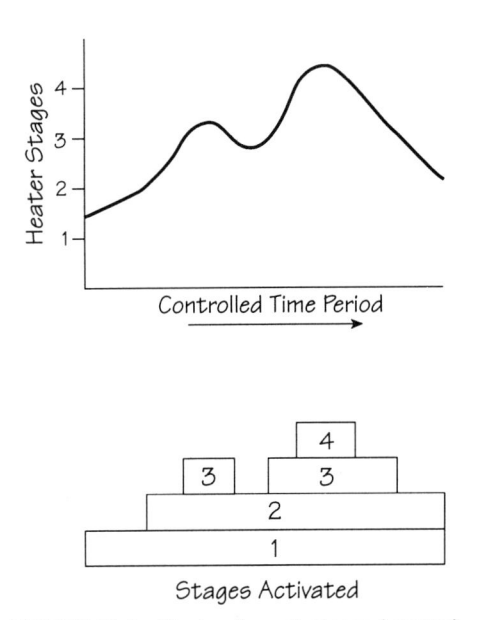

FIGURE 23.4 Staging through stepped control.

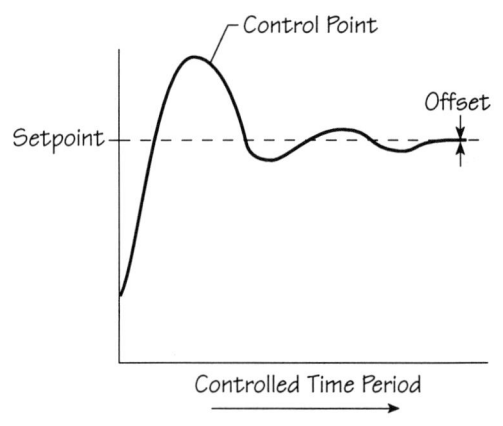

FIGURE 23.6 Proportional-integral (PI) control.

(derivative) of the deviation of the controlled variable. PID control generally shortens response time and reduces overshoot as shown in Figure 23.7. PID control is also known as "rate-reset" or "three-mode" control. Most HVAC loads and needs change at a relatively slow pace which does not require the sophistication provided by more costly PID control.

23.3 SENSORS, CONTROLLERS, AND ACTUATORS

Activation or adjustment of a controlled variable (e.g., temperature) requires the use of sensors that determine the existing condition of the controlled variable, controllers that compare the existing condition to what is desired, and actuators that adjust system performance as needed to achieve the desired condition.

Sensors

Sensors measure the value of a variable being controlled and notify the controller (e.g., a thermostat measures temperature). Other variables typically controlled and/or monitored in HVAC systems are relative humidity, flow, and pressure.

Temperature Sensors. One very common type uses a bimetallic strip made up of two layers of different types of metals which have different rates of expansion as shown in Figure 23.8. With changes in temperature, the curvature of the strip changes. This curvature can then be used to open or close electric circuits or regulate pneumatic control systems.

Other types of temperature sensors include sealed bellows in which changing temperatures cause expansion or contraction of air resulting in movement of a controller mechanism, and thin lengths of "special" wire which changes electrical resistance as temperature changes, used in electronic controls.

Humidity Sensors. Mechanical humidity sensors use elements (usually made of nylon) which absorb or release moisture to the surrounding air and in so doing expand or contract and operate a controller mechanism to respond to the humidity

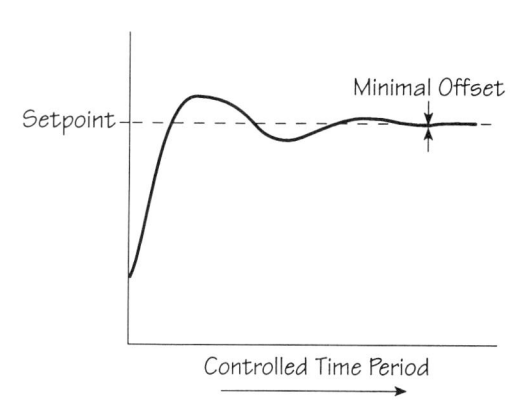

FIGURE 23.7 Proportional-integral-derivative (PID) control.

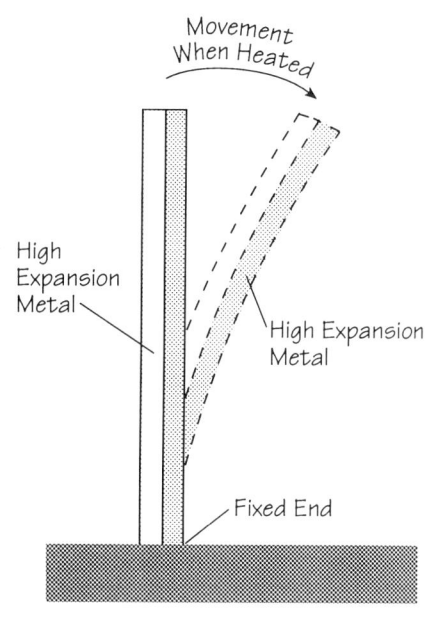

FIGURE 23.8 Bimetallic sensor responds to changes in temperature.

394 CONTROL SYSTEMS

level. Electronic sensors determine the humidity level by changes in the resistance or capacitance of the sensing element.

Flow Sensors. There are a wide variety of flow sensor types. Some use an obstacle (a vane or paddle) which is inserted in the moving medium (air or water) to determine the flow rate of the fluid being measured. Ultrasonic flow detectors which simply clip on to piping are a particularly useful diagnostic tool. Flow sensors are often used to meter the use of steam, water, or fuel.

Pressure Sensors. Pressure sensors use bellows, diaphragms, and other pressure-sensitive devices including solid state sensors. Pressure sensors are used by HVAC systems to monitor and control the operation of fans and pumps, compressors and boilers.

Controllers

The controller receives the measured variable information from the sensor. It then compares this information with the control point (set point, humidity level, flow rate, etc.). Some controllers are integral with the sensor, as in a house thermostat. In larger building systems, sensors and controllers may be located far apart. Controller selection depends upon the control modes required.

Electric Controllers. Primarily used where two-position control such as on/off or open/closed is required.

Electronic and Pneumatic Controllers. Used where modulating analog control is needed.

Microprocessor Controllers. Used where complex control schemes are required (e.g., combination ventilation and smoke control). Activation comes after the information from the sensor is input into a computer which determines the proper system response based upon programmed logic. Easy to reprogram.

Actuators

Actuators take commands from the controller and cause the controlled device to perform as required to bring about the desired condition. They turn equipment on or off or change fluid flows and operating devices such as valves and dampers. The following types of actuators are commonly used in HVAC systems:

Electric Actuators. Motors or solenoids typically used for two-position or stepped operation.

Electronic Actuators. Used to achieve a proportional response (especially for small buildings where a pneumatic system cannot be justified).

Pneumatic Actuators. Commonly used in larger systems where modulating action is required.

23.4 SUMMARY OF CONTROL SYSTEMS

A variety of control methods and systems are available to the HVAC designer. Considerations that should influence control selection include:

- Anticipated load changes. Many HVAC system loads change relatively slowly and can be met by relatively simple and inexpensive controls.
- Size and type of building.
- The degree of accuracy required by the application and the amount of offset which is considered acceptable. For example, in a hospital, operating rooms require extremely close control of temperature and humidity, whereas

TABLE 23.1 CONTROL SYSTEM CHARACTERISTICS

Electric	Electronic	Pneumatic	DDC
Most common for two-position (on/off) control	Solid state reliability	Offers simple and reliable modulating control	Computer-based. Integration with building management system
Integral sensors/controller	Used where precise control is needed	Used for large systems	High order control (P, PI, or PID)
Modulating control is more complex than pneumatic	Complex controllers and actuators	Requires a supply of clean dry air	Inherent energy management
Generally inexpensive for simple systems but can become expensive and complicated for complex control schemes	Higher cost than electric	Requires more maintenance	Compatible with pneumatic actuators
		Commonly used for simple zone control (i.e., VAV boxes) in large buildings	Now available for small buildings and systems at reasonable cost

many other spaces, such as waiting rooms, do not.

Economic considerations, both initial and operational.

Summarized in Table 23.1 are the basic characteristics of electric, electronic, pneumatic, and DDC (microprocessor-based direct digital control) control systems.

23.5 ELECTRIC CONTROL SYSTEMS

Electric controls can be operated using line voltage (120 V) although most systems use low voltage (usually 24 V) produced through the use of a transformer.

Most electric controls are simple two-position controls which use either a snap-acting or mercury switch to turn equipment on or off based upon monitored conditions as shown in Figure 23.9. Perhaps the most common type of low-voltage electric control is a simple thermostat which controls space temperature. Sometimes electric con-

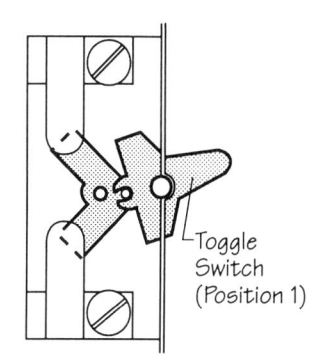

a) Circuit Connection Made

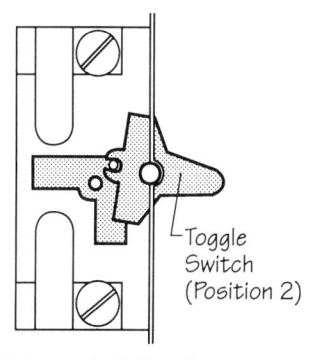

b) Interrupted Circuit

FIGURE 23.9 Snap-acting switch for two-position control.

trols are wired in series. For example, electric controls wired in series can prevent the fan in a fan-coil unit from operating until some temperature of circulating fluid is first attained.

Modulating Control

Electric controls can also provide modulating control. This is accomplished by using two-position controls which have a center off position and a reversible actuator. They work by responding to a change in the controlled variable by driving the actuator to an intermediate position (between the limits of the controller) before opening the circuit. Motorized valves (two-way and three-way) and dampers can be operated in this manner.

23.6 ELECTRONIC CONTROL SYSTEMS

Electronic control systems use solid-state devices which usually operate on low voltage direct current. For conditions that are subject to continual change (such as temperature and pressure), analog sensors signal the controller with a varying low-voltage DC signal. The controller then signals a modulating actuator (e.g., flow control valve) which causes the adjustments needed in the operation of the controlled item.

Items that need two-position signals use digital sensors. The controller then notifies actuators such as relays, solenoid valves, and motor starters of the required operation (either on or off).

Some sensors (e.g., temperature) transmit signals directly to electronic controllers based upon resistance to flow in special wiring. Pressure sensors, though, need transducers or transmitters to convert pressure changes into a variable which can be used by an electronic controller such as voltage, current, or resistance.

23.7 PNEUMATIC CONTROL SYSTEMS

Pneumatic controls use a supply of compressed air and mechanical means, such as a temperature-sensitive bimetal or bellows, to perform control functions.

The basic components of a pneumatic control system are illustrated in Figure 23.10. The compressor cycles on and off (on perhaps 30% of the time) to maintain a storage tank air pressure (typically between 70 and 100 psi). After leaving the storage tank, the compressed air is dried to remove any moisture and filtered to remove any compressor oil and other impurities. The air then passes through a pressure-reducing valve (prv) which lowers the pressure to that to be maintained in the supply main (typi-

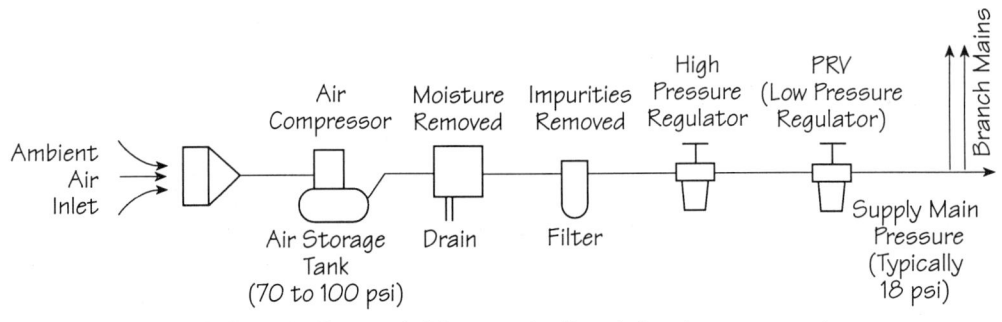

FIGURE 23.10 Pneumatic (air pressurized) control system components.

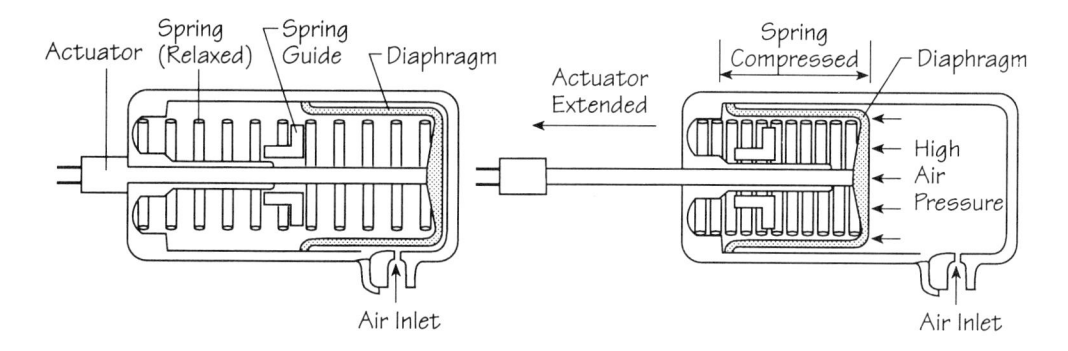

FIGURE 23.11 Damper actuator for modulating control.

cally 18 psi). After the PRV, the compressed air flows through the main to the various branches of the system.

Controlled Devices

Pneumatic systems are particularly useful in providing control which is continuously modulating such as damper actuators (in air supply systems) and flow control valves (in water supply systems). Illustrated in Figure 23.11 is a spring-type piston damper actuator. The piston of the actuator remains in its normal retracted position when the diaphragm pushes against the spring with a very low air pressure (i.e., 5 psig or less). Then as pressure increases against the diaphragm, the spring becomes compressed and the piston extends partially to the degree determined by the sensor and controller. If the controller calls for full extension of the damper actuator, the pressure is increased to a higher air pressure (i.e., 10 psig), at which the spring will be fully compressed.

Such a damper actuator can be used for many applications, such as the ''face and bypass'' damper arrangement shown in Figure 23.12 to direct the required airflow over a (heating or cooling) coil.

Pneumatic actuators can control liquid flow through valves in a similar manner. In-

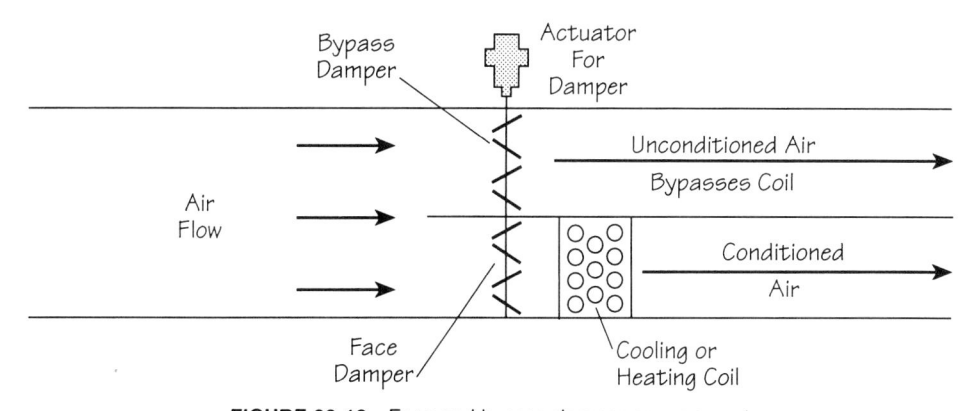

FIGURE 23.12 Face-and-bypass damper arrangement.

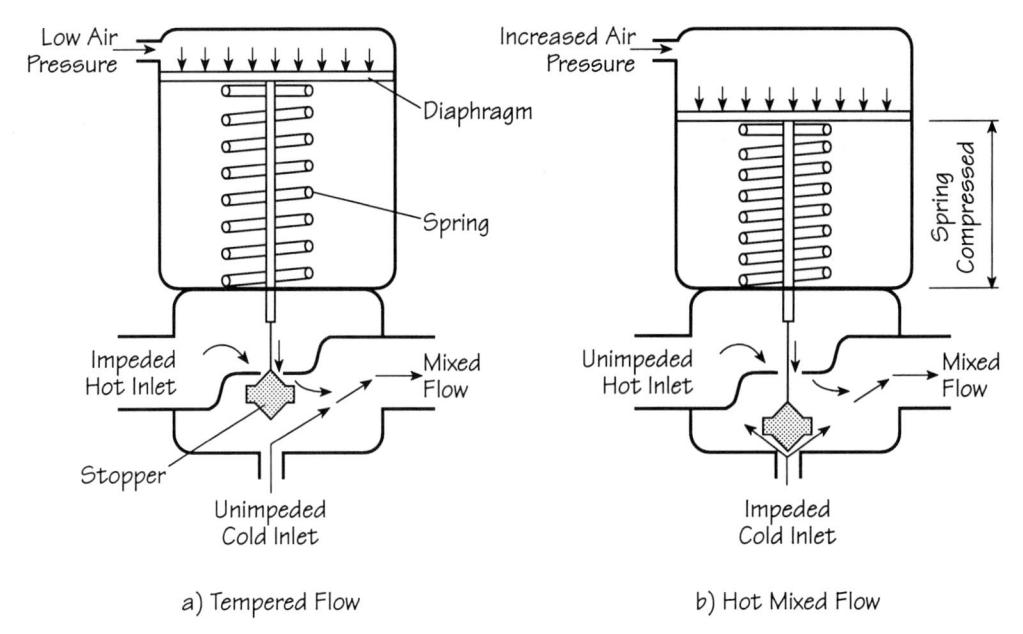

FIGURE 23.13 Pneumatically controlled mixing valve.

dicated in Figure 23.13 is a "mixing valve" which has two inlets (for fluids of different temperatures) and one common outlet (for the mixed fluid of desired temperature). Air pressure is applied to a diaphragm which pushes against a spring which controls the position of the valve stem. This pressure then either opens flow from one inlet while closing the other, or results in the valve stem taking an in-between position as needed to satisfy system requirements.

23.8 DIRECT DIGITAL CONTROL (DDC) SYSTEMS

Computers are now almost everywhere, including in most new automobiles which have numerous microcomputers for various control functions. The heart of a DDC control system is the microprocessor, which can be programmed to perform a wide range of basic control operations, along with more sophisticated functions which optimize operations in an energy efficient manner. DDC control systems generally use electronic sensors and pneumatic actuators where proportional control is required.

The computer can also be used to perform building management functions including operation of lights, fire system monitoring, security, record keeping, and so on, all at one location at a display console.

DDC systems are the most flexible control system because the sequence of control can be modified through software with no need to move wires or pneumatic tubing. Over the past 10 years or so, technological advances have made possible the cost-effective use of DDC controls for many building applications.

23.9 SAFETY CONTROLS

HVAC systems employ safety controls to protect against equipment damage and to assure the safety of building occupants. They include:

1. "Proof of ignition" controls in boilers and furnaces. Upon a call for heat, fuel is released into a combustion chamber where ignition should take place and result in a rise in temperature. Should the temperature not rise within a specified time frame (typically about a minute), these controls will interrupt the supply of fuel.

2. "High limit" controls are found on warm air furnaces to stop combustion should temperatures rise well beyond the normal operating range. Such an occurrence would normally indicate failure of the supply fan.

3. "High temperature" controls are used on hot water systems to sense a water temperature which is above the intended operating range and to shut off the burner. Activation of such a control would indicate that the normal system controls are not functioning properly.

4. "Low water cutoff" controls will interrupt boiler operation should the water level drop to a point where the heat exchanger may overheat and be damaged. Such controls are an absolute necessity on steam boilers where the water line will fluctuate. Activation of a low water cutoff on a steam boiler would be an indication that condensate is not being returned to the boiler.

23.10 ENERGY-CONSERVING CONTROLS

An important goal of most building operators is to minimize the cost of operations. Control systems ranging from simple time clocks to microprocessor-based systems can help to accomplish operational energy savings. For large buildings, the following design strategies are among those that often result in substantial improvements in operational efficiency:

1. Reliance on a microprocessor-based control system, rather than equipment oversizing, to address summer and winter design conditions. For example, when peak cooling conditions occur, the control system can increase the set point temperature slightly, dim or deenergize certain lighting circuits, and so on.

2. Concern for indoor air quality can also be addressed by normally introducing more outside air (perhaps twice that required by code). Controls can then reduce this quantity to code minimums during the few hours per year when outdoor design conditions are approached.

3. Optimized morning warmup. Sensors can be used to monitor outdoor conditions and determine the length of time needed to bring the inside space temperature up to the desired level for building occupancy.

4. Optimized control of an outside air economizer cycle. Sensors can monitor the dry-bulb and wet-bulb temperatures for possible use of outside air for cooling. In dry southwestern United States climates, outside air can usually be used whenever the outside air temperature is below the return air temperature. The changeover temperature to an economizer cycle is usually about 72°F db in temperate climates and 67°F db or lower in humid climates.

5. Deactivation ("duty cycling") of nonessential equipment such as fans can be a very effective control strategy to reduce peak electrical demands. Still, such strategies should be employed in a manner which satisfies applicable code requirements.

6. In large heating systems, controls can continuously monitor flue gas temperatures and composition to determine the efficiency of combustion and percent of excess air. The results can then be used to adjust burners for improved efficiency.

7. Operation of chillers can be optimized by the use of controls to monitor the position of chilled water zone supply valves. A wide open valve indicates that peak load conditions exist in that zone. If no valves are wide open, the controls can

raise the chilled supply temperature until one zone valve opens fully. Such an approach will reduce the power required by the compressor, and result in peak demand savings.

23.11 SMOKE CONTROLS

Control of fire and smoke is an important function performed by control systems. This is particularly true in large, high-rise buildings where an emergency situation must be sensed quickly and addressed effectively.

Fire suppression systems (such as sprinklers, which enjoy a remarkable success rate) are not within the scope of this book. Realize though that smoke, and not flames, is responsible for the great majority of fatalities in building fires. The design of HVAC systems must provide for control of smoke. Local codes typically require that HVAC systems in buildings of various types and sizes comply with NFPA 92A (Recommended Practice for Smoke Control Systems).

Building design should provide architectural barriers to confine the flow of smoke such as doors, and dampers within ducts. Air pressure is also used to control smoke in several ways. For example, some high-rise building codes now require pressurization of fire stairs by large fans which become activated only after a fire situation has been sensed. They then supply 100% outside air to create positive pressure within the stair relative to the building floor areas. This prevents smoke entry through exit doors. Such systems can be difficult to engineer so that the positive pressure which is developed is not so great as to prevent building occupants from opening the fire doors.

Control of dampers within supply and exhaust ductwork can also be used to create a "pressure sandwich" which surrounds a zone where smoke exists to aid in its evacuation. Indicated in Figure 23.14 in simple form is how pressure can be used to remove smoke from floor Y of a building. This is accomplished by supplying air to floors X and Z while closing exhaust air dampers. Therefore, floors X and Z increase in pressure. At the same time the supply of air to floor Y is stopped while the exhaust air system remains open.

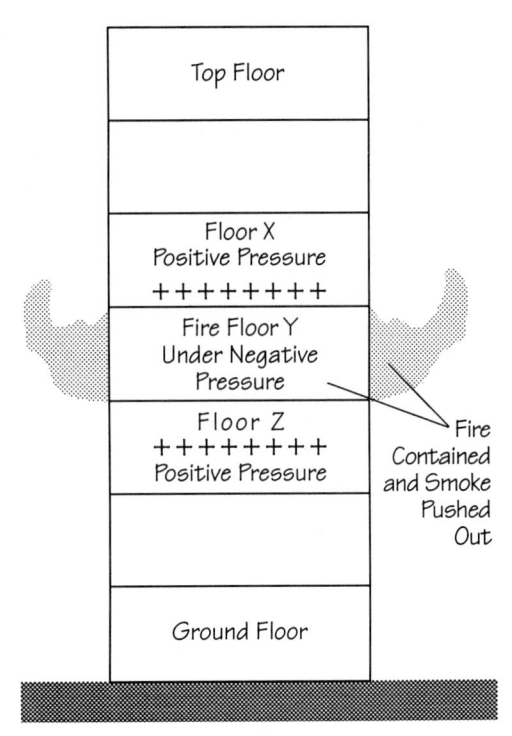

FIGURE 23.14 Building smoke control through pressurization of floors.

23.12 REALISTIC APPLICATIONS

With today's low-cost computer capability there is virtually no limit to the amount of data which can be gathered and functions which can be controlled. Nevertheless, one should be realistic. For example, installation of an optimizing control feature at a cost of $200 is highly questionable if its use results in a savings of only $10 per year. In addition, control system selection should be tailored to the level of oversight. For example, it makes no sense at all to provide a computerized control system with eight modes of cooling system operation if the person in control of the system is untrained in its use.

In theory, control systems allow for efficient, trouble-free operation of buildings. In practice, however, horror stories about maladjusted and misperforming control systems abound. For example, detailed diagnosis of buildings has uncovered motorized dampers which never close, chillers which run all winter (and shed their load through hot-gas bypass), and economizer cycles which are never actually activated due to improper temperature settings. Perhaps the most important lesson is to keep control systems as simple as possible, and never assume that they are working as intended just because they have been turned on.

24

APPLIED ECONOMICS

As in most other aspects of the material world, HVAC design decisions are often driven by economics. Investment criteria vary greatly. For example, operators of institutional facilities typically will make investments which take 10–15 yr to recover since they know that they will be operating these facilities for decades to come. In contrast, homeowners generally look for a payback period of 5–10 yr for conservation measures. And speculative developers often have investment criteria measured in days or months—rather than years.

The time to recover an economic investment is known as the "simple payback." Additional factors are used to account for energy cost escalation and the time value of money which could be invested in some other manner (known as "discounting"). These concepts will be explored and tables provided for quick conversion of a simple payback finding to a result which also accounts for energy cost escalation and discounting.

Perhaps the best way to evaluate energy investments for new houses is in the context of mortgage economics. When fully analyzed, many mortgaged conservation investments will yield "positive cash flow" when the increased yearly loan cost is compared to tax deductions and dollar energy savings.

Also discussed in this chapter are the very important concepts of "diminished return" relative to conservation improvements, and peak electrical demand as it impacts upon the operational cost of many commercial buildings.

It is clear that economics is often a primary factor in HVAC design choices. Yet a certain amount of detachment from pure number-driven decision making is advised when decisions with long-term implications are involved. For example, try to imagine how many buildings remain in operation in relatively cold climates, with single-glazed windows and little or no wall insulation, because the simple paybacks were not attractive when they were built (in particular, buildings built prior to the "energy crisis" of 1973). Such decisions are hard to correct or "retrofit" after the fact in a cost-effective manner.

24.1 COST PER MILLION BTUS (MMBtu)

Because a Btu is a very small quantity of heat (only about the heat content of a wooden kitchen match), heat is normally measured in thousands or millions of Btus. Formula 24A is used to compute the cost of delivered energy in terms of dollars per million Btu ($/MMBtu)—a useful measure for comparing the cost of various fuels and energy sources.

$$\$/\text{MMBtu} = \frac{1{,}000{,}000}{\text{Btu}_{fu} \times \text{Eff}_{sys}} \times \$_{fu}$$

Formula 24A

where

$/MMBtu   is the cost to deliver 1,000,000 Btus of useful energy
Btu_{fu} is the quantity of Btus obtained from a fuel unit (i.e., gallon of oil, therm of gas, etc.)
Eff_{sys} is the efficiency of the system (i.e., Annual Fuel Utilization Efficiency, COP, etc.)
$\$_{fu}$ is the dollar cost per fuel unit (i.e., gallon of oil, therm of gas, etc.)

Example 24A: Determine the delivered cost per million Btus for #2 fuel oil containing 140,000 Btus per gallon (Btu_{fu}) burned in a furnace with an AFUE of 0.80 (Eff_{sys}) when the fuel oil cost is $1.20 per gallon ($\$_{fu}$).

Solution: Use Formula 24A.

$$\$/\text{MMBtu} = \frac{1{,}000{,}000}{\text{Btu}_{fu} \times \text{Eff}_{sys}} \times \$_{fu}$$

(Formula 24A)

$$= \frac{1{,}000{,}000}{140{,}000 \times 0.80} \times \$1.20$$

$$= \$10.71$$

For electricity consumed for lighting, power, or resistance heat, use the simplified Formula 24B to determine the cost of energy.

$$\$/\text{MMBtu} = 293 \times \$_{kWh} \quad \text{Formula 24B}$$

where

$/MMBtu   is the cost to deliver 1,000,000 Btus of electricity consumed for lighting, power or resistance heat
293   is the number of kilowatt hours of electricity needed to provide one million Btus (1,000,000/3413)
$\$_{kWh}$ is the dollar cost per kilowatt hour of electricity

Example 24B: Determine the delivered cost per million Btus for electricity costing twelve cents ($0.12) per kilowatt hour.

Solution: Use Formula 24B.

$$\$/\text{MMBtu} = 293 \times \$_{kWh}$$

(Formula 24B)

$$= 293 \times \$0.12$$

$$= \$35.16$$

Table 24.1 lists the unit costs for common energy sources which are required to yield

TABLE 24.1 DELIVERED ENERGY COST ($/MMBtu)

Energy Source and (Efficiency)	$7.50	$10.00	$12.50	$15.00	$20.00	$30.00	$40.00
	\multicolumn{7}{c}{Dollars per Million Btu}
	\multicolumn{7}{c}{Cost per hundred cubic feet (1 therm)}
Natural gas (75%)	$0.56	$0.75	$0.93	$1.13	$1.50	$2.25	$3.00
Natural gas (80%)	$0.60	$0.80	$1.00	$1.20	$1.60	$2.40	$3.20
Natural gas (82%)	$0.62	$0.82	$1.03	$1.23	$1.64	$2.46	$3.28
Natural gas (90%)	$0.68	$0.90	$1.13	$1.35	$1.80	$2.70	$3.60
	\multicolumn{7}{c}{Cost per gallon}
Propane (75%)	$0.51	$0.69	$0.86	$1.03	$1.37	$2.06	$2.75
Propane (80%)	$0.54	$0.74	$0.92	$1.10	$1.46	$2.20	$2.93
Propane (82%)	$0.56	$0.75	$0.94	$1.13	$1.50	$2.25	$3.01
Propane (90%)	$0.62	$0.82	$1.03	$1.24	$1.65	$2.47	$3.29
	\multicolumn{7}{c}{Cost per gallon}
#2 Fuel oil (75%)	$0.79	$1.05	$1.31	$1.58	$2.10	$3.15	$4.20
#2 Fuel Oil (80%)	$0.84	$1.12	$1.40	$1.69	$2.24	$3.26	$4.48
#2 Fuel Oil (82%)	$0.86	$1.15	$1.43	$1.73	$2.30	$3.44	$4.59
#2 Fuel Oil (90%)	$0.95	$1.26	$1.58	$1.89	$2.52	$3.78	$5.04
	\multicolumn{7}{c}{Cost per cord}
Wood (60%) (assume 24 MMBtu per cord)	$108	$144	$180	$216	$288	$432	$576
	\multicolumn{7}{c}{Cents per kilowatt hour}
Elec. resistance (100%)	2.56	3.41	4.26	5.12	6.83	10.24	13.65
	\multicolumn{7}{c}{Cents per kilowatt hour}
Elect. heat pump							
at 1.5 COP	3.84	5.12	6.40	7.68	10.24	15.36	20.48
at 2.0 COP	5.12	6.83	8.54	10.24	13.65	20.48	27.30
at 2.5 COP	6.40	8.53	10.66	12.80	17.07	25.60	34.13
at 3.0 COP	7.68	10.24	12.80	15.36	20.48	30.72	40.95

the delivered energy cost indicated in dollars per million Btus ($/MMBtu).

Consult Table 24.1 to get a feel for $/MMBtu results for the costs which are common in your area. For example, note that natural gas utilized at an AFUE of 75% and costing $0.56 per hundred cubic feet has a delivered energy cost of $7.50 per million Btu. Whereas the use of electricity costing about $0.10 per kilowatt hour has a delivered energy cost of about $30.00 per million Btu. From a cost point of view, it is clear—all Btus are not created equal!

24.2 SIMPLE PAYBACK AND RATE OF RETURN

The time period required to "break even" or recover the cost of an investment in energy conservation is known as the "simple payback," as calculated using Formula 24C.

$$SP_{yr} = \frac{\$_{invest}}{\$_{sav\text{-}yr}} \quad \text{Formula 24C}$$

where

SP$_{yr}$ is the simple payback in years
$_{invest}$ is the initial investment cost of a conservation investment in dollars
$_{sav-yr}$ is the dollar savings achieved by a conservation investment per year

Example 24C: Additional wall insulation will add \$110 ($_{invest}$) to the initial cost of a house. Determine the simple payback (SP$_{yr}$) if heating costs will be lowered \$20 per year ($_{sav-yr}$).

Solution: Use Formula 24C.

$$SP_{yr} = \frac{\$_{invest}}{\$_{sav\text{-}yr}} \quad \text{(Formula 24C)}$$

$$= \frac{\$110}{\$20}$$

$$= 5.5 \text{ years}$$

Most people understand how to evaluate the annual rate of return they will receive on a savings account, stocks, bonds, or other investment. Formula 24D is used to compute the rate of return for an investment.

$$RofR_\% = \frac{\$_{sav\text{-}yr}}{\$_{invest}} \times 100 \quad \text{Formula 24D}$$

where

RofR$_\%$ is the rate of return earned by a conservation investment in percent
$_{sav-yr}$ is the dollar savings achieved by a conservation investment per year
$_{invest}$ is the initial investment cost of a conservation investment in dollars

Example 24D: Determine the rate of return for an investment in insulation which saves \$20 per year ($_{sav-yr}$) but requires an initial expense of \$110 ($_{invest}$).

Solution: Use Formula 24D.

$$RofR_\% = \frac{\$_{sav\text{-}yr}}{\$_{invest}} \times 100$$

(Formula 24B)

$$= \frac{\$20}{\$110} \times 100$$

$$= .181 \times 100$$

$$= 18.1\%$$

Such a rate of return is very attractive, especially when compared to long-term government-backed savings bonds which in the early 1990s have yielded returns of well below 10%.

24.3 ESCALATION OF ENERGY COSTS

The long-term economics of conservation investments ultimately depends on future energy prices and fuel escalation rates. During the energy crisis of the mid to late 1970s, energy rates increased rapidly, and it was considered reasonable at the time to assume long-term energy escalation rates of 10% or more for evaluating energy conservation options. But the future is uncertain, and hard to predict, as Table 24.2 clearly indicates. In particular, note the volatility of #2 fuel oil prices throughout the decade and the fact that the average retail price was higher in 1980 than it was in 1989.

Table 24.3 can be used to adjust a computed simple payback (0% escalation) for various energy escalation (or deescalation) rates.

406 APPLIED ECONOMICS

TABLE 24.2 ENERGY PRICES IN THE 1980s

	Average Retail[1] Electricity Prices (cents/kWh)	Average Retail[2] Natural Gas Prices ($/1000 ft^3)	Average Retail[1] #2 Fuel Oil Prices (cents/gal)
1980 Average	4.73	2.91	97.4
1981 Average	5.46	3.51	119.4
1982 Average	6.13	4.32	116.0
1983 Average	6.29	4.82	107.8
1984 Average	6.25	4.85	109.1
1985 Average	6.44	4.72	105.3
1986 Average	6.44	4.13	83.6
1987 Average	6.37	4.05	80.3
1988 Average	6.35	4.09	81.3
1989 Average	6.44	4.22	90.0
Average Escalation (Deescalastion)	4.0%	5.0%	(0.8%)

Sources: 1—US DOE/EIA, "Monthly Energy Review;" 2—U.S. DOE/EIA, "Natural Gas Monthly."

Example 24E: Assume that an investment in energy-efficient HVAC fans made in 1980 was estimated to have a simple payback period of 8 yr. Estimate the actual payback period after adjusting for energy cost escalation.

Solution: As indicated in Table 24.2, the cost of electricity had an average escalation rate of 4.0% during the 1980s. Now use Table 24.3 and find the unadjusted simple payback line corresponding to 8 yr. Then go to the right to the column which repre-

TABLE 24.3 ADJUSTED YEARS TO PAYBACK DUE TO ENERGY COST (DE)ESCALATION

Deescalation rate			Unadjusted Simple Payback (Yr)	Escalation Rate				
−10%	−5%	−2%		2%	4%	6%	8%	10%
1.1	1.1	1.0	1	1.0	1.0	1.0	1.0	0.9
2.4	2.2	2.1	2	2.0	1.9	1.9	1.8	1.8
3.9	3.4	3.1	3	2.9	2.9	2.8	2.8	2.7
5.6	4.6	4.2	4	3.9	3.8	3.8	3.7	3.6
7.7	6.0	5.3	5	4.9	4.8	4.7	4.6	4.5
10.4	7.4	6.5	6	5.9	5.8	5.7	5.6	5.5
14.3	9.0	7.6	7	6.9	6.7	6.6	6.5	6.4
20+	10.7	8.8	8	7.8	7.7	7.5	7.4	7.3
20+	12.5	10.0	9	8.8	8.7	8.5	8.3	8.2
20+	14.6	11.3	10	9.8	9.6	9.4	9.3	9.1
20+	16.9	12.6	11	10.8	10.6	10.4	10.2	10.0
20+	19.5	13.9	12	11.8	11.5	11.3	11.1	10.9
20+	20+	15.3	13	12.7	12.5	12.3	12.0	11.8
20+	20+	16.7	14	13.7	13.5	13.2	13.0	12.7
20+	20+	17.4	15	14.7	14.4	14.2	13.9	13.6

sents adjustment to an energy escalation rate of 4%. The answer is 7.7 yr—not a major change.

Browse Table 24.3 a bit more. Note that significant adjustments in simple payback occur only when the time period is long and the energy escalation (or deescalation) rates are high, as illustrated in Example 24H below.

24.4 DISCOUNTED PAYBACK

"Simple" payback (even if adjusted for fuel escalation) is inaccurate if one places a time value on invested money (known as the "discount rate"). An amount of $100 will be worth $107 a year later if invested at 7% interest (100 × 1.07), $114.50 (107 × 1.07) after two years, and so on. Thus, capital accumulates as interest compounds.

A process known as "discounting" allows one to estimate the "present value" of money reflecting future interest accruals. Discount factors for various discount rates for a given number of years in the future are calculated by using Formula 24E.

$$DF = \frac{1}{(1 + d)^n} \quad \text{Formula 24E}$$

where

DF is the discount factor
d   is the discount rate as a decimal (i.e., 10% = 0.10)
n   is the future year being evaluated (i.e., 3 if a three year evaluation is desired)

The use of discount factors enables one to determine the present value of dollar savings which will be realized in the future by using Formula 24F.

$$PV = FV \times DF \quad \text{Formula 24F}$$

where

PV is the present value of savings which will be derived in some future year n
FV is a future amount of dollar savings
DF is the discount factor (computed by Formula 24E or from Table 24.4)

The time period when the accumulated present value of future savings equals the original amount of an investment is known as the "discounted payback period."

Example 24F: Determine the discount factor (DF) for a discount rate (d) of 10% for the third year ($n = 3$).

Solution: Use Formula 24E.

$$DF = \frac{1}{(1 + d)^n} \quad \text{(Formula 24E)}$$

$$= \frac{1}{(1 + .10)^3}$$

$$= \frac{1}{1.331}$$

$$= 0.751, \text{ a result which can also be obtained from Table 24.4}$$

Example 24G: Determine the discounted payback period for a $100 conservation investment which produces energy savings of $25 per year. Assume a discount rate (d) of 10% (0.10).

Solution: First note that an investment of $100 which returns $25 per year has a simple payback of 4 yr. Now use Formula 24F to determine the present value (PV) of the future yearly energy savings of $25 for each year. Obtain discount factors for each year from Table 24.4. The discounted payback period will be when the accumulation of yearly present values equals the investment of $100.

$$PV = FV \times DF \quad \text{(Formula 24F)}$$

408 APPLIED ECONOMICS

After		Savings for year	Accumulated savings
Year 1:	PV = $25 × .909 =	$22.73	$22.73
Year 2:	PV = $25 × .826 =	$20.65	$43.38
Year 3:	PV = $25 × .751 =	$18.78	$62.16
Year 4:	PV = $25 × .683 =	$17.08	$79.24
Year 5:	PV = $25 × .621 =	$15.53	$94.77
Year 6:	PV = $25 × .564 =	$14.10	$108.87

The calculation shows that the accumulated savings equal the original investment of $100 between the 5th and 6th years, resulting in a discounted payback period (by interpolation) of about 5.4 yr.

Listed in Table 24.5 are discounted payback periods in years for a range of simple paybacks and various discount rates. Note the agreement with the answer determined in Example 24G (for a 4-yr simple payback).

24.5 FUEL ESCALATION AND DISCOUNT RATES

In Tables 24.6, 24.7, and 24.8, computed simple payback periods are adjusted for effects of fuel escalation and discount rates (of 5, $7\frac{1}{2}$, or 10%). Note that when the fuel escalation rate and discount rate are identical, the simple payback is unadjusted (see Table 24.8 for 10% discount and 10% escalation rates). Results of more than 20 yr

TABLE 24.4 DISCOUNT FACTORS

For Year in Future	\multicolumn{8}{c}{Discount Rates}
	2%	4%	6%	8%	10%	12%	14%	16%
0	1.000	1.000	1.000	1.000	1.000	1.000	1.000	1.000
1	.980	.962	.943	.926	.909	.893	.877	.862
2	.961	.925	.890	.857	.826	.797	.769	.743
3	.942	.889	.840	.794	.751	.712	.675	.641
4	.924	.855	.792	.735	.683	.636	.592	.552
5	.906	.822	.747	.681	.621	.567	.519	.476
6	.888	.790	.705	.630	.564	.507	.456	.410
7	.871	.760	.665	.583	.513	.452	.400	.354
8	.853	.731	.627	.540	.467	.404	.351	.305
9	.837	.703	.592	.500	.424	.361	.308	.263
10	.820	.676	.558	.463	.386	.322	.270	.227
11	.804	.650	.527	.429	.350	.287	.237	.195
12	.788	.625	.497	.397	.319	.257	.208	.168
13	.773	.601	.469	.368	.290	.229	.182	.145
14	.758	.577	.442	.340	.263	.205	.160	.125
15	.743	.555	.417	.315	.239	.183	.140	.108

24.5 FUEL ESCALATION AND DISCOUNT RATES

TABLE 24.5 DISCOUNTED PAYBACK PERIODS IN YEARS

Unadjusted Simple Payback (yr) (0%)	Adjusted Years to Payback for Discount Rate of:					
	2%	4%	6%	8%	10%	12%
1	1.0	1.0	1.1	1.1	1.1	1.1
2	2.1	2.1	2.2	2.3	2.3	2.4
3	3.1	3.3	3.4	3.6	3.7	3.9
4	4.2	4.4	4.7	5.0	5.4	5.8
5	5.3	5.7	6.1	6.6	7.3	8.1
6	6.5	7.0	7.7	8.5	9.6	11.2
7	7.6	8.4	9.3	10.7	12.6	16.2
8	8.8	9.8	11.2	13.3	16.9	20+
9	10.0	11.4	13.3	16.5	20+	20+
10	11.3	13.0	15.7	20+	20+	20+
11	12.5	14.8	18.5	20+	20+	20+
12	13.9	16.7	20+	20+	20+	20+
13	15.2	18.7	20+	20+	20+	20+
14	16.6	20+	20+	20+	20+	20+
15	18.0	20+	20+	20+	20+	20+

TABLE 24.6 5% DISCOUNT RATE—ADJUSTED YEARS TO PAYBACK DUE TO DISCOUNT RATE AND ENERGY COST (DE)ESCALATION

Deescalation Rate			Unadjusted Simple Payback (yr)	Escalation Rate				
−10%	−5%	−2%		2%	4%	6%	8%	10%
1.2	1.1	1.1	1	1.0	1.0	1.0	1.0	1.0
2.6	2.4	2.2	2	2.1	2.0	2.0	1.9	1.9
4.5	3.8	3.5	3	3.2	3.1	2.9	2.8	2.7
7.1	5.5	4.9	4	4.3	4.1	3.9	3.7	3.6
11.6	7.5	6.4	5	5.5	5.1	4.9	4.6	4.4
20+	10.0	8.1	6	6.7	6.2	5.8	5.5	5.2
20+	13.3	10.1	7	8.0	7.3	6.7	6.3	5.9
20+	18.4	12.3	8	9.3	8.4	7.7	7.1	6.7
20+	20+	14.9	9	10.6	9.5	8.6	7.9	7.4
20+	20+	18.2	10	12.0	10.6	9.5	8.7	8.1
20+	20+	20+	11	13.5	11.7	10.4	9.5	8.7
20+	20+	20+	12	15.0	12.8	11.3	10.2	9.4
20+	20+	20+	13	16.6	14.0	12.2	10.9	10.0
20+	20+	20+	14	18.3	15.1	13.1	11.7	10.6
20+	20+	20+	15	20+	16.3	14.0	12.4	11.2

TABLE 24.7 7.5% DISCOUNT RATE—ADJUSTED YEARS TO PAYBACK DUE TO DISCOUNT RATE AND ENERGY COST (DE)ESCALATION

Deescalation Rate			Unadjusted Simple Payback (yr)	Escalation Rate				
−10%	−5%	−2%		2%	4%	6%	8%	10%
1.2	1.1	1.1	1	1.1	1.0	1.0	1.0	1.0
2.8	2.5	2.3	2	2.2	2.1	2.0	2.0	1.9
4.9	4.1	3.7	3	3.4	3.2	3.1	3.0	2.9
8.5	6.0	5.3	4	4.6	4.4	4.1	4.0	3.8
20+	8.7	7.2	5	6.0	5.6	5.2	4.9	4.7
20+	12.6	9.4	6	7.4	6.8	6.3	5.9	5.6
20+	20+	12.3	7	9.0	8.1	7.4	6.9	6.4
20+	20+	16.2	8	10.7	9.5	8.6	7.8	7.3
20+	20+	20+	9	12.6	10.9	9.7	8.8	8.1
20+	20+	20+	10	14.8	12.4	10.9	9.8	8.9
20+	20+	20+	11	17.1	14.0	12.0	10.7	9.7
20+	20+	20+	12	19.8	15.6	13.2	11.7	10.5
20+	20+	20+	13	20+	17.4	14.5	12.6	11.3
20+	20+	20+	14	20+	19.2	15.7	13.5	12.0
20+	20+	20+	15	20+	20+	17.0	14.5	12.8

TABLE 24.8 10% DISCOUNT RATE—ADJUSTED YEARS TO PAYBACK DUE TO DISCOUNT RATE AND ENERGY CUT (DE)ESCALATION

Deescalation Rate			Unadjusted Simple Payback (Yrs)	Escalation Rate				
−10%	−5%	−2%		2%	4%	6%	8%	10%
1.3	1.2	1.1	1	1.1	1.0	1.0	1.0	1.0
2.9	2.6	2.4	2	2.3	2.2	2.1	2.1	2.0
5.5	4.4	4.0	3	3.6	3.4	3.2	3.1	3.0
10.9	6.8	5.8	4	5.0	4.7	4.4	4.2	4.0
20+	10.6	8.2	5	6.6	6.1	5.6	5.3	5.0
20+	20+	11.5	6	8.4	7.6	6.9	6.4	6.0
20+	20+	16.9	7	10.5	9.2	8.3	7.6	7.0
20+	20+	20+	8	13.1	11.0	9.7	8.7	8.0
20+	20+	20+	9	16.2	13.1	11.2	9.9	9.0
20+	20+	20+	10	20+	15.3	12.8	11.2	10.0
20+	20+	20+	11	20+	17.9	14.5	12.4	11.0
20+	20+	20+	12	20+	20+	16.3	13.7	12.0
20+	20+	20+	13	20+	20+	18.2	15.0	13.0
20+	20+	20+	14	20+	20+	20+	16.4	14.0
20+	20+	20+	15	20+	20+	20+	17.7	15.0

are noted as simply "20+" to underscore the unattractiveness of the investment.

Example 24H: Improvements to a gas-fired heating system were made in 1980, and were estimated to have a simple payback period of 6 yr. Determine the payback period after adjusting for actual energy prices over the last decade, and a discount factor of 10%.

Solution: As indicated in Table 24.2, the average retail price of natural gas actually escalated by an average rate of +5.0 percent during the 1980s. Now use Table 24.8 and find the unadjusted simple payback line corresponding to 6 yr. The answer is about 7.3 yr (the average of 7.6 yr for an escalation rate of 4% and 6.9 yr for an escalation rate of 6%).

24.6 MORTGAGE ECONOMICS

For new house construction, additional conservation costs will normally be included in the mortgage. Mortgage economics can provide a "real world" economic basis which considers tax benefits of borrowed money, fuel escalation, yearly cash flow, and cumulative cash flow over the term of the loan.

Illustrated below is the year-by-year cash flow to repay $1,000 on a 15-yr, 10% fixed-rate mortgage for an energy conservation investment which will save $200 per year (in constant dollars). The analysis is based upon a 28% income tax bracket (assumed to be the minimum 1993 federal rate of most homeowners). This means that 28% of the cost of the mortgage will be tax deductible. The yearly loan cost and values in columns B and C are based upon readily available amortization tables for a mortgage of the in-

SUMMARY: YEAR-BY-YEAR CASH FLOW FOR 15-YEAR MORTGAGE

Mortgage Term:	15 yr		Investment Amount:	$1000
Interest Rate:	10%		Annual Fuel Savings:	$200
Yearly Loan Cost:	$129		Simple Payback:	5 yr
Income Tax Bracket:	28%		Annual Fuel Escalation:	0%

A	B	C	D	E	F	G	H
After Year	Prin. Paid	Int. Paid	$ Tax Savings (.28 × C)	$ Fuel Savings	Total $ Savings (D + E)	Yearly Cash Flow (F − $129)	Cumulative Cash Flow to Date (Sum G)
1	$30	$99	$28	$200	$228	$99	$99
2	$33	$95	$27	$200	$227	$98	$196
3	$37	$92	$26	$200	$226	$97	$293
4	$41	$88	$25	$200	$225	$96	$389
5	$45	$84	$23	$200	$223	$95	$483
6	$50	$79	$22	$200	$222	$93	$577
7	$55	$74	$21	$200	$221	$92	$668
8	$61	$68	$19	$200	$219	$90	$758
9	$67	$62	$17	$200	$217	$88	$847
10	$74	$55	$15	$200	$215	$86	$933
11	$82	$47	$13	$200	$213	$84	$1017
12	$91	$38	$11	$200	$211	$82	$1099
13	$100	$29	$8	$200	$208	$79	$1178
14	$111	$18	$5	$200	$205	$76	$1254
15	$122	$7	$2	$200	$202	$73	$1327

dicated term and rate. This analysis can also be considered to be conservative since no fuel escalation was assumed.

The 15-yr summary on the previous page indicates that a "positive cash flow" of $99 occurs in year 1. This means that the combined tax and energy savings are greater than the yearly loan cost by $99 (228 − 129) in the first year. One way to look at this is that "positive cash flow" means that with this mortgage approach there really is no additional outlay for improved energy conservation. Also note that a total of $1327 of positive cash flow will accrue over the entire 15-yr term of the mortgage. And after the loan is repaid in 15 yr the investment will continue to save $200 (tax free!) annually.

Illustrated below is a similar analysis for a 30-yr mortgage with a fixed interest rate of 10% using the same assumptions: a conservation investment which saves $200 per year in energy savings (5-yr simple payback) with a 28% income tax bracket.

SUMMARY: YEAR-BY-YEAR CASH FLOW FOR 30-YEAR MORTGAGE

Mortgage Term:	30 yr			Investment Amount:	$1000		
Interest Rate:	10%			Annual Fuel Savings:	$200		
Yearly Loan Cost:	$105			Simple Payback:	5 yr		
Income Tax Bracket:	28%			Annual Fuel Escalation:	0%		

A	B	C	D	E	F	G	H
After Year	Prin. Paid	Int. Paid	$ Tax Savings (.28 × C)	$ Fuel Savings	Total $ Savings (D + E)	Yearly Cash Flow (F − $105)	Cumulative Cash Flow to Date (Sum G)
1	$6	$100	$28	$200	$228	$123	$123
2	$6	$99	$28	$200	$228	$122	$245
3	$7	$99	$28	$200	$228	$122	$367
4	$7	$98	$27	$200	$227	$122	$489
5	$8	$97	$27	$200	$227	$122	$611
6	$9	$96	$27	$200	$227	$122	$733
7	$10	$95	$27	$200	$227	$121	$854
8	$11	$94	$26	$200	$226	$121	$975
9	$12	$93	$26	$200	$226	$121	$1096
10	$14	$92	$26	$200	$226	$120	$1216
11	$15	$90	$25	$200	$225	$120	$1336
12	$17	$89	$25	$200	$225	$119	$1455
13	$18	$87	$24	$200	$224	$119	$1574
14	$20	$85	$24	$200	$224	$118	$1693
15	$22	$83	$23	$200	$223	$118	$1811
16	$25	$81	$23	$200	$223	$117	$1928
17	$27	$78	$22	$200	$222	$116	$2044
18	$30	$75	$21	$200	$221	$116	$2160
19	$33	$72	$20	$200	$220	$115	$2275
20	$37	$68	$19	$200	$219	$114	$2389
21	$41	$65	$18	$200	$218	$113	$2501
22	$45	$60	$17	$200	$217	$112	$2613
23	$50	$56	$16	$200	$216	$110	$2723
24	$55	$50	$14	$200	$214	$109	$2832
25	$61	$45	$13	$200	$213	$107	$2939
26	$67	$38	$11	$200	$211	$105	$3045
27	$74	$31	$9	$200	$209	$103	$3148
28	$82	$24	$7	$200	$207	$101	$3249
29	$90	$15	$4	$200	$204	$99	$3348
30	$100	$6	$2	$200	$202	$96	$3444

As shown in the summary, positive cash flow of $123 occurs in year 1 for this 30-yr loan. Also note that a positive cash flow total of $3444 will accrue over the entire 30-yr term of the mortgage. Many financial institutions, utilities, government agencies, and homebuilders have come to understand how beneficial mortgage economics can be in producing positive cash flow for energy conservation measures. As a result, there are now many home-rating systems to encourage such investments, helping both homeowner and society at large.

24.7 CONSERVATION AND "THE LAW OF DIMINISHING RETURNS"

The "law of diminishing returns" applies to investments in additional levels of conservation. To understand how it works let's begin by considering the uninsulated 2 × 10 wood joist ceiling construction in Figure 24.1.

First let's calculate the area-weighted total U-factor for the uninsulated ceiling construction to serve as a "base case." Formula 5C is used to area weight the total R-values where wood framing occurs ("A"), and between the framing ("B"). The joists are installed 12 in. on center and constitute about 0.13 (F_A) of the ceiling area (see Table 5.4). Therefore, the fraction of the ceiling between framing (F_B) is 0.87 (1 − 0.13).

$$U_{aw} = \left(\frac{1}{R_{t-A}} \times F_A\right) + \left(\frac{1}{R_{t-B}} \times F_B\right)$$

(Formula 5C)

$$= \left(\frac{1}{9.86} \times 0.13\right) + \left(\frac{1}{1.31} + 0.87\right)$$

$$= 0.013 + 0.664$$

$$= 0.677 \text{ Btu}/\text{ft}^2 \cdot °\text{F} \cdot \text{h}$$

This U-factor of 0.677 for the uninsulated construction will be considered as a base case. Shown in Table 24.9 are the results of similar area-weighted calculations for ceiling constructions which include the R-values and thicknesses of batt insulation indicated. Assumed for the evaluation is a climate with 7000 heating degree-days per year relative to the house heating balance point temperature.

Examination of Table 24.9 reveals that the first increment of ceiling insulation (R-11) reduces heat loss dramatically (by 88%). Then each additional increment provides less and less of an improvement, as illustrated in Figure 24.2.

In the case of ceiling insulation, installation costs for all thicknesses should be about the same (assume 15 cents per square foot in 1993) while material costs increase approximately 25 cents per square foot for each increment (i.e., from R-0 to R-11, from R-11 to R-19, etc.).

At Framing ("A")	R-value
Attic air film	0.25
9 1/2" wood joist	8.55
1/2" gyp. board	0.45
Inside air film	0.61
R_{t-A} =	9.86

Between Framing ("B")	R-value
Attic air film	0.25
1/2" gyp. board	0.45
Inside air film	0.61
R_{t-A} =	1.31

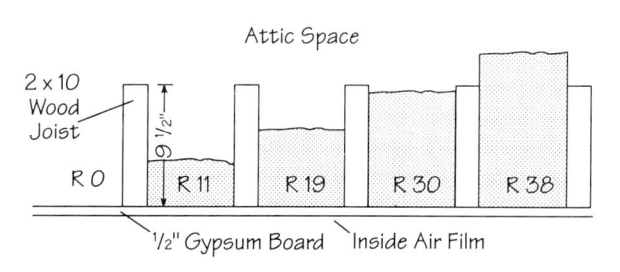

FIGURE 24.1 Wood joist ceiling construction.

TABLE 24.9 INSULATION ALTERNATIVES AMOUNT OF HEAT LOSS

Ceiling Construction Description	Area-Weighted U-factor	Improvement in U-factor Compared to Previous Insulation Amount	Annual Heat Loss per ft² in 7000 HDD Climate (Btu/yr)	Percent Improvement for Increment
R-0 Uninsul.	0.677	N.A.	113,740	N.A.
R-11 (3½ in.)	0.084	0.593	14,110	88%
R-19 (6 in.)	0.056	0.028	9,408	33%
R-30 (9 in.)	0.041	0.015	6,890	27%
R-38 (12 in.)	0.035	0.006	5,880	15%

Yearly savings resulting from an incremental improvement can be computed by using Formula 24G.

$$Q_{\text{sav-yr}} = U_{\text{imp}} \times \text{HDD}_x \times 24$$

Formula 24G

where

$Q_{\text{sav-yr}}$ is the quantity of heating energy saved in Btus per year
U_{imp} is the incremental improvement in U-factor
HDD_x is the quantity of heating degree-days below heating balance point temperature X
24 is the conversion of 24 hr in 1 day

The yearly dollar value of this heating energy is then computed using Formula 24H.

$$\$_{\text{sav-yr}} = Q_{\text{sav-yr}} \times \$/\text{MMBtu}$$

Formula 24H

where

$\$_{\text{sav-yr}}$ is the yearly dollar savings
$Q_{\text{sav-yr}}$ is the quantity of heating energy saved in Btus per year
$\$/\text{MMBtu}$ is the cost to deliver 1,000,000 Btus of useful energy (see Formula 24A)

Example 24I: Determine the simple payback associated with an upgrade of ceiling insulation from R-19 to R-30 in a climate with 7000 heating degree-days (HDD). The increase in insulation cost is 25 cents ($0.25) per square foot, and the cost of delivered heating energy is $10 per million Btus ($/MMBtu).

Solution: As indicated in Table 24.9, in going from R-19 to R-30 ceiling insulation, the improvement in U-factor will be 0.015 Btu/ft²·°F·h (U_{imp}). First use Formula 24G to compute the yearly energy savings in Btus.

$$Q_{\text{sav-yr}} = U_{\text{imp}} \times \text{HDD}_x \times 24$$

(Formula 24G)

$$= 0.015 \times 7000 \times 24$$

$$= 2520 \text{ Btu}$$

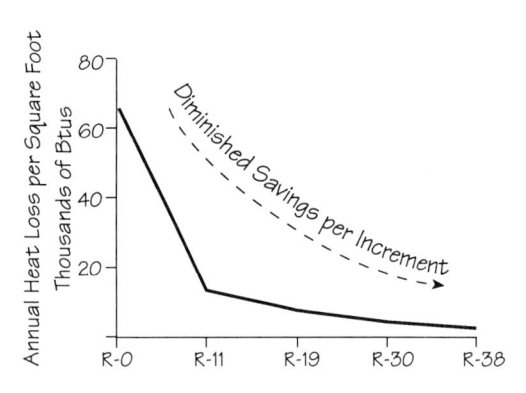

FIGURE 24.2 Diminished incremental savings.

Then determine the dollar value of these savings using Formula 24H.

$$\$_{\text{sav-yr}} = Q_{\text{sav-yr}} \times \$/\text{MMBtu}$$

(Formula 24H)

$$= 2520 \times \frac{\$10}{1,000,000}$$

$$= \$0.025 \text{ or } 2\frac{1}{2} \text{ cents per year per ft}^2$$

The simple payback can then be computed by using Formula 24C, based upon the per square foot cost increase of the additional investment and yearly dollar savings.

$$SP_{yr} = \frac{\$_{\text{invest}}}{\$_{\text{sav-yr}}} \quad \text{(Formula 24C)}$$

$$= \frac{\$0.25}{\$0.025}$$

$$= 10 \text{ yr}$$

A simple payback of 10 yr (also a rate of return of 10%) for R-30 insulation is attractive as a residential conservation investment. In contrast, a similar calculation to evaluate the economics of the diminished improvement from R-30 to R-38 insulation yields a lengthy simple payback result of 25 yr, which most would consider to be too long.

By the way, the simple payback for the very first increment of insulation (from R-0 to R-11) is only about 9 months (0.72) while the next increment (from R-11 to R-19) requires 5 yr to achieve a simple payback.

24.8 UTILITY RATES FOR ELECTRICITY

By law, utilities must maintain sufficient generating capacity to serve the demands of their customer base. In many areas, the demand for power is quite uneven between day and night and summer to winter, as shown in Figure 24.3. This frequently leaves generating equipment underutilized or standing idle, equating to bad economics for the utility.

In an attempt to level their load profiles, many utilities have now adopted pricing structures (known as "time-of-day" rates) to encourage the use of electricity during off-peak periods. They also have pricing structures which impose "peak demand charges" on those users who place the greatest demands upon the system.

Large commercial customers typically have two types of electric meters. They

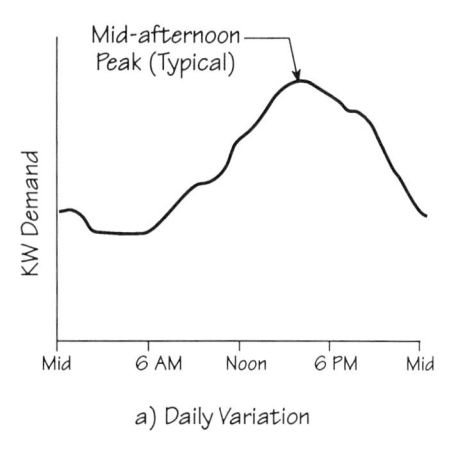

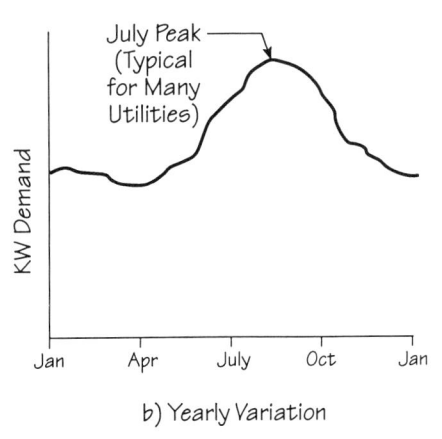

a) Daily Variation

b) Yearly Variation

FIGURE 24.3 Illustration of utility load profiles.

have a kilowatt hour meter (just like a house will) to measure consumption of power. In addition, they will have an additional meter which measures the maximum demand for electricity in kilowatts (kW) during a short period of time (usually 15 or 30 min) depending upon the utility. Then after each period (e.g., 15 min) the meter will reset to measure another short period while retaining the highest demand recorded during that billing period to date. The commercial customer's bill is then calculated based on the highest demand for electricity in kilowatts consumed during any 15- or 30-min period. Demand charges vary greatly from over $20 per kilowatt for some utilities to very low or even zero demand charges in some areas where a relatively constant climate produces a fairly level demand year round. Some utilities base their peak demand charges for many months on the highest recorded demand. This is called "ratcheting." For example, bills in September, October, and November will assume for their basis the same demand for electricity that occurred on the hottest day of the summer.

Design strategies which can take advantage of time-of-day rate structures, and reduce peak kilowatt electrical demand include computerized energy management systems and ice storage. Use of electric boilers is also encouraged in some cold regions where the greatest demand for power is during the winter.

24.9 DOLLARS AND SENSE

When evaluating the economics of HVAC options it is easy to "get lost in the numbers." This is particularly true when a project is running over budget and quick decisions and design adjustments must be made. Yes—respond to the pressures of the moment. But always try to also keep in mind implications on long-term operational cost, ease and minimization of maintenance, system dependability, and durability.

INDEX

Absorber plate, solar collector, 385
Absorption refrigeration cycle, 175, 349–350
Actuators, 394, 397
Additional equivalent length of duct fittings, 146
Adiabatic process, 304
Aesthetics, 177
AFUE, *see* Annual fuel utilization efficiency
Air:
  cleaners, 209
  conditioning processes, 316–317
  constant (1.08), 224
  curtains, 187
  distribution systems, houses, 215
  films, 65
  filtration, 104
  flow, 304
    patterns, 296–297
    warm air heating, 217, 224
  handlers, 104
  inlets and outlets, 296–297
  movement:
    cooling, 296–301
    decreasing comfort, 6
    improving comfort, 5
  outlets, 368–372, 220–223, 225
  pressure, 142
  properties, 138
  purger, 256
  spaces, 65
  supply system description, 360–363
  system types, 363–368
  temperature, warm air systems, 212
  velocity, 358
  venting:
    hydronic systems, 264
    steam systems, 238, 241
Air change:
  method, infiltration heat loss, 76
  rates, houses, 311
Air-circulation fans, 300
Air-cooled:
  condensers, 182
  condensing units, 333
Airfoil fan blades, 151
Air-handling system, 360–363
Airline terminals, 184
Air-to-air:
  heat exchangers, 101
  heat pumps, 276
Altitude angles, 37–43
  noon, 41
American Society of Heating Refrigeration and
    Air Conditioning Engineers, *see* ASHRAE
Analog control, 389
Angles:
  active solar collectors, mounting, 386–387
  altitude, 37–43
  azimuth, 37–43
Annual fuel utilization efficiency (AFUE), 286
Annual heating:
  needs, 86, 91
  requirements, 196, 207
Antifreeze solutions, 252
Apartment buildings, 179
Apparatus dewpoint (Adp), 321
Apparent solar motion, 41

418 INDEX

Appliances, heat gain, 54–56
Approach, cooling towers, 339
Argon-filled windows, 73
ASHRAE:
   calculation procedure:
      CLTD/CLF, 126
      residential cooling loads, simplified, 305–315
   comfort zone, 2
   effective temperature (ET), 3
   Standard 52, 104
   Standard 62, 95–96, 100
   Standard 90.1, 57, 214, 257, 348
   Standard 103, 285
Atmospheric clearness, 31–32
Auxiliary heat requirements, 196, 207
Axial fans, 149
Azimuth angles, 37–43
   solar-wall, 43

Backward-curved fans, 151
Balance point temperatures, 17, 60–62
Balcomb, Dr. J. Douglas, 191
Baseboard heaters:
   electric, 274
   hot water, 264–268
   sizing, 266, 271
Basements:
   floors, heat loss, 83
   unheated, 85
Base-mounted pumps, 164
Base temperature, 16
Bedroom loop, 260
Below-grade:
   floors, heat loss, 83
   walls, heat loss, 81–83
Bimetallic sensors, 393
Bins, temperature, 23
Bioclimatic chart, 6
   building, 7–9
BLC, *see* Building load coefficient
Blowdown, cooling towers, 336
Blower, 208, 210
Blown off, 232
Blow through fans, 361
Boilers:
   cast-iron, 229–231
   electric, 283
   firetube, 231
   horsepower, 233
   hot water, 254–257
   modular, 257
   oil-fired, 255
   related steam piping, 234
   sizing, 257
   steam, 228–234
   steel, 231
   wall-mounted, 257
   watertube, 231

Boiling, water, 157
Brake horsepower (bhp), 152, 168
Breezes, summer, 30
Brick, heat storage capacity, 190
British thermal unit, *see* BTU
BRI, *see* Building-related illness
BTU (British thermal unit), 63
   capacity, piping, 253
   conversion, 55
   cost per million, 403
   heat content of fuels, 286
Building:
   bioclimatic chart, 7–9
   heat loss, 63
   height and infiltration, 79
   materials, R-values, 66
   type, implications for system selection, 173
Building load coefficient (BLC), 193
Building-related illness, 100
Bypass factor (BF), 321

Calculation:
   cooling load, "The Office," 128–132
   heating balance points, 60
   heat loss, "The House," 86–92
   infiltration:
      air change method, 76, 81
      effective leakage area method, 77–80, 89–90
   profile angles, 43–45
   room heat loss, 80
   stack effect, 103
   sun angles, 42
   ventilation:
      heat gain, 97
      heat loss, 96
Capacity control, compressors, 347–348
Carry over, 232
Casablanca fans, 187
Cavitation, 165
CDD, *see* Cooling degree-days
Ceiling fans, 300
Center of glass, 73
Central air conditioning, houses, 301–303
Centrifugal:
   fans, 150
   pumps, 164
CFC, *see* Chlorofluorocarbons
Cfm (cubic feet per minute):
   air flow, 304
   requirements:
      cooling systems, residential, 313–315
      graphic solution, 322–325
      warm air systems, residential, 224
Checklist, heat loss calculations, 92
Chillers:
   absorption, 350
   electric, 342–344
   package, water, 342–344

Chimneys, heating systems without, 179
  vent sizing, 290–292
Chlorofluorocarbons (CFC), 328
Circular equivalent method, duct sizing, 144–147
Circulator pumps, 164, 260
Clean Air Act amendments, 328
Clear day radiation, 45
Clearness, 31–32
CLF, see Cooling load factors
Climate:
  charting, 6–11
  interior, 48
  system selection, implications, 174, 177
Closed, piping systems, 156, 165
Closed loop:
  heat pump, 188
  solar hot water systems, 385–386
CLTD, see Cooling load temperature difference
Coanda effect, 370
Coefficient of performance (COP), 330
  absorption cycle, 350
  heat pumps, 278
Coefficients, infiltration, 78, 89
Coils, heating and cooling, 361
Cold climates, 177
Color correction (CLTD values), 120, 124
Combination heating systems, 263
Combustion, sealed, 210
Comfort zone, 2
Commercial cooking appliances, 56
Component heat loss, 71
Compression cycle, 328–329
Compressors, 329
  capacity control, 347–348
  centrifugal, 346–347
  helical, 345–346
  multiple, 348
  open type, 344
  orbital, 345–346
  reciprocating, 344
  rotary, 344
  screw, 345–346
  scroll, 345–346
Computer rooms, 182
Concrete, heat storage capacity, 190
Condenserless cooling units, 330
Condensers, 329
  air-cooled, 302, 328, 333
  double-bundle, 335, 383
  entering air temperature, 332
  evaporative, 335
  heat recovery, 335
  water-cooled, 335
Condensing, heating appliances, 179, 213
Conditioning coils, 361
Conduction, 64
Conductivity, 190
Conservation investments, 413–414

Constant volume systems, 363
Construction assemblies, heat loss, 65, 67
Controlled variable, 389
Controllers, 394
  microprocessor, 394
Control point, 389
Controls:
  fire, 400
  high limit, 399
  high temperature, 399
  low voltage, 396
  low water cutoff, 399
  modulating, 396
  optimizing, 399
  proof of ignition, 399
  safety, 398–399
  smoke, 400
  terminology, 389–391
  warm air heating systems, 212
  warmup, 399
Control systems, 389–401
  characteristics, 395
  DDC, 398
  electric, 395–396
  electronic, 396
  energy-conserving, 399
  pneumatic, 396
Convection, 64
Convectors, 268
Converter, domestic hot water, 383
Cooking appliances, heat gain, 56
Cool deck, 368–369
Cooling:
  air motion, 296–301
  balance points, 62
  coils, 302, 320–321, 361
  degree-days (CDD), 27–28, 294
  load factors (CLF), 51, 54, 57
    ASHRAE CLF/CLTD procedure, 106
  load hours (CLH), 281–283
    map, United States, 281
  load temperature difference (CLTD), 115–124
    corrections, 120–124
  loads:
    internal heat gains, 128
    latent, 54
    residential, calculation, 305–313
    roofs and walls, 115–124
    sensible, 51
    solar, 105
  moving air, 299
  ponds, 342
  supply air, 355–358
  towers, 336–340
    corrosion control, 336
    fogging, 336, 338
Copper piping, 159–160
COP, heat pumps, 278

Correction factors:
    CLTD values, 120–124
    hot water temperature, 267
    water flow rate, 267
Cost (initial), HVAC systems, 174
Counterflow heat transfer, 101
    cooling coils, 321
Crack method, 77
Crossover point, 283
Cycling, 389
    compressors, 347
    duty, 399
Cylinder unloading, 347

Daily range, temperature, 26
Dampers:
    face and bypass, 397
    motorized, 360
Damper, actuators, 397
Daytime loop, 260
DDC, *see* Direct digital control
Deadband, 390
Decks, multizone unit, 368–369
Declination, 34
Deep earth temperature, 81
Degree-days, 13, 16–20
Dehumidification, dessicant, 320
Delivered energy cost, 403
Density of materials, 190
Design conditions, selection, 25
Design temperatures, 215
    summer, 25–26
    wet-bulb, 26
    winter, 13–16
Design with Climate, 6–8, 296
Dessicant dehumidification, 320
Destratification fans, 187
Desuperheating (heat recovery), 329, 383
Deviation, 390
Dew point temperature, 320
DHW, *see* Domestic hot water
Differential, 391
Diffusers, 220, 370
    laminar flow, 370–371
    linear, 370–371
    noise, 371–372
    round, 370
    supply air, 362
Digital control, 390
Dilution air, 290
Dimension A, 236
Direct digital control (DDC), 398
Direct-fired domestic hot water heaters, 380–381
Direct gain:
    reference designs, 194
    solar heating systems, 189, 205–207
Direct-return hydronic systems, 262

Discounted payback, 407–411
Distribution, supply air, 358
Diurnal temperature, 51
Diversion fittings, 260–261
Diversity, 137
Domestic hot water (DHW), 373–388
    converters, 383
    direct-fired systems, 380–381
    energy requirements, 380–381
    heat exchangers, 383
    heat pump heaters, 384
    heat recovery, 383
    indirect-fired heaters, 382–383
    solar heaters, 385–388
    storage tanks, 383
Double-bundle condensers, 383
Draft, 290
Draw through fans, 361
Drift, 336
Droop, 390
Drop, 221
Dry:
    return, 236
    steam, 227
Dry-bulb temperature, 2
Dual duct systems, 368
Ducts, 144–148
    supply air, 361
    layout, warm air supply, 225
    risers, 358–360
    shafts, sizing, 359
Duct sizing:
    methods, 148
    warm air systems, 218
Dust:
    elimination, 209
    spot ratings, 104
Duty cycling, 399
DX systems, 329, 342

Earth:
    factor, heat loss, 81
    facts, 33
    modified heat loss, 81
Economics, applied, 402–416
    mortgage, 411–413
Economizer cycle, 62, 360
    water-side, 340
Edge of glass, 73
EDR, *see* Equivalent direct radiation
EER, *see* Energy efficiency ratio
Effective:
    leakage area method, 77–80, 89–90
    surface temperature (EST), 321
    temperature (ET), 3
Efficiency:
    electricity, 273

measuring improvement, 288
pumps, 167
steady-state, 285
Electric:
control systems, 395
heat, 273-275
radiant panels, 275
water heaters, 382
Electromagnetic devices, 390
Electronic control systems, 396
Electrostatic air cleaner, 209
Emittance, 66
Empirical correction factor, 19
Energy:
conserving controls, 399
crisis, 94, 188
efficiency ratio (EER), 279
escalation, 405
prices, 405-406
requirements, domestic hot water systems, 380
savings, solar systems, 196
Enthalpy, air, 317, 325
ventilation air, 99
Envelope dominated buildings, 105
Equal friction method, 149
Equinox, 34
Equipment, heat gain, 54
Equivalent:
direct radiation (EDR), 233
length of duct, fittings, 146
pipe length, 163, 236
Escalation, energy prices 405
ET, see Effective temperature
Evaporation, water surfaces, 340, 342
Evaporative cooling, 304-305
map, United States, 305
Evaporators, 329, 342
absorption cycle, 349
entering wet-bulb temperature, 332
flooded coil, 342
Excess air, 286
Exfiltration, 75
Exhaust:
air, 362
gases, 213
Expansion tanks, 256
Extended plenum, air distribution systems, 215, 225
External:
heat gains, 127
static pressure, warm air systems, 210

F2 factors, 84
Face and bypass dampers, 397
FAI, see Fresh air intake
Families of pumps, 167
Fan-assisted appliances, 291-292
Fan-coil units, 181, 353-355

Fans, 149-151
air-circulating, 300
blow through, 361
draw through, 361
laws, 152, 154-155
operation, 152
performance, 151
supply air, 361
Feet of water, 157
Fenestration, 106
Fill (cooling towers), 336
Filters, 104, 361
permanent (washable), 212
Finned-tube radiation, 264-268
Fire:
controls, 400
safety, 177
Flame-retention burner, 254-255
Flat plate solar collectors, 385
Floor-mounted package cooling units, 330
Floor registers, 221
Floors below grade, heat loss, 83
Flow:
control devices, 329
rate correction factors, 267
sensors, 394
Foaming, 232
Forced draft, 290-292
Forward-curved fans, 151
Fouling, 232
Frame construction (heat loss), 70
Free:
air flow, 297
area, outlets, 221
field (or delivery), 149, 152
Freeze protection, water, 252
Freon, 328
Fresh air intake (FAI), 360
Friction:
allowance, 359
ducts, 144-148
hydronic systems, 259
losses, 147
piping and fittings, 161-164
Fuels:
availability, 175
escalation, 405, 408-411
heat content, 285
Furnaces:
arrangements, 210-211
early, 209
efficiency, 212
federal law, 287

Gas-fired cooling, 351
Generator, 349
Givoni, Baruch, 7

Glass load factor (GLF), 306
Glazing:
    conductive heat gain, 119, 122
    heat loss, 72–75
    productivity, solar, 202–204
    sloped, 192–193
    solar cooling loads, 106–110, 127, 129
    surface numbering, 73
Grains of moisture, 6, 139
Graphing HVAC processes, 322–325
Gravity warm air furnace, 209
Grilles, 220
Ground water temperature, 374

Hard water, 374
Hartford loop, 235
HCFC, see Hydrochlorofluorocarbons
HDD, see Heating degree-days
HDH, see Heating degree-hours
Header, steam, 234
Head loss, 163, 166
Health, human, 105
Heat:
    anticipator, 391
    capacity:
        air, 138
        contents, fuels, 285
        materials, 190
        pipe size, hot water heating, 253
    exchanger:
        boiler, 254
        domestic hot water, 383
    gain:
        appliances, 54–56
        equipment, 54–56
        lighting systems, 57–59
        people, 49–54
    loss:
        adjacent unheated spaces, 85
        below-grade floors, 83
        below-grade walls, 81–83
        building, 63
        calculations, checklist, 92
        complex assemblies, 69
        house, 86–92
        infiltration, 75
        room-by-room, 88
        slabs-on-grade, 84
        through building elements, 65–72
        transmission, 71
    of compression, 330
    output, baseboard, 264–268, 271
    recovery:
        domestic hot water, 383
        ventilation, 101, 178
    removal potential, air movement, 299
    sink, heat pump, 276
    source, heat pump, 276

storage, 49, 51
    capacity, 106, 189
    water, 158
transfer:
    counterflow, 101
    processes, 64
traps, 387–388
Heaters:
    hot water (hydronic), 264–268
    steam, 244–248
Heating:
    balance points, 60
    coils, 361
    degree-days (HDD), 13, 16–20
        map, United States, 17
    degree-hours (HDH), 20–23
    load hours (HLH), 280–282
        map, United States, 280
    seasonal performance factor (HSPF), 278
    systems, sizing, 80
Heat pumps, 276–283
    applications, 176, 178
    closed-loop, 355
    cooling, 303
    COP, 278
    domestic hot water heater, 384
    operational cost, 279–283
    sizing, 283
HEPA (high efficiency particulate air) filters, 104
Hermetic, 344
High:
    efficiency air filters, 104
    limit controls, 399
    side, refrigerator cycle, 329
    temperature:
        controls, 399
        hot water, 250
    velocity induction systems, 368
High-rise office buildings, 180
Hooded appliances, 56
Horizontal:
    glazing, heat loss, 74
    projections, 110–115
Horsepower, boilers, 233
Hospitals, 184
Hot air heating, 208
Hot deck, 368–369
Hotels and motels, 182
Hot-gas bypassing, 347
Hot water:
    flow, 250–252
    heaters, instantaneous, 383
    heating system components, 250
    quantity needed, 376
    recirculating, 388
Hour angle, 42
House, see "The House"

HRV, *see* Heat recovery ventilation
HSPF, *see* Heating seasonal performance factor
Human comfort zone, 2
Humidifier, 209
Humidity:
  control, 319
  low winter, 212
  ratio, 319
  relative, 2, 27
  sensors, 393
Hunting, 389
HVAC:
  processes, graphing, 322-325
  system selection, 173-187
Hydrochlorofluorocarbons (HCFC), 328
Hydronic heating systems, 249-272
Hydronics Institute, tables, 265-267

IAQ, *see* Indoor air quality
Ice storage, 176
ICS, *see* Isolated combustion system
Impellers:
  fans, 149
  pumps, 164
Inches of water gauge, 157
Incremental units, 353
Indirect-fired domestic hot water heaters, 382-383
Individual cooling systems, 352-355
Indoor:
  air quality (IAQ), 95, 100
  temperature, air, 198
Induction systems, 368
Inert gas filled windows, 73
Infiltration, 75-80
  air change method, 76
  crack method, 77
  effective leakage area method, 77, 89
Inlets, air, 296-297, 362
Inlet vanes, 152
In-line pumps, 164, 260
Instantaneous hot water heaters, 383
Insulation:
  pipe, 254
  R-values, 66
  warm air ducts, 220
Integrated part-load value (IPLV), 348-349
Interior climate, 48
Internal heat gains:
  cooling loads, 128, 130
  houses, 312
Interpolation, solar savings fraction (SSF), 197, 207
Investments, conservation, 413-414
IPLV, *see* Integrated part-load value
Isogonic chart, 37
Isolated Combustion System (ICS), 210, 287

K factors, 18

Kinetic energy, 346
Krypton-filled windows, 73
KW demand, 416

Lag, 390
Latent:
  cooling load, 54
  heat capacity (LHC), 331
  heat gain, ventilation air, 98
  heat loss, ventilation air, 97
Latitude, 33, 35
  map, United States, 14
Law of diminishing returns, 413-414
  heating equipment, AFUE improvements, 288-290
  passive solar, 203
Laws:
  fan, 152, 154-155
  pump, 170-172
LCR, *see* Load collector ratio method
Leakage area method, infiltration heat loss, 77-80
Leap year, 33
Legionnaires' disease, 100, 373
Libraries, 185
Lighting:
  power density, 57
  systems, heat gain, 57-59
Liquid cooler, 342
Lithium bromide, 349
Living room, cooling load calculation, 305-313
Load collector ratio (LCR) method, 191-198
Load dominated buildings, 105
Loading docks, 187
Local cooling systems, 352-355
Longitude, 35
Loop perimeter air distribution systems, 215-216
Low:
  emissivity (Low-E) glazing, 73, 198-199
  side, refrigeration cycle, 329
  temperature, hot water, 250
  voltage controls, 396
  water cutoff, 399
Low-rise office buildings, 181

Magnetic deviation, 37
  map, United States, 37
Maintenance requirements of systems, 176
Manufacturing plants, 186
Mass, thermal, 194
Master venting, 238
Mean:
  coincident wet-bulb, 24
  daily range, 26
  radiant temperature (MRT), 3-5
Mechanical:
  cooling of houses, 301-304, 313-315

Mechanical (*Continued*)
    efficiency, 169
    equipment floors, 181
Microloads, 179
Microprocessor controllers, 394
Mid-efficiency heating appliances, 212
Milne, Murray, 7
Minimum recovery rate, 377–379
Mixed air, 323
Mixing valves, 398
Modified degree-day method, 19
Modular boilers, 257
Modulating controls, 395
Moisture in air, 139
Monoflow systems, 260–261
Mortgage economics, 411–413
Motorized dampers, 360
Motors, pumps, 168, 170
Movable insulation, 176, 189
MRT, *see* Mean radiant temperature
Mud, 232
Multiple loops, hydronic systems, 260
Multi-tenant occupancy, 176
Multizone systems, 368–369

National Fenestration Rating Council (NFRC), 73
National Fire Protection Association, *see* NFPA
National Fuel Gas Code (NFPA 54), 290, 292
National Oceanic and Atmospheric Administration, 12
Natural:
    draft, 291
    ventilation, 95, 296–300
NC, *see* Noise criteria
Negative pressure, 296
Net:
    energy benefit, glazing, 202–204
    positive suction head (NPSH), 165
Neutral pressure level (NPL), 103
NFPA 54, 290, 292
NFRC, *see* National Fenestration Rating Council
Night insulation, 194
NOAA, *see* National Oceanic and Atmospheric Administration
Noise, systems, 177
Noise criteria (NC), diffusers, 371–372
Nomograph, supply air, 357–358
Non-vertical solar glazing, 192–193
Noon:
    altitude angles, 41
    solar, 37, 41
Normally:
    closed (NC), 390
    open (NO), 390
NPSH, *see* Net positive suction head

OAI, *see* Outside air intake

Occupancy diversity, 137
Occupant density, 51
Office, *see* "The Office"
Office buildings, 180
Off-peak electricity, 284
Offset, 390
Olgyay, Victor, 6–8, 296
One-pipe steam systems, 238
On/off control, 391
Opaque walls, cooling loads, 119–121
Open piping systems, 156, 166
Operational cost, 176
    electric heat, 275
    heat pump, 279–283
Optimization controls, 399
Orientation:
    guidelines, solar, 189
    zoning, 125
Orificing, 262–263
Outdoor:
    reset, 256
    temperature, 12
Outlets, air, 296–297
Outside air:
    intake (OAI), 360
    100%, 184
    requirements, 94
Overhangs, shading from, 307–309
Overheating, passive solar houses, 198
Overshoot, 391
Oversizing, heating equipment, 214

Packaged:
    chillers, 328
    terminal air conditioners (PTAC), 352
    terminal heat pumps (PTHP), 353
Panel heaters, radiant, 275
Partitions, heat gain through, 127
Part-load performance, 348–349
Passive solar heating, 188–207
Passive Solar Industries Council (PSIC), 191
Payback, simple, 404
Peak demand:
    charges, 415
    reductions, 176, 180
Peak hour cooling loads, 125, 132–134
People, heat gain, 49–54
Performance:
    passive solar heating, 191–198
    pumps, 167
Perimeter heaters, 177
Personal environment module (PEM), 1
PI, *see* Proportional-integral control
Pickup factors, 234
PID, *see* Proportional-integral-derivative control
Pipe:
    data, 159–160

hangers, 159, 161
 size, hot water heating capacity, 253
 sizing, steam systems, 242-245
 types, 252
Pipeless furnace, 209
Piping:
 arrangements, hydronic systems, 260-264
 guidelines, steam systems, 243
 pickup, 234
Plastic pipe, 159
Plenum, return air, 362
Plotting weather data, 10-11
Plume abatement, cooling towers, 336, 338
Pneumatic control systems, 396
Point-of-use hot water heaters, 383
Pontiac fever, 100
Positive:
 cash flow, 412
 displacement:
  compressors, 344
  pumps, 165
 pressure, 296
Pound of steam, 233
Power ventilators, 102
Preheat:
 coils, air handler, 361
 tank, DHW, 385
Present value, 407
Pressure:
 absolute, 142
 air, atmospheric, 142
 around buildings, 296-297
 drop, 235
 equivalents, 142
 gauge, 143
 sandwich, fire safety, 400
 sensors, 394
 terminology, 143
 vacuum, 143
 velocity, 143
 water, 156
Pressurization for fire safety, 177
Primary air, 220, 368
Profile angles, 43-45, 110-115
Projected collector area, 192-193
Projections, horizontal and vertical, 110-115
Proof of ignition controls, 399
Propeller fans, 149
Properties, water, 156
Proportional (P) control, 392
Proportional-integral (PI) control, 392
Proportional-integral-derivative (PID) control, 392
PSIC, see Passive Solar Industries Council
Psychrometer, sling, 2
Psychrometric:
 chart, 2, 141, 318
 processes, 139-141, 316-325

 graphing, 322-325
PTAC, see Packaged terminal air conditioners
PTHP, see Packaged terminal heat pumps
Puddling, 370
Pulse combustion, 213
Pumps, 164-172
 capacity curves, 156, 165-167
 efficiency, 167
 laws, 170-172
 motors, 168
 performance, 167
 selection, to the point, 168
 speed, 170
PVC piping, 159

Racheting, 416
Radial perimeter air distribution systems, 215-216
Radiant:
 barriers, 178
 heat, 177
 panel heaters, 275
 temperature effects, 3
Radiation, 64
 surfaces, clear day, 45-47
Radiators, steam, 244-248
Radon, 100
Range, cooling towers, 339
Rate of return, 405
Rate-reset control, 393
Recirculating hot water, 388
Recovery rate, 377-379
Reference house, solar, 191
Refrigerants, 328
Refrigeration effect, 330
Registers, 220
Reheat, 364
Relative humidity, 2, 27
Relay, 390
Reliability, systems, 176
Relief air, 362
Religious buildings, 184
Replacement heating systems, 178
Reset control, 256, 364
Residential:
 cooling load calculations, 305-313
 system selection, 177
Resistance:
 electric heat, 273-275
 values, 65-68
Response time of systems, 176
Restaurant equipment, 56
Retail buildings, 180
Return:
 air:
  ducts, 218-219
  inlets, 362
  intakes, 221-223

426 INDEX

Reverse:
  air conditioners, 276
  return hydronic systems, 262
Risers, duct, 358–360
Roll filters, 104
Roofs (cooling loads), 115–119
Rooftop units, 330
Room-by-room heat loss, 88
Room sensible heat factor (RSHF), 322
R-values, 65–68

Safety controls, 398–399
Salt solution, lithium bromide, 349
SBS, *see* Sick building syndrome
SC, *see* Shading coefficient
Scale deposits, 374
Schedule 40, 159–160
SCL, *see* Solar cooling load
Sealed combustion, 210
Seasonal energy efficiency ratio (SEER), 279
Secondary air, 220
SEER, *see* Seasonal energy efficiency ratio
Selection:
  HVAC systems, 173–187
  unitary air conditioners, 331
Sensible:
  cooling load, 51
  heat:
    capacity (SHC), 331
    factors (SHF), 323
    gain, ventilation air, 98
    loss, ventilation air, 96
Sensors, 393–394
  bimetallic, 393
  flow, 394
  humidity, 393
  pressure, 394
Separation, above windows, 308
Series loop hydronic systems, 260
Service hot water, 373–388
Set-back temperatures, 177
Set point, 390
Shade:
  angles, 43–45
  line factors (SLF), 307
Shading:
  coefficient (SC), 106, 109
  devices, 110–115
  overhangs, 307–309
  projections, 43
Shaft size, risers, 358–360
SHGF, *see* Solar heat gain factor
Shielding class, infiltration, 79
Shutoff static pressure, 152
Sick building syndrome (SBS), 100
Sine curve, 35

Sizing:
  baseboard heaters:
    hot water, 266
    electric, 274
  boilers, 257
  domestic hot water systems, 377–380
  ducts, 144–149
    warm air systems, 218, 225
  heating systems, 80
    in solar houses, 197
  heat pumps, 283
  solar hot water systems, 386
  warm air furnaces, 214, 224
Slab-on-grade construction, 84
Sling psychrometer, 2
Sloped glazing, 192–193
  heat loss, 74
Smoke controls, 400
Soil temperature, 81–83
  map, United States, 82
Solar:
  collectors:
    absorber plates, 385
    flat plate, 385
    mounting angle, 386–387
  constant, 36
  cooling loads, 106–110
  declination, 34
  domestic hot water heaters, 385–388
  glazing, productivity, 202–204
  heat gain factor, 106
  heating systems:
    active, 188
    hybrid, 188
    passive, 188–207
  hot water systems:
    closed loop, 385
    drainback, 385
    draindown, 385
  indoor air temperatures, 198, 200–202
  noon, 37, 41
  orientation guidelines, 189
  savings fraction (SSF), 195–198, 203
  time, 42, 105
Sol-air:
  heat gain, 115–124, 127, 129
  cooling loads, houses, 309–312
Solar-wall azimuth angles, 43
Solenoid valve, 390
Solstice, 34
Source:
  control, indoor air quality, 102
  water temperature, 374–375
Space:
  cooling loads, total, 132
  heaters, 269

requirements, 176
Specific heat of materials, 190
Speed change, compressors, 348
Split:
    loop, hot water heating, 261, 272
    system, air conditioning, 327
Spot heaters, 275
Sprays, cooling, 342
Spread, 221
Square feet per ton, 136
SSF, see Solar savings fraction
Stable operation, 152
Stack:
    coefficient, 78
    effect, 103
Stair, pressurization, 177
Standard air, 138
States of water, 157
Static:
    head, 236
    pressure, 152
    regain method, duct sizing, 149
Steam:
    boilers, 228-234
    header, 234
    heating systems:
        one-pipe, 238
        two-pipe, 239
    pound, 233
    properties, 227
    quality, 231
    traps, 240
    utility supplied, 175
    velocity, 237
    venting of systems, 238, 241
Steel piping, 159-160
Stepped control, 392
Still air, 5
Storage:
    tanks, 383
    water heaters, 381
Strainer cycle, 340-341
Summer:
    breezes, 30
    design conditions, 25-26
    solstice, 34
Sun:
    angles, calculation, 42
    control, 110-115
    path diagrams, 40
Sunspaces, 190
Suntempered houses, 189
Superheat, 329
Superinsulation, 177
Supply air:
    cooling, 355-358

requirements, 322-325
diffusers, 362
ducts, warm air, 218-219
ductwork, 361
fans, 361
flow rates, 357
temperature, warm air systems, 212
warm air systems, 217, 224
Surface effect, 370
Surfaces, radiation on, 45
Swamp coolers, 304
System:
    curves, pumps, 165
    head loss, 163

Tankless coils, 382
Temperate climates, 178
Temperature:
    bins, 23
    control:
        hot water systems, 256
        steam systems, 248
    dry-bulb, 2
    humidity data, plotting, 10-11
    mean radiant (MRT), 3-5
    swing, 198
    water, source, 374-375
    wet-bulb, 2
"The House:"
    electric heat, 275
    heat loss calculation, 86-92
    hot water heating system, 270-272
    mechanical cooling, 313-315
    passive solar heating, 205-207
    sizing heating system, 91
    warm air heating system, 223-226
"The Office:"
    cooling load:
        calculation, complete, 128-132
            equipment, 55
            lights, 58
            people, 53
    solar gain through windows, 109
Thermal:
    break, windows, 74
    bridges, 71
    conductivity, 190
    loading, cooling towers, 339
    mass, 189
    storage:
        capacity, 194
        walls, 190
    zoning, 125
Thermosyphon hot water system, 385
Three-mode control, 393
Throw, 220, 370

Tightness of construction, 76
Tilted solar glazing, 192–193
Time:
   military, 105
   proportioning, 391
   solar, 42
   zones, 36
Timed two-position control, 391
Time-of-day electric rates, 415
Tin knockers, 352
Tons:
   cooling capacity, 133, 135–136
   refrigeration, 157
Total:
   cooling capacity (TC), 331
   heat (enthalpy), 317
   piping system head, 166
   resistance, 67
   space cooling loads, 132
Transducer, 390
Transmission heat loss, 71, 89
Traps, steam, 240
Triple glazing, 73
Trombe-Michel walls, 190
Tropic:
   of Cancer, 33
   of Capricorn, 34
Tubeaxial fans, 150
Turbine-driven chillers, 351
Two-pipe:
   hot water (hydronic) systems, 260
   steam systems, 239
Two-position control, 391
Type B vent, 290

UA (total heat loss), 86
U-factors, 68–70
Undershoot, 391
Unheated basements, 85
Unit:
   heaters, 186, 268
   ventilators, 269
Unitary air conditioners, 328, 330–332
Usable storage capacity, 377
Utility rates, electricity, 415

Vacation houses, 176
Vacuum systems, steam, 241
Valves, mixing, 398
Vaneaxial fans, 150
Vapor compression cycle, 328–329
Variable:
   base, degree-days, 18
   frequency drives (VFD), 152
   pitch motor sheaves, 152
   speed pumps, 170
Variable air volume (VAV):
   boxes, 364–368
   systems, 364–368
VAV, *see* Variable air volume systems
Velocity:
   air, 358
   reduction method, 149
   steam, 237
   warm air supply, 223
Ventilation:
   cooling load, 127
   heat:
      gain, 97–99
      loss, 96
      recovery, 101
   natural, 95
   rates, minimum, 95
   total heat, 99
Ventilators, power, 102
Venting:
   combustion heating appliances, 290
   side wall, 213
   steam systems, 238, 241
Vertical projections, 110–115
Very high-efficiency heating appliances, 213
VFD, *see* Variable frequency drives
Vibration, 177
   pulse combustion, 213
Viscosity, fluids, 161
Volume control, fans, 152
Volume/surface area ratio, 77
Volute type pumps, 164

Walls:
   air outlets, 221–223
   cooling loads, 119
Warehouses, 186
Warm:
   air:
      heating systems, 208
      requirements, 217
      returns, ducts and plenums, 218–219
      supply ducts, 218–219
   climates, 178
Warmup:
   controls, 399
   cycle, 177
Water:
   expansion, 256
   properties, 156
   quality (boiler), 232
   requirements, DHW, 376
   storage, 158
   temperature:
      needed, 373
      source, 374–375
   velocity in piping, 161
   walls, 190

Water-eliminating baffles, 336
Waterline, boiler, 232
Water-side economizer cycle, 340
Weekend houses, 176
Well-water temperature, 374
   map, United States, 375
Wet-bulb:
   degree hours, 29, 295
   design temperature, 26
   temperature, 2
Wet return, 236
Wet-time, 290
Whole-house fans, 300–301
Wind, 30
   chill temperatures, 6
   coefficient, 78
   speed, effect on heat loss, 75
Windows, heat loss, 72–75
Winter:
   design temperatures, 13–16
   solstice, 34
Wood:
   frame construction, heat loss, 69
   heat storage capacity, 190

Zero energy band, 390
Zone type designations, 49
Zoning, 125, 176